Lecture Notes in Mathematics

Edited by J.-M. Morel, F. Takens and B. Teissier

Editorial Policy
for the publication of monographs

1. Lecture Notes aim to report new developments in all areas of mathematics – quickly, informally and at a high level. Monograph manuscripts should be reasonably self-contained and rounded off. Thus they may, and often will, present not only results of the author but also related work by other people. They may be based on specialized lecture courses. Furthermore, the manuscripts should provide sufficient motivation, examples and applications. This clearly distinguishes Lecture Notes from journal articles or technical reports which normally are very concise. Articles intended for a journal but too long to be accepted by most journals, usually do not have this "lecture notes" character. For similar reasons it is unusual for doctoral theses to be accepted for the Lecture Notes series.

2. Manuscripts should be submitted (preferably in duplicate) either to one of the series editors or to Springer-Verlag, Heidelberg. In general, manuscripts will be sent out to 2 external referees for evaluation. If a decision cannot yet be reached on the basis of the first 2 reports, further referees may be contacted: the author will be informed of this. A final decision to publish can be made only on the basis of the complete manuscript, however a refereeing process leading to a preliminary decision can be based on a pre-final or incomplete manuscript. The strict minimum amount of material that will be considered should include a detailed outline describing the planned contents of each chapter, a bibliography and several sample chapters.
Authors should be aware that incomplete or insufficiently close to final manuscripts almost always result in longer refereeing times and nevertheless unclear referees' recommendations, making further refereeing of a final draft necessary.
Authors should also be aware that parallel submission of their manuscript to another publisher while under consideration for LNM will in general lead to immediate rejection.

3. Manuscripts should in general be submitted in English.
Final manuscripts should contain at least 100 pages of mathematical text and should include
- a table of contents;
- an informative introduction, with adequate motivation and perhaps some historical remarks: it should be accessible to a reader not intimately familiar with the topic treated;
- a subject index: as a rule this is genuinely helpful for the reader.

Continued on inside back-cover

Lecture Notes in Mathematics 1765

Springer
Berlin
Heidelberg
New York
Barcelona
Hong Kong
London
Milan
Paris
Tokyo

Thomas Kerler Volodymyr V. Lyubashenko

Non-Semisimple Topological Quantum Field Theories for 3-Manifolds with Corners

Springer

Authors

Thomas Kerler
Department of Mathematics
The Ohio State University
231 West 18th Avenue
Columbus, Ohio 43210, USA

E-mail: kerler@math.ohio-state.edu

Volodymyr V. Lyubashenko
Institute of Mathematics
National Academy of Sciences of Ukraine
3, Tereshchenkivska st.
Kyiv-4, 01601 MSP, Ukraine

E-mail: lub@imath.kiev.ua

Cataloging-in-Publication Data applied for
Die Deutsche Bibliothek - CIP-Einheitsaufnahme

Kerler, Thomas:
Non-semisimple topological quantum field theories for 3-manifolds with corners / Thomas Kerler ; Volodymyr V. Lyubashenko. - Berlin ; Heidelberg New York ; Barcelona ; Hong Kong ; London ; Milan ; Paris ; Singapore ; Tokyo : Springer, 2001
(Lecture notes in mathematics ; 1765)
ISBN 3-540-42416-4
Mathematics Subject Classification (2000):
16W30, 18D05, 18D10, 18E10, 57N10, 57N13, 57N70

Physics and Astronomy Classification (1999):
11.10.Cd, 11.10.Kk, 11.25.Hf, 11.25.Sq

ISSN 0075-8434
ISBN 3-540-42416-4 Springer-Verlag Berlin Heidelberg New York

Springer-Verlag Berlin Heidelberg New York
a member of BertelsmannSpringer Science+Business Media GmbH

http://www.springer.de

Printed in Germany

Typesetting: Camera-ready $\TeX$ output by the authors

SPIN: 10847519 41/3142-543210 - Printed on acid-free paper

Contents

0. Introduction and Summary of Results

In the last decade quantum field theory and string theory have strongly impacted many areas of mathematics, especially the geometry and topology of low dimensional manifolds. In particular, a wealth of intriguing mathematical structures were discovered to be inherent to so called *topological quantum field theories* (TQFT's) and *conformal field theories* (CFT's). Originally, these notions refer to a class of concrete physical quantum field theories, among which three dimensional Chern-Simons theory and two dimensional rational conformal field theory are some of the most prominent ones. It was soon realized that the abstract setting of category theory makes it possible to efficiently organize the zoo of data and structures of these field theories. Eventually, TQFT's evolved into purely mathematical notions, defined axiomatically in the language of categories and functors. Axiomatic TQFT's and similar theories are, therefore, in nature rather similar to other functors in algebraic topology, such as homology. Atiyah was the first mathematician to cast the notion of TQFT's into an axiomatic framework in his seminal work [Ati88]. Independently and at about the same time G. Segal [Seg88] formulates a mathematical definition of CFT's, which very similarly based on categories and functors. The notion of *extended TQFT's* that we will introduce here and on which our constructions will be based involves higher category theory, namely double categories and double functors. It thus contains both Atiyah's notion of a TQFT in dimension three and Segal's notion of CFT as special cases, though they appear on different categorical levels. The definition will not only be a natural and conceptual unification of previous theories, but further abstractions will allow us to construct new classes of TQFT's, namely *non-semisimple TQFT's*, that are manifestly different from other combinatorially defined ones and in some cases describe TQFT's based on classical gauge theories.

In order to explain the main results we give next our definition of an *extended TQFT*. Let us start with a recollection of Atiyah's axioms for manifolds with smooth boundaries. This will be subsequently generalized, using double categories to describe manifolds with corners and double functors to define an extended TQFT.

0.1 Atiyah's TQFT Axioms via Categories

Following to the axioms of Atiyah [Ati88], a TQFT $\mathcal{V}$ in dimension $d+1$ assigns to a d-dimensional oriented manifold Σ^d a vector space $\mathcal{V}(\Sigma^d)$, and to an oriented

$d+1$-dimensional manifold, whose boundary is a disjoint union of d-dimensional manifolds $-\Sigma_0^d$ and Σ_1^d, a linear map $\mathcal{V}(M^{d+1}) : \mathcal{V}(\Sigma_0^d) \to \mathcal{V}(\Sigma_1^d)$. The manifold $-\Sigma_0^d$ is Σ_0^d with the opposite orientation. The *gluing axiom* in [Ati88] requires that if we glue two such $d+1$-manifolds together along a common (closed) d-submanifold of in their boundaries, the linear map for the composite has to be the composition of the linear maps of the individual $d+1$-manifolds.

Using the language of categories and functors, as in [Mac88], we can state Atiyah's axioms very concisely as follows:

Definition 0.1.1 ([Ati88]). *A topological quantum field theory in dimension d is a functor between* symmetric monoidal categories *[Mac88] as follows:*

$$\mathcal{V} \; : \; \mathbf{Cob}_{d+1} \longrightarrow \Bbbk\text{-vect}\,.$$

Here $\Bbbk$-vect denotes the category, whose objects are finite dimensional vector spaces over a field $\Bbbk$, which we assume to be perfect, for instance, a field of characteristic 0. The set of morphisms between two vector spaces is simply the set of linear maps with the usual composition. The category $\mathbf{Cob}_{d+1}$ has as objects closed oriented d-dimensional manifolds. A morphism between two such d-manifolds Σ_0^d and Σ_1^d is a $d+1$-cobordism, meaning an oriented $d+1$-dimensional manifold, M^{d+1}, whose boundary $\partial M^{d+1} = -\Sigma_0^d \sqcup \Sigma_1^d$ is the disjoint union of the two d-manifolds. (Strictly speaking we consider as morphisms cobordisms modulo relative homeomorphisms or diffeomorphisms). Given another cobordism N^{d+1}, between Σ_1^d and Σ_2^d in the above sense, we define the composite by $M^{d+1} \circ N^{d+1} = M^{d+1} \sqcup_{\Sigma_1^d} N^{d+1}$. The union $\sqcup_{\Sigma_1^d}$ stands for the quotient space of the disjoint union, in which we have glued the two $d+1$-manifolds along the common (closed) surface Σ_1^d in their boundaries. The identity on a d-manifold Σ^d in $\mathbf{Cob}_{d+1}$ is easily identified as the (class of) the cylinder $\Sigma^d \times [0,1]$ with canonical boundary identifications. Atiyah's gluing axiom is now implied by functoriality: $\mathcal{V}(M \circ N) = \mathcal{V}(M) \cdot \mathcal{V}(N)$.

The term *monoidal* in Definition 0.1.1 means that $\mathcal{V}$ respects the natural tensor structures on the two categories. The tensor product on $\mathbf{Cob}_{d+1}$ is given by disjoint union, and the product on $\Bbbk$-vect – by the usual tensor product $\otimes_{\Bbbk}$. These conditions allow us to infer the remaining set of axioms from [Ati88].

Note, that in their original form Atiyah's axioms associate to any $d+1$-manifold M with boundary a *vector* $\mathcal{V}(M)$ in $\mathcal{V}(\partial M)$. The assignment of linear maps and the composition rule follow from the additional axioms for tensor products and duals via the identifications $\mathcal{V}(\partial M) = \mathcal{V}(-\Sigma_0 \sqcup \Sigma_1) \simeq \mathcal{V}(\Sigma_0)^* \otimes \mathcal{V}(\Sigma_1) \simeq \mathrm{Hom}(\mathcal{V}(\Sigma_0), \mathcal{V}(\Sigma_1))$.

The axioms in [Ati88] were, in particular, inspired by Witten's investigation [Wit89] of the Chern-Simons field theory, giving rise to a TQFT with 3-dimensional cobordisms ($d = 2$). Although the functional integral formulation used by Witten is a priori not rigorous, it lends itself nicely to illustrate the implied properties of a TQFT as outlined in Appendix A. Witten also writes down formulae for the heuristically defined partition sums of the Chern-Simons theory for some closed manifolds obtained via surgery, but the arguments used to identify them as topological invariants are purely physical and far from rigorous. Almost simultaneous to Witten's work Reshetikhin and Turaev, in their ground breaking paper [RT91], gave a rigorous definition of the 3-manifold invariants using quantum groups as well as a systematic procedure for their computation. These invariants can be considered as a mathematical realization of Witten's Chern-Simons theory. The generalization of their constructions to cobordisms and TQFT's is developed in detail in Turaev's book [Tur94].

At around the same time Segal, Moore and Seiberg found similar categorical structures for conformal field theories on surfaces, see [MS89] and references therein. Witten [Wit89] also realized (again on the heuristic level of physical quantum field theories) that the restriction of a Chern-Simons theory on a 3-manifold M to its boundary ∂M yields precisely such a CFT. The important new ingredient in CFT is that one considers surfaces with boundaries so that surfaces themselves can be "sewn together" along the circles in their boundaries. In the physical interpretation the holes or punctures in the 2-dimensional surface give the locations where charges are inserted into the theory. The corresponding observables in Chern-Simons theory are currents in 3-dimensions that run along Wilson lines, thus creating a charge at their end points on the bounding surface. See Appendix A for a more detailed exposition. This leads us to consider 3-cobordisms, from which we excise tubular neighborhoods of embedded lines. The excision at an end point of a line on the surface thus results in the removal of a disc from the surface at this location.

In order to generalize Atiyah's axioms to include also the sewing operations of the related CFT's besides the gluing axioms of Chern-Simons theory, we need to extend them to the 3-manifolds with corners obtained by tubular excisions along lines. As a consequence, the notion of a TQFT can no longer be formulated by ordinary categories and functors, but we have to pass to higher category theory. The definition of an extended TQFT as a double functor of double categories is the subject of the next paragraphs.

0.2 Double Categories

The fact that a unification of Atiyah's TQFT axioms and the axioms of CFT requires the use of some sort of higher category theory was realized by many people independently. The simplest generalization is that of a *2-category*, which allows us to talk about morphisms between morphisms. In this section let us give the basic definitions of 2-categories and double categories. The outlined definitions will be sufficient for the constructions given in this book. For more details on the theory of 2-categories we refer the reader to the original papers [Bén67], [KS74], or [KV94]. See also Appendix B.1.

A 2-category $\mathfrak{C}$ firstly consists of an ordinary category $\mathfrak{C}_1$, with objects and 1-morphisms between them, as well as a composition operation between 1-morphisms if the target and source objects are matching. In addition, we associate to any two 1-morphisms $A_i : O_s \to O_t$ and $A_f : O_s \to O_t$ with the same source and target a set $\mathrm{Hom}_2(A_i, A_f)$ of 2-morphisms, denoted $B : A_i \Rightarrow A_f$. We have a (vertical) composition operation between two 2-morphisms if the target 1-morphism of the one coincides with the source 1-morphism of the other. The composition of 1-morphisms extends to a second type of (horizontal) composition of 2-morphisms over an intermediate object. Finally, the two compositions are required to be mutually distributive.

One standard example is given by the 2-category **Cat** of categories. The objects of **Cat** are *essentially small categories* (such that are equivalent to categories whose objects form a set), the 1-morphism class $\mathrm{Hom}_1(\mathcal{C}_s, \mathcal{C}_t)$ consists of functors from $\mathcal{C}_s$ to $\mathcal{C}_t$, and $\mathrm{Hom}_2(\mathcal{F}_i, \mathcal{F}_f)$, for functors between the same pair of categories, is the set of natural transformations from $\mathcal{F}_i$ to $\mathcal{F}_f$. The vertical and horizontal compositions of natural transformations are given in standard ways, see [Mac88]. We shall be interested in its 2-subcategory $\mathbf{AbCat}$ for a given field $\Bbbk$ of essentially small, $\Bbbk$-linear, *abelian* categories with additional finiteness conditions (see Chapter 4), *left exact* functors, and natural transformations.

Another 2-category of interest is that of *relative cobordisms*, $\mathfrak{C} = \mathbf{Cob}^{rel}_{d+1}$. The underlying category $\mathfrak{C}_1$ of objects and 1-morphisms is identical to the cobordism category $\mathbf{Cob}_d$ from above. For two d-dimensional cobordisms M^d and N^d between $d-1$-manifolds X^{d-1} and Y^{d-1}, we can consider the *closed* d-manifold $S = M^d \cup (Z^{d-1} \times [0,1]) \cup N^d$, where Z^{d-1} is the disjoint union $X^{d-1} \sqcup Y^{d-1}$. In the definition of S we make identifications $\partial M^d \cong Z^{d-1} \times 0$ and $\partial N^d \cong Z^{d-1} \times 1$. The 2-morphisms $M \to N$ are given by $d+1$-dimensional manifolds W with the boundary $\partial W \cong S$.

The (vertical) composition of 2-morphisms over 1-morphisms is given by gluing the $d+1$-manifolds together over the bounding d-manifolds. For the (horizontal) composition over objects we glue the $d+1$-manifolds together along the cylinders over the respective source and target $d-1$-manifolds.

In this book we define TQFT's using certain generalizations of 2-categories and 2-functors, namely *double categories* and *double functors*. Double categories were introduced by Ehresmann in [Ehr63a]. Let us give an equivalent version of his definition.

Definition 0.2.1. *A double category $\mathfrak{D}$ consists firstly of a class $\mathfrak{D}_0$ of objects.*

For any pair of objects, X and Y, there are sets $\mathrm{Hom}_1^v(X,Y)$ and $\mathrm{Hom}_1^h(X,Y)$ of vertical and horizontal 1-morphisms or 1-arrows. The objects and the vertical 1-morphisms by themselves form an ordinary category $\mathfrak{D}_1^v$, and an analogous category $\mathfrak{D}_1^h$ for the horizontal 1-morphisms.

We call a square $\mathbf{S}$ a set of four objects X_{ij}, with $i,j = 0,1$, two vertical 1-morphisms $g_j \in \mathrm{Hom}_1^v(X_{0j}, X_{1j})$ and two horizontal 1-morphisms $f_j \in \mathrm{Hom}_1^h(X_{j0}, X_{j1})$ so that they can be arranged in a square diagram as follows:

$$\mathbf{S} \;=\; \begin{array}{ccc} X_{00} & \xrightarrow{\;f_0\;} & X_{01} \\ {\scriptstyle g_0}\big\downarrow & \overset{\alpha}{\swarrow\!\!\!\!\swarrow} & \big\downarrow{\scriptstyle g_1} \\ X_{10} & \xrightarrow[\;f_1\;]{} & X_{11} \end{array} \quad , \qquad \alpha \in \mathrm{Hom}_2(\mathbf{S}). \tag{0.2.1}$$

For any square, $\mathbf{S}$*, one has a set* $\mathrm{Hom}_2(\mathbf{S})$ *of 2-morphisms. We often include an element* $\alpha \in \mathrm{Hom}_2(\mathbf{S})$ *in the diagrammatic notation as above.*

If for two squares, $\mathbf{S}$ *and* $\mathbf{S}'$*, we have* $g_1 = g_0'$ *for the vertical 1-morphisms, then we define the horizontal composite square* $\mathbf{S}' \circ_h \mathbf{S}$ *to be the one with vertical 1-morphisms* g_0 *and* g_1'*, and horizontal 1-morphisms* $f_0' \circ_h f_0$ *and* $f_0' \circ_h f_0$*. A double category is equipped with a horizontal composition*

$$\circ_h : \mathrm{Hom}_2(\mathbf{S}') \times \mathrm{Hom}_2(\mathbf{S}) \to \mathrm{Hom}_2(\mathbf{S}' \circ_h \mathbf{S}) : (\alpha, \beta) \mapsto \alpha \circ_h \beta.$$

Analogously, there is a vertical composition $\gamma \circ_v \alpha$ *declared if the target horizontal 1-morphism of* α *coincides with the source horizontal 1-morphism of* γ*.*

We require both compositions to give rise to categories $(\mathfrak{D}_2^h, \circ_h)$ *and* $(\mathfrak{D}_2^v, \circ_v)$*, whose objects are the vertical and horizontal 1-morphisms respectively. In particular,* $\circ_h$ *and* $\circ_v$ *are associative.*

Finally, the interchange law for double categories states that two composition are mutually distributive. More precisely, suppose we have twelve 1-morphisms arranged in a square of squares as depicted below, and α*,* β*,* γ*, and* δ *are 2-morphisms in the* Hom_2*-sets of the four squares of 1-morphisms:*

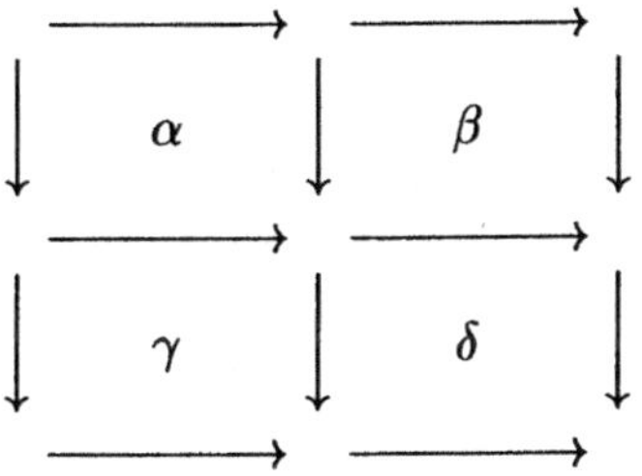

We require that the operations of performing the horizontal and vertical compositions can be interchanged:

$$(\delta \circ_h \gamma) \circ_v (\beta \circ_h \alpha) = (\delta \circ_v \beta) \circ_h (\gamma \circ_v \alpha). \tag{0.2.2}$$

In Appendix B.2 we recall the original definition of Ehresmann [Ehr63a], which does not require the distinction between 0-, 1-, and 2-morphisms.

From a double category $\mathfrak{D}$ we can readily extract a 2-category if we consider only squares, for which we have $X_{0j} = X_{1j}$ and g_j is the identity for $j = 0, 1$. Conversely, if we have a 2-category $\mathfrak{C}$ we can construct a double category $\mathfrak{D} = Q\mathfrak{C}$, called the double category of *quintets* of $\mathfrak{C}$ in [Ehr63b], as follows.

We choose both the horizontal and vertical categories to be identical to the category underlying $\mathfrak{C}$, that is, $(Q\mathfrak{C})_1^v = \mathfrak{C}_1$ and $(Q\mathfrak{C})_1^h = \mathfrak{C}_1$. For a square $\mathbf{S}$ of 1-morphisms as in (0.2.1) the associated 2-morphism sets are

$$\mathrm{Hom}_2^{\mathfrak{D}}(\mathbf{S}) = \mathrm{Hom}_2^{\mathfrak{C}}(g_1 \circ f_0, f_1 \circ g_0).$$

The horizontal composition of 2-morphisms in $\mathfrak{D} = Q\mathfrak{C}$ is the obvious composite $g_1' \circ f_0' \circ f_0 \Rightarrow f_1' \circ g_0' \circ f_0 = f_1' \circ g_1 \circ f_0 \Rightarrow f_1' \circ f_1 \circ g_0$.

0.3 Extended TQFT's

This quintet construction yields the first example relevant to our definition of a TQFT, namely the double category $\mathcal{Q}\mathbf{AbCat}$, of $\mathbb{k}$-linear abelian categories, left exact functors and natural transformations. The precise definition of the topological double category $\widetilde{\mathcal{C}ob}^{\cap}$ used in our definition of a TQFT is more involved. In outline it is as follows.

The set of objects is given as $\{S^{\sqcup a} : a \in \mathbb{Z}_{\geqslant 0}\}$, where $S^{\sqcup a}$ is a fixed, numbered, disjoint union of a oriented circles. The set of horizontal 1-morphisms $\mathrm{Hom}_1^h(a,b)$ between two such 1-manifolds consists of *connected* oriented surfaces, whose boundary is parametrized by (and is homeomorphic to) $-S^{\sqcup a} \sqcup S^{\sqcup b}$. All such surfaces of the same genus g are homeomorphic, so we may leave only one representative in each homeomorphism class, parametrized by $g \in \mathbb{Z}_{\geqslant 0}$. We shall, however, find it technically more convenient to keep several isomorphic copies of horizontal morphisms, namely $\frac{1}{g+1}\binom{2g}{g}$ of them for the class with genus g. To be more concrete, the horizontal morphisms are in one to one correspondence with the set of combinatorial plane graphs that consist of an interval at the boundary of the half-plane and of g non-intersecting arcs in the half plane with endpoints in the interval. See Sect. 7.1.1 for precise definitions and illustrations. The standard surface Σ_G corresponding to the graph G is then obtained as the boundary of a thickening of G in $\mathbb{R}^3$.

The vertical 1-morphism set $\mathrm{Hom}_1^v(a,b)$ of $\widetilde{\mathcal{C}ob}^{\cap}$ is empty if $a \neq b$. We identify the endomorphism set $\mathrm{Hom}_1^v(a,a) = S_a$ with the symmetric group of an a-element set. Hence, in a square $\mathbf{S}$ the two horizontal 1-morphisms Σ_G and Σ_H always lie in the same set $\mathrm{Hom}_1^h(a,b)$ as shown in the following diagram.

$$\mathbf{S} \;=\; \begin{array}{ccc} a & \xrightarrow{\ \Sigma_G\ } & b \\ {\scriptstyle\alpha}\downarrow & \overset{\widetilde{M}}{\swArrow} & \downarrow{\scriptstyle\beta} \\ a & \xrightarrow[\ \Sigma_H\ ]{} & b \end{array} \quad .$$

The a source and b target bounding circles of each of the standard surfaces are numbered due to parametrization homeomorphism. We now sew the surfaces together by connecting the j-th source circle of Σ_G to the $\alpha(j)$-th source circle in the boundary of Σ_H by a cylinder $S^1 \times [0,1]$. Here $\alpha \in S_a$ is the left vertical 1-morphism of the diagram. Doing the same for the target circles, we obtain a closed surface $\Sigma_{\mathbf{S}}$. The cylinders are interpreted as Wilson lines in Chern-Simons theory.

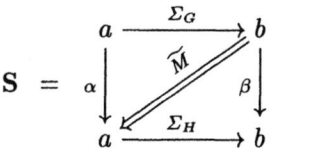

A 2-morphism for a given square $\mathbf{S}$ is now a homeomorphism class of triples $\widetilde{M} = [(M,\phi,\alpha)]$, where each triple (M,ϕ,α) consists of a compact, oriented 3-manifold with corners, M, a homeomorphism, $\phi : \partial M \xrightarrow{\sim} \Sigma_{\mathbf{S}}$, and a 2-framing of its tangent bundle, $\alpha : TM \oplus TM \xrightarrow{\sim} \mathbb{R}^6 \times M$. The additional structure of a 2-framing is motivated by the Chern-Simons gauge theory. In this book we choose an equivalent description as an extensions by signatures of bounding 4-manifolds. In [BHMV95] yet another equivalent definition of this extension is given using so called p_1-structures. Moreover, Walker [Wal] and later Turaev [Tur94] define extensions by enlarging the set of objects to pairs (Σ, L), where Σ is a surface and $L \subset H_1(\Sigma)$ a Lagrangian subspace to indicate a bounding handlebody. The standard surfaces in our approach are thus equipped with fixed Lagrangians. Moreover, the cocycle involving Lagrangians that is put into their definitions of linear or cobordism categories is implicit in our definition via Wall's signature formula.

If we disregard the signature extension of the 3-manifolds, we have natural and well defined vertical and horizontal compositions obtained by gluing two 3-manifolds together along the horizontal pieces Σ_G in their boundary or the cylindrical pieces respectively. One readily verifies that these compositions define a double category $\mathcal{C}ob^{\cap}$. The compositions can be extended to 3-manifolds with 2-framings so that the axioms of a double category are fulfilled. In analogy to group theory we can thus view $\widetilde{\mathcal{C}ob}^{\cap}$ as a central extension

$$1 \to \Omega_4 \hookrightarrow \widetilde{\mathcal{C}ob}^{\cap} \twoheadrightarrow \mathcal{C}ob^{\cap} \to 1\,, \qquad (0.3.1)$$

where Ω_4 is the smooth 4-dimensional cobordism group. Note that Ω_4 is a free cyclic group generated by $[\mathbf{CP}^2]$ and the signature map $\Omega_4 \xrightarrow{\cong} \mathbb{Z} : [W] \mapsto \mathrm{sign}(W)$ is an isomorphism (see, for example [GS99] Sect. 9.1).

There is a natural notion of a strict double functor between double categories. Analogous to ordinary functors it is defined as a map between classes of objects, 1-morphisms and 2-morphisms of two double categories, which is a functor with respect to both horizontal and vertical (ordinary) category structure. Its weak version – double pseudofunctor – is defined in Appendix B.2.

For objects of the main two double categories that we consider, there is a natural tensor multiplication. For $\widetilde{\mathcal{C}ob}^{\cap}$ it is given by the disjoint unions of circles, and for $\mathcal{Q}\mathbf{AbCat}$ by Deligne's tensor product $\boxtimes$ of abelian categories [Del91]. We are now in a position to give a definition of an extended topological quantum field theory.

Definition 0.3.1. *An extended TQFT over a field* $\Bbbk$ *is a double pseudofunctor*

$$\mathcal{V} : \widetilde{\mathcal{C}ob}^{\cap} \longrightarrow \mathcal{Q}\mathbf{AbCat}\,,$$

between double categories as above, which is compatible with tensor structures on the level of objects.

Hence,

$$\mathcal{V}(S^{\sqcup a}) \cong \mathcal{C} \boxtimes \ldots \boxtimes \mathcal{C}\,,$$

where $\mathcal{C}$ is the category associated to one circle. Since this category is of obvious interest as a generator, let us give a formal definition as follows.

Definition 0.3.2. *Let* $\mathcal{V}$ *be an extended TQFT in the sense of Definition 0.3.1. The circle category of* $\mathcal{V}$ *is then defined as*

$$\mathcal{C}_{\mathcal{V}} = \mathcal{V}(S^1).$$

0.4 Statement of Main Result on the Class of Extended TQFT's

To state our main result let us note that circle categories always carry a braided monoidal structure. A few additional assumptions lead to the following definition. The various adjectives will be briefly explained after the definition.

Definition 0.4.1. *A modular category over a field $\Bbbk$ is a bounded abelian, rigid, monoidal, braided, balanced (ribbon) category $\mathcal{C}$ with a special Hopf pairing that is non-degenerate. The endomorphism ring of the unit object is supposed to be $\Bbbk$.*

In particular, $\mathcal{C}$ is $\Bbbk$-linear. The precise definitions and formulations of these conditions, which are all rather natural, will be given in Chapter 4. The notion of an *abelian* category allows us to consider subobjects, quotients and decompositions of objects. The properties *monoidal* and *rigid* imply the existence of tensor products $X \otimes Y$ and duals $X^\vee$ of objects. These notions are part of classical category theory as described in [Mac88]. The word *braided* implies a natural isomorphism $c_{X,Y} : X \otimes Y \xrightarrow{\sim} Y \otimes X$, which does not necessarily square to the identity. *Balanced* (or ribbon) refers to a natural isomorphisms $X \xrightarrow{\sim} X^{\vee\vee}$ compatible with the braiding. Braiding and balancing in categories were introduced, for example, by Joyal and Street [JS91], see also [RT90]. We call a category *bounded* if it is equivalent to a category of finite dimensional modules over a finite dimensional algebra. This turns out to be equivalent to the existence of the coend $\mathbb{F} = \int^{X \in \mathcal{C}} X \boxtimes X^\vee$ in $\mathcal{C} \boxtimes \mathcal{C}$. The details on coends can be found in Chapter 5. For example, a semisimple category is bounded precisely when the set of isomorphism classes of simple objects is finite. The new feature of our approach is that an abelian modular category does not have to be semisimple. The Hopf pairing $\omega : F \otimes F \to 1$, which we require to be non-degenerate, is defined in Sect. 5.2.2 for the coend $F = \otimes\mathbb{F} = \int^{X \in \mathcal{C}} X \otimes X^\vee \in \mathcal{C}$, which is a Hopf algebra.

Now we state our main result.

Theorem 0.4.2. *For every modular category $\mathcal{C}$ there exists an extended TQFT $\mathcal{V}_\mathcal{C}$, which has $\mathcal{C}$ as circle category.*

In summary, we have an assignment $\mathfrak{E} \to \mathfrak{M} : \mathcal{V} \mapsto \mathcal{C}_\mathcal{V}$ from the class $\mathfrak{E}$ of extended TQFT's to the class $\mathfrak{M}$ of modular categories, and by Theorem 0.4.2 a map $\mathfrak{M} \to \mathfrak{E} : \mathcal{C} \mapsto \mathcal{V}_\mathcal{C}$ such that the composite on $\mathfrak{M}$ is identity.

Replacing a modular category $\mathcal{C}$ with an equivalent one, and replacing the structure of symmetric monoidal 2-category of **AbCat** with an equivalent one, we can achieve that $\mathcal{V}_\mathcal{C}$ is a strict double functor.

0.4.1 Specializations and Generalizations

1. Disconnected Surfaces. One assumption in our definition of $\widetilde{\mathcal{Cob}}^{\cap}$ is that all surfaces representing 1-morphisms should be connected. This constraint can be, in principle, overcome, by constructing of a TQFT for disconnected surfaces from

TQFT's for connected ones following the procedure described in [Ker98b]. In the case of non-semisimple TQFT's this requires a slight modification of the TQFT axioms, which have been introduced in [Ker98b] as *half-projective* or *non-semisimple* TQFT's. We will not carry out the generalization to disconnected surfaces in this book, since it requires a considerable formal apparatus to describe as a general theory.

2. Conformal Field Theory. In the spirit of CFT one might consider a double category **Surf** of surfaces. The latter has the same objects and 1-morphisms as $\widetilde{\mathcal{Cob}}^{\cap}$, namely circles and surfaces. The 2-morphisms, however, are homeomorphisms between the surfaces instead of cobordisms. In Chapter 1 we identify the mapping class group of a surface with the invertible cobordism classes from this surface to itself. Hence, we can think of $\mathbf{Surf} \subset \widetilde{\mathcal{Cob}}^{\cap}$ as a double subcategory. The restriction $\mathcal{V}_{\mathcal{C}}^{mod} : \mathbf{Surf} \to \mathcal{Q}\mathbf{AbCat}$ of a TQFT double functor, $\mathcal{V}_{\mathcal{C}}$, turns out to be a version of what Segal, Moore and Seiberg call a *modular functor*. See again Appendix A for more context and details. The double functor $\mathcal{V}_{\mathcal{C}}^{mod}$ also implies projective representations of the mapping class groups that are compatible with respect to concatenations of surfaces.

3. Atiyah's TQFT. In the double category $\widetilde{\mathcal{Cob}}^{\cap}$ we can consider the subcategory, in which all objects are empty 1-manifolds. This means we are dealing with closed surfaces and the only relevant composition is the gluing over these surfaces in vertical direction. We thus obtain a central extension $\widetilde{\mathbf{Cob}_3}$ of a version of $\mathbf{Cob}_3$ as used in Atiyah's definition. The functors associated to closed surfaces, seen as cobordisms between empty 1-manifolds, are naturally identified with vector spaces. As a result, we obtain a *projective* TQFT given by a functor $\mathcal{V}_{\mathcal{C}}^{0} : \widetilde{\mathbf{Cob}_3} \to \Bbbk\text{-vect}$. To a 2-framed *closed* manifold, seen as a cobordism between empty surfaces, the projective TQFT thus assigns a number, which up to normalization is the associated invariant of the underlying topological 3-manifold.

4. Reshetikhin-Turaev Theory. Reshetikhin and Turaev gave in [RT91] a construction of a projective TQFT in the sense of Atiyah as in Definition 0.1.1. The details are worked out in Turaev's book [Tur94]. They use a semisimple modular category $\mathcal{C}$ as input data. Modularity is defined differently in [Tur94] and in our book, however, both definitions are equivalent for semisimple categories as we show in Section 7.4.1. When $\mathcal{C}$ is a semisimple modular category, the restriction of the Reshetikhin-Turaev construction to connected surfaces is a TQFT, isomorphic to the above $\mathcal{V}_{\mathcal{C}}^{0}$. Besides, in semisimple case our construction extends without problems to disconnected surfaces as well, giving a complete agreement of the theories. Semisimple categories can be produced, for instance, as semisimple trace quotients from non-semisimple representation categories of quantum groups [RT91, And92, Ker92, TW93].

5. Hennings Theory. In [Hen96] Hennings defines an invariant of closed 3-manifolds directly from a possibly non-semisimple, quasi-triangular ribbon Hopf algebra $\mathcal{A}$. This invariant naturally extends to a TQFT as shown in [Ker97]. The invariants of closed 3-manifolds and TQFT's are again special cases of $\mathcal{V}_{\mathcal{C}}$, if we insert the representation category $\mathcal{C} = \mathcal{A}-\text{mod}$ and restrict ourselves to closed surfaces.

6. General Vertical Surfaces. An obvious question is whether it is possible and sensible to extend our construction to more general classes of surfaces representing the vertical 1-morphisms instead of mere cylinders connecting the boundary pieces of different surfaces. A small modification, that we can easily deal with in our formalism, is to allow a cylinder to connect a boundary component of a surface to another one of the *same* surface. This means that the vertical category is identical with the category of 1-dimensional cobordisms $\mathbf{Cob}_1$ and we can have arbitrary objects in the corners of a square. Thus, a morphism in $\mathrm{Hom}_1^v(a,b)$ is a 1-manifold with a source endpoints and b target endpoints. The vertical surfaces are obtained by taking the Cartesian product with S^1. Clearly, the group of invertible 1-cobordisms $\mathrm{Aut}_1^v(a)$ is identical with S_a, which is the restriction we have used in the definition of $\widetilde{Cob}^{\cap}$. Further generalizations to other vertical surfaces are possible but quickly become impractical.

0.4.2 Strategy of Construction and Summary of Content

The discovery of new 3-manifold invariants was preceded by constructions of new families of knot and link invariants, for example the Jones polynomial in [Jon87] and the ribbon invariants defined via quantum groups by Reshetikhin and Turaev [RT90]. Studying partition functions of the Chern-Simons quantum field theory on closed 3-manifolds Witten wrote down in [Wit89] remarkable linear relations between the partition functions of a 3-manifold M with included Wilson lines along $\mathcal{L}$ with surgery performed along a link and the ones of the original 3-manifold. For $M = S^1 \times S^2$ and $\mathcal{L}$ transverse to the S^2-fibres Witten's treatment suggests to compute the partition functions as traces over the associated braid group element in the representations obtained from the corresponding conformal field theory. Although written at the physical level of rigor, his program for computing partition functions indicated the existence of a new type of 3-manifold invariants.

In their famous article [RT91] Reshetikhin and Turaev succeeded for the first time to construct quantum 3-manifold invariants in a rigorous and mathematically consistent way. Although partially inspired by Witten's heuristic ideas, the success of their construction was based on the discovery and use of two new crucial ingredients. The first is the algebraic data of quantum groups that allows them produce a large family of invariants of framed links in S^3. The other is the use of Kirby's calculus of links, which establishes when two framed links in S^3 describe the same 3-manifold (with empty boundary) via surgery. They discover and prove that a combination of their link invariants is invariant under Kirby's moves and thus constitutes an invariant of 3-manifolds. Moreover, they generalize their constructions to assign linear maps to cobordisms, which are now represented by ribbon graphs and links rather than only links in S^3, so that they obtain a TQFT in the sense of Atiyah. The approach of Reshetikhin and Turaev [RT91] using embedded ribbon graphs was fully realized by Turaev in his book [Tur94] for the case of semisimple categories. The related question of extending Kirby's calculus of links to manifolds with boundary to determine which tangles describe the same cobordisms is addressed in [Tur94] and in some versions requires additional moves [MP94] and [Ker99].

In our construction of the extended TQFT functor, as given in Theorem 0.4.2, we follow an analogous strategy of first producing a combinatorial surgery presentation of the relative 3-cobordisms with corners and then assigning the algebraic data of an abelian modular category to it. The combinatorial data replacing a framed link also needs to encode the two composition structures. Hence, it will be formulated as a double category $\mathcal{T}gl^{\cap}$ itself, whose 2-morphisms are equivalence classes of tangles of certain type. The presentation respects the two compositions structures and, hence, constitutes an invertible double functor $\mathfrak{Surg}$. Likewise, we formulate the assignment of the algebraic data to combinatorial tangles as a double functor $\mathcal{V}^*_{\mathcal{C}}$. In summary, we construct the TQFT functor $\mathcal{V}_{\mathcal{C}}$ as the composite of two double functors as follows:

$$\mathcal{V}_{\mathcal{C}} : \widetilde{\mathcal{C}ob}^{\cap} \xrightarrow{\cong \mathfrak{Surg}^{-1}} \mathcal{T}gl^{\cap} \xrightarrow{\mathcal{V}^*_{\mathcal{C}}} \mathcal{Q}\mathbf{AbCat}. \qquad (0.4.1)$$

The methods employed here are based on the techniques developed in previous work of the authors. In this book the techniques are further developed and refined. The surgery presentation given by $\mathfrak{Surg}$ generalizes the one from [Ker99] for ordinary cobordisms between closed surfaces. In addition to this, we have to include the cylindrical boundary pieces in the definition and presentation of $\widetilde{\mathcal{C}ob}^{\cap}$ and to define the horizontal composition both in $\widetilde{\mathcal{C}ob}^{\cap}$ and $\mathcal{T}gl^{\cap}$, so that we obtain double categories and double functors compatible with the 2-framing extension. The algebraic assignment $\mathcal{V}^*_{\mathcal{C}}$, for possibly non-semisimple $\mathcal{C}$, generalizes the methods used in the construction of 3-manifold invariants and representations of the mapping class groups in [Lyu95c, Lyu96]. In particular, we extend here the coend techniques to also construct functors and natural transformations, instead of just objects like $\mathbb{F}$ and F. The appearance of the symmetric group necessitates more careful investigations of its action on $\boxtimes$-products of abelian categories. We also expand and refine the theory of braided Hopf algebras in braided tensor categories, which allow very conceptual and concise invariance proofs of dictionary type.

We have organized this book by devoting one or two chapters to the construction and investigation of each of the five ingredients of (0.4.1).

The first three chapters of this book, therefore, concern themselves with the combinatorial representation $\mathfrak{Surg} : \mathcal{T}gl \xrightarrow{\sim} \widetilde{\mathcal{C}ob}$. The double categories $\mathcal{T}gl$ and $\widetilde{\mathcal{C}ob}$ differ from $\mathcal{T}gl^{\cap}$ and $\widetilde{\mathcal{C}ob}^{\cap}$ appearing in (0.4.1) only in that they have one horizontal 1-morphism instead of several isomorphic 1-morphisms. In Chapters 1 and 2 we discuss the definitions and characteristics of the double categories $\widetilde{\mathcal{C}ob}$ and $\mathcal{T}gl$ respectively. The double isomorphism functor $\mathfrak{Surg}$ is constructed in Chapter 3.

More specifically, we start in Chapter 1 with the discussion of the double category $\mathcal{C}ob$ of relative 3-cobordisms. We discover that $\mathcal{C}ob$ contains a canonical balanced braided tensor category. The mapping class group of a surface is identified with the group of invertible cobordisms of $\mathcal{C}ob$ on that surface object. As a special subgroup we also discuss the image of the framed braid groups on a surface in the corresponding mapping class groups. In the last part of Chapter 1 we define the 2-framing extension $\widetilde{\mathcal{C}ob}$ of $\mathcal{C}ob$ using bounding 4-manifolds. Gluings are extended to 3-dimensional cylinders over the respective surfaces. We show that these operations factor into homeomorphism and cobordism classes, and verify that the composition structure on the classes fulfills the axioms of a double category.

In Chapter 2 the tangle double category $\mathcal{T}gl$ is introduced. The 2-morphisms are given as equivalence classes of generic projections of *admissible tangles* with several types of strands. The equivalences are expressed in the form of a list of moves. A large part of this chapter is devoted to finding equivalent description of the category $\mathcal{T}gl$. In particular, we show that $\mathcal{T}gl \cong \mathcal{T}gl^{\mathbf{BL}}_{S^2}$, where $\mathcal{T}gl^{\mathbf{BL}}_{S^2}$ are classes of bridged link diagrams, in the sense of [Ker98a], in the thickened sphere $S^2 \times [0,1]$. The version $\mathcal{T}gl^{\mathbf{BL}}_{S^2}$ will be closer to the surgery presentations of cobordisms, while $\mathcal{T}gl$ is more adequate for the assignment of the algebraic data. Finally, we defined vertical and horizontal compositions for $\mathcal{T}gl$ and prove that they do in fact give rise to a double category. The compositions are mild modifications of the usual stacking and juxtaposition operations.

In Chapter 3 the functor $\mathfrak{Surg} : \mathcal{T}gl^{\mathbf{BL}}_{S^2} \twoheadrightarrow \widetilde{\mathcal{C}ob}$ is constructed by doing surgery along a link in a sum of handlebodies obtained from a respective tangle and thus generalizes the presentation [Ker99] for closed surfaces. We recall the standard tools such as surgery manipulations of handle attachments and Morse and Cerf theory, and review the resulting surgery calculi on non-simply connected manifolds. We prove that the surgery operation factors into an isomorphism $\mathfrak{Surg}$ on the equivalence classes of $\mathcal{T}gl^{\mathbf{BL}}_{S^2}$ and $\widetilde{\mathcal{C}ob}$. We also show that $\mathfrak{Surg}$ respects the vertical and horizontal compositions. Functoriality for the latter requires a more detailed analysis of the handle structure of the bounding 4-manifolds.

Chapter 4 through 7 are concerned with the second composite $\mathcal{V}^* : \mathcal{T}gl \to \mathcal{Q}\mathbf{AbCat}$ of the TQFT double functor as it is given in (0.4.1).

In Chapter 4 the algebraic building blocks for the construction of the functors $\mathcal{V}_\mathcal{C}$ are laid. In particular, we give a thorough discussion of the properties of ordinary braided tensor categories (BTC's) such as braided, reflexive balancings, braided Hopf algebras in BTC's and their integrals. Graphical calculi for both BTC's and Hopf algebras are introduced. We study Hopf pairings and find criteria of their non-degeneracy (side-invertibility) in terms of integrals. In the last section of this chapter we recall the basic definitions and properties of Deligne's tensor product $\boxtimes$ for abelian categories. We first consider only the 2-category of categories of modules over finite dimensional algebras inside a strict version of the category of vector spaces. For this strictified category we ensure that the 2-braiding induces a strict action of the symmetric group S_N on the multifold tensor products $\mathcal{C}_1 \boxtimes \mathcal{C}_2 \boxtimes \ldots \boxtimes \mathcal{C}_N$ of categories of modules. As a result, $\mathbf{AbCat}$ inherits the structure of a weak symmetric monoidal 2-category.

In Chapter 5 we begin with a discussion of a large class of coends in abelian tensor categories that are determined by an expression with operations such as $\otimes$, $\boxtimes$, and $\_^\vee$. In particular, the functors associated to horizontal 1-morphisms are obtained as coends of this form. We review the construction of the braided Hopf algebra structure for the special coend $F = \int^{X \in \mathcal{C}} X \otimes X^\vee$ in a *bounded*, abelian BTC $\mathcal{C}$. We construct a special Hopf pairing $\omega : F \otimes F \to 1$ for such a Hopf algebra F. Modularity of a bounded, ribbon category $\mathcal{C}$ means, by definition, non-degeneracy of the form ω. We prove that ω is non-degenerate if and only if integral-functionals factor through ω. In the modular case we prove that integrals of F are two-sided and that the natural transformation of the identity functor induced by the integral in $\mathrm{Hom}_{\mathcal{C}}(F, 1)$ factors through $1 \oplus \ldots \oplus 1$ for every object of $\mathcal{C}$.

In Chapter 6 we construct the double pseudofunctor $\mathcal{V}^* : \mathcal{T}gl \to \mathcal{Q}\mathbf{AbCat}$ on tangles, which represent cobordisms. The proof of topological invariance, meaning the fact that $\mathcal{V}^*$ is well defined on equivalence classes of tangles, is obtained by a dictionary style translation of elementary moves to algebraic axioms. The proof, that the double functor is compatible with the vertical composition, is straightforward. The horizontal composition is, however, respected only up to isomorphism.

In the first part of Chapter 7 we lift the double pseudofunctor $\mathcal{V} : \widetilde{\mathcal{C}ob} \to \mathcal{Q}\mathbf{AbCat}$ to a double pseudofunctor $\mathcal{V} : \widetilde{\mathcal{C}ob}^{\cap} \to \mathcal{Q}\mathbf{AbCat}$ using an analogous presentation via a tangle double category $\mathcal{T}gl^{\cap}$. It can be made strict after replacing the structure of symmetric monoidal 2-category of $\mathbf{AbCat}$ with an equivalent one.

In the remainder of Chapter 7 we consider two special cases for the input category $\mathcal{C}$. The first is the example of a *semisimple* abelian category $\mathcal{C}$, for which our double functor extends the Reshetikhin-Turaev theory. In the second case we consider the Tannakian situation $\mathcal{C} = \mathcal{A}{-}\mathrm{mod}$ for a general quantum group $\mathcal{A}$, which yields an extension of the Hennings invariant. We discuss in detail the form of the braided Hopf algebras and their integrals for both types of categories. The relations to cellular quantum invariants are also outlined.

In Appendix A we discuss the physical and historical background that leads us to defining an axiomatic field theory in terms of double functors. We start with an exposition of the topological aspects of Chern-Simons theory, that were investigated by Witten, and the functorial formulations of conformal field theories. Other various axiomatic frameworks, that attempt to unite and axiomatize these two theories, are presented and their relation to the double functor picture explained. Relations to gauge theoretic TQFT constructions by Frohman, Nicas, Donaldson, Hutchingson, Lee, and Fukaya are outlined.

We recall the Ehresmann definition of double categories from [Ehr63a] in Appendix B.1. In Appendix B.2 we discuss weak versions of double functors – the pseudofunctors. The related notions – horizontal and vertical natural transformations – are also described. Our interest in those is explained by the fact that we first study a version of the TQFT functor $\mathcal{V}$, which is a double functor in the weak sense.

Finally, in Appendix C.1, we give a description of the category of multiple coends, which are associated to higher genus surfaces, together with natural isomorphisms between them. We do it in terms of the monoidal bicategory of thick tangles, which can be thought of as a free bicategory generated by a self-dual object. We obtain a combinatorial presentation of this bicategory in terms of generators and relations, which have graphical presentations. Coherence of the above mentioned functors and their isomorphisms is asserted in the form of a functor from the combinatorial bicategory to $\mathbf{AbCat}$.

Acknowledgments. We are grateful to J. Baez, Yu. Bespalov, A. Casson, F. Cohen, L. Crane, P. Deligne, G. Felder, Z. Fiedorowicz, J. Fröhlich, V. Jones, M. Karowski,

D. Kazhdan, T. Le, F. Quinn, N. Reshetikhin, R. Schrader, J. Stasheff, A. Sudbery, V. Turaev, E. Witten, D. Yetter for attention to our work, fruitful discussions and invaluable advices.

Commutative diagrams in this book are drawn with the help of the Paul Taylor package `diagrams.tex`.

The work of T.K. was partially supported by NSF grant DMS-9305715. Early parts of this work had been completed while T.K. was at the Institute for Advanced Study, Princeton, supported by NSF grant DMS 9304580, and at the University of California at Berkeley, supported by DFG Forscherstipendium Ke 624/1.

V.L. began to work on this project at the University of York, U.K., partially supported by EPSRC research grant GR/G 42976. Part of this work had been completed while V.L. was at Institut de Recherche Mathématique Avancée, Strasbourg, France. Further work of V.L. was partially supported by NSF grant 530666 while he was visiting Kansas State University, Manhattan, U.S.A. Final touches were added while V.L. visited Max–Planck–Institut für Mathematik in Bonn.

1. The Double Category of Framed, Relative 3-Cobordisms

The category of relative cobordisms in dimension $d+1$ is a category, whose objects are d-manifolds and whose morphisms are classes of $d+1$-manifolds, see [ES52]. The definition of this category was recalled in Sect. 0.2.

The picture of cobordisms emerges very naturally in differential topology when we consider a Morse function $f : M \to [0,1]$ with $f^{-1}\{0,1\} = \partial M$ with N distinct critical values on a $d+1$-dimensional manifold M, see [Mil69]. It yields a handle decomposition of M which is thus categorically expressed as a decomposition into a product $M = M_1 \circ M_2 \circ \ldots \circ M_N$, where each M_j is an *elementary* cobordism, on which f has only one critical point. A classical application of this point of view is the h-Cobordism Theorem [Mil65], which may be considered as a generalization of the Poincaré conjecture in higher dimensions.

In this chapter we will discuss in detail the structures inherent to cobordism categories for $d = 2$, meaning the category of 3-dimensional relative cobordisms between 2-dimensional surfaces in the sense of [ES52]. In three dimensions the differential or piecewise linear theory is equivalent to the topological theory [Moi52]. Our focus will be mostly on the following two important generalizations of this category, which will enter the construction of the topological double category $\widetilde{\mathcal{C}ob}^{\cap}$ that we use to define and construct extended topological quantum fields theories.

The first generalization arises from the fact that we want to consider 2-manifolds with boundaries, and 3-manifolds with corners. The usual definitions says that two d-manifolds V_t and V_s with diffeomorphic boundaries $\partial V_t \cong \partial V_s$ are cobordant if there is a $d+1$-dimensional manifold M with

$$\partial M = V_t \bigcup_{\partial V} -V_s .$$

We slightly modify it by thickening out the 1-manifold ∂V into the product $\partial V \times [0,1]$. The boundary of 3-manifold M, therefore, contains three pieces, namely the surfaces V_t, V_s, and $\partial V \times [0,1]$. Given another pair of d-manifolds V_t' and V_s' that are cobordant by a $d+1$-dimensional manifold M', and a diffeomorphism $V_t \xrightarrow{\sim} V_s'$ one can naturally construct a composite cobordism by gluing the boundary piece V_t of M onto the boundary piece V_s' of M' as follows:

$$M \circ M' \;=\; M \bigcup_{V_t \cong V_s'} M' . \qquad (1.0.1)$$

We may define another gluing operation by identifying components of $\partial V \times [0,1] \subset M$ with corresponding components of $\partial V' \times [0,1] \subset M'$. The second gluing operation yields a second composition law between relative 3-cobordisms, which is distributive with respect to the composition over surfaces defined in (1.0.1). Algebraic structures that generalize the notion of a category by incorporating the combination of two operations are, for example, 2-categories. For our purposes we prefer the slightly more general notion of double categories. In fact, relative cobordisms with corners naturally form double categories and thus provide much richer topological and algebraic structures than ordinary cobordism categories.

The second generalization that will consume the larger part of our exposition arises from the circumstance that the data of a topological manifold by itself does not suffice to construct interesting topological quantum field theories. Geometrical and physical models suggest that we have to consider isotopy classes of framings or 2-framings of the cobordisms as additional data. The latter can be encoded in the signature of a 4-manifold that is bounding the 3-cobordism. Algebraically, this implies an integer extension of the cobordism category so that our morphisms are no longer topological manifolds M but pairs (M, σ) with $\sigma \in \mathbb{Z}$.

The cocycle that expresses the non-additivity of these integers under composition over the surfaces $V_t \simeq V_s'$ can be computed from the results of Wall [Wal69] as in [Ker99], see also [Ati90] for a closely related computation. It is obtained by extending the gluing operation to the bounding 4-manifolds in an obvious way. For the second composition over the $\partial V \times [0,1]$-cylinders, the extension to the 4-dimensional setting is, however, not quite as naïve, and involves a series of additional handle attachments. This complication is justified mainly by properties inherent to the physical examples, after which the topological quantum field theories are modeled.

The independence or distributiveness of the two compositions for the extended relative cobordisms, expressed by the *interchange law* of double categories, is now no longer trivial since the cocycles have to match as well. We give a geometric proof of the interchange law.

Summary of Content

In this chapter we will assume the definition of a double category in Section 0.2 and Appendix B.1. The main goal is to construct the double category $\widetilde{\mathcal{C}ob}^{\cap}$ and prove Theorem 1.6.8.

In Section 1.1 we introduce the objects and 1-arrows sets for the cobordism categories that appear in this chapter. In both cases the sets will be choices of representing 1-manifolds and 2-manifolds, one for each homeomorphism class. In particular, we associate to each non-negative integer a an object, given as a 1-manifold homeomorphic to a copies of the circle, and to each triple of non-negative integers $[g, a/b]$ a surface of genus g and $a + b$ boundary components.

Among the holes on the surface we distinguish between a source boundaries and b target boundaries so that the surface itself may be regarded as a cobordism in dimension $d = 1$. The fact that we confine ourselves to only one surface per homeomorphism class makes the technical representations better to handle. However, in order to identify the composition of two surfaces $[g_1, a/b]$ and $[g_2, b/c]$ with a chosen surface we need to define a homeomorphism that is not canonical. In Chapter 7 we will remedy this situation by introducing a larger set of 1-arrows, where we choose a finite number $\frac{1}{g+1}\binom{2g}{g}$ of surfaces for each homeomorphism class.

Another constraint we work with in Chapter 1 is that all surfaces are connected. It is not very difficult to find generalizations of presentations and TQFT functors for disconnected surfaces using results in [Ker99] and [Ker98b] as outlined again in Chapter 7.

The vertical 1-arrows are given as permutations among the holes induced by the cylindrical boundary pieces. We want to allow non-trivial permutations in order to incorporate the full braid group instead of just the pure braid group on surfaces in our description. We are thus forced to consider double categories instead of just 2-categories.

In Section 1.2 we give the detailed definition of what constitutes a homeomorphism class of relative cobordisms associated to a square of horizontal and vertical 1-arrows. We define the horizontal and vertical compositions as gluings. A routine verification shows that this gives rise to a double category ***Cob***.

Section 1.3 deals with some standard consequences of the double category interpretation. Particularly, we elaborate on the correspondence that assigns to surfaces and relative cobordisms functors and natural transformations on the same vertical category. An important application is that we can identify the vertical category of surfaces with exactly one boundary component as a braided tensor category in the sense of [JS91].

The mapping class group $\pi_0\big(\mathcal{D}iff^+(\Sigma)\big)$ of a surface Σ with boundary is identified in Section 1.4 with the group of vertically invertible relative cobordisms in ***Cob***, whose horizontal 1-arrows sets are homeomorphic to Σ. Furthermore, we investigate the structure of the *framed braid groups* $\widehat{\mathcal{B}_n}(\Sigma)$ over a surface Σ, which are defined as the fundamental groups of the configuration spaces of discs on Σ (instead of points for the ordinary braid groups). In particular, the central extension over the ordinary braid groups is determined as well as its kernel under the natural homomorphism into $\pi_0\big(\mathcal{D}iff^+(\Sigma^*)\big)$, where $\Sigma^* = \overline{\Sigma - \sqcup_1^n D^2}$. They are non-trivial in cases of small genus or number of punctures. The disc rotations that generate the central extension of the framed braid group are also topologically related to the ribbon element of the underlying braided tensor category.

In Section 1.5 we provide several standard tools of handlebody theory that allow us to describe and manipulate handle decompositions of cobordisms in arbitrary dimensions. We relate this in more explicit terms to surgery on the coborded manifolds. Special attention is given to surgery presentations of 3-manifolds and handle decompositions of 4-manifolds, in particular, the sliding and cancellation operations between the 1-handles and 2-handles in dimension 4. Finally, we include a couple of technical lemmas that explicitly describe the effect of a 2-surgery which passes through 1-handles that have been added to a 3-manifold.

Equipped with the technical preliminaries from the previous section, we explicitly construct in Section 1.6 the framing extension $\widetilde{\mathcal{C}ob}$ of $\mathcal{C}ob$. For a relative cobordism M we consider a closure $\langle M_o \rangle$ and identify a 2-framing of M with the signature of a of 4-manifold W with corners such that $\partial W = \langle M_o \rangle$. A vertical composition is, as before, naïvely constructed by gluing the 4-manifolds along cylinders $\Sigma \times [0,1]$ over the bounding surfaces. The horizontal composition, however, involves a series of attachments of 4-dimensional 1-handles and 2-handles instead of boundary identifications. The 2-arrow sets of $\widetilde{\mathcal{C}ob}$ are given as equivalence classes of these 4-manifolds, in which only the bounding 3-manifold M and the signature of W are retained as information. We prove that the induced horizontal and vertical composition operations on these sets are well defined and obey the interchange law of double categories. Hence, $\widetilde{\mathcal{C}ob}$ is a double category as stated in Theorem 1.6.8.

1.1 The 0-1-Arrow Category of Surfaces with Boundaries

In this section we define a category of 1+1-dimensional cobordisms, which consists of the objects and 1-morphisms underlying the double category of 2+1-dimensional cobordisms. We present it in terms of natural generators and algebraic relations. We also discuss the choices between standard surfaces and their composites.

An object is an ordered set of oriented circles. Here, as well as in the generalizations in Chapter 6, we admit only one one-dimensional manifold in each homeomorphisms class. This object, of course, depends only on the number of components, and we denote it as follows:

$$S^{\sqcup a} \cong \underbrace{S^1 \sqcup \ldots \sqcup S^1}_{a} \,. \tag{1.1.1}$$

They are also the objects of the cobordism category $\mathcal{C}ob_2$. The morphisms in $\mathcal{C}ob_2$ are homeomorphism classes of 1+1-dimensional cobordisms. To be more precise, we define first the bicategory $\mathcal{COB}_2$ of all cobordisms. Objects are non-negative integers a, identified with the disjoint union $S^{\sqcup a}$ of a circles. A 1-morphism between $S^{\sqcup a}$ and $S^{\sqcup b}$ is a compact, oriented surface, Σ, together with a homeomorphism $\xi : \partial\Sigma \stackrel{\sim}{\longrightarrow} -S^{\sqcup a} \sqcup S^{\sqcup b}$. Its 2-morphisms are homeomorphisms $k : (\Sigma, \xi) \stackrel{\sim}{\longrightarrow} (\Sigma', \xi')$ such that $\xi' \circ k = \xi$. Two 1-morphisms, (Σ, ξ) and (Σ', ξ'), are considered equal in $\mathcal{C}ob_2$ if there is a 2-morphism $k : \Sigma \stackrel{\sim}{\longrightarrow} \Sigma'$ connecting them.

In this book we want to confine ourselves to the subcategory $\mathcal{C}ob_2^{\mathrm{cn}}$, which is generated by connected surfaces. It is clear that the morphisms in $\mathrm{Hom}_{\mathcal{C}ob_2^{\mathrm{cn}}}(S^{\sqcup a}, S^{\sqcup b})$ can be enumerated using the genus, g, of the surface. We denote the g-th morphism by

$$[g, a/b] : S^{\sqcup a} \longrightarrow S^{\sqcup b}. \tag{1.1.2}$$

If $b \geqslant 1$ then the composition law is given by the simple formula $[g_2, b/c] \circ [g_1, a/b] = [g_1 + g_2 + b - 1, a/c]$.

The category $\mathcal{Cob}_2$ is clearly a symmetric tensor category if we define the tensor product to be the ordered, disjoint union, i.e, $\sqcup = \sqcup^{ordered}$. In particular, we obtain a natural embedding of the symmetric group in a letters into the automorphisms of $S^{\sqcup a}$, i.e.,

$$S_a \subset \mathrm{Aut}_{\mathcal{Cob}_2}(S^{\sqcup a}) : \pi \longmapsto \pi^* . \tag{1.1.3}$$

The cobordism, π^*, associated to a permutation π, is as a two-fold the union of cylinders $S^{\sqcup a}\times[0,1] \cong S^1\times[0,1]\sqcup\ldots\sqcup S^1\times[0,1]$. The boundary identifications are such that the j-the cylinder connects the j-th circle from the source one-dimensional manifold to the $\pi(j)$-th circle in the target manifold.

It is easy to see that any morphism in $\mathcal{Cob}_2$ is the composite of a permutation and the tensor product of morphisms from $\mathcal{Cob}_2^{cn}$.

A useful algebraic way of describing $\mathcal{Cob}_2$ as a freely generated tensor category is provided in the following proposition. A detailed proof can be found in [Abr96], see also [Bae97] and [DJ94] for expositions of this fact.

Proposition 1.1.1. *$\mathcal{Cob}_2$ is the free, strict, symmetric tensor category, generated freely by an associative, commutative Frobenius algebra-object, $\mathcal{A}$, with unit.*

Let us explain in more detail what the contents of the freely generated category is. To say that a strict tensor category is freely generated by an object means that every object is of the form $\mathcal{A}\sqcup\ldots\sqcup\mathcal{A}$, and the tensor product $\sqcup$ on the level of objects is given in the obvious way. The fact that we speak of a symmetric tensor category means that the morphism sets contain as generators isomorphisms, which implement the symmetric group as in (1.1.3), and which are compatible with $\sqcup$. We also assume an object 1, which is a unit with respect to $\sqcup$.

When we say that $\mathcal{A}$ is an algebra-object, we imply that we have a morphism $\otimes : \mathcal{A}\sqcup\mathcal{A}\to\mathcal{A}$, and the conditions of associativity and commutativity mean that we impose the respective relations on $\otimes$. Also we assume a morphism $e : 1 \to \mathcal{A}$, which satisfies the relation that characterizes a unit with respect to $\otimes$.

In the same way the restriction to Frobenius algebras implies further generating morphisms and relations among them. The first is a trace, meaning a morphism $\mathrm{tr} : \mathcal{A}\to 1$. For a Frobenius algebra this has also to be non-degenerate. Implicit to this is the existence of yet another generator, namely $\phi : 1\to\mathcal{A}\sqcup\mathcal{A}$, which is the two-sided side-inverse to the pairing $\phi^* = \mathrm{tr}\circ\otimes : \mathcal{A}\sqcup\mathcal{A}\to 1$.

The object $\mathcal{A}$ is identified with a single circle, and 1 with the empty set. As before, the tensor product $\sqcup$ is given in $\mathcal{Cob}_2$ by the disjoint union. The morphisms e and tr correspond to discs, $\Sigma_{0,1}$, the multiplication $\otimes$ is mapped to the three-holed sphere $\Sigma_{0,3}$, viewed as a cobordism $S^1\sqcup S^1\to S^1$, and the symmetric group generators are mapped onto each other.

Thus almost all of the generators of $\mathcal{Cob}_2$, which we single out by this description, are elementary cobordisms in the Morse theoretical sense. The cobordisms of index 0 and 2 are e and tr, and the fusing index 1 cobordism is $\otimes$. The fissing index 1 elementary cobordism is given by the combination

$$\otimes^* = (\otimes\sqcup\mathbb{I}_{\mathcal{A}})\circ(\mathbb{I}_{\mathcal{A}}\sqcup\phi),$$

and, conversely, ϕ is a combination of an index 1 and an index 2 cobordism.

Similarly, the algebraic relations discussed above translate into elementary moves between Morse functions that describe the same cobordism. The elementary generators are also depicted in Figure 1.1. Here the cobordisms map from the left to the right, i.e., the Morse function is the projection on the horizontal x-axis.

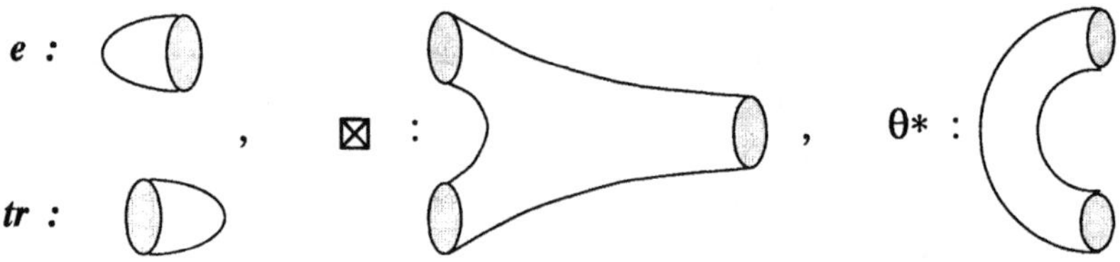

Fig. 1.1. Generating Cobordisms of $\mathcal{C}ob_2$

There is a fundamental problem when we wish to extend a category of 1+1 cobordisms to a category of 1+1+1 cobordisms. In order to describe cobording 3-manifolds we need to be able to talk about specific surfaces and not just equivalence classes. The natural candidate to construct a double category from would be COB_2 – category of all compact oriented surfaces.

Thus, we should select a subset of surfaces in each morphism set, which is finite and which is closed under composition and thus defines a subcategory $COB_2^{finite} \subset COB_2$. In Chapter 7 we describe a way of picking such a subcategory. Roughly speaking, first, we choose specific surfaces for the generators $\otimes$, θ^*, e, tr, and the permutations, and then we describe every other surface as a word in these generators.

Although, the definition of COB_2^{finite} can be chosen such that the isomorphism classes are even finite sets, we shall not insist on having a subcategory under compositions anymore, and select for the purpose of finding presentations of the 3-dimensional cobordisms only *one* surface in each isomorphism class, i.e., for each morphism in $\mathcal{C}ob_2$. If we also confine ourselves to connected cobordisms, this can be formally described by a 2-pseudofunctor

$$\left(\Sigma, \alpha\right) : \mathcal{C}ob_2^{\mathrm{cn}} \hookrightarrow COB_2. \tag{1.1.4}$$

On the level of objects this functor is the identity. It associates a *standard* surface $\Sigma_{g,a/b}$ to the morphism $[g, a/b]$. In place of some formal definition we give $\Sigma_{g,a/b}$ in the form of a picture, as, e.g., the left surface in Figure 1.2.

The labeling of the source holes is from the left to the right, $1_s, \ldots, a_s$ for the first a holes. For the b target holes we choose reversed ordering, $b_t, \ldots, 1_t$. In between we have g handles that are aligned in horizontal direction, and of which we depicted the first and the last.

Besides being a map on the sets of objects and morphisms, a pseudofunctor also contains an isomorphism between the composite of the images of two morphisms

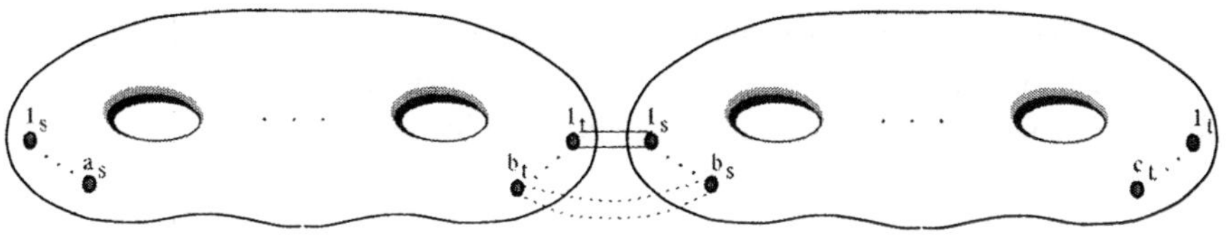

Fig. 1.2. Standard Cobordism, $\Sigma_{g,a/b}$

and the image of the composite in the original category. The isomorphism will be contained in another category, which in the case of 1+1-dimensional cobordisms, consists of the homeomorphisms between surfaces. Specifically, the additional data α for the functor in (1.1.4) consists of a system of homeomorphisms, as follows:

$$\alpha_{[g_1,a/b:g_2,b/c]} : \boldsymbol{\Sigma}\big([g_2,b/c]\big) \cdot \boldsymbol{\Sigma}\big([g_1,a/b]\big) \equiv \Sigma_{g_2,b/c} \sqcup_{S^{\sqcup b}} \Sigma_{g_1,a/b} \xrightarrow{\sim} \Sigma_{g_1+g_2+b-1,a/c}. \quad (1.1.5)$$

The notion of a pseudofunctor also requires that α satisfies an associativity condition of the form $\alpha_{[1:2,3]}(\alpha_{[2:3]} \sqcup_{S^{\sqcup b}} \mathbb{1}_1) = \alpha_{[1,2:3]}(\mathbb{1}_3 \sqcup_{S^{\sqcup c}} \alpha_{[1:2]})$, if we consider the situation, where we compose with a third surface, $\Sigma_{g_3,c/d}$. If we confine ourselves to surfaces with $g > 0$ or $a, b > 1$ these can indeed be found such that the equality holds in the strict sense. For our purposes it suffices to verify associativity up to isotopy, which can be done for all surfaces.

Specifically, we construct the isomorphism in (1.1.5) as follows: instead of the gluing over circles we can also consider the surface, where we have inserted small cylinders between the boundary components that are to be identified, as indicated in Figure 1.2. Hence the composite is homeomorphic to the surface, which in the middle has the form as depicted on the left in Figure 1.3. Moreover, the obvious homeomorphism is unique up to isotopies.

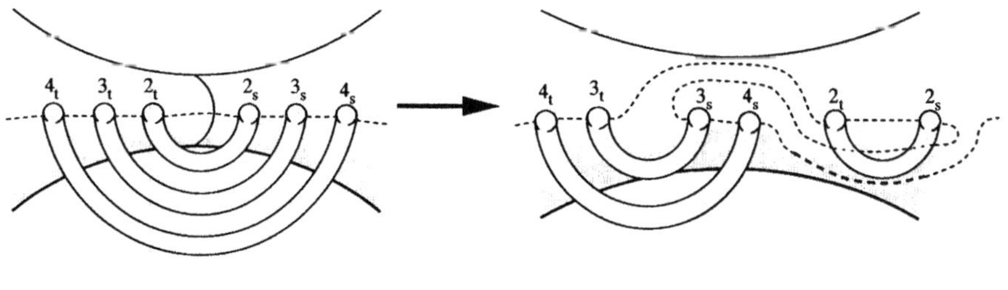

Fig. 1.3. Inductive Definition of α

The slide of the tube connecting 2_s to 2_t, which is indicated on the right hand side of Figure 1.3, defines a homeomorphism:

$$\Sigma_{g_2,b/c} \sqcup_{S^{\sqcup b}} \Sigma_{g_1,a/b} \quad \xrightarrow{\sim} \quad \Sigma_{g_2+1,b-1/c} \sqcup_{S^{\sqcup(b-1)}} \Sigma_{g_1,a/b-1}.$$

Here, we have identified the resulting middle part as a gluing over $b-1$ circles, and the pushed out cylinder as an additional, g_2+1-st handle in the second surface. In Figure 1.4, we explicitly illustrate the latter identification.

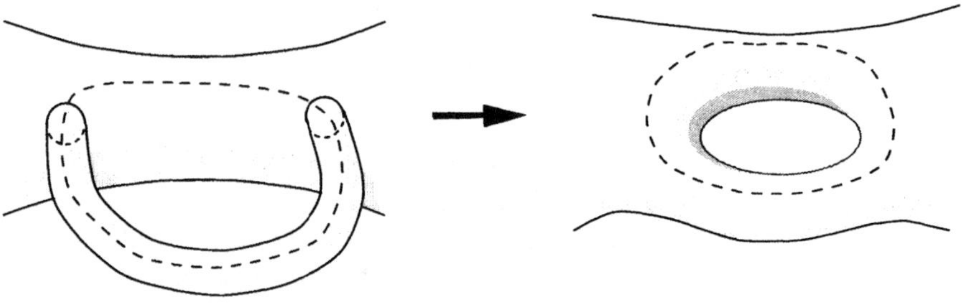

Fig. 1.4. Cylinder-Handle Deformation

The definition of α is now completed by induction. The convention chosen in Figure 1.3, thus, determines the homeomorphism uniquely up to isotopies. The latter will be irrelevant if we consider homeomorphism classes of 2+1-dimensional cobordisms. Also, α does not affect the holes outside of the middle part so that the associativity condition is easily realized.

Let us also give a slightly different construction of α, which defines the same homeomorphism. It will facilitate the description of the compositions of cobordisms or tangle diagrams in horizontal direction. In essence, it is given by the following identity between composites of operations, which produce the standard manifold $\Sigma_{g_1+g_2+b-1,a/c}$ starting from $\Sigma_{g_1,a/b}$ and $\Sigma_{g_2,b/c}$.

$$\begin{aligned}\alpha \circ \{\text{Glue } j_s \text{ to } j_t \text{ for } 1 \leqslant j \leqslant b\} \\ = \{\text{Glue neighbors } j_s \text{ to } j_t \text{ for } 2 \leqslant j \leqslant b\} \circ \Upsilon \circ \{\text{Glue } 1_s \text{ to } 1_t\}. \quad (1.1.6)\end{aligned}$$

The left hand side of this equation has been explained in detail above. The first operation on the right hand side produces a connected, generalized standard surface, $\Sigma_{g_1/g_2,a/2(b-1)/c}$, which is defined in analogy to the surfaces in Figure 1.2 by a horizontal alignment of holes and handles. Specifically, it contains, in order from left to right, a holes with labels $1_s, \ldots, a_s$, g_1 handles, $2(b-1)$ holes with labels $b_t, (b-1)_t, \ldots, 2_t, 2_s, \ldots, b_s$, g_1 handles, and c holes with labels $c_t, \ldots, 1_t$.

Now, the next operation is a composite of homeomorphisms

$$\Upsilon = \Upsilon_{(b-1)} \circ \ldots \circ \Upsilon_2 \subset \mathcal{Diff}^+\left(\Sigma_{g_1/g_2,a/2(b-1)/c}\right). \qquad (1.1.7)$$

Each Υ_j is obtained from the ambient isotopies of $2(b-1)$ discs on $\Sigma_{g_1+g_2,a/c}$, which describe the sliding of a cylinder as in Figure 1.3. As we shall discuss in more detail in Section 1.6, their isotopy classes are, thus, elements of a central extension, $\widehat{\mathcal{B}}_{2(b-1)}\left(\Sigma_{g_1+g_2,a/c}\right)$, of the braid group on the surface with $a+c$ holes.

For example, $\boldsymbol{\Upsilon}_2$ will braid the discs so that the labels of the middle boundary components are permuted into the order $b_t, (b-1)_t, \ldots, 3_t, 3_s, \ldots, b_s, 2_t, 2_s$. An application of the total isomorphism $\boldsymbol{\Upsilon}$ eventually results in the order $b_t, b_s, (b-1)_t, (b-1)_s, \ldots, 2_t, 2_s$.

Until now we have postponed the gluing of the boundary components other than 1_s and 1_t. The gluing, which may be expressed by attaching a cylinder, is performed here between neighboring holes. This will be helpful in later applications. A configuration of cylinders connecting two holes as on the left hand side of Figure 1.4 is deformed as before into the handle on the right side of the same figure.

1.2 2-Arrows from Cobordisms with Corners

In this section we extend the category $\mathcal{C}ob_2^{\mathrm{cn}}$ to a *double category*, $\mathcal{C}ob$, of classes of 3-dimensional manifolds with corners. The notion of a double category is explained in details in the introduction and Appendix B.1. As opposed to two- or bi-categories, we distinguish in this formalism between the category of *horizontal* 1-arrows, which will be given by $\mathcal{C}ob_2^{\mathrm{cn}}$, and the category of *vertical* 1-arrows, which for our purposes will be given by the cobordisms implementing the symmetric group as in (1.1.3). The topological definitions of horizontal and vertical compositions and their compatibility are explained.

The 2-arrows of $\mathcal{C}ob$ are no longer associated simply to a pair of 1-arrows between the same objects, but to a square, $\mathcal{S}q$, of 1-arrows, of which the two horizontal and the two vertical have to be in the respective categories.

If we take the arrows from $\mathcal{C}ob_2$, as described above, we use the map $\boldsymbol{\Sigma}$ from (1.1.4) in order to associate to every square, $\mathcal{S}q$, a closed surface $\boldsymbol{\Sigma}_{\mathcal{S}q}$ as below. It contains embedded curves, where the individual cobordisms have been glued together.

$$\mathcal{S}q - \begin{array}{ccc} S^{\sqcup a} & \xrightarrow{[g_{sc},a/b]} & S^{\sqcup b} \\ {\scriptstyle \alpha^*}\big\downarrow & & \big\downarrow{\scriptstyle \beta^*} \\ S^{\sqcup a} & \xrightarrow{[g_{tg},a/b]} & S^{\sqcup b} \end{array} \longmapsto \boldsymbol{\Sigma}_{\mathcal{S}q} = -\Sigma_{g_{sc},a/b} \bigsqcup_{S^{\sqcup a} \sqcup S^{\sqcup b}} \alpha^* \sqcup \beta^* \bigsqcup_{S^{\sqcup a} \sqcup S^{\sqcup b}} \Sigma_{g_{tg},a/b}.$$

Here, $\alpha \in S_a$ and $\beta \in S_b$, and α^* and β^* are the corresponding cobordisms, see (1.1.3). Instead of inserting the vertical cylinders, we can also construct the surface $\boldsymbol{\Sigma}_{\mathcal{S}q}$, by gluing for each j the boundary component of the source surface $\Sigma_{g_{sc},a/b}$ with label j_s to the boundary circle of the target surface $\Sigma_{g_{tg},a/b}$ with label $\alpha(j)_s$, and similar identifications for the boundary components with labels j_t.

A 2-arrow for the square $\mathcal{S}q$ is then obtained from a compact, oriented three-manifold with corners. In addition, we have to consider an orientation preserving homeomorphism

$$\psi : \boldsymbol{\Sigma}_{\mathcal{S}q} \xrightarrow{\sim} \partial M, \qquad (1.2.1)$$

which also preserves the strata. This means that the special circles in Σ_{Sq} are mapped precisely to the 1-strata of the boundary, and that ∂M contains no 0-strata.

The elements of ***Cob*** are now equivalence classes of pairs (M, ψ). We consider two cobordisms, $(M, \psi) \sim (M', \psi')$, equivalent if there is a homeomorphism $\rho : M \xrightarrow{\sim} M'$, such that $\rho \circ \psi = \psi'$. Abusing notation, we will often use the same letter M to denote the homeomorphism class, and depict it in diagrams by a tilted, double arrow in the middle of the square *Sq*, see, e.g., the diagrams in (1.2.3).

We denote the set of 2-arrows that belong to a square, in which the vertical permutations are α and β, by $\boldsymbol{Cob}(\alpha, \beta)$. Let us also use the (slightly ambiguous) notation $\boldsymbol{Cob}(a, b) = \bigcup_{\alpha \in S_a, \beta \in S_b} \boldsymbol{Cob}(\alpha, \beta)$ for the set of cobordisms, with $S^{\sqcup a}$ and $S^{\sqcup b}$ as objects.

Sometimes we write a cobordism $M \in \boldsymbol{Cob}(a, b)$ also in the form $M : \Sigma_{g_{sc}, a/b} \to \Sigma_{g_{tg}, a/b}$ and define

$$M \longmapsto \Pi(M) = \Pi_s(M) \times \Pi_t(M) = (\alpha, \beta) \in S_a \times S_b \qquad (1.2.2)$$

as the map that associates to a cobordism M its vertical permutations. That is, if we define $[S_a]$ as the category with one object $*$ and $\mathrm{End}(*) = S_a$, we have functors $\Pi_s : \boldsymbol{Cob}(a, b) \to [S_a]$, etc.

It is now easy to see that for fixed $a, b \in \mathbb{Z}^{+,0}$ the cobordisms in $\boldsymbol{Cob}(a, b)$ form a category, whose objects are the standard surfaces $\{\Sigma_{g,a/b} : g = 0, \ldots\}$. The composite of two cobordisms, $M : \Sigma_{g_{sc}, a/b} \to \Sigma_{g_{int}, a/b}$ and $N : \Sigma_{g_{int}, a/b} \to \Sigma_{g_{tg}, a/b}$, is given in the obvious way, namely, by gluing the cobordisms together along the boundary pieces homeomorphic to the intermediate surface with boundary $\Sigma_{g_{int}, a/b}$, using the respective restriction of the boundary maps from (1.2.1).

Clearly, the resulting space is again a manifold with corners. We have canonical boundary maps on the horizontal pieces $\Sigma_{g_{sc}, a/b}$ and $\Sigma_{g_{tg}, a/b}$, and the components of the vertical boundary pieces are naturally parametrized by $S^1 \times [0, 2] = S^1 \times [0, 1] \sqcup_{S^1 \times 1 \sim S^1 \times 0} S^1 \times [0, 1]$. Rescaling the unit interval the latter can be replaced with a parametrization by $S^1 \times [0, 1]$ so that we have a parametrization of the boundary of the composite by a surface associated to a square and, hence, an element $N \circ_v M \in \boldsymbol{Cob}$. This we define to be the *vertical* composition of N and M.

The horizontal 1-arrows of the composite are, as in the ordinary 2- or bi-category situation, simply the horizontal source 1-arrow of the first 2-arrow, and the horizontal target 1-arrow of the second 2-arrow. It is also clear that the permutation of holes, that is defined by the composition of the vertical boundary pieces, is the composite of the respective permutations. More precisely, $\Pi(N \circ_v M) = \Pi(N) \circ \Pi(M)$, where the second composition is in $S_a \times S_b$. Vertical compositions are conveniently described by a pasting of the respective squares as indicated in the next equation:

$$\begin{array}{ccc} S^{\sqcup a} & \xrightarrow{[g_{sc},a/b]} & S^{\sqcup b} \\ {\scriptstyle \alpha_N^* \circ \alpha_M^*} \big\downarrow & \swarrow\!\!\!\swarrow\; N \circ_v M & \big\downarrow {\scriptstyle \beta_N^* \circ \beta_M^*} \\ S^{\sqcup a} & \xrightarrow[{[g_{tg},a/b]}]{} & S^{\sqcup b} \end{array} \quad = \quad \begin{array}{ccc} S^{\sqcup a} & \xrightarrow{[g_{sc},a/b]} & S^{\sqcup b} \\ {\scriptstyle \alpha_M^*} \big\downarrow & \swarrow\!\!\!\swarrow\; M & \big\downarrow {\scriptstyle \beta_M^*} \\ S^{\sqcup a} & =[g_{int},a/b]\Longrightarrow & S^{\sqcup b} \\ {\scriptstyle \alpha_N^*} \big\downarrow & \swarrow\!\!\!\swarrow\; N & \big\downarrow {\scriptstyle \beta_N^*} \\ S^{\sqcup a} & \xrightarrow[{[g_{tg},a/b]}]{} & S^{\sqcup b} \end{array} \quad . \tag{1.2.3}$$

It is obvious that the composition $\circ_v$ on the level of manifolds with corners factors into the homeomorphism classes and, thus, gives a well defined composition on $\mathcal{C}ob(a,b)$. In order for $\circ_v$ to make $\mathcal{C}ob(a,b)$ into a category we also have to verify associativity. Clearly, the bounding three-manifold and the parametrizations of the source and target surfaces do not depend on the order in which we glue. But the parametrization of the cylindrical boundary parts is a priori different, since we have different homeomorphisms that identify $S^1 \times [0,3]$ with $S^1 \times [0,1]$. As in the proof of associativity of fundamental groups, they are, however, isotopic. Since isotopies of the boundary maps do not change the homeomorphism class of a pair (M,ψ) (see Lemma 1.4.1 below) we infer associativity of $\mathcal{C}ob(a,b)$ with respect to $\circ_v$. The analogous problem for horizontal compositions is not quite as easily dealt with, since the mapping class groups of the cobordisms $\Sigma_{g,a/b}$ can be very complicated.

The *horizontal* composition $M_2 \circ_h M_1$ of two cobordisms, $M_1 \in \mathcal{C}ob(\alpha,\beta)$ and $M_2 \in \mathcal{C}ob(\beta,\gamma)$, is defined similarly to the vertical composition. But now, we glue the two manifolds together along the cylindrical boundary pieces $\beta^* \cong S^1 \times [0,1] \sqcup \ldots \sqcup S^1 \times [0,1]$ via the respective restrictions of the boundary parametrizations. At the 1-dimensional corners, where the manifolds meet, we have to make sure that we identify the circles in the first surface $\Sigma_{g_{1,sc},a/b}$ with the circles with the *same* label in the second source surface $\Sigma_{g_{2,sc},a/b}$. Similarly for the target surfaces.

The situation is illustrated in Figure 1.5 in the case of $b = 3$ and $\beta = (3,2,1)$. It is obvious that the ordered identification of the circles between both the source and the target surfaces can be extended to an identification of the cylinders if and only if the vertical permutations of the sides of the square over which we compose are the same, i.e., $\beta_1 = \beta_2$.

Thus, the reason why we use the formalism of double categories to describe cobordisms with corners is that we need to keep track of the permutations $S_a \cong \pi_0(\mathcal{D}iff^+(S^{\sqcup a}))$, after we have chosen standard 1-dimensional manifolds and allow 3-dimensional cobordisms with non-trivial permutations.

For the resulting manifold with corners $M_2 \circ_h M_1 = M_2 \sqcup_{S^{\sqcup b} \times [0,1]} M_1$, we have canonical parametrizations of the remaining, vertical boundary pieces. The horizontal boundary pieces are also canonically parametrized, but by the naïve composite of the surfaces in $\mathcal{COB}_2$ as on the left side of (1.1.5). However, in order to stay in the category $\mathcal{C}ob$ the composite has to belong to the square, where we take the composites in $\mathcal{C}ob_2^{cn}$ as horizontal 1-arrows. We, therefore, redefine the boundary map, e.g., on the source surface as

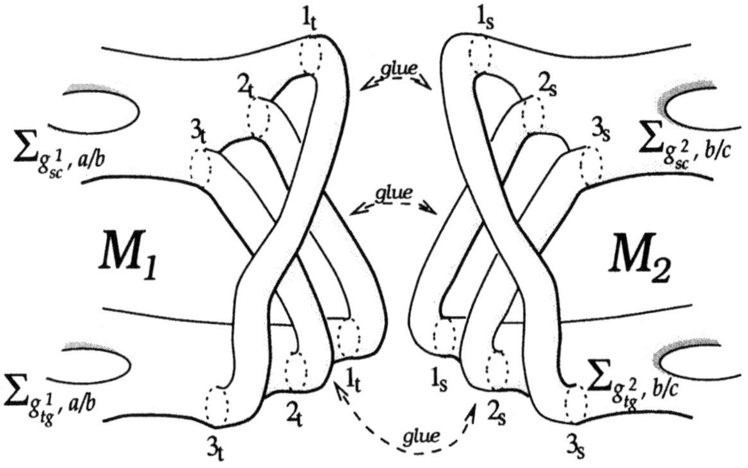

Fig. 1.5. Horizontal Composition as Gluing of Cylinders

$$\psi_2 \circ_h \psi_1 = (\psi_1 \sqcup \psi_2) \circ \alpha^{-1}_{[g_{1,sc},a/b:g_{2,sc},b/c]} : \Sigma_{g_{1,sc}+g_{2,sc}+b-1,a/c}$$
$$\hookrightarrow \partial(M_2 \circ_h M_1) = (\partial M_1 - S^{\sqcup b} \times [0,1]) \sqcup_{S^{\sqcup b}} (\partial M_2 - S^{\sqcup b} \times [0,1]).$$

The composition on the level of manifolds with parametrized boundaries is then defined by the pair $(M_2 \circ_h M_1, \psi_2 \circ_h \psi_1)$, which is easily seen to factor into a product on the homeomorphism classes. Abusing notation, we often write $M_2 \circ_h M_1$ for the composition of equivalence classes.

We can describe horizontal composition $\circ_h$ in terms of a pasting of squares, as in the diagram in Fig. 1.6. Note that we also included the boundary reparametrizations $\alpha_{\ldots}$, since not even their isotopy classes are canonically defined but depend on special choices.

The associativity of the horizontal product $\circ_h$ can be inferred from the associativity property of the boundary maps α. In the degenerate cases, where the system of α's is associative only up to isotopy, we use again Lemma 1.4.1 below. We summarize our discussion in the next proposition.

Proposition 1.2.1. *Let* **Cob** *be the set of classes of manifolds with corners associated to squares with horizontal arrows in* $\mathbf{Cob}_2^{cn}$ *and vertical arrows in the groups of permutations, as defined above. Then* **Cob** *forms a double category in the sense of Ehresmann. That is, we have a vertical composition of squares* $\circ_v$*, and a horizontal pasting* $\circ_h$ *that are both associative. Moreover, the two compositions fulfill the interchange law.*

The interchange law is given in the introduction. For cobordisms it means that if we have four squares that can be pasted into a square of twice the size, the result of first doing two vertical composition and then gluing both composites also in

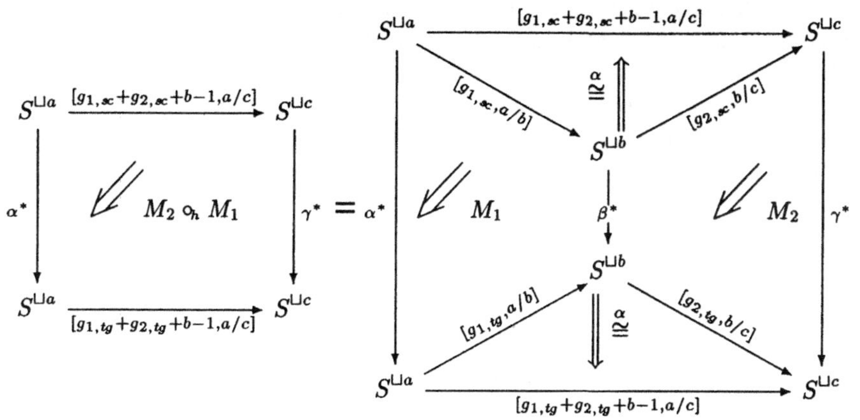

Fig. 1.6. Horizontal composition as a pasting of squares

horizontal direction should be the same as if we started with two horizontal compositions. In short, we have the equation

$$(N_2 \circ_h N_1) \circ_v (M_2 \circ_h M_1) = (N_2 \circ_v M_2) \circ_h (N_1 \circ_v M_1). \tag{1.2.4}$$

The cobordism category we have defined here is not yet going to be the one we shall actually represent in a TQFT. In Section 1.6 we consider a central extension where the cobordisms contain, besides the 3-dimensional topological information, structures equivalent to framings. Also, in Chapter 7 we shall have to consider a category for the horizontal arrows, which is bigger than $\mathcal{C}ob_2^{\text{cn}}$, if we want to construct a strict functor of double categories. Besides these necessary generalizations, we shall also briefly discuss in the next section a natural extension of the category of vertical arrows.

1.3 Basic Consequences and Generalizations of the Double Category Picture

This section discusses a natural functor, $\mathcal{E}_*^s$, implied by the two compositions, that augments the surfaces and cobordism in *Cob* to functors and natural transformations on the same category. As an application, we show that $\mathcal{C}ob(0,1)$ is a braided tensor category. Further, we describe a generalized category, $\mathcal{C}ob^{\text{vtg}}$, in which we have "singular tangles" instead of permutations as vertical 1-arrows.

If our only aim was to simply give a definition of a double category of cobordisms, *Cob*, which allows for corners, there obviously would be no need to confine ourselves to the case, where we have connected surfaces. If we chose COB_2 to be our vertical category, the definition of the three-dimensional cobordism category would be literally the same as in the connected case, once the word "connected" has been omitted everywhere. In order to find *Cob* we also need to extend the functor $(\boldsymbol{\Sigma}, \alpha)$ from (1.1.4) to $\boldsymbol{Cob}_2$.

As mentioned already in Section 1.1, a general cobordism $\langle\Sigma\rangle \in \boldsymbol{Cob}_2$ with k connected components is given as a composite of a permutation and the union of morphisms $[g_j, a_j/b_j] \in \boldsymbol{Cob}_2^{\mathrm{cn}}$ for $j = 1, \ldots, k$ as follows:

$$\langle\Sigma\rangle = \pi^* \circ_h ([g_1, a_1/b_1] \sqcup \ldots \sqcup [g_k, a_k/b_k]) : S^{\sqcup a_1+\ldots+a_k} \longrightarrow S^{\sqcup b_1+\ldots+b_k}. \tag{1.3.1}$$

For a given $\langle\Sigma\rangle$ the numbers g_j, a_j, and b_j are uniquely determined by this form. Once these are fixed, $\langle\Sigma\rangle$ depends only on and defines uniquely the class $[\pi]$ of permutation π in $S_{b_1+\ldots+b_k} \big/ S_{b_1} \times \ldots \times S_{b_k}$.

In order to find an extension of pseudofunctor $\boldsymbol{\Sigma}$ in (1.1.4) we can, thus, use this form and the definition of $\boldsymbol{\Sigma}$ on the $[g_j, a_j/b_j]$. The representation of the permutation coset, $[\pi]$ in $\boldsymbol{COB}_2$ is accomplished, for instance, if we assign to $[\pi]$ the unique shuffle permutation $\pi_{shuff} \in S_{b_1+\ldots+b_k}$, that is, the permutation, which preserves the order the letters in each subset of size b_j, and which lies in the coset $[\pi]$. This is then realized as a cobordism π^*_{shuff} as in (1.1.3), which was already used in the case of the vertical arrows.

It remains to be understood whether α can be extended to this class of surfaces, too. For many parts of the surfaces the homeomorphisms are canonical up to isomorphism and do not affect associativity as in the connected case. There is, however, a crucial difference, which is due to the S_{b_j} action on the isomorphism class $[g_j, a_j/b_j]$ given by composition. In the simplest case with $g = 0$, $a = 1$ and $b = 2$ we have to choose an isomorphism:

$$\alpha_c : (1,2)^* \sqcup_{S^{\sqcup 2}} \Sigma_{0,1/2} \xrightarrow{\sim} \Sigma_{0,1/2}. \tag{1.3.2}$$

If we contract the cylinders of $(1,2)^*$ to circles, we may consider α_c as an element of the mapping class group of $\Sigma_{0,1/2}$, which maps the out-going hole with label 1_t to itself. This group is, of course, the braid group in two strands $B_2 \cong \mathbb{Z}$. The morphism in (1.3.2) has to map the in-going hole with label 1_s to the hole with label 2_s, vice versa, so that it has to represent a non-trivial element in B_2. In Figure 1.7 we indicate the action of α_c, in the case where it is one of the generators of $\mathbb{Z}$.

It is now easy to see from the product $\otimes \circ_h (1,2)^* \circ_h (1,2)^* \xrightarrow{\sim} \otimes$ that it is impossible to satisfy associativity in $\boldsymbol{Cob}_2$. For one bracketing we obtain canonical homeomorphisms, for the other we get α_c^2, which is not trivial.

This situation can be remedied if we distinguish between the two surfaces $\otimes$ and $\otimes \circ_h (1,2)^*$, as we shall do in Appendix C.1. In this modification of $\boldsymbol{Cob}_2$, where we enhanced the isomorphism classes of 1+1-dimensional cobordisms, it will also be easier to describe the choices of α in other disconnected situations, which tends to be somewhat tedious even when we have canonical isomorphisms.

The tensor product, given by disjoint unions on $\boldsymbol{Cob}_2$, extends now naturally to a tensor product on the level of 2-arrows. Specifically, we have for all $a', a'', b', b'' \in \mathbb{Z}^{+,0}$ a functor

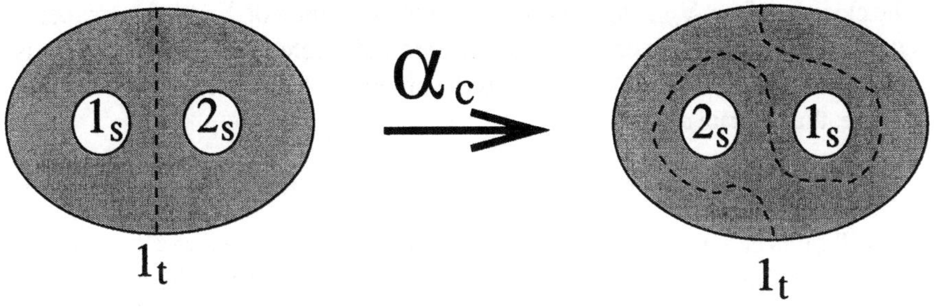

Fig. 1.7. The Braid α_c

$$\sqcup : Cob(a', b') \times Cob(a'', b'') \longrightarrow Cob(a' + a'', b' + b''). \qquad (1.3.3)$$

Note that the vertical arrows of the cobordism in the image of $\sqcup$ all lie in $S_{a'} \times S_{a''}$ and $S_{b'} \times S_{b''}$, respectively. Conversely, it is a simple fact that the vertical permutations can not permute holes which belong to the boundaries of two different connected components of a 2+1-dimensional cobordism.

The fact that we have two types of compositions implies that surfaces can be used to define functors on cobordism categories. Let us begin with a rather general construction.

Let us consider as in (1.2.2) the functors Π_s and Π_t onto the category $[S_b]$ with one object $*$ and $\mathrm{End}(*) = S_b$. Suppose for $b, \overline{b} \in \mathbb{Z}^{+,0}$ we have a functor

$$R_s : [S_b] \longrightarrow Cob(b, \overline{b}), \qquad (1.3.4)$$

such that $\Pi_s \circ R_s = id$ is the identity functor on $[S_b]$. This means that we associate to $*$ a special 1+1-dimensional cobordism $\Sigma_R : S^{\sqcup b} \to S^{\sqcup \overline{b}}$; we have a group homomorphism $S_b \to S_{\overline{b}} : \beta \to \overline{\beta}_\beta$; and to the square with both horizontal arrows equal to Σ_R and vertical arrows β and $\overline{\beta}$ we assign a 2-arrow $R(\beta)$, such that we have consistency with vertical compositions. Then we can construct a functor of cobordism categories as follows:

$$\mathcal{E}^s_R : Cob(a, b) \longrightarrow Cob(a, \overline{b}) : \qquad \begin{array}{l} \Sigma \mapsto \Sigma_R \circ_h \Sigma, \\ M \mapsto R(\Pi_t(M)) \circ_h M. \end{array}$$

The fact that this is a functor is immediate from the interchange law as in (1.2.4). If, instead, we start with a functor R_t, with $\Pi_t \circ R_t = id$, the corresponding extension $\mathcal{E}^t_R$ is obtained by horizontal compositions from the other side.

An easy example can be given for $\overline{b} = 0$ so that $\overline{\beta}$ is also trivial . We choose the surface $\Sigma_{R^{fill}}$ as the union of b discs, seen as a cobordism $S^{\sqcup b} \to \varnothing$. In the language of Proposition 1.1.1 we can also write:

$$\Sigma_{R^{fill}} = \mathrm{tr} \sqcup \ldots \sqcup \mathrm{tr} : \mathcal{A} \sqcup \ldots \sqcup \mathcal{A} \to 1.$$

The closed surface Σ_{Sq}, which is composed of two copies of $\Sigma_{R^{fill}}$ for any permutation β, is clearly homeomorphic to the union of b spheres S^2. Hence, there is a canonical cobordism $R^{fill}(\beta)$, which is the union of b balls D^3.

We, thus, obtain a functor:

$$\mathcal{E}^s_{fill} : \mathcal{C}ob(a,b) \longrightarrow \mathcal{C}ob(a,0). \tag{1.3.5}$$

We will often call this the *fill functor*, since $\Phi(M)$ can be more directly constructed by gluing in the ball with corners $D^2 \times [0,1]$ into the target cylindrical parts of the boundary of M. The discs $D^2 \times \{0,1\}$ are, thus, used to close the holes in a component $\Sigma_{g,a/b}$, yielding $\Sigma_{g,a/0}$.

Another simple example is, of course, the identity R^{id}. Here $b = \overline{b}$, $\overline{\beta}$ is the identity, and $\Sigma_{R^{id}}$ is 1^*, i.e., the union of cylinders with unpermuted boundary identification. The closed surface Σ_{Sq} consists of b tori, each of which is the union of two vertical and two horizontal cylinders. As a surface with corners (the four circles) such a torus bounds in a unique way $S^1 \times [0,1]^2$, where we consider the square $[0,1]^2$ as a manifold with corners. It is clear that the resulting functor $\mathcal{E}^s_{id}$ is the identity.

Unfortunately, these and a few other degenerate cases (e.g., if $b = 1$ etc.) are the only examples of a functor as in (1.3.4), since there are no canonical homomorphisms $S_b \to S_{\overline{\beta}}$ and we can not use braids to realize the $R(\beta)$ as $B_b \to S_b$ does not split. This problem occurs already if we consider a minor modification of the previous examples, namely, if we wish to glue in the discs for the first b' holes, and apply the identity cylinders to the remaining $b'' = b - b'$ holes. That is, we consider the surface

$$\Sigma_R : \mathrm{tr} \sqcup \ldots \sqcup \mathrm{tr} \sqcup \mathbb{I} \sqcup \ldots \sqcup \mathbb{I} : \mathcal{A}^{\sqcup b'} \sqcup \mathcal{A}^{\sqcup b''} \to \mathcal{A}^{\sqcup b''}.$$

It is not clear, which homomorphism $S_b \to S_{b''}$ we should choose, and, although the surface Σ_{Sq} is quite similar to the ones in the previous examples, it is not possible to find bounded manifolds $R(\beta)$ for every β that are compatible with vertical compositions.

Still, there is a way to find a functor which assigns a surface as above, if we consider only a subcategory

$$\mathcal{C}ob(G_s, G_t) \subset \mathcal{C}ob(S_a, S_b) \equiv \mathcal{C}ob(a,b),$$

where we have restricted the category of vertical arrows to subgroups $G_s \subset S_a$ and $G_t \subset S_b$. In the previous example let us choose $G_t = S_{b'} \times S_{b''}$, and $\overline{\beta} : G_t \to S_{b''}$ as the natural projection. Then we only need to consider cobordisms with $\beta = (\beta', \beta'') \in G_t$ so that $R(\beta)$ can be realized as $R^{fill}(\beta') \sqcup R^{id}(\beta'')$. We, thus, obtain a partial fill-functor:

$$\mathcal{E}^s_{part.fill} : \mathcal{C}ob(a, S_{b'} \times S_{b''}) \longrightarrow \mathcal{C}ob(a, S_{b''}). \tag{1.3.6}$$

It is also instructive to consider the case, where we confined ourselves to the trivial subgroup, denoted $I_b = \{1\} \subset S_b$. In this case we may choose any cobordism

$\Sigma : S^{\sqcup b} \to S^{\sqcup \bar{b}}$, and define the cobordism $R(1) = \Sigma \times [0,1]$ as the identity cylinder. We, thus, obtain a functor

$$\mathcal{E}^s_\Sigma : \mathcal{C}ob(a, I_b) \longrightarrow \mathcal{C}ob(a, I_{\bar{b}}).$$

Composition of two such functors is obviously given by the functor associated to the composition of the respective surfaces. Moreover, if we have two surfaces, Σ_1 and Σ_2, together with a cobordism $K : \Sigma_1 \to \Sigma_2$ with trivial vertical arrows, then we can define the system of morphisms $\mathcal{E}^s_K(\Sigma) = K \circ_h (\Sigma \times [0,1]) : \mathcal{E}^s_{\Sigma_1}(\Sigma) \to \mathcal{E}^s_{\Sigma_2}(\Sigma)$, which are easily seen to form a natural transformation from $\mathcal{E}^s_{\Sigma_1}$ to $\mathcal{E}^s_{\Sigma_2}$ using (1.2.4). In summary, we have a natural functor of the form

$$\mathcal{E}^s_* : \mathcal{C}ob(I_*, I_*) \longrightarrow Fun\big(\mathcal{C}ob(a, I_*)\big), \tag{1.3.7}$$

which augments surfaces and cobordisms to functors and natural transformations on a restriction of the same category.

An important application of this correspondence is the following property of the category of connected surfaces with one hole.

Lemma 1.3.1. *The category* $\boldsymbol{\mathcal{C}ob}(0,1)$ *is a strict, braided tensor category with unit. (We denote the respective tensor product also by* $\otimes$*.) The objects are given by the tensor powers* $\phi^{\otimes g}$*, where* $\phi = \otimes \circ_h \phi$ *is the torus with one hole.*

Proof. The tensor product is defined by the composite

$$\otimes : \boldsymbol{\mathcal{C}ob}(0,1) \times \boldsymbol{\mathcal{C}ob}(0,1) \xrightarrow{\ \sqcup\ } \mathcal{C}ob(0, I_2) \xrightarrow{\ \mathcal{E}^s_\otimes\ } \mathcal{C}ob(0,1), \tag{1.3.8}$$

whose image lies in fact in $\boldsymbol{\mathcal{C}ob}(0,1)$. Since in $\boldsymbol{\mathcal{C}ob}_2$ we take cobordisms only up to homeomorphism class, the product $\otimes$ is strictly associative, and we find that

$$\phi = [1, 0/1], \quad \text{and} \quad \phi^{\otimes g} = [g, 0/1].$$

The definition of a braided tensor category also implies a natural transformation c : $\otimes \circ_h P \xrightarrow{\bullet} \otimes$, where P is the flip on the product of categories in (1.3.8). Clearly, $\sqcup \circ_h P = \mathcal{E}^s_{(1,2)^*} \circ_h \sqcup$ so that it suffices to find a transformation $\mathcal{E}^s_\otimes \circ \mathcal{E}^s_{(1,2)^*} \xrightarrow{\bullet} \mathcal{E}^s_\otimes$.

Given the correspondence in (1.3.7) this can be done by specifying a 2-arrow C' from $\otimes \circ_h (1,2)^*$ to $\otimes$, as in the left of the following diagram. Recall from the discussion in the beginning of this section that we have to distinguish between these two surfaces although they are homeomorphic.

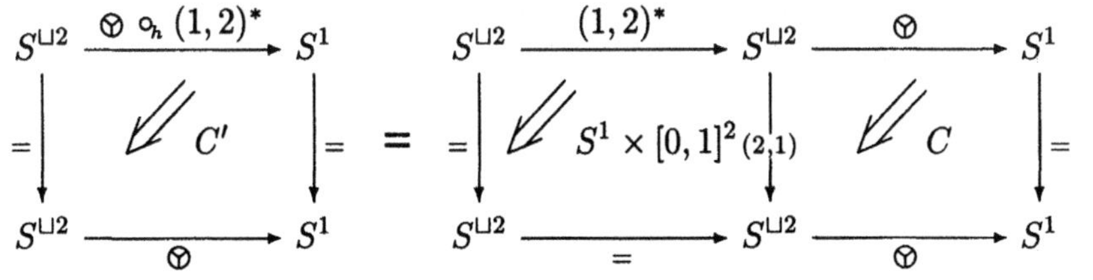

We may rewrite C' as the composite of two cobordisms as on the right of the diagram, where the first is given as in the definition of $\mathcal{E}^s_{id}$ only with the role of vertical and horizontal arrows interchanged. The second C is now a cobordism from the standard three-holed sphere $\Sigma_{0,2/1}$ to itself. Since this is an invertible 2-arrow, we know from Proposition 1.4.2 that it has to be an element of the mapping class group. A natural choice is the braid group generator α_c, which is indicated in Figure 1.7. As a cobordism, C can be embedded into $\mathbb{R}^3$, in the way depicted in Figure 1.5.

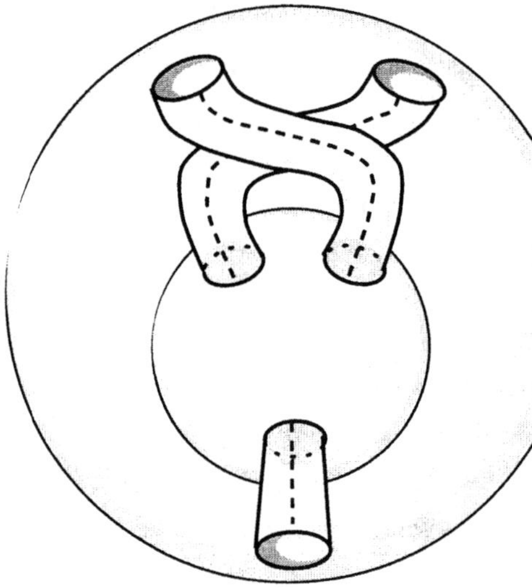

Fig. 1.8. The Braid C as Cobordism

In Fig. 1.8 the target and source copies of $\Sigma_{0,2/1}$ are nested. The cylinder connecting the target holes is the straight piece in the lower part, and the cylinders connecting the source holes are the braided pieces in the upper part. The subset $C \subset \mathbb{R}^3$ consists of the points enclosed by this figure. The triangle equation, that enters the axioms of braided tensor categories, is easily derived from this picture.

The other elementary cobordisms from Proposition 1.1.1 also yield functors with interesting interpretations also on the connected cobordism categories. Let us mention here only two.

The *point splitting functors*

$$\begin{aligned} \mathcal{E}^s_{id\sqcup\otimes^*} &: \boldsymbol{Cob}(a, S_{b-1} \times S_1) \longrightarrow \boldsymbol{Cob}(a, S_{b-1} \times I_2), \\ \mathcal{E}^t_{\otimes\sqcup id} &: \boldsymbol{Cob}(S_1 \times S_{a-1}, b) \longrightarrow \boldsymbol{Cob}(I_2 \times S_{a-1}, b) \end{aligned} \tag{1.3.9}$$

replace the first hole in a surface by a pair of adjacent holes, and a cylinder in a cobordism by a parallel pair of cylinders.

Furthermore, the *recombination functors*

$$\mathcal{E}^s_{id\sqcup\phi^*} : \mathcal{C}ob(a, S_b \times I_2) \longrightarrow \mathcal{C}ob(a,b),$$
$$\mathcal{E}^t_{\phi\sqcup id} : \mathcal{C}ob(I_2 \times S_a, b) \longrightarrow \mathcal{C}ob(a,b) \qquad (1.3.10)$$

glue first two adjacent cylinders to each other, and, as in Figure 1.4, create another handle on a surface, when the adjacent holes are glued to each other. This functor will be discussed in more details in Section 1.4.

Although we focus in this work mainly on the category *Cob* as defined above, let us conclude this section by introducing also a larger double category, to which most of our constructions extend. We consider, as before, two (standard) compact surfaces, Σ_{sc} and Σ_{tg}, and construct a closed manifold, $\boldsymbol{\Sigma}_{Sq}$, by connecting cylinders $S^1 \times [0,1]$ to the boundaries $\partial\Sigma_{sc}$ and $\partial\Sigma_{tg}$ as in Section 1.2. Unlike the definitions we have made so far, we now do not require Σ_{sc} and Σ_{tg} to have the same number of boundary components. Hence, a cylinder $S^1 \times [0,1]$ does not have to start at a component in $\partial\Sigma_{sc}$ and end in a component $\partial\Sigma_{tg}$, but it may also connect $\partial\Sigma_{sc}$ to itself.

Typically, we need this generalization, when we wish to construct pairings, $\Sigma \sqcup \Sigma \longrightarrow \varnothing$, as in the case of closed surfaces. Here, we consider both the upper and the lower part of $\Sigma \times [0,1]$ as a source surface and apply an orientation reversing automorphism $\Sigma \xrightarrow{\sim} -\Sigma$ to one of them.

Another, more generic, example is indicated in the schematic diagram below. The source surface Σ_{sc} has two components Σ^1_{sc} and Σ^2_{sc} with one and two boundary components. The target surface Σ_{tg} has also two components Σ^1_{tg} and Σ^2_{tg} but with three and two boundary components.

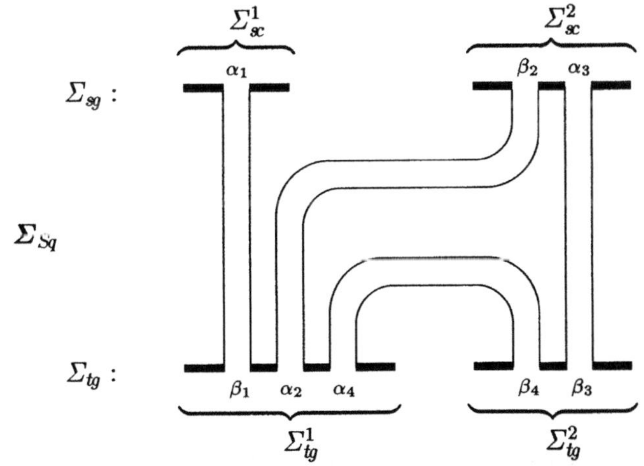

(1.3.11)

The system of these cobordisms can again be embedded in the formalism of double categories, if we consider a larger set of vertical morphisms instead of just the permutations. We should be able to describe with this new set also pairings of labels, which are naturally encoded into the maximum or minimum of a strand. A natural first candidate for the vertical category is, thus, the category of 1+0-dimensional cobordisms, which can be seen as strands or singular tangles immersed in the plane.

It will become clear in the construction of TQFT's that the vertical category is even larger since we may now encounter situations in which we have to make a choice of homeomorphisms α as for the horizontal surfaces. Specifically, maxima and minima are symmetric and side-inverses of each other only up to a canonical isomorphism. This means the following relations between diagrams are not strictly equalities but are given by a specified isomorphism:

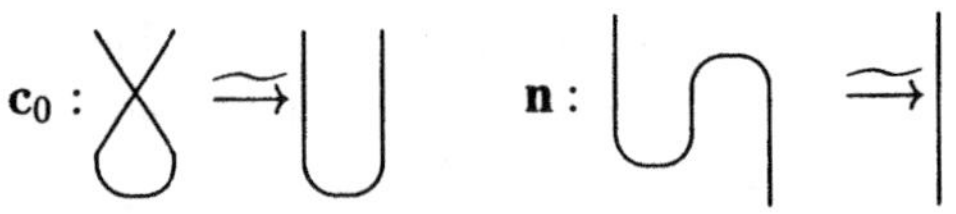

The definition of the double category $\boldsymbol{Cob}^{\mathrm{vtg}}$ with the enhanced vertical category is now in complete analogy to the one for $\mathcal{Cob}$. Yet, the additional structure allows for a few more operations, which reidentify pieces of the vertical surfaces as parts of the horizontal ones.

For example, vertical compositions of singular tangles may result in closed components, which, in the case of singular tangles, can be separated as isolated circles. If the vertical 1-arrow of a cobordism, M, contains a circle, this indicates a torus $\Sigma_{0,2} \cong T \subset \partial M$, which is disjoint from $\Sigma_{sc} \sqcup \Sigma_{tg}$. Since it is composed of pieces $S^1 \times [0,1]$, it has a distinguished long meridian (along the intervals) and a distinguished short meridian (along S^1). Also, with the orientation of the strands of the singular tangle, we have a direction along the long meridian specified, which together with the orientation of T induced by M yields a direction for the short meridian. This means that, up to isotopy, there is a unique homeomorphism from the standard torus to T, once we have specified the long and short meridians on $\Sigma_{2,0}$. As a result, we have a well defined operation that removes a circle from the vertical arrow of a cobordism and adds another $[2,0]$-factor to, e.g., the source surface, as in the following formula:

$$\boldsymbol{Cob}^{\mathrm{vtg}}(*, \zeta_2 \sqcup \bigcirc) \longrightarrow \boldsymbol{Cob}^{\mathrm{vtg}}(*, \zeta_2) \tag{1.3.12}$$

$$\begin{array}{ccc} a & \xrightarrow{\Sigma_{sc}} & b \\ {\scriptstyle \zeta_1}\downarrow & \swarrow\!\!\!\swarrow \; M & \downarrow{\scriptstyle \zeta_2 \sqcup \bigcirc} \\ a' & \xrightarrow[\Sigma_{tg}]{} & b' \end{array} \;\longmapsto\; \begin{array}{ccc} a & \xrightarrow{\Sigma_{sc} \sqcup [2,0/0]} & b \\ {\scriptstyle \zeta_1}\downarrow & \swarrow\!\!\!\swarrow \; M & \downarrow{\scriptstyle \zeta_2} \\ a' & \xrightarrow[\Sigma_{tg}]{} & b' \end{array} .$$

Here, the ζ_j stand for the classes of singular tangles with some additional composition data.

Note, that this operation is not a functor. In particular, the function on the objects depends on the cobordism they belong to. We still mention it, since it will play a practical role in the derivation of presentations of cobordisms.

In the same way we can isolate a strand that connects two adjacent labels in the source, and reinterpret the remaining component $\cong \bigcup$, as a horizontal gluing of the source surface with $S^1 \times [0,1]$ seen as a cobordism $S^{\sqcup 2} \to \varnothing$. Thus, we find an operation of the following type, which is again well defined but not functorial:

$$\mathcal{C}ob^{\mathrm{vtg}}(*, \zeta_2 \sqcup \bigcup) \longrightarrow \mathcal{C}ob^{\mathrm{vtg}}(*, \zeta_2) \tag{1.3.13}$$

$$\begin{array}{ccccccc} a & \xrightarrow{\Sigma_{sc}} & b & & a & \xrightarrow{\Sigma_{\cup} \circ_h \Sigma_{sc}} & b-2 \\ {\scriptstyle \zeta_1}\downarrow & \swArrow M & \downarrow{\scriptstyle \zeta_2 \sqcup \cup} & \longmapsto & {\scriptstyle \zeta_1}\downarrow & \swArrow M & \downarrow{\scriptstyle \zeta_2} \\ a' & \xrightarrow{\Sigma_{tg}} & b' & & a' & \xrightarrow{\Sigma_{tg}} & b' \end{array} ,$$

where $\Sigma_{\cup}$ is the 1+1-cobordism, which has a "returning" cylinder in the respective position, i.e.,

$$\Sigma_{\cup} = \mathbb{1} \sqcup [0, 2/0] \sqcup \mathbb{1}, \quad \text{with} \quad [0, 2/0] \cong S^1 \times [0, 1].$$

The composition of Σ_{sc} with $\Sigma_{\cup}$ has the effect of either connecting two component of Σ_{sc} or adding a handle to one of the components.

1.4 Mapping Class Groups, Framed Braid Groups, and Balancing

The aim of this section is to describe how the mapping class groups and the braid groups on a surface appear as natural constituents of $\mathcal{C}ob$. Specifically, we shall prove in Section 1.4.1 that $\pi_0(\mathcal{D}iff^+(\Sigma, \partial\Sigma))$ is isomorphic to the group of two-sided-invertible cobordisms from Σ to itself. This correspondence yields interpretations of Heegaard diagrams, stable mapping class groups, etc. We also introduce the more adequate definition of the "framed" braid groups $\widehat{\mathcal{B}_b}(\Sigma)$ as the fundamental groups of the configuration spaces of discs rather than points on Σ. In Section 1.4.2 we show that the map $\widehat{\mathcal{B}_b} \longrightarrow \mathcal{B}_b$ onto the ordinary braid group defines a central extension, and that the kernel is generated by a transformation, $\mathbf{v}$, that makes $\mathcal{C}ob(0, 1)$ into a balanced BTC. Finally, we determine in Section 1.4.3 the kernel of the representation of $\widehat{\mathcal{P}}(\Sigma)$ in $\mathcal{C}ob$.

1.4.1 Mapping Class Groups in $\mathcal{C}ob$

The fact that we take the cobordisms only up to homeomorphism also allows us to find equivalences between the boundary maps from (1.2.1). Specifically, we show in the next lemma that for a fixed three-manifold its class in $\mathcal{C}ob$ is unchanged if the boundary map is changed by an isotopy.

Lemma 1.4.1. *Suppose for a manifold with corners M, the boundary maps ψ_0 and ψ_1 for a given square lie in the same path-connected component of*

$$\mathcal{D}iff^+(\Sigma_{Sq}, \partial M | \xi_{1.strata}),$$

which is the space of orientation preserving homeomorphisms with a fixed identification $\xi_{1.strata} : (\Sigma_{Sq})_1 \to (\partial M)_1$ of the 1-strata, that is, the special circles, of the surfaces.

Then the classes of (M, ψ_0) and (M, ψ_1) are the same in $\mathcal{C}ob$.

Proof. With compositions and inversions being continuous on the spaces of functions, the assumptions of the lemma imply that there is an isotopy $f_- : [0,1] \to \mathcal{D}\!i\!f\!f^+(\partial M)$, such that $f_0 = id_{\partial M}$, $f_1 = \psi_1 \circ \psi_0^{-1}$, and f_t is for all t an automorphism of a M, which is identity on the lower dimensional corners. From this we obtain a homeomorphism

$$\Psi : \partial M \times [0,1] \xrightarrow{\cong} \partial M \times [0,1] : (m,t) \longmapsto (m, f_t(m)).$$

Now, the homeomorphism type of a manifold is not changed, if we add a copy of its collar to its boundary. More precisely, we can always find an isomorphism $h_{collar} : M \sqcup_{\partial M \times 0} \partial M \times [0,1] \xrightarrow{\cong} M$, which is identity outside of a collar of M, and which maps the $\partial M \times 1$-piece of the left hand side to the boundary of the right hand side. We can now define an automorphism $\rho : M \xrightarrow{\cong} M$ by the following diagram of isomorphisms:

$$\begin{array}{ccc} M \sqcup_{\partial M \times 0} \partial M \times [0,1] & \xrightarrow{h_{collar}} & M \\ \Big\downarrow {\scriptstyle id_M \sqcup \Psi} & & \Big\downarrow {\scriptstyle \rho} \\ M \sqcup_{\partial M \times 0} \partial M \times [0,1] & \xrightarrow{h_{collar}} & M \end{array}$$

The glued map for the right vertical arrow is well defined, since Ψ is the identity on $\partial M \times 0$. Since it is $\psi_1 \circ \psi_0^{-1}$ on the $\partial M \times 1$ boundary piece, we infer that $\rho \circ \psi_0 = \psi_1$, and, hence, that (M, ψ_0) and (M, ψ_1) are equivalent.

This lemma is used in various places in Section 1.3. The application in which we are interested in this section is to interpret elements of the mapping class group of a surface as cobordisms.

Let us suppose $\chi \in \mathcal{D}\!i\!f\!f^+(\Sigma, S^{\sqcup N})$ to be an automorphism of a compact, oriented and connected surface Σ, which acts by a permutation $\pi_\chi \in S_N$ of the $N = a + b$ components of the boundary of Σ. From this we then obtain a homeomorphism from the square, with both horizontal arrows given by Σ and vertical arrows from π_χ, to the boundary of the cylinder over Σ as follows:

$$\psi_\chi : \boldsymbol{\Sigma}_{Sq} \equiv \Sigma \sqcup \pi_\chi^* \sqcup \Sigma \xrightarrow{id \sqcup id \sqcup \chi^{-1}} \Sigma \sqcup 1^* \sqcup \Sigma \equiv \partial(\Sigma \times [0,1]). \qquad (1.4.1)$$

It is easily checked that the glued map, with χ^{-1} acting on the target piece, is well defined and continuous, since the permutations of the ends of the cylinders match by definition.

We may now consider the cobordism $(\Sigma \times [0,1], \psi_\chi)$, and its class $B_\Sigma^0(\chi) \in \mathcal{C}ob(\pi_\chi)$. It follows from Lemma 1.4.1 that this map factors into a map B_Σ defined on the mapping class group of the surface, as in the following diagram.

$$\begin{array}{ccc} \mathcal{Diff}^+(\Sigma) & \xrightarrow{B^0_\Sigma} & \mathrm{End}_{\mathcal{Cob}(*)}(\Sigma) \\ \downarrow & \nearrow_{B_\Sigma} & \\ \pi_0(\mathcal{Diff}^+(\Sigma)) & & \end{array}$$

It is easy to see that $B^0_\Sigma(\chi_2) \circ B^0_\Sigma(\chi_1) = B^0_\Sigma(\chi_2 \circ \chi_1)$, using the equivalence $id \sqcup_\Sigma \chi_1$, so that also B_Σ is a homomorphism. This implies that its image lies entirely in the set of invertible cobordisms, or, equivalently, the group of automorphisms of Σ in $\mathcal{Cob}(*)$.

A combination of a few classical results from three manifold theory shows that this map is in fact an isomorphism.

Proposition 1.4.2. *The map*

$$B_\Sigma : \pi_0\big(\mathcal{Diff}^+(\Sigma, \partial\Sigma)\big) \xrightarrow{\ \cong\ } \mathrm{Aut}_{\mathcal{Cob}(*)}(\Sigma)$$

is an isomorphism.

Proof. 1) $\hookrightarrow$: In order to show that the map is injective assume that $(\Sigma \times [0,1], id)$ and $(\Sigma \times [0,1], \psi_\chi)$ are mapped to the same cobordism class. This means there is a homeomorphism $\rho : \Sigma \times [0,1] \xrightarrow{\sim} \Sigma \times [0,1]$, which is identity on the lower boundary piece $\Sigma \times 0$ and χ on the upper boundary piece $\Sigma \times 1$. Hence, ρ is a pseudo-isotopy between χ and the identity, which in two dimensions implies isotopy. More precisely, it suffices to consider the homotopy from the identity to χ, which is given by the composite $p_1 \circ \rho : \Sigma \times [0,1] \to \Sigma$ with the natural projection p_1 for the product space. The fact that homotopy implies isotopy has been proven in [Eps66].

2) $\twoheadrightarrow$: Let M be an invertible cobordism, that is, for $M : \Sigma \to \Sigma'$ there is a cobordism $N : \Sigma' \to \Sigma$, such that $\Sigma \cong \Sigma'$, and $M \circ N \cong \Sigma \times [0,1]$, $N \circ M \cong \Sigma' \times [0,1]$. We need to show that each of the cobordisms is homeomorphic to the cylinder $\Sigma \times [0,1]$, as manifolds with corners. It is easy to find a homeomorphisms of $\Sigma \times [0,1]$ that changes an arbitrary boundary map into one of the form ψ_χ so that $M \cong \Sigma \times [0,1]$ does, in fact, imply that it lies in the image of B_Σ.

To begin with, we observe that $\pi_1(\Sigma) \hookrightarrow \pi_1(M)$, as well as $\pi_1(\Sigma') \hookrightarrow \pi_1(M)$, are injective, since the composition $\Sigma \hookrightarrow M \hookrightarrow N \sqcup_{\Sigma'} M \cong \Sigma \times [0,1]$ is a π_1-isomorphism.

It is, in fact, true that these maps of fundamental groups have to be isomorphisms. In order to see this, we have to show that the homomorphism induced from the inclusion $\pi_1(M) \to \Sigma \times [0,1]$ has no kernel. That is, a closed path γ in M, which is contractible in the larger cylinder, $\Sigma \times [0,1]$, should also be contractible in M.

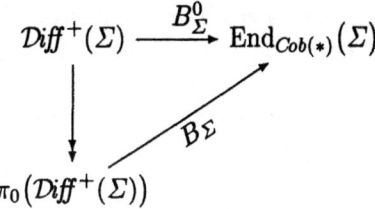

Contractibility of γ may be expressed by the existence a continuous map $h_\gamma : D^2 \to \Sigma \times [0,1]$, whose restriction to the boundary ∂D^2 of the disc parametrizes γ. In the given dimensions we may assume that h_γ is C^1, and also that the submanifold $\Sigma' \subset \Sigma \times [0,1]$, over which M and N had been glued, is C^1. By classical transversality we can achieve by small perturbations of h_γ that $h_\gamma \pitchfork \Sigma'$. Hence, $h_\gamma^{-1}(\Sigma') \subset D^2$ is a compact one-manifold, i.e., a finite disjoint union of Jordan curves. For a review of these results on transversality consult Chapter 3 of [Hir76].

At least one of these curves encloses a contractible region $R \subset D^2$ that contains no other curves. Hence, R is mapped by h_γ entirely into one of the cobordism pieces, M or N, and the bounding curve ∂R maps into Σ'. This implies that the element in $\pi_1(\Sigma')$ represented by ∂R (after choosing an appropriate basepoint) is mapped to the unit in $\pi_1(M)$ or $\pi_1(N)$.

We know, however, that the inclusion is π_1-injective, so ∂R has to be contractible already in Σ'. We can, thus, redefine h_γ on R to be the respective contraction of $h_\gamma(\partial R)$ in Σ', and push this slightly away from Σ' so that $\overline{h_\gamma(R)}$ is disjoint from Σ'. By this process we have, thus, eliminated one of the curves in $h_\gamma^{-1}(\Sigma') \subset D^2$.

Continuing inductively, we can eliminate all curves in a finite number of steps and, thus, obtain a map $h_\gamma : D^2 \to \Sigma \times [0,1]$ that does not meet Σ' and, therefore, maps entirely into M. Thus, γ is contractible also in M, and $\pi_1(\Sigma) \xrightarrow{\sim} \pi_1(M)$ is an isomorphism.

Now we can apply the result in Theorem 10.2 of [Hem76]; see also [Wal68]. It asserts that for $\Sigma \neq S^2$ and the given isomorphism, $M \cong \Sigma \times [0,1] \# P_M$ (also in the case where $\partial \Sigma \neq \varnothing$). Here P_M is a homotopy three-sphere.

From $M \subset M \circ N = \Sigma \times [0,1] \subset S^3$ we obtain an embedding of the homotopy sphere with a ball removed $P_M - D^3 \subset S^3$. But $\partial(P_M - D^3) = S^2$ and every sphere separates S^3 into two 3-balls.

Hence, we have $P_M = S^3$ and, thus, $M \cong \Sigma \times [0,1]$, which completes the proof.

The isomorphism in Proposition 1.4.2 allows us to think of the mapping class group of a two-dimensional surface as a canonical element in the category of three-dimensional cobordisms. Hence, it may be treated on the same footing as, e.g., closed three-dimensional manifolds. In particular, a manifold represented as a Heegaard splitting can be expressed as the composition of three cobordisms

$$\varnothing \xrightarrow{\quad H_g \quad} \Sigma \xrightarrow{\quad B_\Sigma(\chi) \quad} \Sigma \xrightarrow{\quad S^3 - H_g \quad} \varnothing . \tag{1.4.2}$$

Here, $H_g \subset S^3$ is a handlebody that is unknottedly embedded into the three sphere, with boundary $\partial H_g = \Sigma$, and χ is the automorphism of Σ, over which H_g and the complement $S^3 - H_g$ are reglued.

Similarly, we can express the mapping torus $\Sigma \times [0,1] \big/_{\chi : \Sigma \times 0 \sim \Sigma \times 1}$ as the categorical trace $\mathit{Tr}_\Sigma : \mathrm{End}_{\mathit{Cob}(*)}(\Sigma) \to \mathrm{End}_{\mathit{Cob}(*)}(\varnothing)$, evaluated on $B_\Sigma(\chi)$.

Moreover, an interesting point of view on the stable mapping class group can be obtained, if we combine Proposition 1.4.2 with Lemma 1.3.1. If we abbreviate $\mathcal{G} = \mathit{Cob}(0,1)$, and $\Gamma_{g,1}$ the mapping class group of $\Sigma_{g,1}$, we have isomorphisms $\Gamma_{g,1} \cong \mathrm{Aut}_{\mathcal{G}}(\phi^{\otimes g})$ and, hence, for the stable mapping class group

$$\Gamma_{\infty,1} \cong \varinjlim \mathrm{Aut}_{\mathcal{G}}(\phi^{\otimes g}), \tag{1.4.3}$$

where the inclusions in the inductive limit on the right hand side are given by $T \mapsto \mathbb{1}_\phi \otimes T$. A similar categorical approach to $\Gamma_{\infty,1}$ is given in [Mil86]. There, however, three-dimensional cobordisms are not used and the equivalences between morphisms that render the tensor product associative, analogous to Lemma 1.3.1, are more complicated. Closely related to this is the description of $\Gamma_{\infty,1}$ in [Til95], where $\Sigma_{g,1}$ is obtained by horizontal compositions of the two-hole torus $\Sigma_{1,1/1}$. Correspondingly, the limit in (1.4.3) has to be replaced by expressions that contain the special element in $\mathcal{Cob}(1,1)$ $\circ_h$-composed with itself g times.

From Lemma 1.3.1 we also find immediately the action of the infinite braid group $B_\infty = \lim_{\longrightarrow} B_n$ on $\Gamma_{\infty,1}$, which was constructed explicitly by Z. Fiedorowicz[1] using $\Gamma_{g,1} \cong Out(\mathrm{F}_{2g})$, where F_m denotes the free group in m generators.

Let us emphasize here that the group $\mathcal{Diff}^+(\Sigma, \partial\Sigma)$ of automorphisms of Σ that are identity on $\partial\Sigma$ and the group $\mathcal{Diff}^+(\Sigma^{cl}, \{p_j\})$, which consists of maps of the closed surface Σ^{cl} that fix a finite number of points p_j, are *not* homotopy equivalent. With $\Sigma - \partial\Sigma \cong \Sigma^{cl} - \{p_j\}$ the latter group can also be constructed from the maps of Σ to itself that can act non-trivially on $\partial\Sigma$. The simplest example, in which the inequivalence is apparent, is given by a sphere with one hole or puncture. This is manifested in the next lemma, which will be of use also in later discussions.

Lemma 1.4.3. *1. $\mathcal{Diff}^+(D^2, S^1)$ is a contractible space.*
2. Let $R : S^1 \longrightarrow \mathcal{Diff}^+(S^2, p)$ be the map, which assigns to an angle the respective rotation around the axis through p. Then $\pi_j(R)$ is an isomorphism for all $j \geqslant 0$.

Proof. The proof of the first part is basically Alexander's argument [Ale23]: an explicit contraction Ψ_s is given by defining $\Psi_s(f)$ by $R(s) \circ f \circ R(s)^{-1} \in \mathcal{Diff}^+(e^{-s}D^2, \partial)$ on the smaller disc of radius e^{-s} in D^2, where $R(s)$ is as in the proof of Lemma 1.4.7. On the annulus of points, whose distance from the origin is between e^{-s} and 1, we set $\Psi_s(f)$ to be the identity. This yields a continuous function since f leaves S^1 pointwise fixed, and $\Psi_\infty(f) = id_{D^2}$.

For the second part recall that $SO(3) \hookrightarrow \mathcal{Diff}^+(S^2)$ is a homotopy equivalence. This result is due to Kneser, see §2.4 in [Kne26], for homeomorphisms and to Smale [Sma59] for the smooth case. Now, both spaces are naturally fibrations over S^2, and it is easy to see that the inclusion is a map of fibrations as follows:

$$\begin{array}{ccccc} S^1 \hookrightarrow SO(3) & & \hookrightarrow & & \mathcal{Diff}^+(S^2, p) \hookrightarrow \mathcal{Diff}^+(S^2) \\ \downarrow & & & & \downarrow \\ S^2 & & \xrightarrow{\ id\ } & & S^2 \end{array} \qquad (1.4.4)$$

From this the assertion follows by an easy application of the 5-Lemma to the respective map between long exact sequences of the two fibrations.

In [Bir74] it is discussed in detail how the braid groups can be obtained from fibrations of the automorphism groups of punctured surfaces. In view of the previous

[1] Private communication

lemma we expect to find groups that are slightly different from the ordinary braid group if consider instead surfaces with parametrized boundaries.

They arise naturally, when we consider the following fibration of spaces of maps, in analogy to the fibration in Theorem 4.1 in [Bir74]:

$$\begin{array}{ccc} \mathcal{D}\!\mathit{iff}^+(\Sigma_{g,a/b}, S^{\sqcup a+b}) & \hookrightarrow & \mathcal{D}\!\mathit{iff}^+(\Sigma_{g,a/0}, S^{\sqcup a}) \\ & & \downarrow \tau \\ & & \mathcal{E}\!\mathit{mbd}^+(\sqcup^b D^2, \Sigma_{g,a/0}) \big/ S_b \end{array} \tag{1.4.5}$$

Recall that in our convention $\mathcal{D}\!\mathit{iff}^+(\Sigma, S^{\sqcup N})$ is the group of automorphisms of Σ which restricted to the boundary $\partial\Sigma \cong S^{\sqcup N}$ has to coincide with the canonical action of some element of the symmetric group S_N on $S^{\sqcup N}$.

The base space is the space of embeddings of b discs into the surface of genus g and a holes, modulo the canonical action of S_b. The projection τ is defined by restricting an automorphism of $\Sigma_{g,a/0}$ to the b discs in the surface, which have to be deleted in order to obtain $\Sigma_{g,a/b}$.

From this fibration we obtain a long exact sequence of homotopy groups, which ends as follows:

$$\begin{aligned} \ldots \xrightarrow{\tau_*} \widehat{\mathcal{B}}_b(\Sigma_{g,a}) \xrightarrow{d_*} \pi_0\big(\mathcal{D}\!\mathit{iff}^+(\Sigma_{g,a/b}, S^{\sqcup a+b})\big) \\ \xrightarrow{i_*} \pi_0\big(\mathcal{D}\!\mathit{iff}^+(\Sigma_{g,a/0}, S^{\sqcup a})\big) \to 1. \end{aligned} \tag{1.4.6}$$

Here we used that the base space of embeddings is connected, and introduced the notation for the *framed braid group*

$$\widehat{\mathcal{B}}_b(\Sigma_{g,a}) = \pi_1\Big(\mathcal{E}\!\mathit{mbd}^+(\sqcup^b D^2, \Sigma_{g,a/0}) \big/ S_b\Big).$$

As opposed to the ordinary braid group $\mathcal{B}_b(\Sigma_{g,a})$, we use here a symmetrized configuration space of discs instead of just points. For the sake of simplicity, we shall often confine ourselves to the covering of the fibration in (1.4.5), where we assume that the homeomorphisms leave the boundary components pointwise fixed (without permutation), and where we omit the division by S_b of the base space. To this end let us also introduce notations for the pure braid group and the mapping class group with trivial permutation,

$$\begin{aligned} \widehat{\mathcal{P}}_b(\Sigma_{g,a}) &= \pi_1\big(\mathcal{E}\!\mathit{mbd}^+(\sqcup^b D^2, \Sigma_{g,a/0})\big), \\ \mathcal{M}(g, a/b) &= \pi_0\big(\mathcal{D}\!\mathit{iff}^+(\Sigma_{g,a/b}, I_a \sqcup I_b)\big). \end{aligned} \tag{1.4.7}$$

As for the ordinary braid groups, we have for their framed counterparts the short exact sequence,

$$1 \longrightarrow \widehat{\mathcal{P}}_b(\Sigma) \hookrightarrow \widehat{\mathcal{B}}_b(\Sigma) \xrightarrow{\zeta} S_b \longrightarrow 1. \tag{1.4.8}$$

The connecting map in (1.4.6) or its restriction to the pure case $d_* : \widehat{\mathcal{P}}_b(\Sigma_{g,a}) \to \mathcal{M}(g, a/b)$ can be described very explicitly if we identify $\mathcal{M}$ via Proposition 1.4.2 with $\mathrm{Aut}_{\mathcal{C}ob(I_a,I_b)}(\Sigma_{g,a/b})$. For a closed path $t \mapsto f_t : \sqcup^b D^2 \to \Sigma_{g,a/0}$ in the space of embeddings we can define an embedding of solid tubes into cylinder over $\Sigma_{g,a/0}$ by

$$F : \sqcup^b D^2 \times [0,1] \hookrightarrow \Sigma_{g,a/b} \times [0,1] : \quad (d,t) \longmapsto (f_t(d), t).$$

As a three manifold the obvious candidate for the associated cobordism is given by the complement:

$$M_f = \overline{\Sigma_{g,a/0} \times [0,1] - image(F)}.$$

Since f_0 and f_1 are the embeddings of b discs into the standard target positions, the upper and lower boundary piece of M_f is $\Sigma_{g,a/b}$, and we choose for them canonical boundary identifications. The parametrization of the cylindrical pieces of ∂M_f will be given by the restriction of F to $\sqcup^b S^1 \times [0,1] \subset \sqcup^b D^2 \times [0,1]$.

Lemma 1.4.4. *The assignment $f \mapsto M_f$, with boundary identifications as above, realizes the connecting map d_*.*

Proof. We pick a Morse-function $h : M_f \to [0,1]$, such that $h^{-1}(1)$ and $h^{-1}(0)$ are the upper and lower pieces of ∂M_f. Moreover, the gradient ∇h will be parallel to the cylindrical pieces, and pull back of the flow along these pieces to $S^1 \times [0,1]$ will be directed from $S^1 \times 0$ to $S^1 \times 1$. The function h can be modified so that it has no critical points inside M_f, and so that for an extension to $\Sigma_{g,a/0} \times [0,1]$ we have $\langle \nabla \tilde{h}, \nabla p \rangle > 0$, where p is the canonical projection $\Sigma_{g,a/0} \times [0,1] \to [0,1]$ and $\tilde{h}$ is an extension of h to the cylinder $\Sigma_{g,a/0} \times [0,1]$. It is easy to see that the flow $U^!(t)$ of the vector field $\langle \nabla \tilde{h}, \nabla p \rangle^{-1} \nabla h$ maps a slice $p^{-1}(t_1)$ to another slice $p^{-1}(t_1 + t)$, while preserving M_f. Let us denote by $U(t)$ the restriction of $U^!(t)$ to $\Sigma_{g,a/0} \times \{0\} = \Sigma_{g,a/0}$.

It is clear that $p \circ U(t) \in \mathcal{D}iff^+(\Sigma_{g,a/0})$ is a lifting of f_t in the fibration (1.4.5) so that the image of the connecting map is given by the class of $p \circ U(1)$. Moreover, given the isomorphism $U(t) : \Sigma_{g,a/b} \times [0,1] \xrightarrow{\sim} M_f$, the cylinder over $\Sigma_{g,a/b}$ yields the same class in $\mathcal{C}ob(*)$, if we use canonical identifications at the bottom, and cylindrical pieces, but $p \circ U(1)$ at the top-piece. Thus, it is the cobordism associated as in Proposition 1.4.2 to mapping class of $p \circ U(1)$.

It follows also immediately from Lemma 1.4.4 that d_* is a group homomorphism.

1.4.2 Framing Extension of Braid Groups, and the Ribbon Element

Next, let us discuss the connection between the framed and the ordinary braid groups. Clearly, $\mathcal{E}mbd^+(\sqcup^b D^2, \Sigma_{g,a/0})$ maps onto $\mathcal{E}mbd^+(\{c_1, \dots, c_b\}, \Sigma_{g,a/0})$, where c_j is the center point of the j-th disc so that for π_1 of this map we obtain a projection $C_* : \widehat{\mathcal{B}_b} \twoheadrightarrow \mathcal{B}_b$. Surjectivity of C_* is easily seen, but the kernel of C_* is not trivial.

Specifically, for every center point $\overline{c_j} \in \Sigma_{g,a/0}$ of the standard discs in $\Sigma_{g,a/0}$, with $j = 1, \ldots, b$, we can define an element $\delta_j \in \widehat{\mathcal{B}_b}$ by the following path. For the standard disc we can define the rotation $O(\theta) : D^2 \to D^2$ by an angle θ, which keeps the center point fixed. If $i^j : D^2 \hookrightarrow \Sigma_{g,a/0}$ is the standard embedding of the j-th target disc, we get a parametrized embedding $i_\theta : \sqcup^b D^2 \hookrightarrow \Sigma_{g,a/0}$ by using $i^{j'}$ if $j \neq j'$ and $i^j \circ O(\theta)$ on the j-th disc. This yields a closed path in $\mathcal{E}mbd^+(\sqcup^b D^2, \Sigma_{g,a/0})$ and, thus, an element $\delta_j \in \widehat{\mathcal{B}_b}$, which maps to the constant path in $\mathcal{E}mbd^+(\{c_1, \ldots, c_b\}, \Sigma_{g,a/0})$ so that $\delta_j \in \ker(C_*)$.

The additional generators are central both in the pure framed braid group as well as the permutation free mapping class group:

Lemma 1.4.5. *For all $j = 1, \ldots, b$*

$$\delta_j \in center\big(\widehat{\mathcal{P}_b}(\Sigma_{g,a/0})\big) \quad and \quad d_*\delta_j \in center\big(\mathcal{M}(g, a/b)\big).$$

Proof. Let an element $\overline{f} \in \widehat{\mathcal{P}_b}$ be represented by functions $f^k : [0,1] \times D^2 \to \Sigma_{g,a/0}$ such that $f(t, \_) = \sqcup_{k=1}^b f^k(t, \_)$ is an embedding of b discs, and $f^k(1, \_) = f^k(0, \_) = i^k$, i.e., the standard embedding of the k-th disc. If we compose f with a rotation of the j-th disc, i.e., we replace f^j by $f^j \circ (id \times O(\theta))$, we obtain a function

$$H : [0,1] \times [0, 2\pi] \longrightarrow \mathcal{E}mbd^+(\sqcup^b D^2, \Sigma_{g,a/0}).$$

If we restrict H to the path on the boundary of the rectangle going from $(0,0)$ to $(t, \theta) = (1, 0)$, and then to $(1, 2\pi)$ we obtain the composite path for $\delta_j * \overline{f} \in \widehat{\mathcal{P}_b}$. Going from $(0,0)$ to $(1, 2\pi)$ over $(0, 2\pi)$ instead we obtain the path representing $\overline{f} * \delta_j \in \widehat{\mathcal{P}_b}$. From H we easily construct a homotopy between these two paths, which shows that δ_j commutes with $\overline{f}$ in $\widehat{\mathcal{P}_b}$.

The rotation of a disc, D^2, in $\mathcal{E}mbd^+(\sqcup^b D^2, \Sigma_{g,a/0})$ can be lifted to a path, ξ_θ, in $\mathcal{D}iff^+(\Sigma_{g,a/0}, I_a)$. If the disc is of radius one in some local coordinates, ξ_θ will be identity outside of a disc of radius $1 + \varepsilon$ and the rotation $O(\theta)$ inside D^2. In the collar we define in polar coordinates $\xi_\theta(1 + \varepsilon t, \phi) = (1 + \varepsilon t, \phi + (1 - t)\theta)$, with $0 < t < 1$. The element $d_*\delta_j$ is, thus, represented by the Dehn twist $\xi_{2\pi}$ on $\Sigma_{g,a/b}$, which is localized in an ε-thick collar of the j-th target boundary component of $\Sigma_{g,a/b}$. Clearly, any other element in $\mathcal{M}(g, a/b)$ can be represented by an automorphism of $\Sigma_{g,a/b}$, which is identity *inside* of the ε-collar. It is obvious that such an automorphism commutes with $\xi_{2\pi}$ so that, in particular, $d_*\delta_j$ is central in $\mathcal{M}(g, a/b)$.

For the full braid group $\widehat{\mathcal{B}_b}$ we get instead of centrality the relation

$$g\delta_j g^{-1} = \delta_{\zeta(g)(j)}, \qquad \forall g \in \widehat{\mathcal{B}_b}, \tag{1.4.9}$$

where ζ is the projection from (1.4.8), and an analogous equation for $d_*\delta_j$ in $\pi_0\big(\mathcal{D}iff^+(\Sigma_{g,a/b}, S^{\sqcup a+b})\big)$.

By Proposition 1.4.2 the element $d_*\delta_j$ can also be viewed as a cobordism in $\mathrm{Aut}_{\mathcal{C}ob(*)}(\Sigma)$. Let us elaborate this interpretation with a few more details, since it is closely related with the notion of balancing in a braided tensor category.

Following Lemma 1.4.4, the cobordism for $d_*\delta_j$ is given as a three-fold by the cylinder $\Sigma \times [0,1]$, and the boundary identification $\psi : \partial(\Sigma \times [0,1]) \xrightarrow{\sim} \partial(\Sigma \times [0,1])$ from (1.2.1) is canonical on all pieces, except the j-th cylindrical piece. There we set $\psi(\phi, t) = (\phi + 2\pi t, t)$, where ϕ is the coordinate of the standard S^1. Of particular interest is the twist $V_0 = d_*\delta$ for the target hole of the surface $\Sigma_{0,1/1} \cong S^1 \times [0,1]$. In the double category picture the latter surface is the identity 1-arrow id_{S^1} in $\mathrm{End}_{\mathcal{C}ob_2 S^1}$, and $V_0 = d_*\delta$ is a 2-arrow for the square, in which all objects are S^1 and all 1-arrows are id_{S^1}.

For a general surface Σ we can cut out a collar $\cong S^1 \times [1, 1+\varepsilon] \times [0,1]$ around the j-th cylindrical piece. If we choose canonical identifications along $S^1 \times \{1+\varepsilon\} \times [0,1]$ the cobordism $d_*\delta_j$ is, thus, a horizontal gluing (over only the j-th piece) of V_0 with the cylinder over Σ minus the ε-collar. More precisely, with the canonical (up to an isotopy) choice of the homeomorphism α between composed and standard surfaces, we, thus, obtain the following form of $d_*\delta_j$ in the double category language of Section 1.2:

$$d_*\delta_j = \quad V_0^{(j)} \circ_h id_\Sigma = (1 \sqcup \ldots \sqcup \underset{\substack{\uparrow \\ j-\text{th position}}}{V_0} \sqcup \ldots \sqcup 1) \circ_h id_\Sigma \tag{1.4.10}$$

Comparing this with the construction of the functor $\mathcal{E}^s_*$ in (1.3.7) in Section 1.3, we can identify the expression with the natural transformation of functors we obtain by application of $\mathcal{E}^s_*$ to the morphism $V_0^{(j)} : \Sigma^{\sqcup b}_{0,1/1} \to \Sigma^{\sqcup b}_{0,1/1}$. Since we can identify $\mathcal{E}^s_{\Sigma^{\sqcup b}_{0,1/1}}$ with the identity functor on $\mathcal{C}ob(a,b)$, we recover $d_*\delta_j$ by evaluating the following natural automorphism on the particular surface Σ:

$$\mathcal{E}^s_{V_0^{(j)}} : id_{\mathcal{C}ob(a,b)} \xrightarrow{\cong\bullet} id_{\mathcal{C}ob(a,b)} : d_*\delta_j = \mathcal{E}^s_{V_0^{(j)}}(\Sigma) : \Sigma \xrightarrow{\sim} \Sigma.$$

Observe that since $d_*\delta_j$ is a natural transformation it automatically has to be central in the group $\mathrm{Aut}_{\mathcal{C}ob(a,b)}(\Sigma)$. Note also, that this proof of Lemma 1.4.5 uses implicitly the same argument as the one we gave in the original proof, only in a more formal language.

The natural transformation also has a particular meaning for the category $\mathcal{G} = \mathcal{C}ob(0,1)$, for which we extracted the algebraic structure of a braided tensor category. Let us introduce the special notation for the case, where $a = 0$ and $b = 1$

$$\mathbf{v} = \mathcal{E}^s_{V_0} : id_\mathcal{G} \xrightarrow{\cong\bullet} id_\mathcal{G}, \qquad \Sigma \mapsto \mathbf{v}_\Sigma \in \mathrm{Aut}_\mathcal{G}(\Sigma).$$

This natural transformation relates to the category $\mathcal{G}$ as follows.

Lemma 1.4.6. *The transformation* $\mathbf{v}$ *is the ribbon-element for* $\mathcal{G}$ *for the braided structure defined in Lemma 1.3.1 so that* $\mathcal{G}$ *is a* balanced *(ribbon) BTC, as in Section 4.1.3.*

We recall that the notion of a ribbon-element in a BTC is a natural isomorphism v of the identity functor, such that

$$(\mathbf{c} \circ_h P) \circ_v \mathbf{c} = \mathbf{v}_\bullet \otimes_\bullet (\mathbf{v}^{-1} \times \mathbf{v}^{-1}) : \otimes \xrightarrow{\cong\bullet} \otimes : \mathcal{G} \times \mathcal{G} \longrightarrow \mathcal{G}. \qquad (1.4.11)$$

Here $\circ_h$ stands for the composition of functors, and $\circ_v$ for the composition of natural transformations. The transformation $\mathbf{c} : \otimes \xrightarrow{\bullet} \otimes \circ_h P$ is as in the proof of Lemma 1.3.1.

Since these transformations have been obtained by the augmentation $\mathcal{E}^s_*$, the proof of equation (1.4.11) and, thus, Lemma 1.4.6 can be done by verifying the respective relations for the elementary cobordisms V_0 and C. This, however, is implied by the pictorial identity in Figure 1.9.

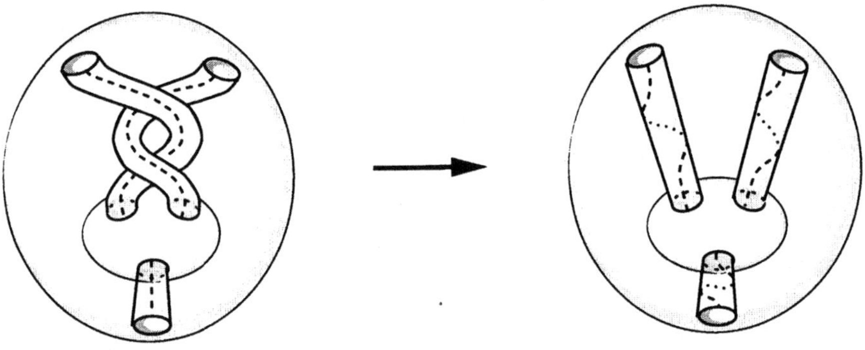

Fig. 1.9. Proof of Ribbon Relation

There the composite of C with itself leads to the cobordism on the left side with a pure braid between the upper two tubes. We can untwist the braid by turning the inner sphere by 2π around its vertical axis. In the course all of the tubes will pick up a 2π-twist themselves, which is indicated by the dashed lines. These correspond to the reparametrization of the cylinder given by V_0. The rotation can be extended ambiently to the space enclosed by the spheres and cylinders so that we can find a homeomorphism between the left and the right picture in Figure 1.9.

More commonly, the defining relation (1.4.11) between the natural transformations is expressed as the corresponding family of relations between the isomorphisms associated to specific objects:

$$\mathbf{c}_{\Sigma_2,\Sigma_1} \circ \mathbf{c}_{\Sigma_1,\Sigma_2} = \mathbf{v}_{(\Sigma_1 \otimes \Sigma_2)} \circ \mathbf{v}^{-1}_{\Sigma_1} \otimes \mathbf{v}^{-1}_{\Sigma_1} \in \mathrm{Aut}_{\mathcal{G}}(\Sigma_1 \otimes \Sigma_2) \qquad \forall \Sigma_1, \Sigma_2.$$

For most surfaces the generators δ_j from Lemma 1.4.5 freely generate the kernel of the map from $\widehat{\mathcal{P}_b}$ onto $\mathcal{P}_b$. The precise relation is given as follows.

Lemma 1.4.7. *For $\Sigma \neq S^2$ or $b \geqslant 3$ we have the following short exact sequence:*

$$1 \longrightarrow \mathbb{Z}(\delta_1, \ldots, \delta_b) \hookrightarrow \widehat{\mathcal{P}_b}(\Sigma) \twoheadrightarrow \mathcal{P}_b(\Sigma) \longrightarrow 1.$$

For the remaining special cases with $\Sigma = S^2$ we have exact sequences: for $b=1$:

$$1 \longrightarrow \mathbb{Z}(2\delta_1) \hookrightarrow \mathbb{Z}(\delta_1) \twoheadrightarrow \widehat{\mathcal{P}}_1(S^2) \cong \mathbb{Z}/2 \longrightarrow \mathcal{P}_1(S^2) = 1,$$

and for $b=2$:

$$1 \longrightarrow \mathbb{Z}(2(\delta_1+\delta_2)) \hookrightarrow \mathbb{Z}(\delta_1,\delta_2) \twoheadrightarrow \widehat{\mathcal{P}}_2(S^2) \cong \mathbb{Z}\oplus\mathbb{Z}/2 \longrightarrow \mathcal{P}_2(S^2) = 1.$$

In particular, by Lemma 1.4.5, $\widehat{\mathcal{P}}_b$ is a central extension of $\mathcal{P}_b$ in all cases.

Proof. We claim that the above are the π_1-parts of the long exact homotopy sequence for fibration $\mathbf{F}_{b,\Sigma} \hookrightarrow \mathbf{E}_{b,\Sigma} \twoheadrightarrow \mathbf{B}_{b,\Sigma}$, that follows:

$$\begin{array}{ccc} \mathcal{E}mbd^+\left(\sqcup_j^b(D^2,p_j),(\Sigma,\{\overline{p_1},\dots,\overline{p_b}\})\right) & \hookrightarrow & \mathcal{E}mbd^+\left(\sqcup_j^b D^2,\Sigma\right) \\ & & \Big\downarrow C^{fr} \\ & & \mathcal{E}mbd^+\left(\{p_1,\dots,p_b\},\Sigma\right) \end{array}$$

Clearly, the fundamental groups of $\mathbf{E}_{b,\Sigma}$ and $\mathbf{B}_{b,\Sigma}$ are the framed and the unframed, pure braid group, respectively. The fiber space is given (up to homotopy) by the space of embeddings of b discs into Σ, such that the center point of the j-th standard disc, p_j, is mapped to the fixed point $\overline{p_j}$ on Σ, defined, e.g., by the image of p_j for the standard embedding of D^2.

The topology of $\mathbf{F}_{b,\Sigma}$ is really independent of Σ. To see this let us pick disjoint vicinities, $W_j \cong D^2$, of the points, $\overline{p_j}$, and introduce the subspace $\mathbf{F}^0_{b,\Sigma} \subset \mathbf{F}_{b,\Sigma}$ of embeddings, whose image lies in these vicinities:

$$\mathbf{F}^0_{b,\Sigma} = \mathcal{E}mbd^+\left(\sqcup_j^b(D^2,p_j),\sqcup_j^b(W_j,\overline{p_j})\right) = \mathop{\times}_{j=1}^{b} \mathcal{E}mbd^+\left((D^2,p_j),(W_j,\overline{p_j})\right).$$

Let $R(s) : D^2 \hookrightarrow D^2$ be the one-parameter family of embeddings of the disc into itself, which in polar coordinates, where $p_j = (0,0)$, is defined by $(r,\phi) \mapsto (e^{-s}r,\phi)$, for $0 \leqslant s < \infty$. This defines a semigroup action, Φ_s, of $\mathbb{R}^+$ on $\mathbf{F}_{b,\Sigma}$, given by $\Phi_s(f) = f \circ (\sqcup_j^b R(s))$.

A simple compactness argument shows that for every embedding $f \in \mathbf{F}_{b,\Sigma}$ the dilated element $\Phi_s(f)$ will lie in $\mathbf{F}^0_{b,\Sigma}$ for a sufficiently large s. This argument is easily extended to paths in $\mathbf{F}_{b,\Sigma}$, homotopies of paths, etc., that can, thus, be deformed with Φ_s into paths and homotopies in $\mathbf{F}^0_{b,\Sigma}$. It follows that the inclusion $\mathbf{F}^0_{b,\Sigma} \hookrightarrow \mathbf{F}_{b,\Sigma}$ yields isomorphisms for the homotopy groups.

In order to identify the free, abelian group in sequence of Lemma 1.4.7 with $\pi_1(\mathbf{F}_{b,\Sigma})$, it remains to make sure that $\pi_1(\mathbf{F}^0_1)$, with $\mathbf{F}^0_1 = \mathcal{E}mbd^+\left((D^2,p),(W,\overline{p})\right)$, is the free, cyclic group, $\mathbb{Z}(\delta)$, generated by the class, δ, of the 2π-rotation as defined in the paragraph preceding Lemma 1.4.5.

It is somewhat more convenient to compute the groups for the homotopy equivalent space $\mathbf{F}_{1,S^2}$ instead for $\mathbf{F}_1^0$. It appears as the base space of the fibration

$$\mathcal{D}\!\mathit{iff}^+(D^2, S^1) = \mathcal{D}\!\mathit{iff}^+(S^2, D^2) \hookrightarrow \mathcal{D}\!\mathit{iff}^+(S^2, p) \twoheadrightarrow \mathbf{F}_{1,S^2}.$$

The fiber is contractible by Lemma 1.4.3 so that we have isomorphisms

$$\pi_j(\mathcal{D}\!\mathit{iff}^+(S^2, p)) \cong \pi_j(\mathbf{F}_{1,S^2}). \tag{1.4.12}$$

Using the second part of Lemma 1.4.3, we find that $\pi_j(R) : \pi_j(S^1) \xrightarrow{\sim} \pi_j(\mathbf{F}_1^0)$ is an isomorphism for all $j \geqslant 0$, where R assigns to an angle the respective rotation of the disc.

Finally, we have to make sure that the π_1–piece in the long exact sequence for the fibration of $\mathbf{E}_{b,\Sigma}$ does in fact form a short exact sequence. The fact that $\mathbf{F}_{b,\Sigma}^0$ is connected takes care of the lower π_0-end. For the upper π_2-piece, we know from Corollary 2.2 in [FN62] that $\pi_2(\mathbf{B}_{b,\Sigma}) = 0$ is $\Sigma \neq S^2$. Also, we know from the Corollary in Section II in [FV62] that $\pi_2(\mathbf{B}_{b,S^2}) = 0$ for $b \geqslant 3$.

In the discussion of the special cases with $\Sigma = S^2$ let us start with $b = 1$. Clearly, the restriction of an automorphism of S^2 to a disc yields a map of fibrations as follows:

$$\begin{array}{ccccc}
\mathcal{D}\!\mathit{iff}^+(S^2, p) \hookrightarrow \mathcal{D}\!\mathit{iff}^+(S^2) & \xrightarrow{\text{restr.}} & \mathbf{F}_{1,S^2} & \hookrightarrow & \mathbf{E}_{1,S^2} \\
\downarrow & & & & \downarrow \\
S^2 & & \xrightarrow{id} & & S^2
\end{array} \tag{1.4.13}$$

Since $\mathcal{D}\!\mathit{iff}^+(S^2, D^2) \hookrightarrow \mathcal{D}\!\mathit{iff}^+(S^2) \twoheadrightarrow \mathbf{E}_{1,S^2}$ has contractible fiber, and by (1.4.12) from the previous discussion, the map between the long sequences of the fibrations is an isomorphism. Using that the same is true for the map in (1.4.4), we obtain the following isomorphism of short exact sequences:

$$\begin{array}{ccccccccc}
1 = \pi_2(SO(3)) & \longrightarrow & \pi_2(S^2) & \longrightarrow & \pi_1(S^1) & \longrightarrow & \pi_1(SO(3)) & \longrightarrow & \pi_1(S^2) = 1 \\
 & & \downarrow\cong & & \downarrow\cong & & \downarrow\cong & & \\
1 = \pi_2(\mathbf{E}_{1,S^2}) & \longrightarrow & \pi_2(\mathbf{B}_{1,S^2}) & \longrightarrow & \pi_1(\mathbf{F}_{1,S^2}) & \longrightarrow & \pi_1(\mathbf{E}_{1,S^2}) & \longrightarrow & \pi_1(\mathbf{B}_{1,S^2}) = 1
\end{array}$$

The sequence for $\Sigma = S^2$ and $b = 1$ in Lemma 1.4.7 follows immediately.

Combining the fact that $\pi_2(\mathbf{E}_{1,S^2}) = 1$ with the long, exact homotopy sequence for the fibration $\mathbf{F}_1^0 \hookrightarrow \mathbf{E}_{2,S^2} \twoheadrightarrow \mathbf{E}_{1,S^2}$, where the projection is the restriction to one of two discs, we obtain that also $\pi_2(\mathbf{E}_{2,S^2}) = 1$. Hence, the restriction yields the following map between exact sequences:

$$\begin{array}{ccccccccc}
1 = \pi_2(\mathbf{E}_{2,S^2}) & \longrightarrow & \pi_2(\mathbf{B}_{2,S^2}) & \hookrightarrow & \mathbb{Z}(\delta_1, \delta_2) & \twoheadrightarrow & \pi_1(\mathbf{E}_{2,S^2}) & \longrightarrow & 1 \\
 & & \downarrow\cong & & \downarrow p_1 & & \downarrow & & \\
1 = \pi_2(\mathbf{E}_{1,S^2}) & \longrightarrow & \mathbb{Z}(2\delta_1) & \hookrightarrow & \mathbb{Z}(\delta_1) & \twoheadrightarrow & \pi_1(\mathbf{E}_{1,S^2}) & \longrightarrow & 1
\end{array}$$

The first vertical map is an isomorphism since the restriction map on the base spaces is the fibration $D^2 \hookrightarrow \mathbf{B}_{2,S^2} \twoheadrightarrow \mathbf{B}_{1,S^2}$ with contractible fiber. In order for the first square to commute, the generator of $\pi_2(\mathbf{B}_{2,S^2})$ has to be mapped to an element of the form $2\delta_1 + n\delta_2$. Since we can also consider the restriction to the second disc, we have that also $n = 2$. This completes the proof for the last sequence with $b = 2$ and $\Sigma = S^2$.

The central extensions are (non-canonically) split, if Σ is not closed. The reason for this is that with $\partial\Sigma \neq \varnothing$ we can define a non-vanishing vector field on Σ. We can then construct a global section $\mathcal{E}mbd^+(\{p_1, \dots, x_b\}, \Sigma) \to \mathcal{E}mbd^+(\sqcup^b D^2, \Sigma)$, since at every point $p \in \Sigma$ a direction is specified, which determines uniquely an isotopy class of an isometrical embeddings $(D^2, 0) \hookrightarrow (\Sigma, p)$ for a given a metric on Σ. Using this we could also specify the extension in the case of closed surfaces.

1.4.3 Framed Braid Groups in $\mathcal{C}ob$

With the sequence in (1.4.6) there is a map d_* from the framed braid group $\widehat{\mathcal{P}_b}(\Sigma)$ to the mapping class group of $\Sigma - \sqcup^b D^2$. With a few exceptions this map is in fact an inclusion.

Lemma 1.4.8. *Let us suppose that Σ is neither S^2, D^2, $S^1 \times [0,1]$ nor $S^1 \times S^1$, i.e., $(g,a) \notin \{(0,0),(1,0),(0,1),(0,2)\}$, and $b \geqslant 1$. Then there is a short exact sequence*

$$1 \longrightarrow \widehat{\mathcal{P}_b}(\Sigma_{g,a}) \overset{d_*}{\hookrightarrow} \mathcal{M}(g, a/b) \twoheadrightarrow \mathcal{M}(g, a/0) \longrightarrow 1.$$

Proof. First, we show for all surfaces $\Sigma = \Sigma_{g,a}$ that the kernel of d_* lies in the center of $\widehat{\mathcal{P}_b}(\Sigma)$, following the proof of the analogous to Theorem 4.1 in [Bir74] for ordinary braid groups. Since (1.4.6) is exact we know that an element in $\ker(d_*)$ can be represented by the restriction of a closed path $t \mapsto H_t \in \mathcal{D}iff^+(\Sigma)$ with $H_1 = H_0 = id$ to the b standard discs in Σ. Let us suppose another element in $\widehat{\mathcal{P}_b}(\Sigma)$ be represented by a path $s \to A_s \in \mathcal{E}mbd^+(\sqcup^b D^2, \Sigma)$, such that $A_0 = A_1$ are the standard embeddings, then we can consider $K : [0,1] \times [0,1] \longrightarrow \mathcal{E}mbd^+(\sqcup^b D^2, \Sigma) : (t,s) \mapsto H_t \circ A_s$. If we restrict K to the paths along the edges of $[0,1]^2$ from $(0,0)$ to $(1,1)$, we get the composite paths in $\widehat{\mathcal{P}_b}(\Sigma)$ of the elements represented by H and A in both orders. Thus K yields a homotopy, proving $[\tau_* H] * [A] = [A] * [\tau_* H]$.

For $\Sigma \neq S^2, D^2, S^1 \times [0,1], S^1 \times S^1$ we know that the center of $\mathcal{P}_b(\Sigma)$ is trivial. For closed surfaces with $g \geqslant 2$ this is proven in Lemma 4.2.2 in [Bir74]. The proof for punctured surfaces, $\Sigma_{g,a}$ with $2g + a \geqslant 3$, follows by exactly the same arguments from Theorem 1.4 of the same book. It follows that $d_* H$ lies in the group $\mathbb{Z}(\delta_1, \dots, \delta_b)$ of Lemma 1.4.7. Thus, there is an isotopy that deforms the braid $t \mapsto H_t \in \mathcal{E}mbd^+(\{p_1, \dots, p_b\}, \Sigma)$ into a constant braid. Extending this to an ambient isotopy we can also deform $t \to H_t \in \mathcal{D}iff^+(\Sigma)$, such that the resulting isotopy leaves the points $\overline{p_j}$ fixed. It is, therefore, sufficient to consider the fibration

$$\mathcal{D}\!\mathit{iff}^+(\Sigma,\sqcup^b D^2) \hookrightarrow \mathcal{D}\!\mathit{iff}^+(\Sigma,\{\overline{p_j}\}) \xrightarrow{\tau'} \mathcal{E}\!\mathit{mbd}^+((\sqcup^b(D^2,p_j)),(\Sigma,\{\overline{p_j}\})),$$

and show that $\pi_1(\tau')$ is trivial. (Recall $\mathcal{D}\!\mathit{iff}^+(\Sigma,X)$ means that X is point-wise fixed).

Now, if $a+b \geqslant 2$ we can choose a path γ between a given point, say $\overline{p_1} \in \Sigma$, and either another point $\overline{p_j}$ or a point on $\partial\Sigma$, all of which are fixed by a closed path $t \mapsto H_t \in \mathcal{D}\!\mathit{iff}^+(\Sigma,\{\overline{p_j}\}$. To the cut defined by γ we consider the covering space $p^\gamma : \Sigma^\gamma \longrightarrow \Sigma - \{\overline{p_j}\}$, where the covering group $\cong \mathbb{Z}$ is cyclically generated by $c \in \mathcal{D}\!\mathit{iff}^+(\Sigma^\gamma)$.

Now, the isotopy H_t can be lifted to an isotopy H_t^γ on Σ^γ, such that $H_0^\gamma = id$, $p^\gamma \circ H_t^\gamma = H_t \circ p^\gamma$, and H_1^γ is a covering transformation, i.e., $H_1^\gamma = c^{w_1}$ with $w_1 \in \mathbb{Z}$.

We claim that $w_1 = 0$. In order to see this let us give an explicit construction of Σ^γ. To this end we consider the surface $\Sigma^c = \Sigma - (\gamma \cup \{\overline{p_j}\})$, which we obtain by cutting the punctured surface along γ. The result has the same genus as Σ but only $a+b-1$ holes, one of which, S^c, contains two copies of γ in its boundary. The covering space is now given by gluing copies of Σ^c together along the γ-pieces in S^c:

$$\Sigma^\gamma = \ldots \sqcup_\gamma \Sigma^c \sqcup_\gamma \Sigma^c \sqcup_\gamma \ldots \quad .$$

Let γ_0 be one of the paths with $p^\gamma \circ \gamma_0 = \gamma$. Then H_t^γ deforms γ_0 into a shifted copy $c^{w_1} \circ \gamma_0$, where the endpoints of $H_t^\gamma \circ \gamma_0$ remain in two components, C_1 and C_2, of $\partial\Sigma^\gamma$, which cover the respective holes in $\Sigma - \{\overline{p_j}\}$ connected by γ. Since the components $C_1 \cong C_2 \cong \mathbb{R}$ in $\partial\Sigma^\gamma$ are contractible, the deformation H_t^γ actually gives rise to a contraction of the closed path $s^c = b_1 * (-\gamma_0) * b_2 * (c^{w_1} \circ \gamma_0)$. Here b_j denotes a small segment in C_j.

Let us suppose that $x^{(k)}$ is the path along the boundary component S^c of the k-th copy of Σ^c in the above presentation of Σ^γ, then s^c is easily seen to be homotopic to $x^{(k_0+w_1)} * \ldots * x^{(k_0+1)} * x^{(k_0)}$. Since $\pi_1(\Sigma^\gamma)$ is the corresponding, infinite, free product of the $\pi_1(\Sigma^c)$ it follows that s^c is contractible in Σ^γ only if x is contractible in Σ^c. If $w_1 > 0$, this is only possible if $\Sigma^c \sim D^2$, which, however, is not compatible with the assumption that Σ is neither D^2 nor S^2. Hence, $w_1 = 0$ and the image of $\pi_1(\tau')$ is trivial.

For the remaining case, with $b = 1$ and $a = 0$, we have to consider a refinement of the above argument, that involves non-abelian coverings. For simplicity let us (again) replace $\mathcal{D}\!\mathit{iff}^+(\Sigma_{g,0} - \{p_1\})$ by the automorphisms of the compactification $\Sigma_{g,1}$, which preserve the boundary, although not pointwise. Instead for $\pi_1(\tau')$ we may show triviality of $\pi_1(\tau'')$, where $\tau'' : \mathcal{D}\!\mathit{iff}^+(\Sigma_{g,1}) \longrightarrow \mathcal{D}\!\mathit{iff}^+(S^1)$ is the restriction to the boundary. We are going to consider the *universal* covering $p^\infty : \Sigma_{g,1}^\infty \longrightarrow \Sigma_{g,1}$, whose covering group is the free group $\mathrm{F}(a_1, b_1, \ldots, a_g, b_g)$ in $2g$ generators. As before we lift the isotopy H_t to an isotopy $t \mapsto H_t^\infty \in \mathcal{D}\!\mathit{iff}^+(\Sigma_{g,1}^\infty)$. Now, the restriction $p^\infty : \partial\Sigma_{g,1}^\infty \longrightarrow S^1$ has components $\cong \mathbb{R}$, on each of which the covering group is $\cong \mathbb{Z}$, generated by $c = \prod[a_j, b_j]$. Since H_t preserves the boundary it also must preserve each component of this sub-covering, and, hence, H_1^∞ must be a power c^{w_1}.

Select now any point $q \in \partial\Sigma^{\infty}_{g,1}$ and choose a closed path γ in $\Sigma_{g,1}$, that starts and ends in the respective point in $\partial\Sigma_{g,1}$. In $\Sigma^{\infty}_{g,1}$ this lifts to a path γ_0 from q to $[\gamma](q)$. The isotopy H^{∞}_t deforms this path into the path $\gamma'_0 = H^{\infty}_1 \circ \gamma_0$ that joins $c^{w_1}(q)$ to $c^{w_1} \circ [\gamma](q)$. But in the projection γ'_0 is mapped again to γ since $H_1 = id$ on $\Sigma_{g,1}$. It, therefore, also joins $c^{w_1}(q)$ to $[\gamma] \circ c^{w_1}(q)$. By uniqueness of liftings and the fact that the covering group acts freely, we infer from this $[[\gamma], c^{w_1}] = 1$, i.e., that c^{w_1} lies in the center of a free group. Hence, $w_1 = 0$ and $\pi_1(\tau'') = 1$.

The right side of the sequence from Lemma 1.4.8 can, in fact, be expressed in terms of elements of *Cob* using Proposition 1.4.2, as well as the fill functor from (1.3.5) from Section1.3. We have the following commutative diagram:

$$\begin{array}{ccc} \mathcal{M}(g, a/b) & \longrightarrow\!\!\!\!\rightarrow & \mathcal{M}(g, a/0) \\ \cong\downarrow & & \downarrow\cong \\ \mathrm{Aut}_{Cob(a,b)}(\Sigma_{g,a/b}) & \xrightarrow{\mathcal{E}^s_{fill}} & \mathrm{Aut}_{Cob(a,0)}(\Sigma_{g,a/0}) \end{array} . \tag{1.4.14}$$

Hence, for the above cases we can define the framed braid group as the kernel of $\mathcal{E}^s_{fill}$ on the respective group of invertible cobordisms.

In the next two lemmas we describe the maps from $\widehat{\mathcal{P}_b}$ to $\mathcal{M}$ for the remaining special cases, for which the braid group $\mathcal{P}_b$ has non-trivial center. We begin with the case of a sphere with two or less holes.

Lemma 1.4.9. *For $g = 0$ we have the following short exact sequences:*

1. $1 \longrightarrow \pi_1(\mathcal{D}\!\mathit{iff}^+(S^2)) \cong \mathbb{Z}/2 \hookrightarrow \widehat{\mathcal{P}_b}(S^2) \longrightarrow\!\!\!\!\rightarrow \mathcal{M}(0, 0/b) \longrightarrow 1.$
2. $1 \longrightarrow \widehat{\mathcal{P}_b}(D^2) \xrightarrow{\cong} \mathcal{M}(0, 1/b) \longrightarrow 1.$
3. $1 \longrightarrow \widehat{\mathcal{P}_b}(\Sigma_{0,2}) \hookrightarrow \mathcal{M}(0, 2/b) \longrightarrow\!\!\!\!\rightarrow \mathcal{M}(0, 2/0) \cong \mathbb{Z} \longrightarrow 1.$

Proof.

1. For the first sequence observe that $\pi_1(\mathcal{D}\!\mathit{iff}^+(\Sigma_{0,b}, \partial\Sigma_{0,b})) = 0$ of $b \geqslant 1$. This follows by induction from the long sequence for the fibration

$$\mathcal{D}\!\mathit{iff}^+(S^2, \sqcup^{b+1}D^2) \hookrightarrow \mathcal{D}\!\mathit{iff}^+(S^2, \sqcup^{b}D^2) \longrightarrow\!\!\!\!\rightarrow \mathcal{E}\!\mathit{mbd}^+(D^2, \Sigma_{0,2}),$$

using that $\pi_2(\mathcal{E}\!\mathit{mbd}^+(D^2, \Sigma_{0,b}) = 1$. Moreover, we have that $\pi_0(\mathcal{D}\!\mathit{iff}^+(S^2)) = 1$ and that $\pi_1(\mathcal{D}\!\mathit{iff}^+(S^2)) = \mathbb{Z}/2$. The latter group is obtained from the inclusion $i_S : SO(3) \hookrightarrow \mathcal{D}\!\mathit{iff}^+(S^2)$ of fibrations, for which $\pi_1(i_S)$ is an isomorphism. The claim follows now from the long sequence for the fibration $\mathcal{D}\!\mathit{iff}^+(\Sigma_{0,b}, \partial\Sigma_{0,b}) \hookrightarrow \mathcal{D}\!\mathit{iff}^+(S^2) \longrightarrow\!\!\!\!\rightarrow \mathcal{E}\!\mathit{mbd}^+(\sqcup^b D^2, S^2)$. Note, that for $b > 2$ the kernel for the map from the braid to the mapping class group is exactly the same as for the corresponding sequence for the ordinary braid group and punctured surfaces, where it coincides with the center of the braid group (see Lemma 4.2.3 in [Bir74]). We, thus, have precisely the same exact sequence as in Lemma 1.4.7, if we replace the groups $\widehat{\mathcal{P}_b}(S^2)$ and $\mathcal{P}_b(S^2)$ by their respective images in the mapping class groups. The cases $b = 1, 2$ are consistent with Lemma 1.4.7 since $\mathcal{M}(0,0/1) = 1$ and $\mathcal{M}(0,1/1) = \mathbb{Z}$, generated by the Dehn-twist around one hole.

2. The next sequence follows immediately from the fact, that in the fibration $\mathcal{Diff}^+(\Sigma_{0,1/b}, \partial\Sigma_{0,1/b}) \hookrightarrow \mathcal{Diff}^+(\Sigma_{0,1/0}, \partial\Sigma_{0,1/0}) \twoheadrightarrow \mathcal{Embd}^+(\sqcup^b D^2, D^2)$ the entire space $\mathcal{Diff}^+(D^2, S^1)$ is contractible by part *1* of Lemma 1.4.3. The map on the framed braid group should not be confused with the respective map into the mapping class group of the Euclidean plane E^2 with holes as in [Bir74]. The collective 2π-rotation of the b points or discs in the braid group is realized in the mapping class group as a Dehn twist around the hole in either E^2 or D^2. In $\pi_0(\mathcal{Diff}^+(E^2 - \{p_j\}))$ this is trivial since we can rotate the points around the puncture freely so that the map from $\mathcal{P}_b$ has a kernel. But for $\pi_0(\mathcal{Diff}^+(D^2 - \sqcup^b D^2, S^1))$ the 2π-twist is a non-trivial element, since the points on the boundary can not be moved.
3. We have $\pi_1(\mathcal{Diff}^+(\Sigma_{0,2}, \partial)) = 1$ and $\pi_0(\mathcal{Diff}^+(\Sigma_{0,2})) = \mathbb{Z}$, using the fibration $\mathcal{Diff}^+(\Sigma_{0,2}, \partial) \hookrightarrow \mathcal{Diff}^+(D^2\partial) \twoheadrightarrow \mathcal{Embd}^+(D^2, D^2)$ and $\pi_2(\mathcal{Embd}^+(D^2, D^2)) = 1$ and $\pi_1(\mathcal{Embd}^+(D^2, D^2)) = \mathbb{Z}$. The claim then follows from the long sequence for the fibration

$$\mathcal{Diff}^+(\Sigma_{0,2/b}, \partial) \hookrightarrow \mathcal{Diff}^+(\Sigma_{0,2/0}, \partial) \twoheadrightarrow \mathcal{Embd}^+(\sqcup^b D^2, \Sigma_{0,2}).$$

Finally, the map from the framed braid group of the torus to the respective mapping class group is given as follows

Lemma 1.4.10. *We have the following exact sequence*

$$1 \to \pi_1(\mathcal{Diff}^+(S^1 \times S^1)) \cong \mathbb{Z} \oplus \mathbb{Z} \hookrightarrow \widehat{\mathcal{P}_b}(S^1 \times S^1) \to$$
$$\xrightarrow{d_*} \mathcal{M}(1,0/b) \twoheadrightarrow \mathcal{M}(1,0) \cong SL(2,\mathbb{Z}) \to 1,$$

where the framing extensions are injected into the mapping class group, in particular,

$$im(d_*) = \mathbb{Z}(\delta_1, \ldots, \delta_b) \oplus \frac{\mathcal{P}_b(S^1 \times S^1)}{\text{center}}.$$

Proof. It is implied by [Bir69b, Bir69a] that $\pi_1(\mathcal{Diff}^+(\Sigma_{1,0/b}))$ is trivial for $b > 0$ and $\mathbb{Z}(g_1, g_2)$ for $b = 0$, where the generators g_1 and g_2 are represented by uniform rotations along one of the S^1-directions. The exact sequence then follows from the same fibration $\mathcal{Diff}^+(S^1 \times S^1)$ as before.

Now, the flow of g_1 defines a vector field on $S^1 \times S^1$, and, thus, a section

$$\chi : \mathcal{Embd}^+(\{p_1, \ldots, p_b\}, S^1 \times S^1) \longrightarrow \mathcal{Embd}^+(\sqcup^b D^2, S^1 \times S^1),$$

which gives a splitting of the sequence in Lemma 1.4.7. By construction χ is also covariant with respect to the action of the g_j so that the image of the subgroup $\mathbb{Z}(g_1, g_2) \subset \mathcal{Diff}^+(S^1 \times S^1)$ in $\mathcal{Embd}^+(\sqcup^b D^2, S^1 \times S^1)$ lies entirely in the image of χ. Thus, also the image of $\pi_1(\mathcal{Diff}^+(S^1 \times S^1))$ in $\widehat{\mathcal{P}_b}(S^1 \times S^1)$ lies entirely in the $\mathcal{P}_b(S^1 \times S^1)$-part of the splitting. For the fact that there it coincides with the center see again [Bir69b, Bir69a].

1.5 Some Facts about Handle Decompositions

The purpose of this section is to provide some basic technical tools that are needed for Section 1.6 as well as for the proof of the main presentation theorem in Chapter 3. The notions of handlebody decompositions and surgery are introduced in Section 1.5.1. In particular, the method of simplifying handle decomposition by cancellation is described. This is applied in Section 1.5.2 to describe 4-dimensional handle decompositions and 3-dimensional surgery explicitly in terms of links and, more generally *bridged links*. Cancellations and handle slides are, thus, given by diagrammatic equivalences. Finally, in Section 1.5.3, we explain the effect of a surgery on a 3-manifold along a curve which passes parallelly through an attached 3-handle.

1.5.1 Handles, Surgery, Isotopies and Cancellation

In this section let us review some elementary facts about handle attachments and surgery in arbitrary dimensions as well as the particular features of dimensions 3 and 4. The two notions are closely intertwined and yield basic procedures to build arbitrary manifolds. The existence and equivalences of handle decompositions or surgery presentations is based on the theory of Morse functions. This will be discussed in greater detail in Chapter 3. Here let us only use this as a motivation and continue our discussion using only the notion of handle attachments.

Assume for a compact, connected, differentiable $k+1$-dimensional manifold N that $\partial N = -V_0 \cup (\partial V \times [0,1]) \cup V_1$, where the pieces are fit along $\partial V \times \{j\} \cong \partial V_j$. We consider differentiable functions from N to the unit interval of the form

$$f : (N, V_0, V_1) \longrightarrow ([0,1], \{0\}, \{1\}) \qquad \text{with } V_a = f^{-1}(a) \text{ for } a = 0, 1 ,$$

which restrict to the projection on the second factor on the $\partial V \times [0,1]$-part of the boundary. In this situation the following is a basic fact of Morse theory.

Lemma 1.5.1 (e.g., [Mil69]). *Let f be a Morse function on N as above, which has exactly one non-degenerate singularity of Morse index j. Then there is a homeomorphism between N and the cylinder over V_0 with an attachment of a handle of index j:*

$$N \cong V_0 \times [0,1] \bigcup_{\mathcal{G}} \mathrm{e}_j^{k+1} .$$

Here the *j-handle* e_j^{k+1} is a $k+1$-ball written as $D^j \times D^{k+1-j}$ so that its boundary is given as the union

$$\partial \mathrm{e}_j^{k+1} = S^{j-1} \times D^{k+1-j} \bigcup_{S^{j-1} \times S^{k-j}} D^j \times S^{k-j} .$$

Moreover, $\mathcal{G}$ denotes an embedding

$$\mathcal{G} : S^{j-1} \times D^{k+1-j} \hookrightarrow V_0 .$$

The quotient space in Lemma 1.5.1 is then obtained by identifying every point $x \in S^{j-1} \times D^{k+1-j} \subset \mathrm{e}_j^{k+1}$ with the corresponding point $(\mathcal{G}(x), 1) \in V_0 \times \{1\} \subset V_0 \times [0,1]$. In other words, the $k+1$-ball is glued to the upper k-dimensional side of the cylinder along a tubular neighborhood of an embedded $j-1$-sphere.

The following basic lemma shows that isotopic handle attachments ought to be considered as equivalent.

Lemma 1.5.2. *If two embeddings $\mathcal{G}_0$ and $\mathcal{G}_1$ of $S^{j-1} \times D^{k+1-j}$ into the boundary ∂N of a $k+1$-manifold N are isotopic, then there is a homeomorphism between two manifolds $N_{\mathcal{G}_0}$ and $N_{\mathcal{G}_1}$ obtained by a j-handle attachment as in Lemma 1.5.1, which is supported in a vicinity of ∂N and the handle e_j^{k+1}.*

Proof. It is a standard fact that an isotopy $t \mapsto \mathcal{G}_t : [0,1] \to \mathcal{E}mbd^+(S^{j-1} \times D^{k+1-j}, \partial N)$ can be extended to an ambient isotopy. See Chapter 8 (Theorem 1.3) in [Hir76]. This means that there is a diffeotopy $t \mapsto \Phi_t : [0,1] \to \mathcal{D}iff^+(\partial N)$, such that $\mathcal{G}_t = \Phi_t \circ \mathcal{G}_0$ and Φ_0 is the identity on ∂N. This allows us to construct a diffeomorphism on Ψ' on $N \cup_{\partial N \times \{0\}} \partial N \times [0,1]$ as follows.

Ψ' maps the N-part to itself by identity. It also maps the $\partial N \times [0,1]$-part to itself by $\Psi'(y,t) = (\Phi_t(y), t)$ for $y \in \partial N$. In particular, $\Psi'(y,0) = y$ so that it is continuously defined on the union $N \cup_{\partial N \times \{0\}} \partial N \times [0,1]$. Since one can choose the isotopies $\mathcal{G}_t$ and Φ_t to be constant for t in a vicinity of 0 and 1, $\Psi'(y,t)$ can be assumed a smooth diffeomorphism.

Using a collar of N one constructs straightforwardly a diffeomorphism $\xi : N \cup_{\partial N \times \{0\}} \partial N \times [0,1] \xrightarrow{\sim} N$, such that $\xi : \partial N \times \{1\} \to \partial N$ is the identity, and ξ is also identity outside of a collar vicinity of ∂N on N. The map $\Psi = \xi \circ \Psi' \circ \xi^{-1} \in \mathcal{D}iff^+(N)$ has now the property $\Psi \circ \mathcal{G}_0 = \mathcal{G}_1 \in \mathcal{E}mbd^+(S^{j-1} \times D^{k+1-j}, \partial N)$ and is supported in a collar vicinity of ∂N. Hence, Ψ can be extended to a smooth map $\Psi : N_{\mathcal{G}_0} \xrightarrow{\sim} N_{\mathcal{G}_1}$ with the desired properties.

The notion of *surgery* addresses now the question how a description of the other boundary component, namely, the k-manifold $V_1 \subset \partial N$, can be derived from a presentation of N as in Lemma 1.5.1. The image of $\mathcal{G}$, which is a piece homeomorphic to $S^{j-1} \times D^{k+1-j}$ is identified with a part of the boundary of the j-handle and, hence, it will no longer appear in V_1. We have $\partial(V_0 - im(\mathcal{G})) = \partial V_0 \sqcup S^{j-1} \times S^{k-j}$, where the latter component is from $\partial im(\mathcal{G})$. Yet, the remaining piece $\cong D^j \times S^{k-j}$ in the boundary of the j-handle is now added to the total boundary. This is done by gluing it along its boundary $S^{j-1} \times S^{k-j} = \partial(D^j \times S^{k-j})$ to the extra boundary component created by removing $im(\mathcal{G})$. We, thus, have the homeomorphism:

$$V_1 \cong \overline{V_0 - \mathcal{G}(S^{j-1} \times D^{k+1-j})} \bigcup_{\delta} D^j \times S^{k-j}. \tag{1.5.1}$$

The relation δ for the quotient space identifies a point $x \in S^{j-1} \times S^{k-j} \subset \partial(D^j \times S^{k-j})$ with the corresponding point $\mathcal{G}(x) \in \partial(V_0 - im(\mathcal{G}))$. In particular, $\partial V_0 = \partial V_1$ again. We sometimes also use the notation

$$V_1 = (V_0)_{\mathcal{G}}$$

for a surgered manifold.

The boundary pieces of ∂e_j^{k+1} give, hence, rise to submanifolds in the k-dimensional level manifolds V_0 and V_1 that can both be viewed as the tubular neighborhoods of embedded spheres. Let us introduce the following notions for these spheres themselves in the center of the respective neighborhoods:

ascending critical submanifold $S_{asc}^{k-j} = 0 \times S^{k-j} \subset D^j \times S^{k-j} \subset V_1$,
descending critical submanifold $S_{des}^{j-1} = S^{j-1} \times 0 \subset S^{j-1} \times D^{k+1-j} \subset V_0$.

For these sphere-embeddings we clearly have a natural trivializations of their unit normal bundles in the respective k-manifolds, equipped with some metric.

$\psi_1 : \nu_{V_1}^1(S_{asc}^{k-j}) \overset{\cong}{\longrightarrow} S^{k-j} \times S^{j-1}$,
$\psi_0 : \nu_{V_0}^1(S_{des}^{j-1}) \overset{\cong}{\longrightarrow} S^{j-1} \times S^{k-j}$.

Conversely, it follows from basic tubular neighborhood theorems (see again [Hir76] Chapter 6) that any such trivialization determines a tubular neighborhood up to an ambient isotopy. By Lemma 1.5.2 it is, thus, enough to consider only the isotopy classes of trivialization $[\psi_j]$ for the embedded spheres in order to characterize handle attachments and surgery.

Since all compact differentiable manifolds admit Morse functions, such that there are only finitely many critical points and each of them has a different critical value, we know that every such manifold with boundary as above has a handle decomposition of the form

$$N \cong (\dots(((V \times [0,1]) \cup e_{j_1}^{k+1}) \cup e_{j_2}^{k+1}) \cup \dots \cup e_{j_P}^{k+1}.$$

In this expression we attach successively P handles with indices $j_1, \dots, j_P$ to the $V \times \{1\}$ part of the boundary. Correspondingly, the boundary can be expressed as a series of surgeries

$$\partial N \cong -V \cup (\dots((V)_{\mathcal{G}_1})_{\mathcal{G}_2} \cdots)_{\mathcal{G}_P}.$$

We say that a handle decomposition is in *general position* or *generic* if the corresponding ascending and descending manifolds are transverse to each other:

$$S_{asc}^{k-j_r} \pitchfork S_{des}^{j_s-1} \quad \text{in} \quad V_{int}^k \quad , \quad \text{whenever} \quad r < s. \tag{1.5.2}$$

Here V_{int}^k is some intermediate level k-manifold, as, for example,

$$(\dots((V)_{\mathcal{G}_1})_{\mathcal{G}_2} \cdots)_{\mathcal{G}_{s-1}}.$$

It is a standard consequence of transversality theorems [Hir76] that any handle decomposition can be put into general position by an arbitrarily small isotopy.

Lemma 1.5.3. *In a generic handle decomposition the handles can be renumbered so that*

$$j_1 \leqslant j_2 \leqslant \dots \leqslant j_P$$

and handle attachments of equal index are independent of each other.

Proof. If $r < s$ and $j_r \geqslant j_s$, then the transversality condition (1.5.2) implies that $S^{k-j_r}_{asc} \cap S^{j_s-1}_{des} = \varnothing$, since in this case $(k - j_r) + (j_s - 1) < k$. Hence, there are also small tubular neighborhoods of these spheres that are also disjoint, and, hence, we may assume that the corresponding handles $\mathrm{e}^{k+1}_{j_s}$ and $\mathrm{e}^{k+1}_{j_r}$ are disjoint in the decomposition. Hence, if these are successive handles, their order of attachment can be exchanged. In general, by applying all such exchanges we arrive at an order, where a handle with smaller index is attached before any handle with greater index.

In a generic handle decomposition that is ordered in this way we can now consider handle attachments with indices differing by 1. That is, we assume that $r < s$ and $j_s = j_r + 1$. By dimension counting we see that the transversality condition (1.5.2) implies that the intersection $S^{k-j_r}_{asc} \cap S^{j_s-1}_{des} = \{p_1, \dots, p_K\}$ is a finite set of points in some V^k_{int}, at each of which the spheres intersect transversally. The following lemma allows us to *cancel* two such handles against each other if the number of intersection points is $K = 1$.

Lemma 1.5.4. *Let us suppose that the handle decomposition of N contains two handles $\mathrm{e}^{k+1}_{j_s}$ and $\mathrm{e}^{k+1}_{j_r}$ with $r < s$ and $j_s = j_r + 1$ such that the critical manifolds intersect in exactly one point (i.e., $K = 1$).*

Then if we remove the two handles from the handle decomposition of N the resulting handle complex is still homeomorphic to N.

Proof. This lemma is also known as "Smale Cancellation" and is originally proven by using Morse functions directly. Let us here give a prove entirely within the language of handles.

Set $j = j_r$ and $j_s = j + 1$. Let us begin with the observation that $\mathrm{e}^{k+1}_{j_s}$ and $\mathrm{e}^{k+1}_{j_r}$ meet at the intersection of the thickened critical manifolds $D^j \times S^{k-1}$ and $S^j \times D^{k-j}$. Notice that the intersection of the subsets $0 \times S^{k-1}$ and $S^j \times 0$ is by assumption one point p. Since the intersection is also transverse, the normal bundles of the two submanifolds span the entire tangent space, so that for small enough tubular neighborhoods the intersection $D^j \times S^{k-j} \cap S^j \times D^{k-j}$ is a contractible neighborhood of p and, thus, is homeomorphic to D^k. Hence, the union of the two handles in the decomposition is $\mathrm{e}^{k+1}_j \cup \mathrm{e}^{k+1}_{j+1} \cong D^{k+1} \cup_{D^k} D^{k+1} \cong D^{k+1}$.

Next let us investigate in which way this combined handle is attached. The lower handle e^{k+1}_j is attached along the piece $S^{j-1} \times D^{k+1-j}$ to an intermediate manifold V^k_{int}. The upper handle e^{k+1}_j is attached along a piece $S^j \times D^{k-j}$. This, however, intersects the boundary of the lower handle in a bi-disc, $D^j \times D^{k-j}$. If we write $D^{k+1-j} = [0,1] \times D^{k-j}$ we identify this bi-disc in the lower handle $D^j \times D^{k+1-j}$ as the subset $D^j \times \{0\} \times D^{k-j}$. We also denote the disc $D^j_\bullet \subset S^j \subset D^{j+1}$ such that the bi-disc is $D^j_\bullet \times D^{k-j}$ in the boundary piece $S^j \times D^{k-j}$ of the upper handle. The attaching area for the combination of the two handles to V^k_{int} is therefore given as

$$S^{j-1} \times [0,1] \times D^{k-j} \cup (S^j - \mathring{D}^j_\bullet) \times D^{k-j} \cong \left(S^{j-1} \times [0,1] \cup_\sim (S^j - \mathring{D}^j_\bullet)\right) \times D^{k-j}.$$

The identification $\sim$ here is between the $j-1$-spheres $S^{j-1} \times \{0\}$ and $\partial D^j_\bullet \cong \partial(S^j - \mathring{D}^j_\bullet)$ and adds a collar to a j-disc. Hence, the identification area itself combines to a disc $D^j \times D^{k-j} \cong D^k$.

Thus, the combined attachment of the two handles is equivalent to attaching a disc D^{k+1} along another disc D^k into the level manifold V^k_{int}. This, obviously, does not change the homeomorphism class of the manifold so that the two handles can be removed from the decomposition.

From the view point of Morse theory a cancellation is described by a generic one-parameter family of functions $t \mapsto f_t$, such that each f_t with $t \neq 0$ is a Morse function and f_0 is a "co-dimension one function" characterized by a second order jet-transversality condition. More precisely, the function f_0 has exactly one degenerate singularity. Around this we can find local coordinates and a parametrization such that $t \mapsto f_t$ is given by

$$f_t(x) = -x_1^2 - x_2^2 - \ldots - x_j^2 + x_{j+1}^2 + \ldots + x_k^2 \\ + tx_{k+1} + x_{k+1}^3 + \mathcal{O}_3(\|x\|) + \text{const.} \qquad (1.5.3)$$

For $t < 0$ this function has in the given vicinity two singularities of indices j and $j+1$, and has no singularities for $t > 0$. It is a standard conclusion from jet-transversality that the space of Morse functions together with this space of co-dimension one functions is connected. See, for example, Exercise 5, Section 1, Chapter 6 in [Hir76]. In picture of handle decompositions this means the following.

Lemma 1.5.5. *Any handle decomposition of a compact, differentiable manifold can be related to another by applying isotopies and cancellations.*

Finally let us comment on the genericity of handle decompositions and their isotopies.

Lemma 1.5.6. *Any handle decomposition of a compact differentiable manifold can be changed to a generic one as defined in (1.5.2) by an arbitrarily small isotopy of the handles.*

Furthermore, any isotopy of a handle can be deformed into such isotopy that the resulting isotopy of the descending (ascending) manifold is transverse to the ascending (descending) manifolds of the other handles.

By a transverse isotopy of, for example, the s-th handle attachment over the lower handles we understand here that there is an isotopy of the attaching data

$$A : [0,1] \times S^{j_s - 1}_{des} \longrightarrow V^k_{int}$$

such that we have transversality

$$A \pitchfork S^{k-j_r}_{asc} \quad , \qquad \text{whenever} \quad r < s. \qquad (1.5.4)$$

The proof of Lemma 1.5.6 is again a direct consequence of standard transversality results as given in [BJ73] or [Hir76].

1.5.2 4-dim Handles, 3-dim Surgery, and "Bridged Links"

In this section let us discuss in more detail the presentation of attachments of 4-dimensional handles of index 1 or 2, and the respective index 1 and 2 surgery on the 3-dimensional bounding manifolds.

An attachment of a handle e_1^4 to a 4-manifold W with boundary $M = \partial W$ occurs along the image of an embedding $\mathcal{P} : S^0 \times D^3 \hookrightarrow M$. For $S^0 = \{+,-\}$ the image is a pair of 3-balls $D^3_\pm = \mathcal{P}(\{\pm\} \times D^3)$ inside the 3-manifold M.

The boundary of the new 4-manifold is then the 3-manifold after surgery:

$$\partial(W \cup e_1^4) = M_{\mathcal{P}} = \overline{(M - (D^3_+ \sqcup D^3_-))} \bigcup_{\partial D^3_\pm = \{\pm\varepsilon\} \times S^2} [-\varepsilon,\varepsilon] \times S^2 \cong \frac{M - (D^3_+ \sqcup D^3_-)}{\partial D^3_+ = \partial D^3_-}.$$

In the last identity the reglued $D^0 \times S^2$ for the index 1 surgery is shrunk with $\varepsilon \to 0$ so that, using collars of the balls $D^3_\pm$, we may, equivalently, glue the S^2-boundary components created by removing the 3-balls directly to each other.

Correspondingly, a handle of index 2 is attached to a 4-manifold W along the image of an embedding:

$$\mathcal{L} : S^1 \times D^2 \hookrightarrow M = \partial W$$

and the surgered 3-dimensional boundary is given as

$$\partial(W \cup e_2^4) = M_{\mathcal{L}} = (M - \mathcal{L}(S^1 \times \mathring{D}^2)) \bigcup_{\partial} D^2 \times S^1 .$$

The equivalence ∂ again identifies a point $(x,y) \in S^1 \times S^1 = \partial D^2 \times S^1$ in the boundary of the full torus on the right with a corresponding point $\mathcal{L}(x,y) \in M - \mathcal{L}(S^1 \times \mathring{D}^2)$ in the torus-boundary component that is created when the image of $\mathcal{L}$ is removed.

This operation of removing a standard torus $S^1 \times D^2 \cong im(\mathcal{L})$ from a manifold and regluing the opposite torus in the 3-sphere $D^2 \times S^1 \subset D^2 \times S^1 \cup_{S^1 \times S^1} S^1 \times D^2 = \partial e_2^4 = S^3$ along the joint boundary is known as *integral Dehn surgery* along a knot $\mathcal{L}(S^1 \times 0)$. See [Rol76] for details and generalizations to fractional Dehn surgery.

The embedding $\mathcal{L}$ parametrizes the gluing torus $S^1 \times S^1$ as a sub-manifold in both M and $M_{\mathcal{L}}$. Using this we can define curves in the form of embeddings ν and λ of an S^1 into both M and $M_{\mathcal{L}}$ by the formulas $\lambda(x) = \mathcal{L}(x,1)$ and $\nu(y) = \mathcal{L}(1,y)$. The curve λ, thus, runs along the long meridian of the embedded and thickened knot so that $\lambda : S^1 \hookrightarrow M_{\mathcal{L}}$ is contractible but $\lambda : S^1 \hookrightarrow M$ may not be contractible. Conversely, ν winds as a short meridian around the knot at a given point so that $\nu : S^1 \hookrightarrow M$ is contractible but not necessarily $\nu : S^1 \hookrightarrow M_{\mathcal{L}}$.

By Lemmas 1.5.3 and 1.5.6 we may assume that in a 4-dimensional handle decomposition with only 1-handles and 2-handles all 1-handle are attached before the 2-handles, and that the decomposition is in general position. This means that the attaching manifolds of handles with the same indices are disjoint and that every ascending critical S^2 of a 1-handle intersects the descending critical S^1 of a 2-handle transversally.

In order to describe the combined surgery on a given 3-manifold, generated by attaching 1- and 2-handles in this way, we can first consider the 3-manifold

$$\check{M} = (\ldots((M_{\mathcal{P}_1})_{\mathcal{P}_2})\ldots)_{\mathcal{P}_K} = \partial(W \cup \mathrm{e}_1^4 \cup \ldots \cup \mathrm{e}_1^4),$$

on which all of the surgeries of index 1 have been performed. The subsequent index 2 surgeries are now described by disjoint embeddings of tori into the 1-surgered manifold, as in

$$\mathcal{L}_1 \sqcup \ldots \sqcup \mathcal{L}_M : \underbrace{S^1 \times D^2 \sqcup \ldots \sqcup S^1 \times D^2}_{M} \hookrightarrow \check{M}$$

so that the total surgered manifold is given by

$$M_{surg} = (\ldots(((\ldots(M_{\mathcal{P}_1})_{\mathcal{P}_2}\ldots)_{\mathcal{P}_K})_{\mathcal{L}_1})_{\mathcal{L}_2}\ldots)_{\mathcal{L}_M} = (\ldots(\check{M}_{\mathcal{L}_1})_{\mathcal{L}_2}\ldots)_{\mathcal{L}_M}.$$

By genericity, embeddings of the thick links $\mathcal{L}_j$ can be made transverse to the K spheres S^2_{asc} in $\check{M}$ from the index-1 surgeries. This means that the restriction of the embedding to the circle $S^1 \times 0 \subset S^1 \times D^2 \hookrightarrow \check{M}$ is transverse to any of the S^2_{asc}'s. Now M can be obtained by removing a vicinity $S^2_{asc} \times [-\varepsilon, \varepsilon]$ for each of the ascending spheres from the index-1 surgeries $\check{M}$ and fill the S^2-boundary components with 3-balls. Hence, the $\mathcal{L}_j$'s passing transversally through these spheres will, in the picture of M, enter a surgery 3-ball at one point and emerge at a corresponding point on the corresponding other 3-ball.

This surgery picture in M can be further simplified if we take into account that every trivialization of ψ_0 of the normal bundle as above determines the parametrized tubular neighborhood up to isotopy. Moreover, a trivialization of an orientable 2-dimensional bundle is already determined, up to isotopy, by choosing a non-vanishing section in it.

Such a section, in turn, is determined by the restriction of the embeddings $\mathcal{L}_j$ to $S^1 \times R$, where $R \subset D^2$ is an arbitrarily small line segment in the 2-disc starting at the origin 0. Hence, instead of isotopy classes of embeddings of tori we can, equivalently, consider isotopy classes of embeddings of 2-dimensional annuli $S^1 \times [0, 1]$, whose images are ribbon links and knots in $\check{M}$ and, thus, ribbons in M interrupted by transverse passages through pairs of surgery balls.

This type of surgery data has been introduced as *bridged links* in [Ker99]. A typical picture of a bridged link is given next, where ϕ and ϕ' are corresponding pairs of surgery balls. In the precise definition of a bridged link given in [Ker99] there is also an orientation reversing diffeomorphism between the spheres of a pair, which is implicit to the the description above of this surgery data. For example, the strand a entering ϕ at one interval emerges at the corresponding as a strand with the same label a at the image of the interval under this diffeomorphism on the other sphere.

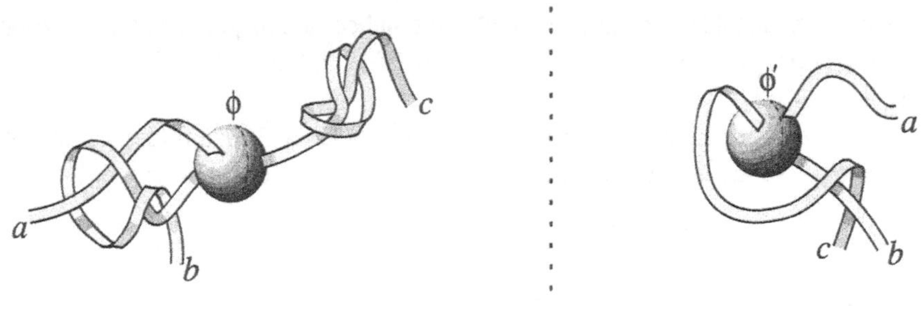

Fig. 1.10. Bridged Link

In Lemma 1.5.6 we described generic isotopies of handles over each other. If we consider the case of a 2-handle being slid over a 1-handle this means that an isotopy of a loop $S^1_{des,j} = \mathcal{L}_j(S^1 \times 0)$ is transverse to any $S^2_{asc} \subset \check{M}$ it intersects. More precisely, let $l : [0,1] \times S^1 \to \check{M} : (s,\theta) \mapsto l_s(\theta)$ be such an isotopy, and denote $D_\theta l : [0,1] \times S^1 \to T\check{M} : (s,\theta) \mapsto (l_s(\theta), \frac{\partial l_s}{\partial \theta}(\theta))$. The transversality condition may then be formally stated as

$$D_\theta l \pitchfork TS^2_{asc} .$$

The fact that any isotopy is arbitrarily close and, hence, isotopic to such an isotopy follows from first order jet-transversality, since $S^1 \times D_\theta l = j^1_\theta l_s(\theta)$, and $J^1(S^1, \check{M}) \cong S^1 \times T\check{M}$ as in [Hir76]. By dimension counting we see that $D_\theta l$ meets TS^2_{asc} only in a finite number of points. That is, only at a finite number of points $\{(s_k, \theta_k)\}_{k=1}^K$, at which the embedded circle is tangential to a sphere S^2_{asc}. Transversality implies for the component l_ν of l which is normal to S^2_{asc} that $\frac{\partial^2 l_{s_k}}{\partial^2 \theta}(\theta_k) \neq 0$ and $\frac{\partial l_{s_k}}{\partial s}(\theta_k) \neq 0$. The first property ensures that at such a point the strand is a loop with non-degenerate parabolic form, and the second guarantees that the apex of the loop is pushed through the sphere transversally. A picture of a generic isotopy of 2-handles over 1-handles as it appears in M is shown below. Both the loops a and b are being moved through the sphere pair of ϕ and ϕ'. During all other parts of the isotopy S^1_{des} and S^2_{asc} are in general position by themselves.

The attaching data of handles with the same index is generically disjoint. During a generic isotopy, however, the ascending critical manifold of one handle can run through the descending of another transversally in one point. For 1-handles we have, for example, points S^0_{des} passing through spheres S^2_{asc}. In the bridged link diagram in M this means that one surgery sphere Hi is moved through a pair of different surgery spheres Lo and Lo' as indicated in Figure 1.12.

Finally, in order to explain the sliding of 2-handles over each other, let us consider a generic isotopy of a line segment $t \mapsto A_t$, where for each $t \in [-\varepsilon, \varepsilon]$ we have that $A_t \subset M_{surg}$ is a one-dimensional submanifold in the manifold M_{surg}, on which we have already performed index 2-surgery. Hence, there we have opposite

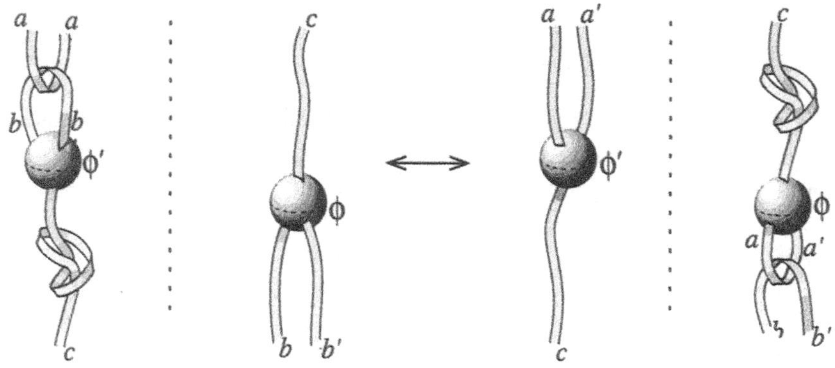

Fig. 1.11. Isotopy of S^1_{des} over S^2_{asc}

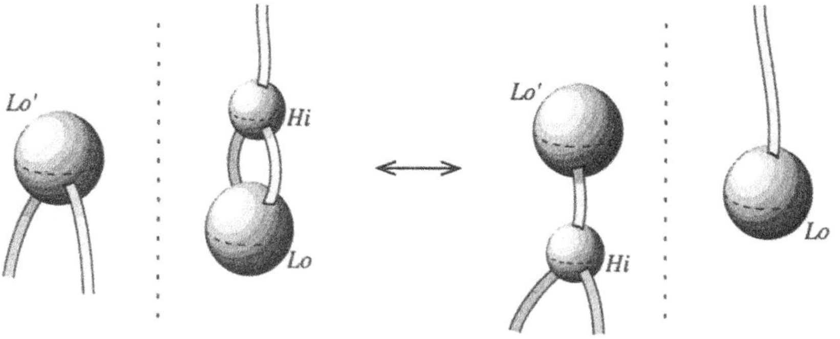

Fig. 1.12. Isotopy of S^0_{des} through S^2_{asc}

tori $D^2 \times S^1_{asc} \subset M_{surg}$ of the index-2 surgeries, and we want the isotopy $t \mapsto A_t$ to be transverse to any $\{0\} \times S^1_{asc}$, where $0 \in D^2$ is the center point of the disc.

Now, if a line segment is disjoint from an ascending critical manifold it is not necessarily yet disjoint from the whole boundary $D^2 \times S^1_{asc}$ of the attached 2-handle. In order to obtain a good separation between the 2-handle attaching data this is, however, desirable, so that we need to find a retraction from $D^2 \times S^1_{asc}$ without S^1_{asc} to its outside.

To this end let us pick a collar neighborhood $\cong [1, 1+\delta] \times S^1_{des} \times S^1_{asc}$ of the torus $D^2 \times S^1_{asc}$ in M_{surg} so that

$$D^2_\delta \times S^1_{asc} = D^2 \times S^1_{asc} \bigcup_{1 \times S^1_{des} \times S^1_{asc} = \partial D^2 \times S^1_{asc}} [1, 1+\delta] \times S^1_{des} \times S^1_{asc} \subset \check{M}$$

is a slightly thicker torus. Since we can write the punctured disc as $D^2 - \{0\} \cong (0,1] \times S^1_{des}$ we see that also the thickened disc component is naturally $D^2_\delta - \{0\} \cong$

$(0, 1+\delta] \times S^1_{des}$. Using any monotone map $(0, 1+\delta] \to (1, 1+\delta]$ we, thus, obtain a homeomorphism

$$\Delta : D^2_\delta \times S^1_{asc} - \{0\} \times S^1_{asc} = (0, 1+\delta] \times S^1_{des} \times S^1_{asc} \xrightarrow{\cong} (1, 1+\delta] \times S^1_{des} \times S^1_{asc},$$

which pushes the complement of the ascending critical manifold from the inside to the collar of the torus. Hence, by this map any line segment as well as any isotopy of a line segment that is disjoint from S^1_{asc} can be made disjoint from the whole piece $D^2 \times S^1_{asc}$ in a uniform way.

For a generic isotopy we know that the line segment may pass transversally through the ascending critical manifold during a finite number of times t_k and at a finite number of points $p_k \in S^1_{asc}$. It is not hard to see that in the vicinity of the passage point such an isotopy can always be deformed, so that the relevant part of the line segment lies entirely within the section $D^2_\delta \times \{p_k\} \subset D^2 \times S^1_{asc}$. There looks like the one indicated in the first row of Figure 1.13.

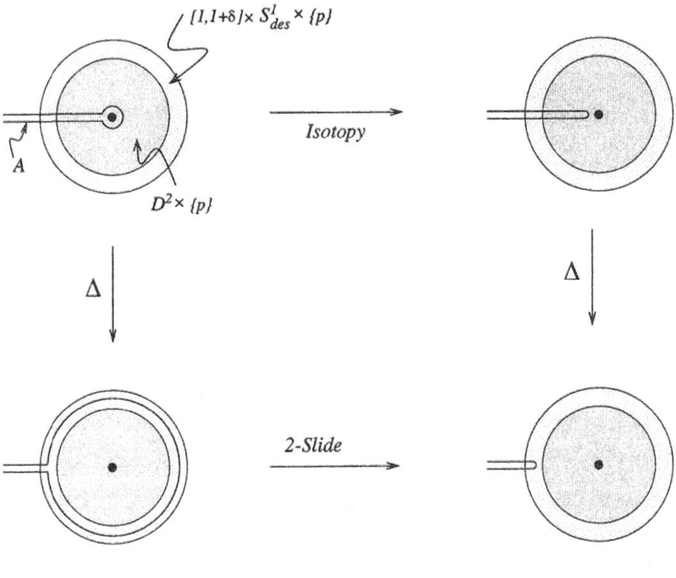

Fig. 1.13. Sectional Isotopy of A over S^1_{asc}

For all times before the crossing t_k and all times after the crossing we can apply the deformation Δ to make the isotopy disjoint from S^1_{asc}. The resulting isotopy of the segment in the complement of $D^2 \times S^1_{asc}$ is now discontinuous at the crossing time t_k. As it can be seen in the last row of Figure 1.13 the transition implies that the loop $\{1+\varepsilon\} \times S^1_{dec} \times \{p_k\}$ (with $\delta > \varepsilon \to 0$) is connected to the segment that is tangential to the torus $S^1_{des} \times S^1_{asc}$ at some point $\{q_k\} \times \{p_k\}$.

If we use framed strands or ribbons instead of the full tori to describe surgery, this loop will be a push-off along the framing of S^1_{asc} or given by splitting the ribbon

$R \times S^1_{asc}$, that is, dividing the interval R into two parts and using one component as the extra loop and identifying the other with the original R. The ribbon slicing also implies the case, where we isotope a line segment A_t, which is by itself framed. This follows, for example, from the observation that a framing of a segment is, in turn, determined by a push-off of itself so that sliding the two copies one after the other determines the framing of the segment after sliding it over.

In summary, a 2-handle slide is given as follows. If two ribbons Lo and Hi are in a tangential position as on the left of Figure 1.14 then we can obtain the ribbon configuration after the slide by cutting Lo along its longitude into two parallel ribbons Lo' and Lo'', and form the connected sum between Hi and the one component Lo' or Lo'' that is closest to Hi. The result is depicted on the right side of Figure 1.14.

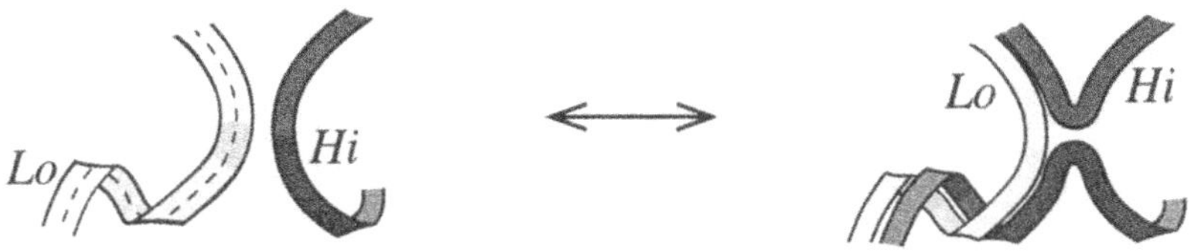

Fig. 1.14. Isotopy of S^1_{asc} over S^1_{des}

The only type of Smale cancellation, as described in Lemma 1.5.4, in our situation is between a 1-handle and a 2-handle, whose critical manifolds intersect in exactly one point. In a surgery diagram this manifests itself by a ribbon C_2 that passes through a pair of surgery balls C_1 and C_1' exactly once. Since the configuration consisting of the open strand C_2, with the two balls attached to it at the ends, is contractible, we can isotope it into a position as indicated on the left hand side of Figure 1.15 using, possibly, 1-slides as in Figure 1.12. In this situation we may still have ribbons $a, b, c \ldots$ running through the surgery pair C_1 and C_1'. They can be slid over the ribbon C_2 and removed parallelly from the configuration. Hence, after the ribbon C_2 is cancelled against the pair of surgery balls these strands are connected in the way indicated on the right hand side of Figure 1.15.

We shall come back to handle decompositions and surgery in Chapter 3. For more details and properties of bridged links see [Ker99].

1.5.3 3-dim Handles and Surgery

In the construction of the extended categories as in Section 1.6 as well as for the tangle presentations in the subsequent chapters, we encounter situations in which 3 dimensional 1-handles are attached to a given 3-manifold, and we carry out a surgery, as described in Section 1.5.2, along attaching data running through the 3-dimensional handles. We consider in this subsection some situations for which such

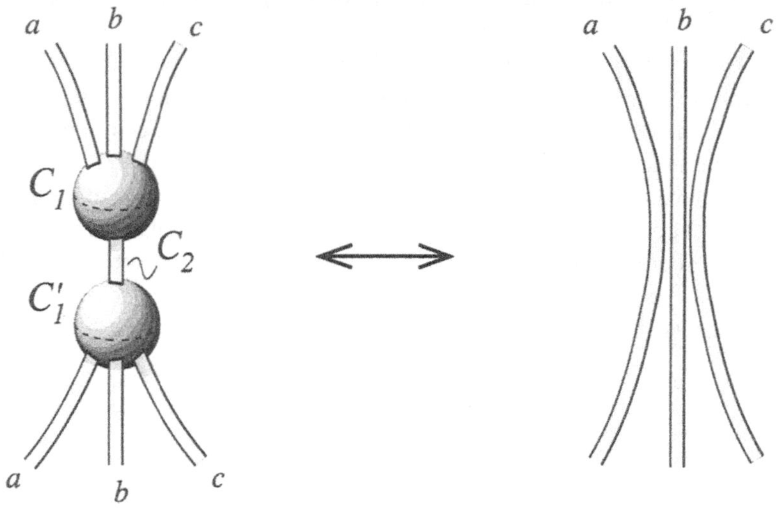

Fig. 1.15. Cancellation between 1- and 2-handles

combinations of handle attachments and surgery in mixed dimensions 3 and 4 are equivalent to much simpler operations.

As a first configuration we consider a 3-manifold M with boundary ∂M to which we attach a 3-dimensional 1-handle $\mathrm{e}^3_1 \cong [0,1] \times D^2_a$. On the combined manifold $M \cup \mathrm{e}^3_1$ we consider an index 2 surgery with attaching torus $\mathcal{L} : S^1 \times D^2_s \hookrightarrow M \cup \mathrm{e}^3_1$ with the following properties.

We require the preimage $\mathcal{L}^{-1}(\mathrm{e}^3_1)$ of the 1-handle to be precisely a torus segment $C \times D^2_s$, where $C \subset S^1$ is a closed, connected piece of the circle. On this segment let the restriction of the map $\mathcal{L}$ be of the form

$$\mathcal{L} : C \times D^2_s \xhookrightarrow{p_I \times j_D} [0,1] \times D^2_a = \mathrm{e}^3_1 \,, \tag{1.5.5}$$

where $p_I : C \xrightarrow{\cong} [0,1]$ is a diffeomorphism of 1-dimensional manifolds, and $j_D : D^2_s \hookrightarrow \mathring{D}^2_a \subset D^2_a$ is a fixed embedding of discs, such that one disc lies in the interior of the other.

This condition means that the image of the surgery torus in e^3_1 is precisely given by slightly shrinking e^3_1 into itself as depicted on the right hand side of Figure 1.16.

The manifold after handle attachment and surgery is denoted as

$$M^{\&} = (M \cup \mathrm{e}^3_1)_{\mathcal{L}}$$

so that its boundary is given by the usual index-1 surgery on ∂M:

$$\partial M^{\&} = \overline{\partial M - \{0,1\} \times D^2_a} \bigcup_{\{0,1\}\times\partial D^2_a = \partial C \times S^1_x} C \times S^1_x \,,$$

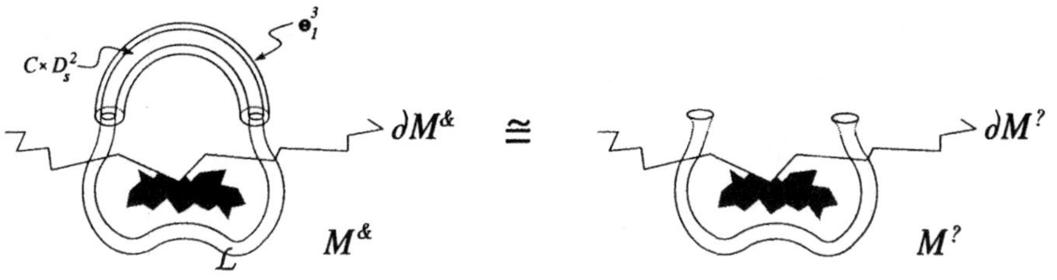

Fig. 1.16. Surgery with 1-handle

where we identify the boundary of the cross-sectional disc D_a^2 with the short meridian S_x^1 of e_1^3 and ∂C with $\{0,1\}$ by p_I as above. Another manifold to consider in this situation is

$$M^? = \overline{M - im(\mathcal{L})} ,$$

where the part $\cong C^{opp} \times D_s^2$, with $C^{opp} = \overline{S^1 - C}$, of the embedded surgery torus intersecting the manifold M is removed. Its boundary is given similarly by an "opposite" index 1 surgery as follows:

$$\partial M^? = \overline{\partial M - \{0,1\} \times D_s^2} \quad \bigcup_{\{0,1\}\times\partial D_s^2=\partial C^{opp}\times S_x^1} \quad C^{opp} \times S_x^1 ,$$

By shrinking the D_a^2 holes out into the D_s^2 holes and choosing an identification $C \xrightarrow{\sim} C^{opp}$, which is the identity on the endpoints $C \cap C^{opp}$ we obtain a natural diffeomorphism

$$\rho_\partial : \partial M^{\&} \xrightarrow{\sim} \partial M^? ,$$

which is unique up to isotopy. The claim is now that also the 3-manifolds $M^?$ and $M^{\&}$ themselves are homeomorphic.

Lemma 1.5.7. *There is a diffeomorphism*

$$M^{\&} \xrightarrow{\sim} M^? ,$$

which is the identity on M outside of a vicinity of the union $\mathcal{L} \cup e_1^3$, and which restricts on the boundaries to the map ρ_∂.

Proof. The first step in performing the surgery along $\mathcal{L}$ on $M \cup e_1^3$ is to remove the image, i.e., the attaching torus $\cong (C \cup C^{opp}) \times D_s^2$ with $S^1 = C \cup C^{opp}$ the partition of the circle into the two segments as defined above. The manifold with 1-handles attached and surgery torus removed can, thus, be written in the form:

$$(M \cup e_1^3) - im(\mathcal{L}) = (M - \mathcal{L}(C^{opp} \times D_s^2)) \bigcup_{\partial C\times(D_a^2-D_s^2)} C \times (D_a^2 - D_s^2).$$

The closure of the first piece in this presentation is, clearly, nothing else but $M^?$. In the second piece the set $D_a^2 - D_s^2$ is up to isotopy the unique boundary collar of D_a^2

so that with $S^1_x = D^2_a$ we can write $D^2_a - D^2_s = J \times S^1_x$, where $J \cong [0, \tau)$. Hence, the identity from above can be rewritten as

$$\overline{(M \cup e^3_1) - im(\mathcal{L})} = M^? \bigcup_{\partial C \times \overline{J} \times S^1_x} C \times \overline{J} \times S^1_x .$$

Before we complete the surgery by regluing the opposite torus $D^2_o \times S^1_x$, let us note that a closed collar vicinity of the removed torus in M is of the form $N = [0, \varepsilon] \times C \times \partial D^2_s$. The collar interval in normal direction of the torus can be identified with $\overline{J}$ and by radial assignment we identify $\partial D^2_s \cong S^1_x$. Hence, a combined vicinity of the deleted torus N and the attached 1-handle is identified with the thick torus $N \cup (C \times \overline{J} \times S^1_x) = (C \cup C^{opp}) \times \overline{J} \times S^1_x$.

To this we now glue the opposite torus along $(C \cup C^{opp}) \times \{0\} \times S^1_x \cong \partial D^2_o \times S^1_x$ so that we find

$$N \cup (C \times \overline{J} \times S^1_x) \cup D^2_o \times S^1_x \quad \cong \quad D^2_T \times S^1_x ,$$

where $D^2_T = (C \cup C^{opp}) \times \overline{J} \cup_{(C \cup C^{opp}) \times \{0\} = \partial D^2} D^2$ is a disc with an annulus attached to its boundary, and, hence, again a disc. This piece $D^2_T \times S^1_x$ is now attached to the complement $M - N$ along the boundary piece $\cong C^{opp} \times S^1_x$. Clearly, this attaching operation can also be seen as the attachment of the disc D^2_T to a local cross-sectional surface along a connected boundary piece C^{opp} multiplied with S^1_x. Since the 2-dimensional attachment is trivial so is its product with the circle. This shows that the total result of surgery is diffeomorphic to $M - N \cong M^?$.

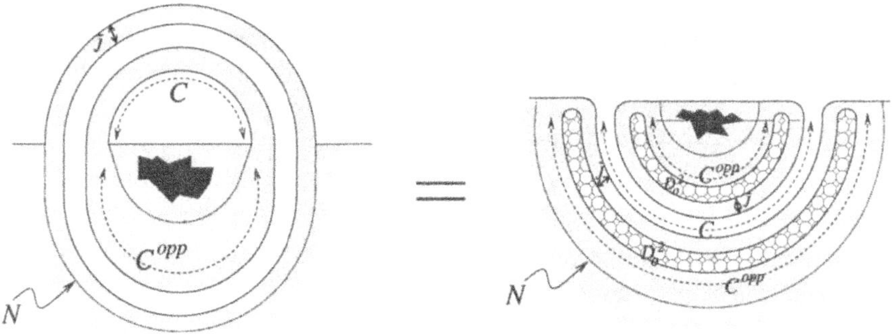

Fig. 1.17. Surgery with 1-handle

The situation can also be summarized in the cross sectional picture given in Figure 1.17. On the right side the manifold $\overline{(M \cup e^3_1) - im(\mathcal{L})}$ is depicted. The two upper arcs are the cross sectional strips through the 1-handle $C \times \overline{J} \times \{1, -1\}$ with $\pm 1 \in S^1_x$. In the lower part the vicinity N around the removed torus is depicted, which is in the cross sectional presentation also of the form $C^{opp} \times \overline{J} \times \{1, -1\}$.

This configuration is obviously equivalent to the one, where we let the thick cylinder $C \times \overline{J} \times S^1_x$ run parallelly to $C^{opp} \times \overline{J} \times S^1_x$, without changing the attaching annuli $\partial C \times \overline{J} \times S^1_x \subset \partial M$. In the cross sectional view this yields the picture on the right of Figure 1.17. The two strips combine with the components of the neighborhood N to two "U-shaped" annuli, which are precisely the space $(C^{opp} \cup C) \times \overline{J} \times \{1, -1\} \cong S^1 \times \overline{J} \sqcup S^1 \times \overline{J}$. The annuli, correspondingly, surround two "U-shaped" discs $D^2_o \times \{1, -1\}$, which are indicated in the picture as the patterned areas. The regluing of $D^2_o \times S^1_x$ in the surgery procedure, thus, fills up these areas in the cross sectional view. It is then obvious that this results in a manifold that is naturally diffeomorphic to $M^?$.

A slightly more involved situation is given when a surgery torus $\mathcal{L}$ passes through two attached 1-handles. This situation is described in Figure 1.18. The two handles e^3_1 are attached to manifold at discs in disjoint neighborhoods $U_A, U_B, U_C, U_D \subset \partial M$ in the boundary. The link $\mathcal{L}$ also splits up into four pieces of the form $D^2 \times [0, 1]$. Two of them $\mathcal{L}_{up/lo} = e^3_1 \cap im(\mathcal{L})$ are given by the intersections with the 1-handles, and the remaining two pieces $\mathcal{L}_1 \sqcup \mathcal{L}_2, = M \cap im(\mathcal{L})$ are the intersections of the surgery torus with M. The situation is depicted on the left side of Figure 1.18.

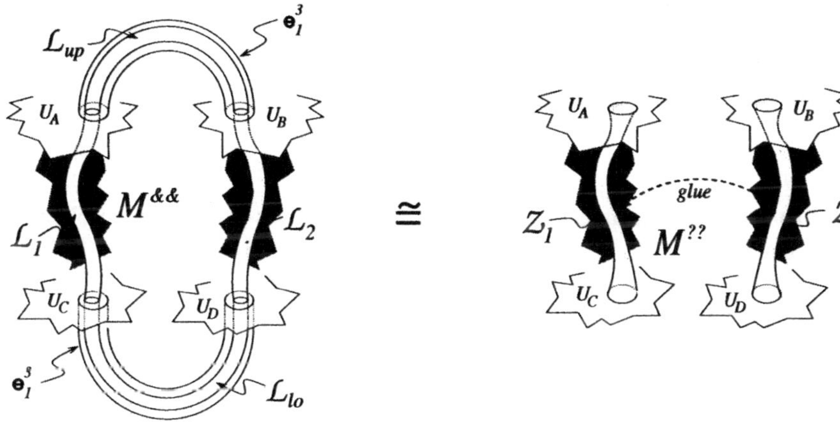

Fig. 1.18. Surgery with 1-handle

The manifold with the two 1-handles e^3_1 and the surgery performed on it is given by

$$M^{\&\&} \quad = \quad (M \cup e^3_1 \cup e^3_1)_{\mathcal{L}} \, .$$

Its boundary is given by removing discs $D^2_A \subset U_A, \ldots, D^2_D \subset U_D$ and regluing cylinders so that

$$\partial M^{\&\&} =$$

$$(\partial M-(D^2_A \sqcup D^2_B \sqcup D^2_C \sqcup D^2_D)) \bigcup_{\partial D^2_{A/B}=S^1_{AB}\times\{0/1\},\partial D^2_{C/D}=S^1_{CD}\times\{0/1\}} S^1_{AB}\times[0,1]\sqcup S^1_{CD}\times[0,1].$$

A manifold with a homeomorphic boundary can be obtained as follows. First, we remove the cylindrical parts $\mathcal{L}_1 \sqcup \mathcal{L}_2 \subset M$ from it so that we obtain two additional boundary pieces $\mathcal{Z}_1 = \partial\mathcal{L}_1$ and $\mathcal{Z}_2 = \partial\mathcal{L}_2$, both of which are naturally $\cong S^1 \times [0,1]$. The manifold $M^{??}$ is then obtained from this by identifying $\mathcal{Z}_1$ with $\mathcal{Z}_2$ so that the circle boundary piece ending in U_A is identified with the corresponding one in U_B:

$$M^{??} = \frac{\overline{M - (\mathcal{L}_1 \sqcup \mathcal{L}_2)}}{\mathcal{Z}_1 = \mathcal{Z}_2}.$$

Correspondingly, the boundary of this manifold is obtained by removing the end pieces of the $\mathcal{L}_j$ from the boundary, which are precisely the discs $D^2_A, \ldots, D^2_D$, and gluing the respective endings of the $\mathcal{Z}_j$ pieces together, which are just the circles $\partial D^2_A, \ldots, \partial D^2_D$

$$\partial M^{??} = \frac{\overline{M - (D^2_A \sqcup D^2_B \sqcup D^2_C \sqcup D^2_D)}}{\partial D^2_A = \partial D^2_B; \partial D^2_C = \partial D^2_D}.$$

Comparing the expressions for the boundaries $\partial M^{\&\&}$ and $\partial M^{??}$, we see that we can push the cylindrical pieces $S^1_{AB/CD} \times [0,1]$ into collars of the discs $D^2_A, \ldots, D^2_D$ in ∂M and thereby construct a natural diffeomorphism

$$\rho_{\partial\partial} : \partial M^{\&\&} \xrightarrow{\sim} \partial M^{??}.$$

We can in fact use Lemma 1.5.7 to show that this extends to a homeomorphism on the 3-manifolds itself.

Lemma 1.5.8. *There is a diffeomorphism*

$$M^{\&\&} \xrightarrow{\sim} M^{??},$$

which is the identity on M outside of a vicinity of the union $\mathcal{L} \cup \mathrm{e}^3_1 \cup \mathrm{e}^3_1$, and which restricts on the boundaries to the map $\rho_{\partial\partial}$.

Proof. The configuration from above is a special case from the one described in Figure 1.16, if we substitute for the 3-manifold M there the 3-manifold $M \cup \mathrm{e}^3_1$ that we are considering here with the lower 1-handle between U_C and U_D attached. To the added upper 1-handle between U_A and U_B and the surgery torus $\mathcal{L}$ we then apply Lemma 1.5.7.

For a more precise description let us introduce also closed tubular neighborhoods N_1 and N_2 of the pieces $\mathcal{L}_1$ and $\mathcal{L}_2$, such that the inclusion $\mathcal{L}_1 \subset N_1$ is given by $[0,1] \times D^2_s \subset [0,1] \times D^2_a$ and $\mathcal{L}_1 \subset N_1$ by $[2,3] \times D^2_s \subset [2,3] \times D^2_a$ in a parametrization extending that of $\mathcal{L}$ as already in the proof of Lemma 1.5.7.

Moreover, the end discs $N_j \cap \partial M$ will coincide with the attaching discs $e_1^3 \cap \partial M$ of the 1-handles. Hence, the union of the 1-handles and the neighborhoods is naturally identified with

$$N_1 \cup N_2 \cup e_1^3 \cup e_1^3 \quad \cong \quad D_a^2 \times S_T^1 \,,$$

where each of the four pieces corresponds to a segment of the circle S_T^1. In particular, if the upper 1-handle is omitted we can write $N_1 \cup N_2 \cup e_1^3 \cong D_a^2 \times [0,3]$. The manifold $M \cup e_1^3$ can, thus, be given by gluing $D_a^2 \times [0,3]$ into $M - (N_1 \sqcup N_2)$ by identification of cylindrical pieces $\partial D_a^2 \times [0,1]$ and $\partial D_a^2 \times [2,3]$ with ∂N_1 and ∂N_2 respectively. The image of $\mathcal{L}$ in the above parametrization of $N_1 \cup N_2 \cup e_1^3 \cup e_1^3$ is by construction identified with $D_s^2 \times S_T^1$, where $D_s^2 \subset \overset{\circ}{D}{}_a^2$ is the fixed embedding of discs as in Lemma 1.5.7. Also, as before, we write $D_a^2 - D_s^2 \cong [0,\tau) \times S_x^1$ so that we can express the manifold with the upper 1-handle and the surgery torus removed as

$$\overline{(M \cup e_1^3) - im(\mathcal{L})} = \overline{M - (N_1 \sqcup N_2)} \bigcup_{\partial N_1 = \{0\} \times S_x^1 \times [0,1];\, \partial N_2 = \{0\} \times S_x^1 \times [2,3]} [0,\tau] \times S_x^1 \times [0,3].$$

A cross sectional view of this construction is given by the diagram on the left side of Figure 1.19.

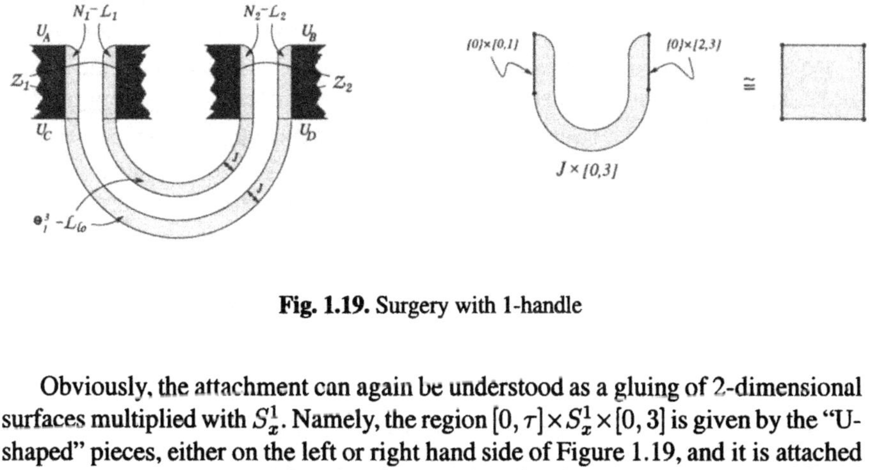

Fig. 1.19. Surgery with 1-handle

Obviously, the attachment can again be understood as a gluing of 2-dimensional surfaces multiplied with S_x^1. Namely, the region $[0,\tau] \times S_x^1 \times [0,3]$ is given by the "U-shaped" pieces, either on the left or right hand side of Figure 1.19, and it is attached along the line segments $\{0\} \times [0,1]$ and $\{0\} \times [2,3]$. Since this is topologically nothing but a disc there is a diffeomorphism that identifies it with a square, such that the two line segments become opposite edges.

However, by shrinking the horizontal coordinate of the square it is obvious that the gluing of the square is equivalent to identifying the two edges it is glued to. In the circle product this means that instead of gluing in $[0,\tau] \times S_x^1 \times [0,3]$ we may as well identify the pieces $\{0\} \times S_x^1 \times [0,1]$ and $\{0\} \times S_x^1 \times [2,3]$. This is, hence, equivalent to identifying the boundary pieces ∂N_1 with ∂N_2 in $\overline{M - (N_1 \sqcup N_2)}$.

But the tubular pieces N_1 and N_2 are nothing but thickenings of the pieces $\mathcal{L}_1$ and $\mathcal{L}_2$. Hence, the manifold that we obtain is diffeomorphic in the desired way to $M^{??}$.

1.6 The Central Extension $\Omega_4 \to \widetilde{\mathbf{Cob}} \to \mathbf{Cob}$

Most of the relevant physical topological quantum field theories are not truly functors of the category $\mathbf{Cob}$. The largest class are (projective) TQFT's that can be understood as a representation not of the ordinary cobordism category but an extension thereof by $\mathbb{Z}$. Specifically, we consider pairs (M, n), where M is a homeomorphism class of 3-cobordisms with corners, and $n \in \mathbb{Z}$ is an integer, which indicates an additional structure.

In this section we extend the definitions of the two types of compositions so as to make the $\mathbb{Z}$-extended 3-cobordisms into a double category $\widetilde{\mathbf{Cob}}$. The constructions of the compositions involve bounding 4-manifolds, and the integers are naturally interpreted as Ω_4, the Thom-cobordism group in dimension four determined by the signatures of the 4-manifolds. Hence, the relation between the categories can be viewed as central extension as follows:

$$1 \longrightarrow \Omega_4 \hookrightarrow \widetilde{\mathbf{Cob}} \twoheadrightarrow \mathbf{Cob} \longrightarrow 1. \tag{1.6.1}$$

In the remaining sections of this chapter we construct the extended double category $\widetilde{\mathbf{Cob}}$ explicitly. We start with the definition of the 2-arrow sets of the large category $\widetilde{\mathbf{COB}}$, which are four manifolds W with corners. One 3-stratum of ∂W contains a cobordism from $\mathbf{Cob}$ and the others have a prescribed standard form. The 2-arrow sets of $\widetilde{\mathbf{Cob}}$ are then given by taking homeomorphism classes of the 3-manifolds and cobordism classes of the 4-manifolds. We obtain a natural projection $\widetilde{\mathbf{COB}} \twoheadrightarrow \widetilde{\mathbf{Cob}}$. In $\widetilde{\mathbf{COB}}$ vertical and horizontal compositions are defined via gluings and handle attachments. They are shown to fulfill the double category axioms modulo surgery on the 4-manifolds W so that we obtain a strict double category for $\widetilde{\mathbf{Cob}}$. Furthermore, the restriction $\widetilde{\mathbf{Cob}} \twoheadrightarrow \mathbf{Cob}$ preserves both compositions, i.e., it is a double functor.

The cobordism categories appearing in this section are summarized in the following diagram:

$$\begin{array}{ccc}
\widetilde{\mathbf{COB}} & \xrightarrow{\ \partial\ } & \mathbf{COB} \\
{\scriptstyle \text{mod surgery \& homeomorphism}}\ \big\downarrow\ {\scriptstyle Sg} & \searrow^{\triangleleft} & \big\downarrow\ {\scriptstyle \text{mod homeomorphism}} \\
\widetilde{\mathbf{Cob}} & \xrightarrow{\ \pi_C\ } & \mathbf{Cob}
\end{array} \tag{1.6.2}$$

Let us first justify the particular extension we choose from physical models such as Chern-Simons Theory in Section 1.6.1. Then in Section 1.6.2 we construct the almost double category $\widetilde{\mathbf{COB}}$ of 4-manifolds bounding 3-dimensional relative cobordisms. Compositions and the double category $\widetilde{\mathbf{Cob}}$ are then constructed in Section 1.6.3. In particular, Section 1.6.3.V describes the vertical composition; Section 1.6.3.H1 – the first step of the horizontal composition, in which a series of

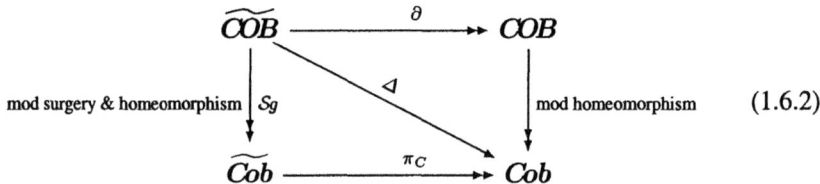

4-dimensional 1-handles is attached; Section 1.6.3.H2 – the final 2-handle attachments for the horizontal composition; and in Section 1.6.3.H3 we verify that this composition on $\widetilde{\mathcal{COB}}$ is also well defined on $\widetilde{\mathcal{C}ob}$. Finally, in Section 1.6.4 we show, using 5-dimensional handle attachments, that the interchange law holds on $\widetilde{\mathcal{C}ob}$, although it is not valid on the auxiliary category $\widetilde{\mathcal{COB}}$. This proves that $\widetilde{\mathcal{C}ob}$ is indeed a double category

1.6.1 2-Framings, and Closure of 3-Cobordisms

An exciting and important observation of Witten's in the construction of the Chern–Simons theory [Wit89], see Appendix A, is that, although it is purely topological as a classical theory, the quantized theory will depend on additional structures of the 3-manifold, namely framings or 2-framings. In the quantum group constructions of Reshetikhin and Turaev [RT91] one also encounters a dependence on additional data – in this case the signature of the linking matrix of a representing surgery diagram, which is at the same the signature of a bounding 4-manifold. These two extension are in fact equivalent. Indeed, the possible set of topological quantum field theories is dramatically reduced if one insists on ones that do not depend on framings or signatures. At the same time their nature is rather controllable since these are merely $\mathbb{Z}$-extension of out TQFT's.

Specifically, when S_{CS} is quantized, functional integration has to be restricted to a set of representatives of gauge classes. This means one has to impose a "gauge fixing" condition, which in turn requires a choice of a metric ρ. Witten [Wit89] computes that the dependence on this metric is given by an overall phase factor obtained from the η-invariant $\eta(\rho, 0)$ of an associated Dirac operator, see [APS75a]. The fact shown in [APS75b] that $S_{CS}(\omega_\rho) = 3\eta(\rho, 0) \mod \mathbb{Z}$, where ω_ρ is the Levi-Civita on the tangent bundle TM of the 3-manifold M, makes this functional a candidate for a counter term to the ordinary Chern–Simons action in [Wit89].

One needs to keep in mind, however, that although $S_{CS}(B)$ is defined in $\mathbb{R}/\mathbb{Z}$ for any connection B on TM any lift to a functional with values on $\mathbb{R}$ depends on a choice of a framing, that is an isotopy class of trivializations $\alpha : TM \xrightarrow{\sim} \mathbb{R}^3 \times M$. That is, we obtain a branch $S^{\alpha}_{CS}(B) \in \mathbb{R}$. As a result the invariants constructed in [Wit89] depend on such framings, but, by the relations in [APS75a] and [APS75b] no longer on the metric itself. For non-abelian Chern–Simons theory the invariants do, in fact, depend only on a 2-framing, meaning an isotopy class of trivializations $\alpha^{\#} : TM \oplus TM \xrightarrow{\sim} \mathbb{R}^6 \times M$. The relation between framings and 2-framings can, for example, be understood from the exact sequence $1 \to \mathrm{H}^3(M, \mathbb{Z}) \to [M, SO(3)] \to \mathrm{H}^1(M, \mathbb{Z}/2) \to 1$. A class of framings is (non-canonically) given by an element in $[M, SO(3)]$, while a 2-framing only sees the mapping degree of a trivialization and thus corresponds to an element in $\mathrm{H}^3(M, \mathbb{Z})$. The term $\mathrm{H}^1(M, \mathbb{Z}/2)$ then counts the spin structures in M.

Atiyah defines in [Ati90] a canonical 2-framing for a 3-manifold M with a bounding 4-manifold W by the fact that the Hirzebruch formula

$$\mathrm{sign}(W) = \frac{1}{6} p_1(TW \oplus TW, \alpha^{\#}_W) \qquad \text{with} \qquad M = \partial W$$

holds, where p_1 is the relative first Pontrjagin class with respect to given trivializations at the boundaries. It turns out [Ati90] that this canonical framing is precisely the one that lifts the action on the Levi-Civita connection to the original η-invariant, i.e., $S^{\alpha}_{CS}(\omega_\rho) = \eta(\rho, 0)$ in $\mathbb{R}$, given that W is a 4-manifold to which the metric ρ extends. Thus this framing is precisely the desired one for the renormalization of the Chern–Simons functional.

Instead of 2-framings or metrics it thus suffices to consider the signatures of bounding 4-manifolds, to which such framings and metrics can be extended. We still need to extend the choices of canonical 2-framings from closed manifolds to manifolds with boundaries and corners, as well as explain the composition structure of the extended data. Especially the latter will occupy the larger part of this section.

To begin with let us consider an ordinary cobordism,

$$M_o : \Sigma_{g_{sc}} \longrightarrow \Sigma_{g_{tg}}, \tag{1.6.3}$$

of *closed* surfaces. When we impose a 2-framing on every such cobordism, we have to make sure that compositions give us again 2-framed cobordisms. This forces us to impose boundary conditions, i.e., we can allow only such framings on M_o that the restriction to every boundary component

$$\xi_g : 2TM_o\Big|_{\Sigma_g \times [0,\varepsilon]} \longrightarrow \mathbb{R}^6 \times \Sigma_g \times [0, \varepsilon].$$

is a fixed trivialization that depends only on Σ_g. If we consider for simplicity trivializations in the simple tangent bundles (ordinary framings) a choice can be made by a standard embedding of $\Sigma_g \times [0, \varepsilon] \subset \mathbb{R}^3$ and using the induced trivialization of the Euclidean space as indicated in the figure below:

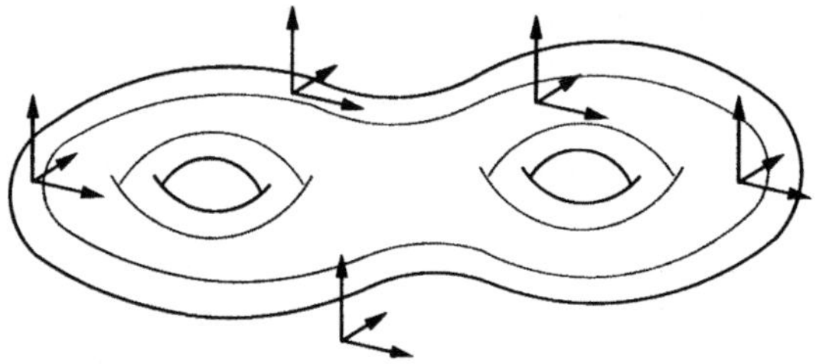

For each $g \geqslant 0$ let us consider an unknotted, untwisted embedding of the handlebody of genus g into the 3-sphere,

$$\mathcal{H}_g^+ = \mathrm{e}_0^3 \cup \underbrace{\mathrm{e}_1^3 \cup \ldots \cup \mathrm{e}_1^3}_{g} \subset S^3, \tag{1.6.4}$$

where $\mathrm{e}_j^d \cong D^j \times D^{d-j}$ denotes a j-handle of dimension d. We identify $\Sigma_g = -\partial\mathcal{H}_g^+$. Moreover, we shall use the complementary handlebody $\mathcal{H}_g^- = S^3 - \mathcal{H}_g^+$, with $\partial\mathcal{H}_g^- = \Sigma_g$. Obviously, the framing and associated 2-framing ξ_g for the choice of collar framing depicted above also extend to the standard handlebodies as subsets of $\mathbb{R}^3$ or S^3.

Moreover, we may choose embeddings of spheres $S^2_\pm \hookrightarrow S^3$, such that $S^2_\pm \subset \mathcal{H}^\pm_g$ for every g, and the restriction of the 2-framing on S^2_+ coincides with that of S^2_-. Let us also assume that the standard sphere is inside the initial 0-handle in the handle decomposition (1.6.4), i.e., $S^2_+ \subset e^3_0$. This allows us to extend the standard framings and 2-framings to the connected sum of two such handlebodies, one standard and one complementary, for which the cut out balls are bounded by S^2_+ and S^2_- respectively. We can write them naturally as the *standard connected cobordism*:

$$\mathcal{H}^{g_{tg}}_{g_{sc}} = \mathcal{H}^+_{g_{tg}} \# \mathcal{H}^-_{g_{sc}} : \Sigma_{g_{tg}} \longrightarrow \Sigma_{g_{sc}}. \tag{1.6.5}$$

Now, instead of requiring a 2-framing on a general cobordism $M_o : \Sigma_{g_{sc}} \longrightarrow \Sigma_{g_{tg}}$ to be compatible with ξ_g, we may as well consider 2-framings on the *standard closure* $\langle M_o \rangle$ that restrict to the standard ones on the handlebody pieces. This is obtained by gluing M_o to the corresponding standard cobordism from (1.6.5) along their common surface, which is thickened by an interval $[f_0, f_1]$. More precisely, we have the following identification space:

$$\langle M_o \rangle = \mathcal{H}^{g_{tg}}_{g_{sc}} \bigsqcup_{(-\Sigma_{g_{sc}} \sqcup \Sigma_{g_{tg}}) \times -\{f_0\}} (-\Sigma_{g_{sc}} \sqcup \Sigma_{g_{tg}}) \times [f_0, f_1] \bigsqcup_{(-\Sigma_{g_{sc}} \sqcup \Sigma_{g_{tg}}) \times \{f_1\}} M_o, \tag{1.6.6}$$

which we can consider as a formal map between classes:

$$\langle . \rangle : \{\text{3-cobordisms of } \textit{closed} \text{ surfaces}\} \to \{\textit{closed} \text{ 3-manifolds with 2-strata}\} : M_o \mapsto \langle M_o \rangle.$$

The 2-framings on $\langle M_o \rangle$ that restrict properly to $\mathcal{H}^{g_{tg}}_{g_{sc}}$ are, thus, in one-to-one correspondence with the 2-framings on M_o, which yield the prescribed ones at the boundaries. Using the correspondence from [Ati90] this structure can thus be further substituted by the signature of a 4-manifold that bounds $\langle M_o \rangle$. This will be our point of view in the following.

Since we have fixed 2-framings on collars of surfaces the 2-framings of the 3-manifolds extend to the gluing of the manifold over surfaces. Correspondingly, the framings extend to 4-manifolds bounding the respective standard closures if we glue these 4-manifolds along cylinders over the surfaces that appear in their boundary. Thus, the composition and extension of framings can be translated into the gluing of 4-manifolds.

Now, the cobordisms M_o between closed surfaces are only a very special subset of the 2-arrow sets of the category ***Cob***. For the relative cobordisms M between connected surfaces with boundaries we shall consider only 2-framings that are obtained by restricting a 2-framing on the corresponding ordinary cobordism $M_o \cong M \sqcup_{nS^1 \times [0,1]} nD^2 \times [0,1]$, where we have "filled up" the $\cong S^1 \times [0,1]$ boundary pieces between the holes in the surfaces.

This is justified from the geometric picture of Chern-Simons theory, where the cylindrical pieces $\cong S^1 \times [0,1]$ in the boundary $\Sigma_{S_q} \cong \partial M$ play a different rôle than the horizontal surface pieces. They correspond to "currents" or "Wilson-lines" inside the three-manifold, and the holes have the meaning of "charges" that are inserted in the physical state space. Hence, they correspond to a choice of *observables* that can be evaluated against states in a given quantum field theory. But the construction of the field theory itself should not depend on this choice. Thus, the 2-framing needed for the construction should extend over tubular fillings of cylinders, such that it is compatible with the standard framing of $D^2 \times [0,1]$.

Finally, we need to explain how the 2-framings will extend under the horizontal compositions. Note that part of the horizontal composition is the gluing of neighborhoods of holes in corresponding surfaces to each other. This needs to be done such that the two standard 2-framings extend to the glued surface collars and yield the standard 2-framing of the resulting surface.

One way of achieving this is to align two given (collars of) surfaces in R^3 and attach a tubular piece $A \times [0,1]$ to the annuli $\cong A \times \{0,1\}$ around corresponding holes. As indicated in the top piece of Fig. 1.20 the embedding in $\mathbb{R}^3$ then induces the desired framing. In the horizontal composition this procedure is applied both to the source and the target side of the cobordism.

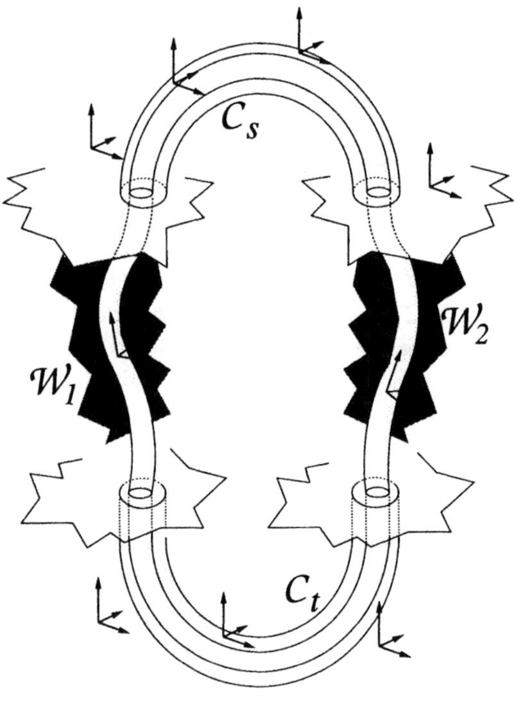

Fig. 1.20. Standard 2-framings of surfaces

We can consider the resulting space as the 3-manifold, where we attach 3-dimensional 1-handles to the cobordisms, in which the tubular pieces are filled, and then remove a torus $D^2 \times S^1$ along the curve $\mathcal{T} = \mathcal{W}_1 \cup \mathcal{C}_1 \cup \mathcal{W}_2 \cup \mathcal{C}_2$. As the framing is assumed to compatible with the standard $D^2 \times [0,1]$ framing along the lines $\mathcal{W}_j$ it, thus, extends over the said torus $D^2 \times S^1$, where it coincides with the induced framing of a standard unknotted embedding of $D^2 \times S^1$ into S^3.

But this means that it also extends to the opposite torus $\overline{S^3 - D^2 \times S^1}$. Hence, we may instead glue the opposite torus in and still be able to extend the framing. The 3-manifold we obtain then differs from the original one by a surgery along $\mathcal{T}$. We find from the results in Section 1.5.3 that this yields precisely the gluing of the cylindrical boundary parts that give rise to the horizontal gluing.

For a bounding 4-manifold we can now translate 1-handle attachments in dimension 3 into 1-handle attachments in dimension 4 and surgery along a curve into 2-handle attachments, to which the respective framings extend. To see that also the spin structure extends to the bounding 4-manifold with this surgery description we note that the framing is "incompatible" with the normal framing on the torus $D^2 \times S^1$ and refer to Lemma 0 in [MK89]. Thus in the following sections we shall define the horizontal composition with the extended structure as a composition of 4-manifolds with handle attachments so that we obtain the gluing procedure as described above.

1.6.2 Bounding 4-manifolds and 2-Arrows of $\widetilde{COB}$

In this section we replace the admissible 2-framings on M_o by four-manifolds, W, which bound the standard closure, i.e., $\partial W \cong \langle M_o \rangle$. Moreover, we introduce additional structures on W, which allow us to construct from the cobordism M_o of closed surfaces a relative cobordism of surfaces with boundaries as they are used in the definition of *Cob*.

More precisely, as 2-arrows of $\widetilde{COB}(a,b) \subset \widetilde{COB}$ we consider tuples consisting of a smooth four manifold W, an oriented, compact, connected cobordism of closed surfaces M_o, a surface coordinate chart ψ_o, a homeomorphism ρ, and an embedding $\mathbf{b}$, $(W, M_o, \psi_o, \rho, \mathbf{b})$, where $\rho : \langle M_o \rangle \xrightarrow{\sim} \partial W$, $\psi_o : \Sigma_{tg} \sqcup -\Sigma_{sc} \xrightarrow{\sim} \partial M_o$ and

$$\mathbf{b} : \bigsqcup^a D^2 \times [p_2, p_3] \sqcup \bigsqcup^b D^2 \times [p_0, p_1] \xhookrightarrow{\mathbf{b}^s \sqcup \mathbf{b}^t} M_o .$$

The 2-strata of $\langle M_o \rangle$, thus, become, via the homeomorphism ρ, 2-dimensional corners of W. Specifically, these are the four connected surfaces $\Sigma_{g_{sc}} \times f_0$, $\Sigma_{g_{sc}} \times f_1$, $\Sigma_{g_{tg}} \times f_0$, and $\Sigma_{g_{tg}} \times f_1$.

The condition on the embedding $\mathbf{b}$ of $(a+b)$ copies of the full cylinders $D^2 \times I$ with $I = [p_0, p_1]$ or $I = [p_2, p_3]$ into the 3-manifold M_o is that $\mathbf{b}$ maps the discs at the ends of the cylinders into the boundaries.

More precisely, we require that $\mathbf{b}^t(\bigsqcup^b D^2 \times p_0) \subset \psi_0(\Sigma_{g_{tg}})$, such that

$$\psi_0^{-1} \circ \mathbf{b}^t : \bigsqcup^b D^2 \times p_0 \quad \hookrightarrow \quad \Sigma_{sc}$$

maps the discs into the positions of the *target* holes on $\Sigma_{g_{sc},a/b}$ in the given order. Similarly, we want that $\psi_0^{-1} \circ \mathbf{b}^t : \bigsqcup^b D^2 \times p_1 \hookrightarrow \Sigma_{g_{tg}}$ maps b discs onto the target holes of $\Sigma_{g_{tg},a/b}$, but with order permuted by the vertical arrow $\beta \in S_b$.

The mapping properties of $\mathbf{b}^s$ are now completely analogous. That is, $\psi_0^{-1} \circ \mathbf{b}^s$ maps the a discs $D^2 \times p_3$ to the source hole positions in $\Sigma_{g_{sc}}$ in natural order, and the discs $D^2 \times p_2$ to the source hole positions in $\Sigma_{g_{sc}}$ in order permuted by $\alpha \in S_a$.

Finally, we want that $\mathbf{b}$ maps the inner part of the tubes $D^2 \times \overset{o}{I}$ without the end discs to the interior $\overset{o}{M}_o$ of the 3-manifold.

In the picture to the right the way the pieces of $\langle M_o \rangle$ fit together to form the boundary of ∂W is depicted schematically as the surface of a cube (all depicted dimensions are one lower than the actual ones). Also the embedding $\mathbf{b}$ with all its boundary conditions as given above is indicated for at least one $D^2 \times I$ component.

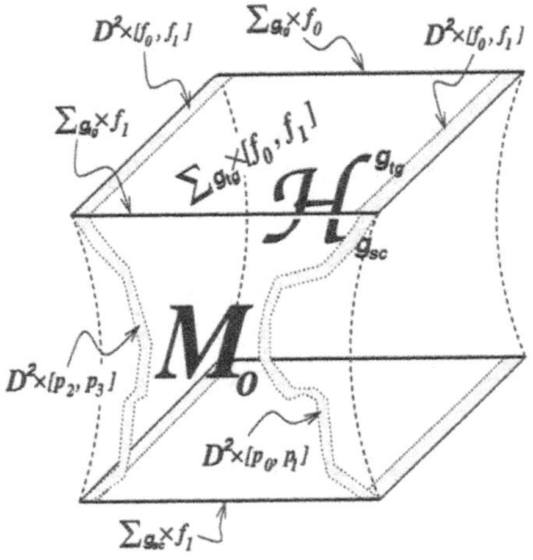

Recall that the $D^2 \times [p_0, p_1]$ and $D^2 \times [p_3, p_2]$ components of $im(\mathbf{b}) \subset M_o$ in the cobordism M_o have the meaning of the Wilson line observables in the quantum field theory picture. In the topological picture they fill out the $S^1 \times [0,1]$ boundary parts of the relative cobordisms that we considered in the previous sections of this chapter.

As we already sketched in the end of the previous Section 1.6.1, the relevant cobordisms of ***Cob*** are recovered by cutting away the $D^2 \times [p_{0/3}, p_{1/2}]$ pieces from M_o. More formally, we define the following map Δ from the 2-arrows sets of $\widetilde{\mathbf{\mathit{COB}}}$, in the sense defined without double category structure so far, and the 2-arrow sets of the double category ***Cob***:

$$\Delta : \widetilde{\mathbf{\mathit{COB}}} \longrightarrow \mathbf{\mathit{Cob}} : Q = (W, M_o, \psi_o, \rho, \mathbf{b}) \mapsto \Delta(Q) = \left[\overline{M_o - im(\mathbf{b})}, \psi\right]. \tag{1.6.7}$$

Here $[\ldots]$ denotes the homeomorphism class of cobordism classes used in the definition of ***Cob*** in Section 1.2, and ψ is obtained by extending ψ_o in the canonical way to the new $S^1 \times [0,1]$ boundary components that are created by removing $im(\mathbf{b})$ from M_o. Thus ψ is, in fact, defined on Σ_{S_q} as in (1.2.1).

Clearly, if a relative cobordism, $M \cong \overline{M_o - im(\mathbf{b})}$, is obtained by cutting away cylindrical pieces from a cobordism M_o of closed surfaces, then M_o can be recovered from M by simply regluing the $D^2 \times [0,1]$ pieces as indicated already at the end of the previous Section 1.6.1. The assignment $\phi : M \mapsto M_o$ can be described in formal language using the terminology of fill functors as in (1.3.5) or, alternatively, the elementary cobordisms from Proposition 1.1.1. The resulting map is denoted as follows:

$$\phi : \mathcal{C}ob(a,b) \longrightarrow \mathcal{C}ob(0,0) \qquad (1.6.8)$$
$$M \longmapsto \phi(M) = \mathcal{E}^t_{fill} \circ \mathcal{E}^s_{fill}(M) = \mathbb{1}^{\sqcup b}_{\mathbf{tr}} \circ_h M \circ_h \mathbb{1}^{\sqcup a}_{\mathbf{e}}.$$

In this notation we then have the following obvious identity:

$$\phi\big(\Delta(W, M_o, \psi_o, \rho, \mathbf{b})\big) = [M_o, \psi_o]. \qquad (1.6.9)$$

Recall that the purpose of introducing these four manifolds with corners is to construct a double category (2-framing) extension of $\mathcal{C}ob$. More precisely, every cobordism M of $\mathcal{C}ob$ is equipped, in addition, with a 2-framing that extends to the filling $M_o = \phi(M)$. As explained in Section 1.6.1 and in [Ati90] any such framing can be obtained from a four manifold W bounding the standard closure of the filling of M, i.e., $\partial W \cong \langle\phi(M)\rangle$. Since the isotopy classes of 2-framings are one-to-one with the signature of W, the 2-arrow sets of the extended category $\widetilde{\mathcal{C}ob}$ without structure are simply given as

$$\widetilde{\mathcal{C}ob}(a,b) \;=\; \mathcal{C}ob(a,b) \times \mathbb{Z} \qquad \forall a, b \in \mathbb{N} \cup \{0\}.$$

Let us denote also the obvious projection by

$$\pi_C : \widetilde{\mathcal{C}ob} \longrightarrow \mathcal{C}ob : (M, n) \mapsto M. \qquad (1.6.10)$$

The map Δ from (1.6.7) can be, thus, modified to give the corresponding projection onto the extended category:

$$\mathcal{S}g : \widetilde{\mathcal{COB}} \longrightarrow \widetilde{\mathcal{C}ob} : Q = (W, M_o, \psi_o, \rho, \mathbf{b}) \mapsto \big(\Delta(Q), \mathrm{sign}(W)\big). \quad (1.6.11)$$

Clearly, we have $\pi_C \circ \mathcal{S}g = \Delta$ so that the diagram in (1.6.2) indeed commutes. Let us also give the precise reason that all of these maps are onto as already suggested in the notation:

Lemma 1.6.1. *The maps Δ from* (1.6.7) *and $\mathcal{S}g$ from* (1.6.11) *are surjective on each 2-arrow set.*

Proof. For a given M in $\mathcal{C}ob$ we construct a preimage of Δ in $\widetilde{\mathcal{COB}}$ as follows. Clearly, we choose $M_o = \phi(M)$ and denote the filled in tubes by $\mathbf{b}$ so that $M = M_o - \mathbf{b}$. Given now M_o, we construct the closed 3-manifold $\langle M_o\rangle$. By Rohlin's Theorem [Roh51] we can find a 4-manifold W such that $\partial W \cong \langle M_o\rangle$. Together with the appropriate homeomorphism this, thus, defines a tuple that is mapped to M.

In order to prove that $\mathcal{S}g$ is surjective we have to find a preimage in $\widetilde{\mathcal{COB}}$ of $[M, \sigma]$ for M in $\mathcal{C}ob$ and $\sigma \in \mathbb{Z}$. Suppressing the homeomorphisms we first find a tuple $(W, M_o, \mathbf{b})$ that is mapped to M by Δ, using the fact that this is already surjective. The signature of W may now differ from σ. Hence, if $m = \sigma - \mathrm{sign}(W) \geqslant 0$ we substitute W by $W' = W\#\mathbb{CP}^2\#\ldots\#\mathbb{CP}^2$, where we have attached m copies of the projective plane $\mathbb{CP}^2$. As each $\mathbb{CP}^2$ has signature 1 and signatures are additive under connected summing we find that $\mathrm{sign}(W') = \mathrm{sign}(W) + m = \sigma$. In case

$m < 0$ we connect $|m|$ copies of $\overline{\mathbb{CP}^2}$, the projective plane with opposite orientation and signature -1. Therefore, the tuple $(W', M_o, \mathbf{b})$ is mapped by $\mathcal{S}g$ precisely to $[M, \sigma]$ so that we have found the desired preimage.

Another way of viewing the 2-arrow sets of $\widetilde{\mathcal{C}ob}$ can be extracted from the observation that the smooth 4-dimensional cobordism group $\Omega_4 \cong \mathbb{Z}$ is given by the signature. More precisely, two four manifolds, W_1 and W_2 with $\partial W_1 \cong \partial W_2$ are 5-cobordant, that is there is a 5-manifold Q with $\partial Q = W_1 \sqcup_{\partial W} (-W_2)$, if and only if $\mathrm{sign}(W_1) = \mathrm{sign}(W_2)$. In particular, this explains (1.6.1) on the level of sets.

In this sense the 2-arrows $\widetilde{\mathcal{C}ob}$ can be characterized topologically as smooth 4-cobordisms with structured corners modulo 5-cobordisms. It follows from basic Morse theory that if W_1 and W_2 are cobordant via a compact, differentiable 5-manifold they can be obtained from each other by doing surgery on the 4-manifolds. Thus, the 2-arrow sets can also be viewed as follows:

$$\widetilde{\mathcal{C}ob}(a,b) \quad = \quad \frac{\widetilde{\mathcal{COB}}(a,b)}{\text{4-dim surgery \& homeomorphism}}.$$

1.6.3 Compositions in $\widetilde{\mathcal{COB}}$ and $\widetilde{\mathcal{C}ob}$

Let us now define a vertical and a horizontal composition operation on the 2-arrows of $\widetilde{\mathcal{COB}}$. These compositions will not make $\widetilde{\mathcal{COB}}$ into a double category. However, on the equivalence classes used as 2-arrows of $\widetilde{\mathcal{C}ob}$ they will, in fact, be well defined as well as fulfill the axioms of a double category. We begin with the easier definition of a vertical product on $\widetilde{\mathcal{COB}}$.

1.6.3.V) Vertical Compositions : For simplicity let us suppress the homeomorphisms ρ and ψ_o entering the 2-arrow tuples of $\widetilde{\mathcal{COB}}$. For the vertical composition of two cobordisms we, thus, have to declare what $\circ_v$ is on each part of a triple,

$$(V, N_o, \mathbf{d}) \circ_v (W, M_o, \mathbf{b}) = (V \circ_v W, N_o \circ_v M_o, \mathbf{d} \circ_v \mathbf{b}),$$

where $M_o : \Sigma_{sc} \to \Sigma_{int}$ and $N_o : \Sigma_{int} \to \Sigma_{tg}$, where $\Sigma_{int} \neq \varnothing$, are ordinary cobordisms of $\mathcal{C}ob$.

Clearly, the $\Sigma_{g_{int}} \times [f_0, f_1]$ 3-stratum appears in the boundaries of both W and V, so that we can define the composition of the 4-manifolds simply but gluing over this part:

$$V \circ_v W = V \bigsqcup_{\Sigma_{g_{int}} \times [f_0, f_1]} W. \tag{1.6.12}$$

This yields again a 4-manifold with corners. The f_1-stratum of its boundary is then obviously given by $N_o \sqcup_{\Sigma_{g_{int}} \times f_1} M_o$, which is nothing but the composite $N_o \circ_v M_o$ of representing cobordisms in $\mathcal{C}ob(0,0)$. For the f_0-stratum observe that

$$\mathcal{H}^{g_{tg}}_{g_{int}} \sqcup_{\Sigma_{g_{int}}} \mathcal{H}^{g_{int}}_{g_{sc}} = \mathcal{H}^{+}_{g_{tg}} \# \mathcal{H}^{-}_{g_{int}} \sqcup_{\Sigma_{g_{int}}} \mathcal{H}^{+}_{g_{int}} \# \mathcal{H}^{-}_{g_{sc}} = \mathcal{H}^{+}_{g_{tg}} \# S^3 \# \mathcal{H}^{-}_{g_{sc}} = \mathcal{H}^{g_{tg}}_{g_{sc}}. \tag{1.6.13}$$

The other remaining boundary pieces of $V \circ_v W$ are $\Sigma_{g_{sc}} \times [f_0, f_1]$ and $\Sigma_{g_{tg}} \times [f_0, f_1]$, which are just the cylinders over the source and target surface of $N_o \circ_v M_o$. Hence, we have, indeed, that $\partial(V \circ_v W) \cong \langle\langle(N_o \circ_v M_o)\rangle\rangle$.

The definition of $\mathbf{d} \circ_v \mathbf{b}$ is also straightforward. For example, if $\mathbf{b}$ embeds a component $D^2 \times [p_0', p_1']$ and $\mathbf{d}$ a component $D^2 \times [p_0'', p_1'']$, then their images meet in $N_o \circ_v M_o \subset \partial(V \circ_v W)$ in $\mathbf{b}(D^2 \times p_1')$ and $\mathbf{d}(D^2 \times p_0'')$, in the respective incoming or out going hole of $\Sigma_{g_{int}}$. Thus, any monotonous function $[p_0', p_1'] \cup_{p_1'=p_0''} [p_0'', p_1''] \xrightarrow{\sim} [p_0, p_1]$ yields an embedding of the four cube with the properties required in Section 1.6.2.

Note, that the gluing of the embeddings $\mathbf{d}$ and $\mathbf{b}$ is completely parallel to the gluing of the $S^1 \times [0,1]$ pieces in the definition of the vertical composition for Cob in (1.2.3). Hence, if we remove $im(\mathbf{d} \circ_v \mathbf{b})$ from $N_o \circ_v M_o$, the result is homeomorphic to the vertical composite of $N_o - im(\mathbf{d})$ and $M_o - im(\mathbf{b})$ as relative cobordisms. In summary, the product $\circ_v$ has the following properties.

Lemma 1.6.2. *The gluing operation*

$$(P, Q) \mapsto P \circ_v Q$$

closes within the 2-arrow sets of $\widetilde{COB}$ and, thus, defines a binary operation if the intermediate horizontal 1-arrows coincide.

Moreover, for the vertical composite of their images in Cob we have

$$\Delta(P) \circ_v \Delta(Q) = \Delta(P \circ_v Q).$$

One way to describe a vertical composition on $\widetilde{Cob}$ is to choose for two 2-arrows (M, m) and (N, n) corresponding 2-arrows P and Q in $\widetilde{COB}$ with $\mathcal{S}g(Q) = (M, m)$ and $\mathcal{S}g(P) = (N, n)$. Note, that if P and P' are obtained from each other via surgery on 4-manifolds or homeomorphism, then, clearly, so are $P \circ_v Q$ and $P' \circ_v Q$. Hence, the vertical product of (M, m) and (N, n) is well defined by $(N, n) \circ_v (M, m) = \mathcal{S}g(P \circ_v Q)$.

A more explicit construction can be obtained by using the Wall 2-cocycle [Wal69] that expresses the non-additivity or the signature of two 4-manifolds when glued with corners. In [Ker99] this has already been applied in detail to the situation of ordinary cobordisms $M_o, N_o \in Cob(0,0)$ of closed surfaces. Here we consider the rational first homology $V = H_1(\Sigma_{int})$ of the intermediate surface, with a natural symplectic form ω. $\Lambda_{M_o}, \Lambda_{N_o} \subset V$ denote the kernels of the maps into $H_1(M_o \circ_v \mathcal{H}^{+}_{g_{sc}})$ and $H_1(\mathcal{H}^{-}_{g_{tg}} \circ_v N_o)$, and $\Lambda^{\pm} \subset V$ the corresponding kernels for the maps into $H_1(\mathcal{H}^{\pm}_{g_{int}})$. If for a an element $\lambda \in U' = \Lambda_{M_o} + \Lambda_{N_o}$ we denote by $\lambda = \lambda_{M_o} + \lambda_{N_o}$ a corresponding decomposition, and, similarly, $\lambda = \lambda^+ + \lambda^-$ for the $\Lambda^{\pm}$ decomposition, we can define a symmetric bilinear form ν on U' by $\nu(\lambda, \eta) = \omega(\lambda_{M_o}, \eta_{N_o}) - \omega(\lambda^+, \eta^-)$. The Wall cocycle $\mu(N_o, M_o) \in \mathbb{Z}$ is then given by the signature of the form ν. Its value is precisely the anomaly, by which the signature of the composite of two bounding 4-manifolds is different from the sum of their individual signatures. Thus, we find:

Lemma 1.6.3. *If we define a vertical composition on $\widetilde{Cob}$ by*

$$(N,n) \circ_v (M,m) \;=\; (N \circ_v M, n+m+\mu(\phi(N),\phi(M)))$$

then we have

$$\mathcal{S}g(P \circ_v Q) = \mathcal{S}g(P) \circ_v \mathcal{S}g(Q).$$

1.6.3.H1) Horizontal Compositions: 1-Handle-Attachments: Let us now describe the horizontal gluing operation in $\widetilde{COB}$ for two cobordisms

$$Q_1 = (W_1, M_{1,o}, \mathbf{b}_1) \in \widetilde{COB}(a,b), \quad Q_2 = (W_2, M_{2,o}, \mathbf{b}_2) \in \widetilde{COB}(b,c), \tag{1.6.14}$$

with

$$\Delta(Q_1) = M_1 : \Sigma_{g_1,sc,a/b} \to \Sigma_{g_1,tg,a/b}, \quad \Delta(Q_2) = M_2 : \Sigma_{g_2,sc,b/c} \to \Sigma_{g_2,tg,b/c}$$

in $\mathcal{C}ob$ and coinciding intermediate vertical arrow β^*.

The first step in the construction of $\circ_h$ is a binary gluing operation $\diamond_h$ of the 4-manifolds W_1 and W_2. It consists of attaching $2b$ 4-dimensional 1-handles e_1^4 between the two 4-manifolds, such that one end of each handle is glued to W_1 and the other to W_2. Specifically, b of the handles $e_1^4 = (D^2 \times [f_0, f_1]) \times [p_1, p_2]$ are attached at the $\Sigma_{g_1,tg,a/b} \times p_1 \times [f_0, f_1]$ stratum of ∂W_1 and the $\Sigma_{g_2,tg,b/c} \times p_2 \times [f_0, f_1]$ stratum of ∂W_2, such that the $(D^2 \times [f_0, f_1]) \times p_j$ are identified with the cylinders over the incoming and outgoing holes of the two target surfaces respectively. The other b 1-handles are attached in the analogous way at the source surfaces. In summary, the $\diamond_h$-gluing is the following identification space:

$$W_2 \diamondsuit_h W_1 \;=\; W_1 \bigsqcup_{b\{p_0,p_1\}\times[f_0,f_1]} \underbrace{e_1^4 \cup \ldots \cup e_1^4}_{2b} \bigsqcup_{b\{p_2,p_3\}\times[f_0,f_1]} W_2. \tag{1.6.15}$$

The boundary of $W_2 \diamondsuit_h W_1$ has three natural pieces, which we shall discuss in the following.

<u>$\diamondsuit$-1) *The $\Sigma \times [f_0, f_1]$-Pieces :*</u>

Since the 1-handles have been added at these pieces of W_1 and W_2, the corresponding piece in $W_2 \diamondsuit_h W_1$ is obtained by an index-1-surgery. Moreover, since the 1-handles themselves are cylinders, $e_1^4 = e_1^3 \times [f_0, f_1]$, we can describe the result also by doing index-1-surgeries on the surfaces, and then taking the cylinder over the result. The surgeries on the surfaces are done by removing the discs, where the e_1^3 are attached, and then gluing in copies of $S^1 \times [p_1, p_2]$.

As a result we obtain the surface from Figure 1.2, with $a = c = 0$. We may, thus, use the homeomorphism from (1.1.5) to identify the surgered $\Sigma \times [f_0, f_1]$-pieces with the cylinder over a standard target surface as follows:

$$\Big(\Sigma_{g_2,b/0} \bigsqcup_{b(S^1\times p_2)} b(S^1 \times [p_1,p_2]) \bigsqcup_{b(S^1\times p_1)} \Sigma_{g_1,0/b}\Big) \times [f_0, f_1]$$

$$\xrightarrow{\alpha_{[g_1,0/b:g_2,b/0]}\times[f_0,f_1]} \Sigma_{g_1+g_2+(b-1)} \times [f_0, f_1]. \quad (1.6.16)$$

Thus, up to this homeomorphism the $\Sigma \times [f_0, f_1]$-pieces in $W_2 \Diamond_h W_1$ are already of the form as required for a 2-morphism in $\widetilde{COB}(a,c)$, whose horizontal 1-arrows are the composites of those of Q_1 and Q_2.

$\Diamond$-2) The $\mathcal{H}$-Pieces:

Up to here we know how the pieces in $\partial(D^2 \times [p_1,p_2]) \times [f_0, f_1] \subset \partial e_1^4$ in the boundary of a 1-handle are glued and placed in $W_2 \Diamond_h W_1$. The other pieces $(D^2 \times [p_1,p_2]) \times \partial[f_0, f_1] \subset \partial e_1^4$ will be added to the handlebodies and 2+1-dimensional cobordisms in ∂W_1 and ∂W_2. To the handlebodies we add the 1-handles $e_1^3 = D^2 \times [p_1,p_2] \times f_0$ at the standard discs $D^2 \times p_j$ in their boundaries.

The result of the handle addition is particularly simple for the $\mathcal{H}_g^+$ handlebodies. We assume the handlebodies to be the interiors of the surfaces depicted in Figure 1.2. It is clear that the definition of α, given by deformations as in Figure 1.3 can be extended to the interior of the combined handlebody. We, thus, obtain a homeomorphism, unique up to isotopy, as follows:

$$\alpha^{\mathcal{H}}_{g_1,b,g_2} : \mathcal{H}^+_{g_1} \cup \underbrace{e_1^3 \cup \ldots \cup e_1^3}_{b} \cup \mathcal{H}^+_{g_2} \rightrightarrows \mathcal{H}^+_{g_1+g_2+b-1}. \quad (1.6.17)$$

This allows us to make the following identifications for the 1-handle attachments to the $\mathcal{H}^{g_{tg}}_{g_{sc}}$ components:

$$\mathcal{H}^{g_{1,tg}}_{g_{1,sc}} \cup \underbrace{e_1^3 \cup \ldots \cup e_1^3}_{b} \cup \mathcal{H}^{g_{2,tg}}_{g_{2,sc}} \cong \mathcal{H}^-_{g_{1,sc}} \# \Big(\mathcal{H}^+_{g_{1,tg}} \cup \underbrace{e_1^3 \cup \ldots \cup e_1^3}_{b} \cup \mathcal{H}^+_{g_{2,tg}}\Big) \# \mathcal{H}^-_{g_{2,sc}}$$

$$\cong \mathcal{H}^-_{g_{1,sc}} \# (\mathcal{H}^+_{g_{1,tg}+g_{2,tg}+b-1}) \# \mathcal{H}^-_{g_{2,sc}} \cong (\mathcal{H}^+_{g_{1,tg}+g_{2,tg}+b-1}) \# (\mathcal{H}^-_{g_{1,sc}} \# \mathcal{H}^-_{g_{2,sc}}).$$

Abstractly, the last homeomorphism is trivial from the properties of # for connected spaces. However, we wish to characterize the deformation underlying the homeomorphism more specifically. To this end note that the first 1-handle $e^{first} \cong e_1^3$, which attaches to the holes with label 1_s and 1_t respectively, in the gluing of the $\mathcal{H}_g^+$ handlebodies as in Figure 1.2, plays a special rôle. We may assume that e^{first} is attached directly to the $D^3_{[1/2]} \cong e_0^3 \subset \mathcal{H}^+_{g_{1/2,tg}}$ pieces of each handlebody that enters the handle decomposition as in (1.6.4). Since the sphere S^2_+ used in each $\mathcal{H}^+_{g_{tg}}$ to connect the corresponding opposite handlebodies lies entirely within the balls $D^3_{[1/2]}$, the $\mathcal{H}^-_{g_{sc}}$ are already attached to the piece

$$D^3_{[1]} \cup e^{first} \cup D^3_{[2]} \subset \mathcal{H}^+_{g_{1,tg}} \cup \underbrace{(e_1^3 = e^{first}) \cup \ldots \cup e_1^3}_{b} \cup \mathcal{H}^+_{g_{2,tg}}.$$

Assuming that a deformation of the S^2_+-attachment of the $\mathcal{H}^-_{g_{sc}}$ handlebodies into the e^{first} handle is entirely within this simply connected part, we see that it is uniquely given up to isotopy. Furthermore, the isotopy will also be such that the alignment of the handlebodies with the first handle as in Figure 1.3 does not change. Now the $\mathcal{H}^-_{g_{sc}}$ are complements of the standard handle bodies in S^3 so that $S^3 - \mathcal{H}^+_{g_{sc}}$ with a 3-ball removed is really $D^3 - \mathcal{H}^+_{g_{sc}}$. In connected sum of $\mathcal{H}^-_{g_{sc}}$ with neighborhood U also a 3-ball is removed from U and $D^3 - \mathcal{H}^+_{g_{sc}}$ glued in so that $U \# \mathcal{H}^-_{g_{sc}} = \overline{U - \mathcal{H}^+_{g_{sc}}}$. Hence, in summary, we obtain the following homeomorphism, which is unique up to isotopy:

$$\mathcal{H}^{g_{1,tg}}_{g_{1,sc}} \cup \underbrace{\mathrm{e}^3_1 \cup \ldots \cup \mathrm{e}^3_1}_{b} \cup \mathcal{H}^{g_{2,tg}}_{g_{2,sc}}$$

$$\cong \mathcal{H}^+_{g_{1,tg}} \cup \left(\overline{\mathrm{e}^{first} - (\mathcal{H}^+_{g_{1,sc}} \sqcup \mathcal{H}^+_{g_{2,sc}})}\right) \cup \underbrace{\mathrm{e}^3_1 \cup \ldots \cup \mathrm{e}^3_1}_{b-1} \cup \mathcal{H}^+_{g_{2,tg}}$$

$$\cong \overline{\mathcal{H}^+_{g_{1,tg}+g_{2,tg}+b-1} - (\mathcal{H}^+_{g_{1,sc}} \sqcup \mathcal{H}^+_{g_{2,sc}})}. \qquad (1.6.18)$$

The region inside e^{first} with the handlebodies $\mathcal{H}^+_{g_{1,sc}}$ and $\mathcal{H}^+_{g_{2,sc}}$ cut out is depicted in the figure below.

F

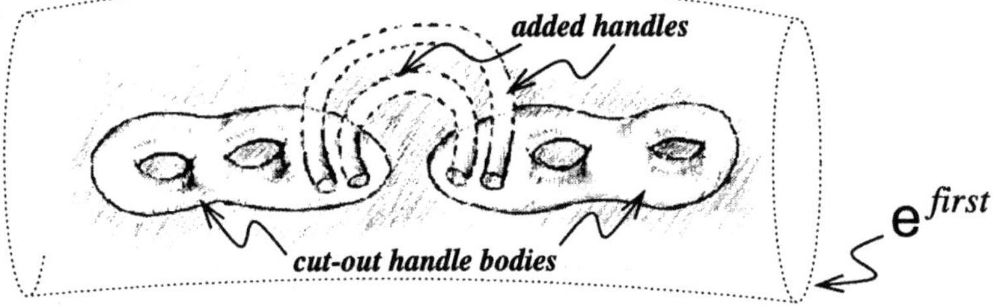

In order to obtain the complete $\mathcal{H}$-part of $W_2 \Diamond_h W_1$ as in (1.6.15) we also need to consider the other b e^4_1-handles that are attached to the bottom side of the four-folds if viewed as in the schematic picture of Section 1.6.2. The restriction to the $\mathcal{H}$-part of the boundary implies again an attachment of b e^3_1-handles, but this time to the $\mathcal{H}^-_{g_{sc}}$-handlebodies in the two $\mathcal{H}^{g_{tg}}_{g_{sc}}$ components. Analogously, a e^3_1-handle is attached at a target disc in $\Sigma^-_{g_{1,sc}} = -\partial\mathcal{H}^-_{g_{1,sc}}$ and the corresponding source disc in $\Sigma^-_{g_{2,sc}} = -\partial\mathcal{H}^-_{g_{2,sc}}$. In the presentation of the $\mathcal{H}$-part after gluing the first b e^3_1-handles as in (1.6.18) these surfaces appear when we cut out the two $\mathcal{H}^+_{g_{sc}}$ from e^{first}. Hence, this is where we have to add the remaining e^3_1-handles. Let us denote the resulting e^{first}-piece with the $\mathcal{H}^+_{g_{sc}}$ cut out and the b e^3_1-handles added as follows:

$$\mathbf{F} = \overline{\mathrm{e}^{first} - (\mathcal{H}^+_{g_{1,sc}} \sqcup \mathcal{H}^+_{g_{2,sc}})} \cup_{\Sigma_{g_{1,sc}} \sqcup \Sigma_{g_{2,sc}}} \underbrace{\mathrm{e}^3_1 \cup \ldots \cup \mathrm{e}^3_1}_{b}.$$

The piece $\mathbf{F}$ is depicted in the diagram above, including the ($b = 2$) added handles that are attached to the inside surfaces of the cut out parts. Obviously, $\mathbf{F}$ cannot be embedded in Euclidean 3-space so that the dashed parts of the latter handles have to be thought of as lying in an additional dimension.

F is now also a piece in the complete $\mathcal{H}$-part of $W_2 \Diamond_h W_1$ with all 1-handles added. Specifically, the way it is connected with handlebodies is given by the following extension of the homeomorphism from (1.6.18):

$$\mathcal{H}_{g_{1,sc}}^{g_{1,tg}} \cup \underbrace{\mathrm{e}_1^3 \cup \ldots \cup \mathrm{e}_1^3}_{2b} \cup \mathcal{H}_{g_{2,sc}}^{g_{2,tg}} \cong \overline{\mathcal{H}^+_{g_{1,tg}+g_{2,tg}+b-1} - (\mathcal{H}^+_{g_{1,sc}} \sqcup \mathcal{H}^+_{g_{2,sc}})} \cup \underbrace{\mathrm{e}_1^3 \cup \ldots \cup \mathrm{e}_1^3}_{b}$$

$$\cong \mathcal{H}^+_{g_{1,tg}} \cup \mathbf{F} \cup \underbrace{\mathrm{e}_1^3 \cup \ldots \cup \mathrm{e}_1^3}_{b-1} \cup \mathcal{H}^+_{g_{2,tg}} \cong \mathcal{H}^+_{g_{1,tg}+g_{2,tg}+b-1} \# \mathbf{F}. \qquad (1.6.19)$$

The last homeomorphism is here by virtue of α_{g_1,b,g_2} as in (1.6.17), and we can assume that the connected sum is such that the 3-ball in $\mathcal{H}^+_{g_{1,tg}+g_{2,tg}+b-1}$-handle bounded by the corresponding standard S^2_+-sphere is replaced precisely by **F**.

Thus, the only reason that we are not able to canonically identify the handlebody in (1.6.19) with $\mathcal{H}^{g_{1,tg}+g_{2,tg}+b-1}_{g_{1,sc}+g_{2,sc}+b-1}$ is that in the construction of **F** the last b e_1^3-handles were added to instead of cut away from e^{first}.

$\Diamond$-3) The M_o-pieces :

The M_o-part of $\partial(W_2 \Diamond_h W_1)$ is obtained completely analogous to the $\mathcal{H}$-part. Again, the e_1^4-handles $\cong D^2 \times [p_1, p_2] \times [f_0, f_1]$ or $\cong D^2 \times [p_0, p_3] \times [f_0, f_1]$ restrict to 3-dimensional 3-handles $\mathrm{e}_1^3 = D^2 \times [p_{1/0}, p_{2/3}] \times \partial f_1$ on the M_o-parts, and, thus, account for the last pieces of $\partial \mathrm{e}_1^4$. As for the handlebodies, an e_1^3-handle, say, of the form $\mathrm{e}_1^3 = D^2 \times [p_1, p_2]$, is attached with one end $\cong D^2 \times p_1$ at a target-disc with label j_t to the target surface $\Sigma_{g_{1,tg}} \subset \partial M_{1,o}$ of the first cobordism and with the other $\cong D^2 \times p_1$ at a source-disc with the same label j_s to the target surface $\Sigma_{g_{2,tg}} \subset \partial M_{2,o}$ in the second cobordism. The 1-handles $\mathrm{e}_1^3 = D^2 \times [p_0, p_3]$ are correspondingly attached between the source surfaces of $M_{1,o}$ and $M_{2,o}$.

With the homeomorphisms of surfaces as in (1.6.16) we can, thus, already identify $M_{1,o}$ and $M_{2,o}$ with the $2b$ 1-handles attached between them as a cobordism between the correct closed surfaces. However, this cobordism is in general different from the one we obtain if we consider the filling of the horizontal product $\phi(M_2 \circ_h M_1)$, with $\circ_h$ as defined in the diagram in Fig. 1.6, ϕ as in (1.6.8), and, correspondingly, M_1 and M_2 the relative cobordisms that are obtained from $M_{1,o}$ and $M_{2,o}$ as in (1.6.14). For this reason let us describe the M_o-part in a slightly different way.

As required in Section 1.6.2 $M_{1,o} \cap im(\mathbf{b}_1^t)$ is identified with b copies of $D^2 \times [p_0, p_1]$ and $M_{2,o} \cap im(\mathbf{b}_2^s)$ is identified with b copies of $D^2 \times [p_2, p_3]$, and top and bottom discs $\cong D^2 \times p_i$ coincide with the source and target discs on the standard surfaces Σ_g. Moreover, the position labels of the discs at the top and bottom of $D^2 \times [p_0, p_1]$ are related precisely by the target vertical 1-arrow $\beta \in S_b$ of Q_1. Since this coincides with source vertical 1-arrow of Q_2, the same permutation β tells us which disc in the source surface of $M_{2,o}$ is connected by a $D^2 \times [p_2, p_3]$-piece to which disc in the target surface.

This correspondence in the union of the $D^2 \times [p_0, p_1]$- and $D^2 \times [p_2, p_3]$-pieces with the 1-handles $e_1^3 \cong D^2 \times [p_1, p_2]$ and $e_1^3 \cong D^2 \times [p_1, p_2]$ results now in b connected pieces, each containing exactly four of these cylindrical pieces, one of each type. Specifically, we have that

$$\mathcal{T}_1 \sqcup \mathcal{T}_2 \sqcup \ldots \sqcup \mathcal{T}_b = \left(M_{1,o} \cap im(\mathbf{b}_1^t)\right) \cup \underbrace{e_1^3 \cup \ldots \cup e_1^3}_{2b} \cup \left(M_{2,o} \cap im(\mathbf{b}_2^s)\right). \tag{1.6.20}$$

Here each connected component $\mathcal{T}_j \cong D^2 \times S_{4p}^1$ is a full torus, where we denote the circle obtained by gluing the four $[p_i, p_{i+1}]$-intervals together by

$$S_{4p}^1 = \frac{[p_0, p_1] \cup_{p_1} [p_1, p_2] \cup_{p_2} [p_2, p_3] \cup_{p_3} [p_3, p_4]}{p_0 \sim p_4}. \tag{1.6.21}$$

The connectedness structure asserted in (1.6.20) becomes apparent when we identify each of the four cross-sectional discs $D^2 \times p_i$ in the j-th torus component $\mathcal{T}_j$ with the target or source holes of given labels on the standard surfaces:

$D^2 \times p_0 \subset \mathcal{T}_j = j_t$ – labeled disc in $\Sigma_{g_{1,sc}}$,
$D^2 \times p_1 \subset \mathcal{T}_j = \beta(j)_t$ – labeled disc in $\Sigma_{g_{1,tg}}$,
$D^2 \times p_2 \subset \mathcal{T}_j = \beta(j)_s$ – labeled disc in $\Sigma_{g_{2,tg}}$,
$D^2 \times p_3 \subset \mathcal{T}_j = j_s$ – labeled disc in $\Sigma_{g_{2,sc}}$.

Our goal is to present the M_o-part of $\partial(W_2 \Diamond_h W_1)$ as a union of these tori with cobordisms. We may define the latter either from the relative cobordisms by filling the source holes of M_1 and the target holes of M_2 using the functors (1.3.5), or, alternatively, by removing the target cylinders from $M_{1,o}$ and source cylinders from $M_{2,o}$:

$$\begin{aligned} M_{\frac{1}{2},1} &= \overline{M_{1,o} - im(\mathbf{b}_1^t)} = \mathcal{E}_{fill}^t(M_1), \\ M_{\frac{1}{2},2} &= \overline{M_{2,o} - im(\mathbf{b}_2^s)} = \mathcal{E}_{fill}^s(M_2). \end{aligned}$$

With this notation, also introducing the cylinders $\mathcal{Z}_{j_s} \cong S^1 \times [p_0, p_1] = \mathcal{T}_j \cap \partial M_{\frac{1}{2},1}$ and $\mathcal{Z}_{j_t} \cong S^1 \times [p_2, p_3] = \mathcal{T}_j \cap \partial M_{\frac{1}{2},2}$, we obtain the presentation:

$$M_{1,o} \cup \underbrace{e_1^3 \cup \ldots \cup e_1^3}_{2b} \cup M_{2,o} = M_{\frac{1}{2},1} \cup_{\mathcal{Z}_{1_t} \ldots \mathcal{Z}_{b_t}} \left(\mathcal{T}_1 \sqcup \ldots \sqcup \mathcal{T}_b\right) \cup_{\mathcal{Z}_{1_s} \ldots \mathcal{Z}_{b_s}} M_{\frac{1}{2},2}. \tag{1.6.22}$$

In Fig. 1.21 we illustrate the 1-handle attachments to obtain the 4-manifold $W_2 \Diamond_h W_1$ for the simplest case $b = 1$ using the same scheme in one dimension lower as in the picture in Section 1.6.2. Here $\mathbf{b}_1$ and $\mathbf{b}_2$ are short for $im(\mathbf{b}_1^t)$ and $im(\mathbf{b}_1^s)$, respectively.

1.6.3.H2) Horizontal Compositions: 2-Handle-Attachments:

The remaining step in the construction of a horizontal binary operation on $\widetilde{\mathcal{COB}}$ is to add, in addition to the 1-handles, also $2b$ 4-dimensional 2-handles e_2^4 to the four folds:

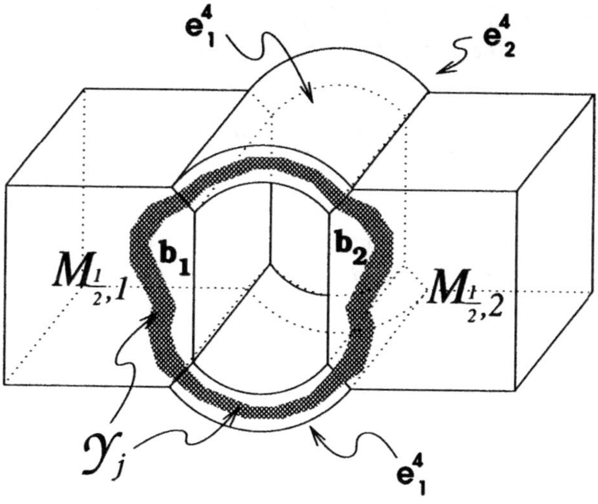

Fig. 1.21. 1-handle attachments

$$W_2 \circ_h W_1 \;=\; (W_2 \Diamond_h W_1) \cup_{\mathcal{Y}_1 \sqcup \ldots \sqcup \mathcal{Y}_b \sqcup \mathcal{L}_1 \sqcup \ldots \sqcup \mathcal{L}_b} \underbrace{(\mathrm{e}^4_2 \cup \ldots \cup \mathrm{e}^4_2)}_{2b}. \qquad (1.6.23)$$

Here the $\mathcal{Y}_j$ and $\mathcal{L}_j$, with $j = 1, \ldots, b$ denote $2b$ full tori $\cong S^1 \times D^2$ that are embedded in the boundary $\partial(W_2 \Diamond_h W_1)$. Recall that an attachment of a 4-handle e^4_2 along a torus is obtained by writing $\mathrm{e}^4_2 = D^2 \times D^2$, so that there is a canonical identification $S^1 \times D^2 = (\partial D^2) \times D^2 \subset \partial(D^2 \times D^2) = \partial \mathrm{e}^4_2$. Hence, a standard torus in $\partial(W_2 \Diamond_h W_1)$ is naturally identified with a piece in $\partial \mathrm{e}^4_2$ along which e^4_2 is glued to $W_2 \Diamond_h W_1$. In order to complete the definition of (1.6.23) we still need to specify the location of the $2b$ embedded tori.

The b tori $\mathcal{L}_j$ all lie in the $\mathcal{H}$-piece of $\partial(W_2 \Diamond_h W_1)$ as described in (1.6.19) of the previous section. In this presentation all of the tori will be contained in the F-part, as depicted in Section 1.6.3.H1, Part $\Diamond - 2$. Half of each torus will run through the "added handles" e^3_1, such that

$$\mathcal{L}_j \cap \mathrm{e}^3_1 = (\tfrac{1}{2} D^2) \times [p_0, p_3] \subset D^2 \times [p_0, p_3] = \mathrm{e}^3_1$$

for the j-th 1-handle e^3_1. Here $\frac{1}{2} D^2 \subset D^2$ denotes the standard disc with half radius. Hence, $\mathcal{L}_j$ intersects the surfaces $\Sigma_{g_{1,sc}}$ and $\Sigma_{g_{2,sc}}$ in **F** in the discs with labels j_t and j_s and half radius. The second half of the j-th torus in **F**, namely the piece

$$\mathcal{L}_j \cap \overline{\mathrm{e}^{first} - (\mathcal{H}^+_{g_{1,sc}} \sqcup \mathcal{H}^+_{g_{2,sc}})} \;\cong\; (\tfrac{1}{2} D^2) \times [q_0, q_3],$$

will be the full cylindrical piece added to the handlebody in Figure 1.2 between discs with labels j_s and j_t, only that we halved the diameter of each. Thus, we have a complete torus $\mathcal{L}_j \cong (\frac{1}{2} D^2) \times S^1_{2p}$, where $S^1_{2p} = [p_0, p_3] \cup_{p_0 \sim q_0, p_3 \sim q_3} [q_0, q_3]$. For a picture of the $\mathcal{L}_j$ in **F** see the discussion of the $\mathcal{H}$-boundary part below.

The other b tori $\mathcal{Y}_j$ lie in the M_o-piece of $\partial(W_2 \Diamond_h W_1)$ as given in (1.6.22) above. Specifically, the j-th torus $\mathcal{Y}_j$ is simply defined as the j-th torus $\mathcal{T}_j$ with halved radius. Furthermore, we have that the inclusion

$$(\tfrac{1}{2}D^2) \times S^1_{4p} \cong \mathcal{Y}_j \subset \mathcal{T}_j \cong D^2 \times S^1_{4p}$$

is described by the standard one of $(\frac{1}{2}D^2) \subset D^2$.

Let us next describe again in more detail how the various pieces of $\partial(W_2 \circ_h W_1)$ differ from the respective ones in $\partial(W_2 \Diamond_h W_1)$ after attaching the 2-handles.

o-1) The $\Sigma \times [f_0, f_1]$-Pieces :

The 2-handles e^4_2 are all disjoint from the $\Sigma \times [f_0, f_1]$-parts of $\partial(W_2 \Diamond_h W_1)$. Hence, they are unchanged and still identified canonically with the correct composite surface as in (1.6.16).

o-2) The $\mathcal{H}$-Pieces :

As explained in Section 1.5.2, attaching a e^4_2-handle to a 4-manifold W along a torus $\mathcal{T} \cong D^2 \times S^1 \subset N = \partial W$ results in an index-2-surgery on N along $\mathcal{T}$. Specifically, $N_{\mathcal{T}} = \partial(e^4_2 \cup W)$, where the surgered manifold $N_{\mathcal{T}}$ is obtained by cutting away the torus $\mathcal{T}$ from N and regluing the opposite torus $S^1 \times D^2$ along $S^1 \times S^1 \cong \partial(N - \mathcal{T})$.

In our case this means that the $\mathcal{H}$-piece of $\partial(W_2 \circ_h W_1)$ is obtained from the corresponding piece in $\partial(W_2 \Diamond_h W_1)$ by b 2-surgeries along the tori $\mathcal{L}_1, \ldots, \mathcal{L}_b$. Since all of these lie in the **F** piece, it suffices to describe the effects of surgery only there. Here each surgery torus passes parallelly exactly once through a 3-dimensional 1-handle. As explained in Section 6 this implies that the surgered manifold can be described by removing both the 3-dimensional 1-handle as well as a neighborhood of the surgery torus. In the following picture it becomes apparent that with our choices of the $\mathcal{L}_1$, this results in cutting away precisely those full cylinders in e^{first} that are in the locations of the 1-handles inserted between the standard handlebodies $\mathcal{H}^+_{g_{1,sc}}$ and $\mathcal{H}^+_{g_{1,sc}}$ as in Figure 1.2.

Hence, with this homeomorphism and the one in (1.6.17) we obtained for the surgered **F**-piece the following:

$$\mathbf{F}_{\mathcal{L}_1,\ldots,\mathcal{L}_b} \cong \overline{e^{first} - (\mathcal{H}^+_{g_{1,sc}} \cup \underbrace{e^3_1 \cup \ldots \cup e^3_1}_{b} \cup \mathcal{H}^+_{g_{2,sc}})}$$

$$\cong \overline{e^{first} - (\mathcal{H}^+_{g_{1,sc}+g_{2,sc}+b-1})} \cong e^{first} \#(\mathcal{H}^-_{g_{1,sc}+g_{2,sc}+b-1}). \quad (1.6.24)$$

The complete $\mathcal{H}$-piece of $\partial(W_2 \circ_h W_1)$ is now obtained by applying the surgeries along the $\mathcal{L}_j$ to the identifications in (1.6.19). Combining this with (1.6.24) we finally obtain a canonical homeomorphism of the $\mathcal{H}$-piece with the correct standard connected cobordism from (1.6.5) with genera $g_{tg} = g_{1,tg} + g_{2,tg} + b - 1$ and $g_{sc} = g_{1,sc} + g_{2,sc} + b - 1$:

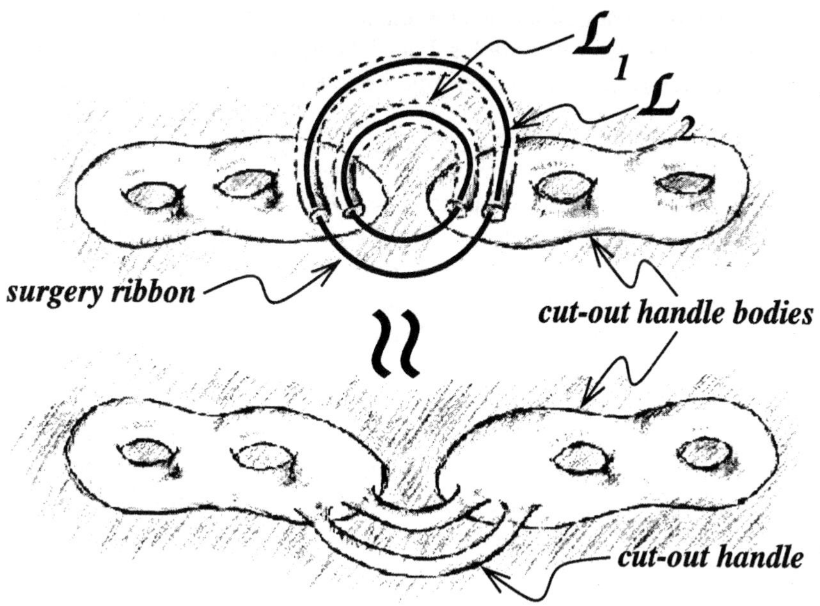

Fig. 1.22. From added to cut out handles via surgery along $\mathcal{L}_j$

$$\Big(\mathcal{H}^{g_{1,tg}}_{g_{1,sc}} \cup \underbrace{e^3_1 \cup \ldots \cup e^3_1}_{2b} \cup \mathcal{H}^{g_{2,tg}}_{g_{2,sc}}\Big)_{\mathcal{L}_1,\ldots,\mathcal{L}_b} \cong \mathcal{H}^+_{g_{1,tg}+g_{2,tg}+b-1} \#(\mathcal{H}^-_{g_{1,sc}+g_{2,sc}+b-1})$$

$$\cong \mathcal{H}^{g_{1,tg}+g_{2,tg}+b-1}_{g_{1,sc}+g_{2,sc}+b-1}. \qquad (1.6.25)$$

In the construction of tangle presentations the 4-manifolds W_2 and W_1 will be considered as 4-dimensional relative cobordisms from the handlebodies to the 3-manifolds $M_{1,o}$ and $M_{2,o}$. For the composite we still have the glued W_2 and W_1 but in addition we now also have added the 4-dimensional 2-handles. These 2-handles can be interpreted as an additional 4-cobordism, which we want to describe in some more detail here. In order to simplify the notation let us write

$$\mathcal{H}_\diamond = \mathcal{H}^{g_{1,tg}}_{g_{1,sc}} \cup \underbrace{e^3_1 \cup \ldots \cup e^3_1}_{2b} \cup \mathcal{H}^{g_{2,tg}}_{g_{2,sc}} \qquad \text{and} \qquad \mathcal{H}_\circ = \mathcal{H}^{g_{1,tg}+g_{2,tg}+b-1}_{g_{1,sc}+g_{2,sc}+b-1}$$

so that (1.6.25) becomes $(\mathcal{H}_\diamond)_{\mathcal{L}_1,\ldots,\mathcal{L}_b}, = \mathcal{H}_\circ$. The extra 4-cobordism is now obtained by considering a collar of $\mathcal{H}_\diamond$ in $W_2 \diamond_h W_1$, homeomorphic to the product of the handlebody with an interval $[0,\varepsilon]$, and add the handles e^4_2 along the tori in the $\mathcal{H}_\diamond \times \varepsilon$ boundary part:

$$P = \big(\mathcal{H}_\diamond \times [0,\varepsilon]\big) \cup_{(\mathcal{L}_1\times\varepsilon)\sqcup\ldots\sqcup(\mathcal{L}_b\times\varepsilon)} \big(\underbrace{e^4_2 \cup \ldots \cup e^4_2}_{b}\big). \qquad (1.6.26)$$

By definition of surgery this implies now that P is a relative 4-cobordism between $\mathcal{H}_\diamond \times 0$ and $(\mathcal{H}_\diamond)_{\mathcal{L}_1,\ldots,\mathcal{L}_b} \times \varepsilon \cong \mathcal{H}_\circ$. As we indicated above, in the cobordism picture for the derivation of tangle presentation we need to consider the standard handlebodies as the source manifolds. For P, however, $\mathcal{H}_\circ$ is the target manifold. Hence, we are really interested in the opposite cobordism

$$P^{opp} = -P : \mathcal{H}_\circ \longrightarrow \mathcal{H}_\diamondsuit .$$

In an opposite 4-cobordism a k handle $e^4_k = D^k \times D^{4-k}$ is reverted to a $4-k$ handle $e^4_{4-k} = D^{4-k} \times D^k$, since the rôles of a piece in the boundary of the handle, serving as the attaching part to the source manifold, and the new piece added to the target manifold are switched. In our case P^{opp} is still composed of $2 = 4 - 2$-handles e^4_2. The boundary of the j-th 2-handle is correspondingly made up of two full tori, one of which is identified with $\mathcal{L}_j$ in the attachment. If we denote the opposite part by $\mathcal{L}_j^{opp}$ we, thus, have $\partial e^4_2 = \mathcal{L}_j \cup \mathcal{L}_j^{opp}$, where $\mathcal{L}_j \cong D^2 \times S^1$ lies in $\mathcal{H}_\diamondsuit$ and $\mathcal{L}_j^{opp} \cong S^1 \times D^2$ is the opposite torus in $\mathcal{H}_\circ$ by which $\mathcal{L}_j$ is replaced in the process of surgery. From our discussion in Section 1.5 of a 2-surgery passing through 1-handles, it is apparent that $\mathcal{L}_j^{opp}$ is a torus surrounding the j-th cut away handle in $\mathcal{H}_\circ$. The tori are depicted in Fig. 1.23.

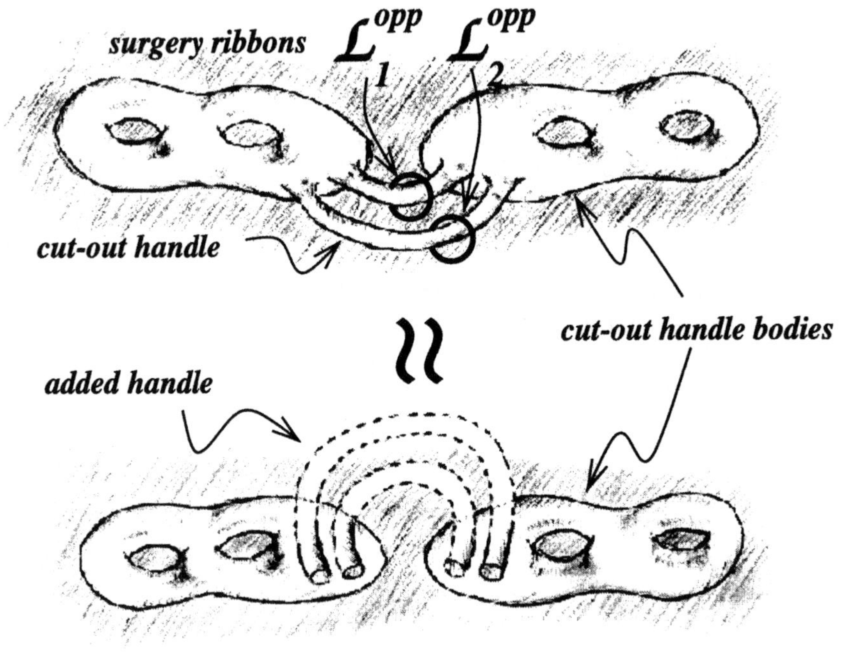

Fig. 1.23. From cut out to added handles via surgery along $\mathcal{L}_j^{opp}$

In Section 1.5 we showed that the effect of surgery along the annuli $\mathcal{L}_j^{opp}$, which could be substituted by an index-1 surgery, is to fill up the cut away cylindrical piece and add a 1-handle at the opposite side of the bounding surface. Thus we do obtain the homeomorphism $(\mathcal{H}_\circ)_{\mathcal{L}_1^{opp}\sqcup\ldots\sqcup\mathcal{L}_b^{opp}} \cong \mathcal{H}_\diamondsuit$ as expected. The handle structure of the opposite cobordism is thus:

$$P^{opp} = \underbrace{(e_2^4 \cup \ldots \cup e_2^4)}_{b} \cup_{(\mathcal{L}_1^{opp}\times 0)\sqcup\ldots\sqcup(\mathcal{L}_b^{opp}\times 0)} (\mathcal{H}_\circ \times [0,\varepsilon]). \qquad (1.6.27)$$

o-3) The M_o-Pieces :

As for the $\mathcal{H}$-piece, this part is obtained by performing index-2 surgery along the tori $\mathcal{Y}_j$ in the M_o-piece of $\partial(W_2 \diamondsuit_h W_1)$. Since they are parallelly contained in the tori $\mathcal{T}_j$, it is enough to study the effects of surgery in a vicinity of every $\mathcal{T}_j$. As explained in Section 1.5, in this situation the surgered full torus is homeomorphic to another full torus, but with S^1 and D^2 factors exchanged. Specifically, we have a homeomorphism

$$(\mathcal{T}_j)_{\mathcal{Y}_j} \xrightarrow{\sim} S^1 \times [0,\delta] \times [p_0, p_1],$$

which restricts to the cylinders $\mathcal{Z}_{j_s}, \mathcal{Z}_{j_s} \subset \partial\mathcal{T}_j$, defined at the end of Part $\diamondsuit$-*3)* from Section 1.6.3H1, as

$$\mathcal{Z}_{j_s} = S^1 \times [p_0,p_1] \xrightarrow{\sim} S^1 \times 0 \times [p_0,p_1], \quad \mathcal{Z}_{j_t} = S^1 \times [p_2,p_3] \xrightarrow{\sim} S^1 \times \delta \times [p_0,p_1],$$

with the obvious factorwise identifications, and a monotonous isomorphism $[p_0,p_1] \cong [p_3,p_2]$. The surgery along the $\mathcal{Y}_j$ on the manifold in the presentation (1.6.22), thus, allows us to give the M_o-piece of $\partial(W_2 \circ_h W_1)$ as a gluing of the $M_{\frac{1}{2},j}$ along their common cylinders:

$$\begin{aligned}
&\Big(M_{1,o} \cup \underbrace{e_1^3 \cup \ldots \cup e_1^3}_{2b} \cup M_{2,o}\Big)_{\mathcal{Y}_1,\ldots,\mathcal{Y}_b} \\
&\qquad \cong M_{\frac{1}{2},1} \cup_{\mathcal{Z}_{1_t}\ldots\mathcal{Z}_{b_t}} b(S^1 \times [0,\delta] \times [p_0,p_1]) \cup_{\mathcal{Z}_{1_s}\ldots\mathcal{Z}_{b_s}} M_{\frac{1}{2},2} \\
&\qquad\qquad \cong M_{\frac{1}{2},1} \cup_{\mathcal{Z}_{1_t}\sim\mathcal{Z}_{1_s},\ldots,\mathcal{Z}_{b_t}\sim\mathcal{Z}_{b_s}} M_{\frac{1}{2},2} \\
&\cong \mathcal{E}^s_{fill}(M_2) \circ_h \mathcal{E}^t_{fill}(M_1) \cong \mathcal{E}^s_{fill} \circ \mathcal{E}^t_{fill}(M_2 \circ_h M_1) \cong \phi(M_2 \circ_h M_1). \qquad (1.6.28)
\end{aligned}$$

Here, the first identity is obtained by inserting the form of $(\mathcal{T}_j)_{\mathcal{Y}_j}$ for each j. We arrive at the next line by shrinking the intermediate cylinder, with $\delta \to 0$, so that the corresponding cylindrical boundary pieces of $M_{\frac{1}{2},1}$ and $M_{\frac{1}{2},2}$ can be thought of as being glued directly onto each other. This is, obviously, just the definition of the horizontal composition $M_{\frac{1}{2},2} \circ_h M_{\frac{1}{2},1}$ of relative cobordisms in ***Cob*** as defined in Section 1.2. We continue to use the presentation of the $M_{\frac{1}{2},j}$ in terms of the partial fill functors, their functoriality, and the definition in (1.6.8).

Finally, let us also denote the relative 4-cobordism given by the b last e_2^4-attachments on the M_o-side of the composite of the 4-manifolds as

$$R = (M_{2,o} \diamondsuit_h M_{1,o}) \times [0,\varepsilon] \cup_{\mathcal{Y}_1\times\varepsilon,\ldots,\mathcal{Y}_b\times\varepsilon} \underbrace{(e_2^4 \cup \ldots \cup e_2^4)}_{b} :$$

$$M_{2,o} \diamondsuit_h M_{1,o} \longrightarrow \phi(M_2 \circ_h M_1), \qquad (1.6.29)$$

where we use the notation

$$M_{2,o}\Diamond_h M_{1,o} = M_{1,o} \cup \underbrace{\mathrm{e}_1^3 \cup \ldots \cup \mathrm{e}_1^3}_{2b} \cup M_{2,o}.$$

1.6.3.H3) Factoring the Horizontal Composition:

The construction of $\circ_h$ for the four-manifolds allows us now to define a binary operation on the 2-arrow sets of $\widetilde{\mathbf{COB}}$ with compatible vertical 1-arrows. We suppress in notation the homeomorphism, and write the operation for 2-arrows Q_1 and Q_2 from (1.6.14) as

$$Q_2 \circ_h Q_1 = \big(W_2 \circ_h W_1, \phi(M_2 \circ_h M_1), \mathbf{b}_2^t \sqcup \mathbf{b}_1^s\big). \qquad (1.6.30)$$

Our observations from the previous section are now summarized as follows.

Lemma 1.6.4. *The composition* $\circ_h$ *of manifolds defined in* (1.6.30) *yields again a tuple of manifolds as characterized in Section 1.6.2) and thereby a horizontal binary operation in the class* $\widetilde{\mathbf{COB}}$ *compatible with the 1-arrow compositions.*

Proof. The $\Sigma \times [f_0, f_1]$-pieces of $\partial(W_2 \circ_h W_1)$ have been identified with

$$\Sigma_{g_{1,tg}+g_{2,tg}+b-1} \times [f_0, f_1] \text{ and } \Sigma_{g_{1,sc}+g_{2,sc}+b-1} \times [f_0, f_1]$$

in Part $\Diamond - 1$ above. These are precisely the surfaces associated to the composites of the corresponding horizontal 1-arrows, $[g_2, b/c] \circ [g_1, a/b] = [g_2 + g_1 + b - 1, a/c]$, as defined in (1.1.2). The identification in (1.6.25) also shows that the $\mathcal{H}$-piece of $\partial(W_2 \circ_h W_1)$ is also the correct handlebody associated to these two composites of 1-arrows. Clearly, the 1-arrow structure of $\phi(M_2 \circ_h M_1)$ that was identified with the M_o-piece of $\partial(W_2 \circ_h W_1)$ is also given by the corresponding composite square. Putting all these parts together in the standard closure as defined in (1.6.6), we finally obtain

$$\langle \phi(M_2 \circ_h M_1) \rangle \quad \stackrel{\sim}{\rightarrow} \quad \partial(W_2 \circ_h W_1),$$

as required in the definition of Section 1.6.2. Now $\mathbf{b}_1^s$ ends precisely in the $[f_0, f_1]$-cylinders over the a source holes of the surfaces Σ_{g_1}, which are also the a source holes of the composed surfaces $\Sigma_{g_1+g_2+b-1}$. In the same way $\mathbf{b}_2^t$ is shown to bound the c target holes in $\Sigma_{g_1+g_2+b-1}$ as required in Section 1.6.2.

In the previous lemma only the 1-arrow structure of $\phi(M_2 \circ_h M_1)$ was relevant. The particular way in which it is constructed is summarized next more formally, using the map Δ from (1.6.7).

Lemma 1.6.5. *The map* $\Delta : \widetilde{\mathbf{COB}} \to \mathbf{Cob}$ *is functorial with respect to the horizontal composition* $\circ_h$*:*

$$\Delta(Q_2) \circ_h \Delta(Q_1) = \Delta(Q_2 \circ_h Q_1).$$

Proof. The only thing that we still need to realize here is that

$$M_2 \circ_h M_1 \quad = \quad \phi(M_2 \circ_h M_1) - (\mathbf{b}_2^t \sqcup \mathbf{b}_1^s).$$

It is, however, obvious from the constructions and definitions of the manifolds M_j, $M_{\frac{1}{2},j}$, and $M_{j,o}$, that the images of the $\mathbf{b}_2^t$ and $\mathbf{b}_1^s$ fill in precisely the target and source cylinders of $M_2 \circ_h M_1$. They also identify the source vertical 1-arrow of $M_2 \circ_h M_1$ with the source vertical 1-arrow α of M_1, and the target vertical 1-arrow of $M_2 \circ_h M_1$ with the target vertical 1-arrow γ of M_2.

Recall that the main purpose of constructing the operation $\circ_h$ on $\widetilde{COB}$ is to extend the horizontal product on Cob to a product on $\widetilde{Cob}$. With Lemma 1.6.5 the construction can now be done completely analogous to the one for the vertical composition. Using again the map $Sg : \widetilde{COB} \to \widetilde{Cob}$ from (1.6.11), we find:

Lemma 1.6.6. *The category $\widetilde{Cob}$ admits a horizontal composition that is defined by*

$$Sg(Q_2) \circ_h Sg(Q_1) = Sg(Q_2 \circ_h Q_1).$$

Proof. Since Sg is surjective by Lemma 1.6.1, the only thing to check is that if $Sg(Q_1') = Sg(Q_1)$ and $Sg(Q_2') = Sg(Q_2)$, then also $Sg(Q_2' \circ_h Q_1') = Sg(Q_2 \circ_h Q_1)$. As explained at the end of Section 1.6.2 we can think of Sg as the map that assigns to a 4-manifold its 4+1-cobordism class. Hence, we know that W_1 and W_1' differ precisely by some collection of 5-dimensional surgeries in their interior. However, the construction of the composite $W_2 \circ_h W_1$ leaves the interior of the manifolds unchanged and only adds handles along the boundaries of their union. Hence, $W_2' \circ_h W_1'$ is obtained from $W_2 \circ_h W_1$ by combining the surgeries on each 4-manifold. It follows that their signatures and, hence, their image under Sg are the same.

1.6.4 Double Category Properties of $\widetilde{Cob}$

It is trivial to verify that Cob satisfies the associativity axioms and the interchange law from Section B.1. Hence, it is a double category, since the involved compositions are given as the same gluings over boundary pieces, only in different orders. For $\widetilde{Cob}$ the associativity axioms are also evidently fulfilled. However, the interchange law is not obvious, since the vertical composition is defined by gluing of 4-manifolds whereas the horizontal composition is defined quite differently by handle attachments to corresponding 4-manifolds with corners. In fact, the different product orders in $\widetilde{COB}$ do in fact yield different 4-manifolds. However, the following lemma shows that they are related by a 4-dimensional surgery and, hence, yield the same 2-morphisms as equivalence classes in $\widetilde{Cob}$.

Lemma 1.6.7. *Let W_1, W_2, V_1, and V_2 be 4-manifolds of four 2-arrows that have the compatible 1-arrow structure as in the assumptions of the interchange law for double categories. Then $(W_2 \circ_v V_2) \circ_h (W_1 \circ_v V_1)$ can be obtained by doing b 4-dimensional index-2 surgeries on $(W_2 \circ_h W_1) \circ_v (V_2 \circ_h V_1)$.*

Equivalently, $(W_2 \circ_v V_2) \circ_h (W_1 \circ_v V_1)$ *is cobordant to* $(W_2 \circ_h W_1) \circ_v (V_2 \circ_h V_1)$*, where the 5-dimensional cobordism is obtained by attaching* b *handles* e_2^5 *to* $(W_2 \circ_h W_1) \circ_v (V_2 \circ_h V_1) \times [0, 1]$.

Proof. We need to show that an index-2 surgery on $(W_2 \circ_h W_1) \circ_v (V_2 \circ_h V_1)$ yields the opposite product order as in the interchange law. To this end let us present this product in a way where the identifications for the individual vertical compositions have almost been made with the exception of the cylinders over the respective intermediate discs in the surfaces. Thus, as a modification of the vertical product for the 4-manifolds defined in equation 1.6.12 of Section 1.6.3.V, let us introduce the following product

$$W \ominus_v V = W \bigsqcup_{\Sigma^*_{g_{int}} \times [f_0, f_1]} V. \tag{1.6.31}$$

Here $\Sigma^*_{g_{int}} = \Sigma_{g_{int}} - \sqcup^b \mathring{D}^2$, where we removed the b open source discs from the intermediate surfaces of the products for the product $W_2 \ominus_v V_2$ and the corresponding b target discs from the intermediate surfaces for $W_1 \ominus_v V_1$. Two corresponding discs in an intermediate surface are still glued together along their S^1-boundary, thus, combining to a sphere S^2. Hence, in the product we obtain additional boundary pieces, namely the cylinders $S^2 \times [f_0, f_1]$ over these spheres. In particular, we have

$$W_j \circ_v V_j = W_j \ominus_v V_j \bigsqcup_{\sqcup^b S^2 \times [f_0, f_1]} D^3 \times [f_0, f_1]. \tag{1.6.32}$$

The product order in the interchange law, in which the horizontal compositions are carried out first, can now be presented also by making the vertical gluing over the first $\Sigma^*_{g_{int}}$ and then adding all the 1-handles and 2-handles for the two horizontal composition. With $2b$ handles of each type added, in each composition we obtain the following formula:

$$\begin{aligned}(W_2 \circ_h W_1) \circ_v (V_2 \circ_h V_1) &= \\ &= (W_2 \ominus_v V_2) \sqcup (W_1 \ominus_v V_1) \cup \underbrace{\mathrm{e}_1^4 \cup \ldots \cup \mathrm{e}_1^4}_{4b} \cup \underbrace{\mathrm{e}_2^4 \cup \ldots \cup \mathrm{e}_2^4}_{4b} \\ &= Q \cup \underbrace{\mathrm{e}_1^4 \cup \ldots \cup \mathrm{e}_1^4}_{2b} \cup \underbrace{\mathrm{e}_2^4 \cup \ldots \cup \mathrm{e}_2^4}_{4b}.\end{aligned}$$

Note, here that not all of the $4b$ 1-handles $\mathrm{e}_1^4 \cong D^2 \times [p', p''] \times [f_0, f_1]$ are glued in the strict sense of a handle attachment, since the b source 1-handles of the $W_2 \circ_h W_1$ composition are glued in the subsequent vertical composition to the corresponding b target 1-handles of the $V_2 \circ_h V_1$ composition along the pieces $K = S^1 \times [p', p''] \times [f_0, f_1] \subset \partial \mathrm{e}_1^4$ in their boundaries. Above we have also denoted by Q the manifold $(W_2 \ominus_v V_2) \circ_h (W_1 \ominus_v V_1)$ to which we have already added these $2b$

1-handles that are attached to the intermediate surface. Each pair of these 1-handles, thus, combines in the vertical composition to

$$\mathrm{e}_1^4 \sqcup_K \mathrm{e}_1^4 \cong S^2 \times [p', p''] \times [f_0, f_1],$$

so that

$$Q = (W_2 \ominus_v V_2) \coprod_{\sqcup^b S^2 \times \{p''\} \times [f_0, f_1]} \sqcup^b S^2 \times [p', p''] \times [f_0, f_1] \coprod_{\sqcup^b S^2 \times \{p'\} \times [f_0, f_1]} (W_1 \ominus_v V_1).$$

Observe next that each $S^2 \times D^2_{\&} = S^2 \times [p', p''] \times [f_0 + \varepsilon, f_1 - \varepsilon]$ lies in the interior $\mathring{Q}$ of Q for some small enough $\varepsilon > 0$. We can use this to perform an index-2 surgery, which corresponds to the 5-dimensional 2-handle attachment described in Lemma 1.6.6. As described in Section 1.5.2 this means removing each $S^2 \times D^2_{\&}$-piece and replacing it by $D^3 \times S^1_{\&}$ along the common $S^2 \times S^1$-boundary. We denote the surgered manifold by

$$Q_{\&} = (Q - \sqcup^b S^2 \times D^2_{\&}) \cup_{\sqcup^b S^2 \times S^1} \sqcup^b D^3 \times S^1_{\&}.$$

Since the removed $S^2 \times D^2_{\&}$-pieces are part of the 1-handles added to $(W_2 \ominus_v V_2) \sqcup (W_1 \ominus_v V_1)$, it is clear that $Q_{\&}$ can also be described by adding only $\mathrm{e}_1^4 \sqcup_K \mathrm{e}_1^4 - S^2 \times D^2_{\&}$-pieces to begin with, as well as the $D^3 \times S^1_{\&}$-pieces. Note that since $D^2_{\&} = [p', p''] \times [f_0, f_1]$ we have that $S^1_{\&} = \{p', p''\} \times [f_0 + \varepsilon, f_1 - \varepsilon] \cup [p', p''] \times \{f_0 + \varepsilon, f_1 - \varepsilon\}$ so that the total piece added to $(W_2 \ominus_v V_2) \sqcup (W_1 \ominus_v V_1)$ can be written as

$$\begin{aligned}
&(\mathrm{e}_1^4 \sqcup_K \mathrm{e}_1^4 - S^2 \times D^2_{\&}) \cup D^3 \times S^1_{\&} = \\
&\quad = S^2 \times [p', p''] \times [f_0, f_0 + \varepsilon] \cup D^3 \times \{p', p''\} \times [f_0 + \varepsilon, f_1 - \varepsilon] \cup \\
&\qquad \cup D^3 \times [p', p''] \times \{f_0 + \varepsilon, f_1 - \varepsilon\} \cup S^2 \times [p', p''] \times [f_1 - \varepsilon, f_1] \\
&\quad = D^3_{0,\varepsilon} \times [p', p''] \cup D^3 \times \{p', p''\} \times [f_0 + \varepsilon, f_1 - \varepsilon] \cup D^3_{1,\varepsilon} \times [p', p''],
\end{aligned}$$

where we denoted
$D^3_{0,\varepsilon} = S^2 \times [f_0, f_0 + \varepsilon] \cup_{S^2 \times \{f_0 + \varepsilon\}} D^3 \times \{f_0 + \varepsilon\}$
and $D^3_{1,\varepsilon} = S^2 \times [f_1 - \varepsilon, f_1] \cup_{S^2 \times \{f_1 - \varepsilon\}} D^3 \times \{f_1 - \varepsilon\}$.

The $D^3 \times \{p', p''\} \times [f_0 + \varepsilon, f_1 - \varepsilon]$-pieces in the above formula are glued into $(W_2 \ominus_v V_2) \sqcup (W_1 \ominus_v V_1)$ as subsets of the corresponding pieces in (1.6.32). We have, for example,

$$\begin{aligned}
(W_1 \ominus_v V_1) &\coprod_{\sim} \sqcup^b D^3 \times \{p'\} \times [f_0 + \varepsilon, f_1 - \varepsilon] \\
&= W_1 \circ_v V_1 - \sqcup^b (D^3 \times [f_0, f_0 + \varepsilon] \cup D^3 \times [f_1 - \varepsilon, f_1]) \\
&\cong W_1 \circ_v V_1.
\end{aligned}$$

The last homeomorphism results from the fact that a 4-ball $D^3 \times [f_0, f_0 + \varepsilon] \subset W_1 \circ_v V_1$ is removed from the collar with $D^3 \times \{f_0\} \subset \partial(W_1 \circ_v V_1)$ so that the gap caused by the removal can be "pushed out". In the course of this the boundary pieces $D^3_{0,\varepsilon}$ are pushed into the positions of the D^3's in $M_1 \circ N_1 \subset \partial(W_1 \circ_v V_1)$ that are obtained by removing a 3-dimensional neighborhood around the holes in the intermediate surface of composition of 3-dimensional cobordisms.

Note that $Q_{\&}$ is now presented by adding the remaining $D^3_{j,\varepsilon} \times [p', p'']$ with $j = 0, 1$ to $(W_2 \circ_v V_2) \sqcup (W_1 \circ_v V_1) - \sqcup^{2b}(D^3 \times [f_0, f_0 + \varepsilon] \cup D^3 \times [f_1 - \varepsilon, f_1])$. With the described "push out" homeomorphism this presentation then becomes an ordinary 1-handle attachment of ${}^*\mathrm{e}^4_1 = D^3_{j,0} \times [p', p'']$ to $(W_2 \circ_v V_2) \sqcup (W_1 \circ_v V_1)$ along the D^3's in the boundary around the holes of the intermediate surface. We, thus, obtain the following homeomorphism, which is canonical up to isotopy:

$$Q_{\&} \cong (W_2 \circ_v V_2) \cup \underbrace{{}^*\mathrm{e}^4_1 \cup \ldots \cup {}^*\mathrm{e}^4_1}_{2b} \cup (W_1 \circ_v V_1).$$

Below the mechanics of the surgery on the interior of Q is depicted schematically again only for the simplest case $b = 1$. The shaded area on the left corresponds to the $S^2 \times D^2_{\&} \subset \overset{\circ}{Q}$ piece that is removed, and the shaded area on the right indicates the replaced $D^3 \times S^1_{\&}$-part. The b^M_1 denotes, as before, the full cylinder $D^2 \times [p_0, p_1]$ in $M_{1,o}$ and b^M_2 – the corresponding piece in $M_{2,o}$. In the horizontal gluing they combine to $\mathrm{b}^M_{1,2}$, which runs transversally through the 1-handle ${}^*\mathrm{e}^4_1$. Analogously, for the thickened strands $\mathrm{b}^N_{1,2}$.

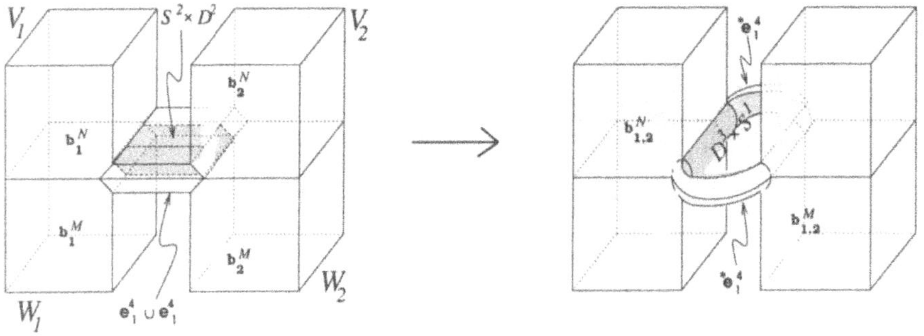

Since Q is a part of $(W_2 \circ_h W_1) \circ_v (V_2 \circ_h V_1)$, the surgery on Q also extends to this total product of the interchange law in the way implied by the statements in Lemma 1.6.6. The general form of the surgered product manifold is now:

$$\begin{aligned}&\big((W_2 \circ_h W_1) \circ_v (V_2 \circ_h V_1)\big)_{\&}\\ &= (W_2 \circ_v V_2) \sqcup (W_1 \circ_v V_1) \cup \underbrace{{}^*\mathrm{e}^4_1 \cup \ldots \cup {}^*\mathrm{e}^4_1}_{2b} \cup \underbrace{\mathrm{e}^4_2 \cup \ldots \cup \mathrm{e}^4_2}_{4b} \cup \underbrace{\mathrm{e}^4_1 \cup \ldots \cup \mathrm{e}^4_1}_{2b}.\end{aligned}$$

In the remainder of this proof we show that $2b$ of the e^4_2-handles can be cancelled against the $2b$ new ${}^*\mathrm{e}^4_1$-handles, such that the remaining $2b$ of the e^4_2-handles and $2b$ of the e^4_1-handles are attached precisely in the way prescribed by the definition of the horizontal composition. The attachments of these handles are, using the terminology of Section 1.6.3.H, all either in the "M_o-pieces" and the "$\mathcal{H}$-pieces" of $\partial(W_2 \circ_v V_2)$ and $\partial(W_1 \circ_v V_1)$.

For the "M_o-pieces" the gluing of a $^*e_1^4$-handle results in an index-1 surgery on $(M_2 \circ_v N_2) \sqcup (M_1 \circ_v N_1)$. Each of the b pairs of surgery balls S' and S'' is given as a neighborhood of corresponding source or target discs $D^2 \subset \Sigma_{g_{int}} \subset interior(M_j \circ_v N_j)$, in the intermediate surfaces $\Sigma_{g_{int}}$ over which we do the vertical gluing. The attaching tori for the e_2^4-handles on the "M_o-side" for either horizontal composition run transversally through the corresponding $^*e_1^4$-handle. In the surgery diagram on $(M_2 \circ_v N_2) \sqcup (M_1 \circ_v N_1) \cup \underbrace{e_1^4 \cup \ldots \cup e_1^4}_{2b}$ (where we have added the 1-handles at the total source and target of the composition) the attaching data for a e_2^4-handle of the $(M_2 \circ_h M_1)$ appears as a surgery ribbon $\mathbf{b}^M$, which enters transversally one surgery sphere $S' \subset M_1 \circ_v N_1$ and emerges at the partner sphere $S'' \subset M_2 \circ_v N_2$ and, furthermore, passes through the e_1^4-handle at the very source. In the picture below its pieces in M_2 and M_1 are denoted by $\mathbf{b}_2^M$ and $\mathbf{b}_1^M$. Analogously, we have a ribbon $\mathbf{b}^N$ for each of the b e_2^4-handles in the $(N_2 \circ_h N_1)$-composition, which runs over the same $^*e_1^4$-handle and, hence, through the same pair of surgery spheres S' and S'' as the $\mathbf{b}^M$-ribbon with the same labels. The surgery diagram is summarized on the right side of the next figure:

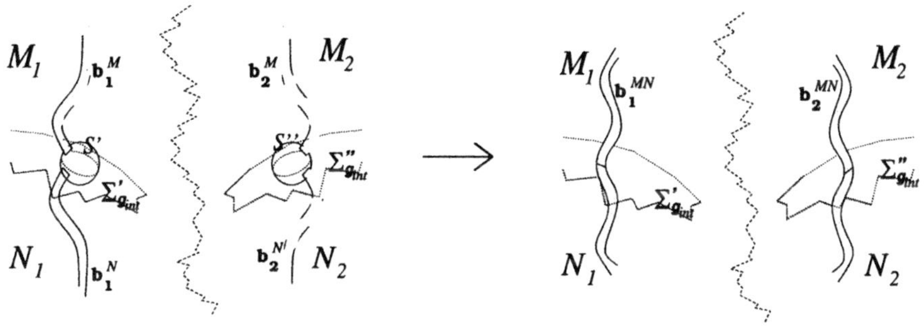

To this configuration we can apply an isotopy that moves one surgery sphere S' along the ribbon piece $\mathbf{b}^M$ through the added 1-handle from M to N until it is right next to its partner surgery sphere S''. In this situation apply the cancellation move as described in Figure 1.15 of Section 1.5.2. As a result, the ribbons $\mathbf{b}^M$ and $\mathbf{b}^N$ are replaced by their connected sums, and the surgery spheres disappear. Thus, instead of the intermediate attachments of the $^*e_1^4$-handles and the $2b$ attachments of the e_2^4-handles we can consider $(M_2 \circ_v N_2) \sqcup (M_1 \circ_v N_1)$ with attaching data for b e_2^4-handles given by joining up the strands $\mathbf{b}_j^M$ with the respective ones $\mathbf{b}_j^N$ at the disc locations of the intermediate surface in the interiors of the vertical composites. This is, however, exactly the "M_o-part" of the attaching prescription for the horizontal composition $(W_2 \circ_v V_2) \circ_h (W_1 \circ_v V_1)$.

For the "$\mathcal{H}$-part" the boundary piece of $(W_2 \circ_v V_2) \sqcup (W_1 \circ_v V_1)$ is given by the union of handlebodies $\mathcal{H}_{g_{1,sc}}^{g_{1,tg}} = \mathcal{H}_{g_{1,sc}}^{g_{1,int}} \circ_v \mathcal{H}_{g_{1,int}}^{g_{1,tg}}$ and $\mathcal{H}_{g_{2,sc}}^{g_{2,tg}} = \mathcal{H}_{g_{2,sc}}^{g_{2,int}} \circ_v \mathcal{H}_{g_{2,int}}^{g_{2,tg}}$ as in (1.6.13). As before, the addition of the b intermediate ${}^{*}e_1^4$-handles on the $\mathcal{H}$-side results an index-1 surgery along a pair of surgery balls S_j' and S_j'' on this combined handlebody, located at the intermediate surfaces in the interior $\Sigma_{g_{int}} \subset \overset{o}{\mathcal{H}}{}^{g_{tg}}_{g_{sc}}$. From the $(W_2 \circ_h W_1)$-composition we also have the e_1^3-handles attached to the $\Sigma_{g_{sc}}$ boundary pieces and running through these surgery ribbons $\mathcal{L}_1^M, \dots, \mathcal{L}_b^M$. Since the original prescription for the horizontal composition was that the ribbons $\mathcal{L}_j^M$ should close through the *first* handle of $\Sigma_{g_{int}}$ they will all run in the composite through the first pair of surgery balls S_1' and S_1''.

Similarly, the surgery ribbons $\mathcal{L}_1^N, \dots, \mathcal{L}_b^N$ for the $(V_2 \circ_h V_1)$-composition run through the e_1^3-handles that are attached to $\Sigma_{g_{int}}$. Hence, in the total composition every one of them runs through another pair of surgery balls S_j' and S_j''. In summary, we obtain a surgery diagram on $\mathcal{H}^{g_{1,tg}}_{g_{1,sc}} \sqcup \mathcal{H}^{g_{2,tg}}_{g_{2,sc}} \cup \underbrace{e_1^3 \cup \dots \cup e_1^3}_{2b}$, where the extra e_1^3-handles are those attached to $\Sigma_{g_{sc}}$ and $\Sigma_{g_{tg}}$. It is depicted in the following figure.

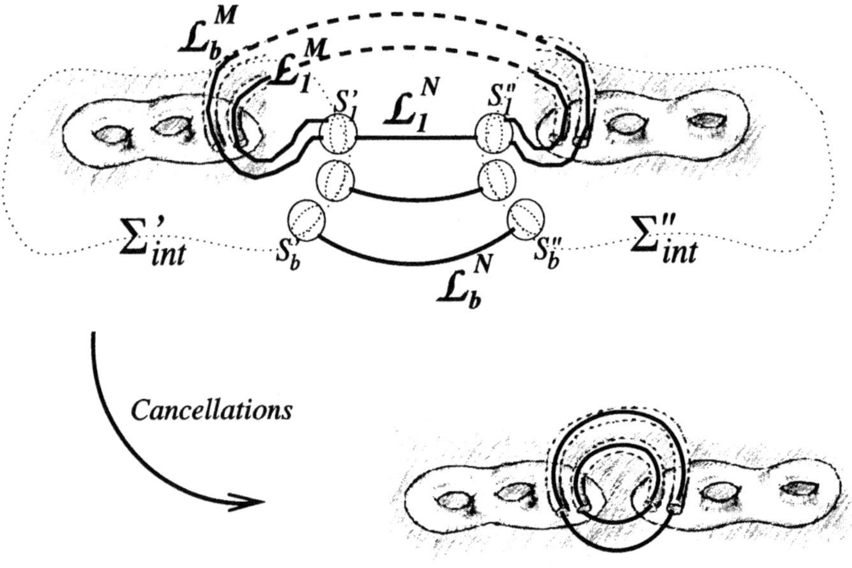

From this picture it is now easy to see that every pair of surgery balls S_j' and S_j'' can be cancelled against a ribbon $\mathcal{L}_j^N$ as described in Section 1.5.2. The resulting picture is then precisely the one for the horizontal composition $(W_2 \circ_v V_2) \circ_h (W_1 \circ_v V_1)$ with the $\mathcal{H}^{g_{tg}}_{g_{sc}}$ handlebodies.

Hence, $(W_2 \circ_v V_2) \circ_h (W_1 \circ_v V_1)$ and $((W_2 \circ_h W_1) \circ_v (V_2 \circ_h V_1))_{\&}$ are homeomorphic as 4-manifolds with corners. This completes the proof.

We obtain now the main result of this chapter by combining the results from Lemma 1.6.3, Lemma 1.6.6, Lemma 1.6.7, the identification in (1.6.29) and the definitions of the maps between 2-morphisms from Section 1.6.2.

Theorem 1.6.8. *The horizontal and vertical composition defined above make* $\widetilde{\mathbf{Cob}}$ *into a strict double category.*

The map $\pi_C : \widetilde{\mathbf{Cob}} \longrightarrow \mathbf{Cob}$ *from* (1.6.10) *is a strict double functor of double categories.*

2. Tangle-Categories and Presentation of Cobordisms

All known rigorous constructions of quantum invariants of 3-manifolds and TQFT's proceed by, first, presenting a given manifold in a finite combinatorial way, and then applying algebraic functors to this combinatorial data. The most basic example of such is a presentation of a manifold as a simplicial complex, taken modulo so called Alexander or Pachner subdivision moves. It is not hard to imagine that these types of presentation become quite complicated when considering cobordisms between closed surfaces and even more cumbersome when we also want to describe relative cobordisms with corners. Among other things, the latter would imply that we simplicially encode the framed braid groups.

The types of presentations we will, therefore, exclusively use in all of our constructions are ones, which arise from what is called *surgery* on manifolds. It was known already to Lickorish and Wallace, (see [Lic62] and [Wal60]), that any 3-manifold without boundary can be obtained by doing surgery along a framed link, $\mathcal{L} \subset S^3$, in the three-sphere. Later, Kirby [Kir78] solved the question, when surgery along two different links results in the same 3-manifold. He extracted two types of modifications of links, called the $\mathcal{O}_1$*-Move* and the $\mathcal{O}_2$*-Move*, that do not change the associated 3-manifold, and, using Cerf-theory, he showed that any two links, that yield the same manifold, could be obtained one from the other by applying a succession of these moves.

The described surgery presentations will have to be substantially modified and extended for our purposes. For example Kirby Calculus only apply to closed manifold as surgery on the simply-connected space S^3, but we need to describe manifolds with boundaries and corners obtained via surgery on non-simply-connected handlebodies. And not only that: besides the presentations for the relative cobordisms, we also must find simple and efficient rules that translate the two types of compositions in $\widetilde{\mathcal{C}ob}$, as defined in Chapter 1, into operations on the level of presentations.

Another necessary modification of Kirby's calculus that we need for our purposes is to make the moves local. This is because the construction of TQFT's proceeds by assigning algebraic data to local pictures. Hence, for the verification that algebraic relations indeed yield topological invariance it is both convenient and instructive to have local, elementary moves.

For cobordisms between closed surfaces the type of surgery presentations, that fit best for our needs, has already been developed in [Ker99]. Instead of a link $\mathcal{L} \subset S^3$ we now consider tangles $\mathcal{T} \subset \mathbb{R}^2 \times [-1, 1]$, or rather their projections into the strip $\mathbb{R} \times [-1,1]$. In this framework the composition of cobordisms over surfaces translates into simply stacking tangles on top of each other. Also, the Kirby moves are replaced there by a set of local moves. They arise from the "bridged link" calculus developed in [Ker99], complemented by additional moves at the boundaries of $\mathbb{R}^2 \times [-1,1]$. In summary, cobordisms in [Ker99] are presented as a functor from the ordinary category $\widetilde{\mathcal{C}ob}(0,0)$ into a special subcategory of a natural tangle category, similar to the ones introduced in [FY92], but taken modulo additional equivalence relations given by the moves.

The restriction to invertible tangle classes yields a representation of the mapping class group of a corresponding closed surface. If we use manifolds with corners we also obtain analogous presentations for surfaces with boundaries. For a surface with one boundary component we reproduce the tangle presentation of Matveev and Polyak [MP94], which was obtained via Wajnryb's presentations of the mapping class groups.

Following the same principles, we want to represent the entire double category $\widetilde{\mathcal{C}ob}$ by constructing an invertible functor on it that maps it to a combinatorial double category. The purpose of this chapter is to define and describe the latter. Its objects and 1-arrows are as those of $\widetilde{\mathcal{C}ob}$, but the 2-morphisms are now generalizations of the previous tangle spaces with additional elements.

A vertical composition is defined as before by stacking tangle diagrams that represent given classes on top of each other. The construction of a horizontal composition starts with the juxtaposition of representing tangles but is then followed by two further operations on the tangle diagram. The two compositions will turn out to be compatible with each other and, thus, define a tangle double category.

Summary of Content

In Sections 2.1 and 2.2 we develop the notions and conventions needed to define the admissible tangles that we shall, eventually, use to represent cobordisms. This included specifying the types of strands and coupons that can occur in a diagram, the possible local pictures, the 1-arrows that define the boundaries of a diagram, and the allowed global properties of strands. From this we define in Section 2.2 the 2-arrow sets of the double category $\mathcal{T}gl$ as equivalence classes of admissible tangle diagrams. The equivalence notion will be defined by introducing a number of elementary *moves*, which serve as a set of generators for all other equivalences.

The purpose of Section 2.4 is to show that the classes of planar tangle projections in $\mathcal{T}gl$ are in natural one-to-one correspondence with classes of tangles that are embeddings in three dimensional space. It is carefully examined with the help of transversality methods how the process of projecting tangles in the plane gives rise to the so called TI-Moves of $\mathcal{T}gl$. In this section we also relate another group of moves, the TD-Moves of $\mathcal{T}gl$, to the addition of auxiliary strands that have no relevance for a surgery presentation of a cobordism but are useful in the construction of the TQFT functor. In the course of this and the following sections we introduce a variety of tangle classes for different types of tangles and different sets of moves.

For example, in Section 2.5 we also introduce the formulations that correspond to the surgery calculi of Kirby and Fenn Rourke as well as the "bridged link calculus". Moving between various types of tangles allows to pass gradually from the double category of tangles equivalent to the category of cobordisms to an equivalent double category of tangles, in terms of which TQFT's are easy to construct. We prefer to do it in many steps, changing the classes of tangles and sets of moves gradually, rather than to do it in one theorem, whose proof would occupy dozens of pages. Instead of a more detailed survey we summarize different although equivalent tangle categories in the table at the end of this introduction.

In Section 2.6 we introduce vertical and horizontal compositions $\circ_v$ and $\circ_h$ for the tangle categories explicitly. It is shown that corresponding combinatorial binary operations on the level of admissible tangles do indeed factor into equivalence classes of tangles. Moreover these compositions fulfill an interchange law so that $(\mathcal{Tgl}, \circ_v, \circ_h)$ does indeed form a double category as desired. In the remainder of this chapter we consider special forms and decompositions of tangles that are of technical relevance for the horizontal composition. We also explain further how the structures of braid groups of surfaces and braided tensor categories are naturally represented within the double category $\mathcal{Tgl}$.

N.B.: In the table on the next page the 2-arrow sets of all the listed tangle categories are equivalent, in the sense that there is natural bijection between the respective classes of tangles, which then also extends to a double isomorphism between double categories.

For each category we characterize the tangles that we consider by the space they are embedded or projected into, by the different types of ribbon strands and other components that constitute the tangle according to their local and global properties, and by the constraints on the allowed embeddings or projections. On these sets an equivalence relation is then generated by the listed moves. For convenience we also provide on the next page the reference in Chapter 2, where the complete definition of a given category is given and where its equivalence to all the other categories is asserted.

2.1 Local Ingredients of Tangle-Diagrams and Horizontal 1-Arrows

In this section we shall compile a list of all local ingredients needed to make up a tangle diagram, i.e., the projection of a ribbon tangle in $\mathbb{R}^2 \times [-1, 1]$. The list will include several more types of ribbons as well as so called coupons, which, in an other context, is called the "Kirby dotted circle".

2.1.1 Horizontal 1-Arrows and Intervals on $\mathbb{R} \times \{\pm 1\}$

In a combinatorial terms the horizontal arrows are the same as the ones of $\widetilde{\mathcal{C}ob}$, namely triples of numbers, $[g, a/b]$, with $g, a, b \in \mathbb{N} \cup \{0\}$. The composition is formally defined for $b \geqslant 1$ as in Section 1.1 by

Category	Space	Compo-nents	Constraints	Moves	Reference
$\mathcal{T}\boldsymbol{gl}$	$\mathbb{R}_x \times [-1,1]$	Strands: external internal: (top, bot, thru, clos) auxiliary: Coupons	Strands: in gen. pos. w.r.t. projection height fct. Coupons&joints: well positioned projected	TI1 - TI11, TD1 - TD5, TS1 - TS4	Section 2.3.4
$\mathcal{T}\boldsymbol{gl}_{\mathbb{R}^2}$	$\mathbb{R}^2 \times [-1,1]$	As for $\mathcal{T}\boldsymbol{gl}$	All embeddings	Isotopy TD1 - TD5, TS1 - TS4	Section 2.4.1 Thm. 2.4.8
$\mathcal{T}\boldsymbol{gl}_{\mathbb{R}^2}^{well-pos}$	$\mathbb{R}^2 \times [-1,1]$	As for $\mathcal{T}\boldsymbol{gl}$	Coupons&Joints well positioned	Isotopy, TI6, TI11, TD1 - TD5 TS1 - TS4	Section 2.4.1 Lemma 2.4.3
$\mathcal{T}\boldsymbol{gl}_{\mathbb{R}^2}^{dec-proj}$	$\mathbb{R}^2 \times [-1,1]$	As for $\mathcal{T}\boldsymbol{gl}$	Coupons&Joints projectable & well positioned	Isotopy,TI6, TI7, TI11, TD1 - TD5 TS1 - TS4	Section 2.4.1 Lemma 2.4.5
$\mathcal{T}\boldsymbol{gl}_{\mathbb{R}^2}^{plan}$	$\mathbb{R}^2 \times [-1,1]$	As for $\mathcal{T}\boldsymbol{gl}$	Strands: in gen. pos. w.r.t. projection Coupons&Joints: projectable & well positioned	Isotopy, TI1 - TI3, TI6, TI7, TI10, TI11, TD1 - TD5 TS1 - TS4	Section 2.4.1 Lemma 2.4.6
$\mathcal{T}\boldsymbol{gl}_{S^2}$	$S^2 \times [-1,1]$	As for $\mathcal{T}\boldsymbol{gl}$	All embeddings	Isotopy, TD1 - TD5, TS1 - TS3	Section 2.4.2 Lemma 2.4.9
$\mathcal{T}\boldsymbol{gl}_{S^2}^{\mathbf{nx}}$	$S^2 \times [-1,1]$	As for $\mathcal{T}\boldsymbol{gl}$ *without* the auxiliary strands	All embeddings	Isotopy, TD1, TD2, TS1 - TS3	Section 2.4.3 Lemma 2.4.10
$\mathcal{T}\boldsymbol{gl}_{S^2}^{\mathbf{s};*}$	$S^2 \times [-1,1]$	Strands: external internal: (top, thru) Coupons	All embeddings	Isotopy, TD1, TD2, TD3*, TD4*, TD5*, TS1*, TS2*, TS3*	Section 2.4.3 Lemma 2.4.11
$\mathcal{T}\boldsymbol{gl}_{S^2}^{\mathbf{BL}}$	$S^2 \times [-1,1]$	Strands: external internal: (top, bot thru, clos) Ball-Pairs	All embeddings	Isotopy, TD1♠, TD2♠, TS1♠, TS2♠, TS3♠	Section 2.5.1 Eqn. 2.5.1

Category	Space	Components	Constraints	Moves	Reference
$\mathcal{T}gl^{\mathrm{Ki}}_{S^2}$	$S^2 \times [-1,1]$	Strands: external internal: (top, bot thru, clos)	All embeddings	Isotopy, Hopf-Link, $\mathcal{O}_2$-Move, Ribbon-TS3	Section 2.5.2 Lemma 2.5.5
$\mathcal{T}gl^{\mathrm{FR}}_{S^2}$	$S^2 \times [-1,1]$	Strands: external internal: (top, bot thru, clos)	All embeddings	Isotopy, Signature, κ-Move, Ribbon-TS3	Section 2.5.2 Lemma 2.5.6
$\mathcal{T}gl^{tg1\to\infty}$	$\mathbb{R}_x \times [-1,1]$	Strands: external: last removed internal: (top, bot, thru, clos) auxiliary: Coupons	Strands: in gen. pos. w.r.t. projection height fct. Coupons&joints: well positioned projected	TI1 - TI11, TD1 - TD5, TS1, TS2, TS3	Section 2.7.1

$$[g_2, b/c] \circ [g_1, a/b] = [g_1 + g_2 + b - 1, a/c]. \tag{2.1.1}$$

Now we wish to describe a tangle diagram in $\mathbb{R}^2 \times [-1,1]$ that represents a 2-arrow between horizontal arrows $[g_{sc}, a/b]$ and $[g_{tg}, a/b]$ (see, e.g., the squares in Section 1.2). To this end we, first, have to specify $a + 2g_{sc} + b$ disjoint intervals on the upper boundary line $\mathbb{R} \times \{1\}$ and $a + 2g_{tg} + b$ disjoint intervals on the lower boundary line $\mathbb{R} \times \{-1\}$. Specifically, to each $[g, a/b]$ we associate a sequence of $a + 2g + b$ closed, connected, mutually disjoint intervals on the real line $\mathbb{R}$ denoted as follows:

$$J_1^s < J_2^s < \ldots < J_a^s < I_1^i < I_1^o < I_2^i < \ldots < I_g^o < J_b^t < J_{b-1}^t < \ldots < J_1^t. \tag{2.1.2}$$

The order $<$ on sets of disjoint intervals is the obvious one, i.e., all elements of the smaller interval are smaller than all elements of the greater interval. If one of the numbers a, b, or g is zero, the respective type of intervals will simply not occur. Besides the notation we use the following terms for the different types of intervals:

The $a + b$ J-intervals are called *external intervals*, of which the first a, namely J_j^s with $j = 1, \ldots, a$, are called *initial* or *left* intervals, and the other b, namely J_j^t with $j = 1, \ldots, b$, the *final* or *right* intervals.

The $2g$ I-intervals will be the *internal* intervals, and the ones labeled I_j^i with $j = 1, \ldots, g$ are the *in-going* ones and those denoted I_j^o are called the *out-going* intervals.

Additionally, at the *top* piece, $\mathbb{R} \times \{+1\}$, of the boundary we shall introduce an infinite sequence of *auxiliary* intervals, $\{K_j\}_{j=1}^{\infty}$, that is fit in between I_g^o and J_b^t, i.e.,

$$I_g^o < K_1 < K_2 < \ldots < J_b^t \,. \tag{2.1.3}$$

In conclusion, we summarize the interval configuration on the real line $\mathbb{R}$ that is associated to a 1-arrow, $[g, a/b]$, in the following picture:

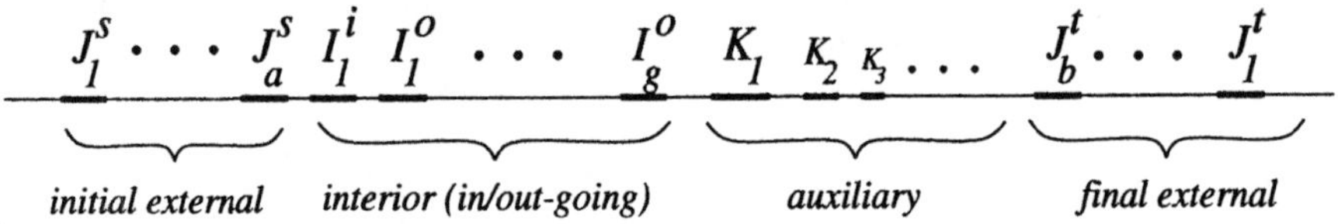

2.1.2 Types of Elements

A tangle is an embedding of a 1-dimensional manifold S (with boundary) into $\mathbb{R}^2 \times [0,1]$. However, for the purpose of doing surgery we also need to pick a framing of the normal bundle of the so embedded submanifold. This is up to isotopy equivalent to embedding strips or ribbons instead of just threads, or, more formally, embedding $S \times [-\varepsilon, \varepsilon]$ into $\mathbb{R}^2 \times [0,1]$ so that the image of $\{p\} \times [-\varepsilon, \varepsilon]$ specifies a section in the normal bundle at the image point of $p \in S$ and, hence, a framing class. The ribbons that are related to surgery will be called *internal* ribbons.

In addition, we also need to describe the tube-like boundary pieces of the relative cobordisms with corners, as well as represent the cut and paste procedure involved in surgery in a more controlled way for the algebraic insertions. As a result, we distinguish two further ribbon types in a tangle diagram, which are called *external* and *auxiliary* ribbons, respectively. Finally, we also introduce the coupons that will stand for local 1-handle surgery besides the 2-handle surgery along ribbons.

The four types of ingredients, i.e., three ribbon types and coupons, are listed next, together with the convention for coloring that we will use henceforth:

1) Internal Ribbon : 

3) Auxiliary Ribbon : 

2) External Ribbon : 

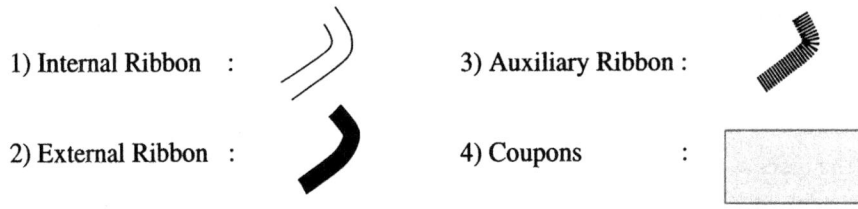

4) Coupons : 

2.1.3 Local Pictures of Elementary Slices

The next set of constraints we impose on a tangle diagram concerns the genericity of the tangle projection and the possible joints between different elements.

We say that the projection of a tangle into $\mathbb{R} \times [0,1]$ is *generic*, if the interval $[0,1]$ can be subdivided into consecutive intervals $[0,t_1], [t_1,t_2], \ldots, [t_N,1]$, where the projection in each $\mathbb{R} \times [t_j, t_{j+1}]$ is an *elementary* ribbon tangle. In this context call an elementary tangle a tangle diagram, which consists entirely of straight vertical ribbons, except an elementary *local picture*, as indicated in the following diagram:

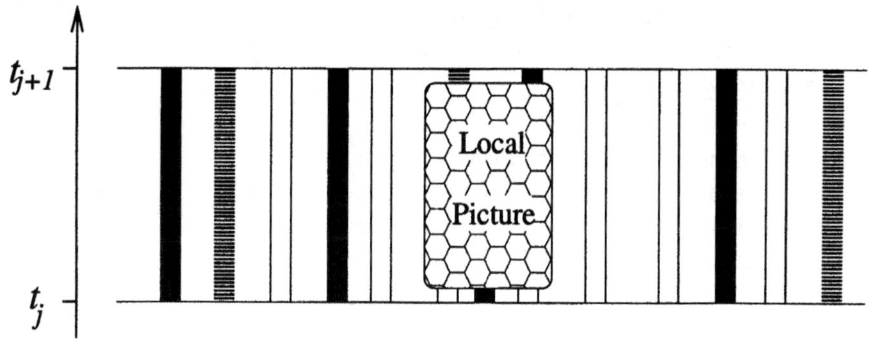

Next we give a list of the possible local pictures:

1. Over- and undercrossings of any type of strands.
   Three of 18 examples are depicted to the right:

2. Maxima and Minima of any type of strand.
   See three of six possibilities:

3. 2π-twists of any type of strand.
   There are two possible directions for the twist and three type of strand. Three of six pictures are on the right:

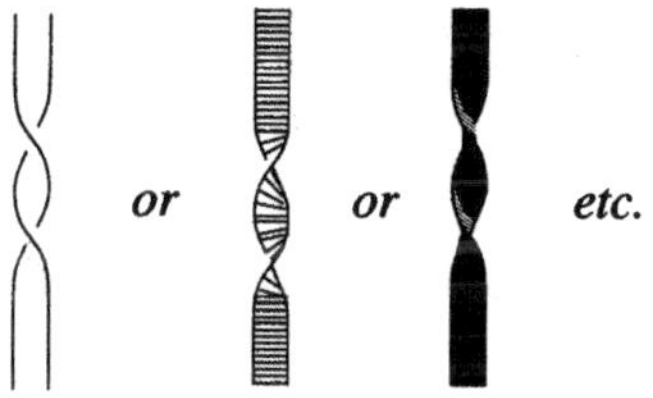

   For convenience we also admit π-twists, but *only for the auxiliary* ribbons, see:

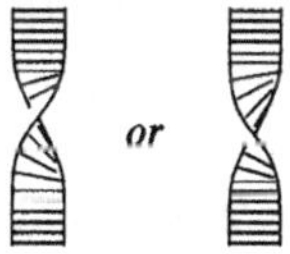

4. They may be ribbons attached to coupons. To each strand joined at the top there has to be a ribbon of the same type at the corresponding bottom interval. I.e., the coupon appears to be placed on top of a series of parallel strands. For example:

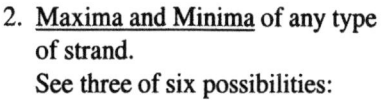

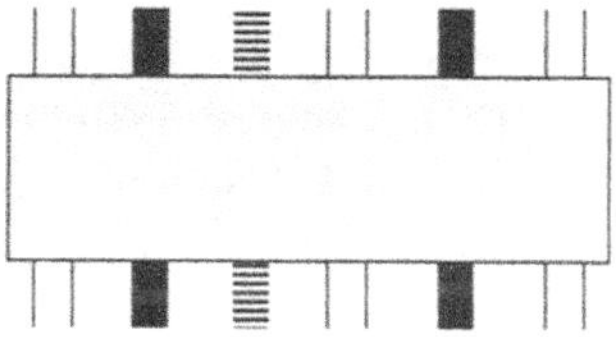

5. An auxiliary strand can end at a right angle in an internal strand from the right or the left. There are exactly two possible pictures of such joints:

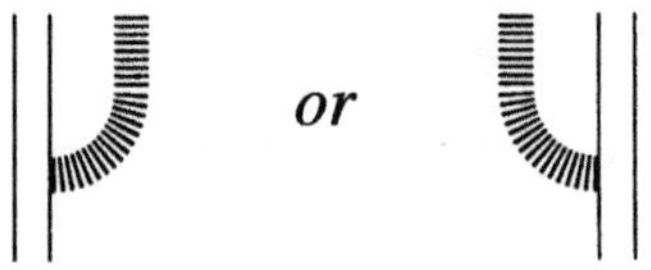

6. All ribbons can end in the boundary $\mathbb{R}\times\{\pm 1\}$ in the respective boundary intervals as defined in Section 2.1.1. I.e., internal ribbons can end in the internal intervals, $I_j^{i/0}$, the external ribbons in the external intervals, $J_j^{s/t}$, and the auxiliary ribbons may end in the auxiliary intervals, K_j, either at the top or bottom line. Some of the possible pictures of ribbon boundaries are:

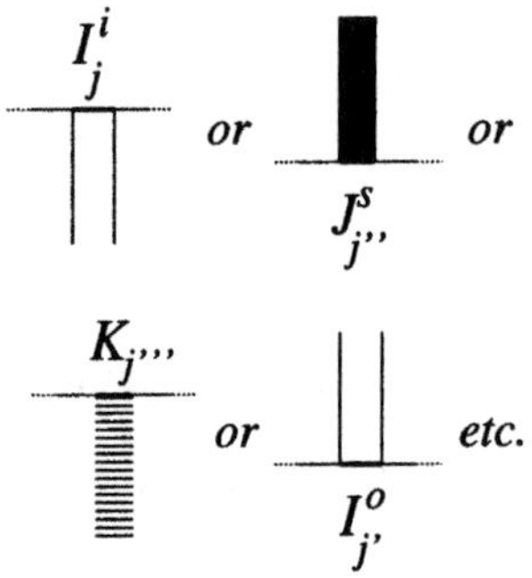

2.2 Admissible Tangles and Vertical 1-Arrows

Not every possible diagram that is composed from the local pictures will make sense as a surgery prescription in the way we are going to present cobordisms. We consider some tangles *admissible* depending on global properties of the strands. We distinguish several types of strands mainly by conditions on where they will end. The global properties will depend in part not only on the horizontal 1-arrows, but also the vertical 1-arrows, which we shall define first.

2.2.1 Vertical 1-Arrows, and Squares

We let the 1-arrow categories of our tangle double categories be isomorphic to those of $\widetilde{\mathcal{C}ob}$. We already introduced the horizontal category in Section 2.1.1 for tangles simply by considering the arrows, $[g, a/b] : a \mapsto b$, as formal triples.

In the same way we shall define a vertical arrow, $\alpha : a \mapsto a$, formally only as an element of the symmetric group, $\alpha \in S_a$. The composition in the vertical 1-category is simply the composition in the symmetric group. There are no vertical morphisms between distinct integers.

Hence, a *square* of 1-arrows in a tangle category is given by four integers a, b, g_{sc}, and g_{tg}, and two permutations $\alpha \in S_a$ and $\beta \in S_b$. As usual, we arrange them in a diagram as follows:

$$\begin{array}{ccc} a & \xrightarrow{[g_{sc},a/b]} & b \\ {\scriptstyle\alpha}\downarrow & & \downarrow{\scriptstyle\beta} \\ a & \xrightarrow[{[g_{tg},a/b]}]{} & b \end{array} \tag{2.2.1}$$

2.2.2 Types of Strands

In Section 2.1.2 we already distinguished between internal, external, and auxiliary strands. For each type there will be additional conditions on where a strand can end. In order to make more precise what is meant by a total strand, let us define:

Definition 2.2.1. *A* connected component *of a tangle diagram made up from the pieces given in Section 2.1 is a maximal (finite) sequence of ribbon pieces, in which successive pieces are* attached *to each other. Such attachments can be:*

1. *Two ribbon pieces that come together at the boundary of an elementary slice. For example, also the ribbon pieces that are joined to each side of an arc belong to the same component.*
2. *The left upper and right lower ribbon piece of a crossing are considered attached to each other. Analogously, the right upper and left lower pieces. (However, e.g., left upper and left lower are not attached in this part of the diagram.)*
3. *A ribbon piece at the top of a coupon and the corresponding piece at the bottom of the coupon are considered attached to each other.*
4. *The pieces entering and emerging from a 2π-twist (or a π-twist in the case of auxiliary ribbons).*

This definition is obviously identical with that of usual topological connected components of a ribbon tangle in $\mathbb{R}^2 \times [-1,1]$ disregarding the coupons.

It is also clear in the formal definition for diagrams in $\mathbb{R} \times [-1,1]$, that the sequence of attached pieces can be continued in a unique way forward or backwards from each given ribbon piece, until a ribbon piece is attached to a boundary piece $\mathbb{R} \times \{\pm 1\}$ or a previous element of the sequence. Thus, connected components are uniquely defined.

Clearly, the pieces of a component need to be all of the same type. Furthermore, for the internal and external types we can always introduce an *orientation* on the ribbon as a 2-dimensional submanifold with boundary from an orientation of the plane they are projected into, since we only allowed 2π-twists, which preserve orientations.

We now list the following possibilities for a connected component, distinguished by their types:

1. **Internal Connected Tangle Components :** They consist only of internal ribbon pieces. From the previous conditions we already know they can end only in the internal intervals $I_j^{i/o}$. There are four configurations that we allow:
   a) *Closed* (internal) tangle components: Here the last piece of a sequence of attached ribbon pieces is itself attached to the first piece of the sequence. Because of orientability the resulting component as a 2-dimensional manifold has to be homeomorphic to $S^1 \times [0,1]$.
   b) *Top* (internal) tangle components: This is a ribbon piece homeomorphic to $[-L,L]\times[0,1]$, where both ends $\{\pm L\}\times[0,1]$ are attached to corresponding *internal* intervals at the *top* line. More precisely, for a given j the end

$\{-L\} \times [0,1]$ is identified with interval I^i_j, and the end $\{+L\} \times [0,1]$ with interval I^o_j.

c) *Bottom* (internal) tangle components: These types of strands are completely analogous to the top components, and, hence, start in an interval, I^i_j, now at the *bottom* line and end in the neighboring interval I^i_j also at the bottom of the diagram.

d) *Through pairs* of (internal) tangle components: There may be also ribbon components that start at the top line and end in the bottom line. However, they always have to occur in *pairs* of components, $[-L, L] \times [0,1] \cup [-L, L] \times [0,1]$. Furthermore, the intervals at the top-line they start from have to be a neighboring pairs, $I^i_{j_t}, I^o_{j_t}$, and likewise the pair of components has to end at the bottom line in corresponding intervals $I^i_{j_b}, I^o_{j_b}$. There are, obviously, two possibilities to interpolate between the j_t-th interval pair at the top and j_b-th pair at the bottom line:

   i. Parallel through pairs: Here one strand goes from $I^i_{j_t}$ to $I^i_{j_b}$, the other strand connects $I^o_{j_t}$ to $I^o_{j_b}$.

   ii. Crossed through pairs: Here one strand goes from $I^i_{j_t}$ to $I^o_{j_b}$, the other strand connects $I^o_{j_t}$ to $I^i_{j_b}$.

We summarize the possible connectivity of internal ribbons in the following schematic diagram:

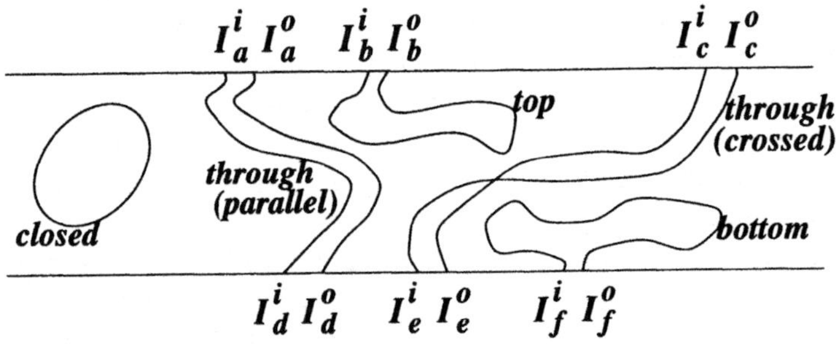

Complications such as thickness of strands, knottings, further components of one type, external and auxiliary components are omitted here.

2. **External Connected Tangle Components:** The global constraints on the external tangle components depend on the permutations, $\alpha \in S_a$ and $\beta \in S_b$, of the vertical arrows in Diagram 2.2.1.
External components consist of external pieces that are globally strands $[-L, L] \times [0,1]$, starting at an external interval at the top-line and ending at a corresponding external interval at the bottom-line. Specifically, the rules are as follows:

   a) *Source external strands* are components that start at an interval J^s_j at the top line and end at the interval $J^s_{\alpha(j)}$ at the bottom line, where α is the first permutation in the 2-arrow diagram of the tangle.

   b) *Target external strands* are components that start at an interval J^t_j at the top line and end at the interval $J^t_{\beta(j)}$ at the bottom line, where β is the second permutation in the 2-arrow diagram.

Below is a schematic picture of how the external strands are connected. Again we omit complications, such as thickness, tangling of the strands with each other and other types of strands:

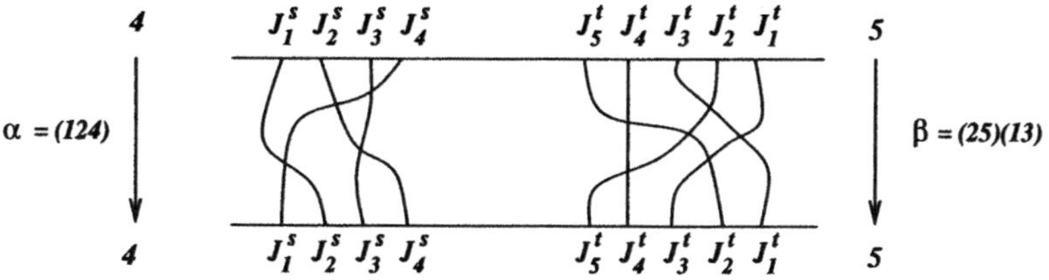

3. **Auxiliary Connected Tangle Components:** They consist of auxiliary ribbon pieces, that if pieced together yield a strand homeomorphic to $[-L, L] \times [0, 1]$. One end of the strand connects to an internal ribbon in a joint as in the fifth of the local pictures above, and the other end of the ribbon connects at one of the auxiliary intervals, K_j, at the top line of the diagram. We admit only two possibilities:
   a) *Closed auxiliary ribbons:* They start at a closed internal ribbon and end at an auxiliary interval at the top line.
   b) *Bottom auxiliary ribbons:* They start at a bottom internal ribbon and end at an auxiliary interval at the top line.

   Here again a schematic picture. The dashed lines represent the internal ribbons to which the respective auxiliary ones are connected to:

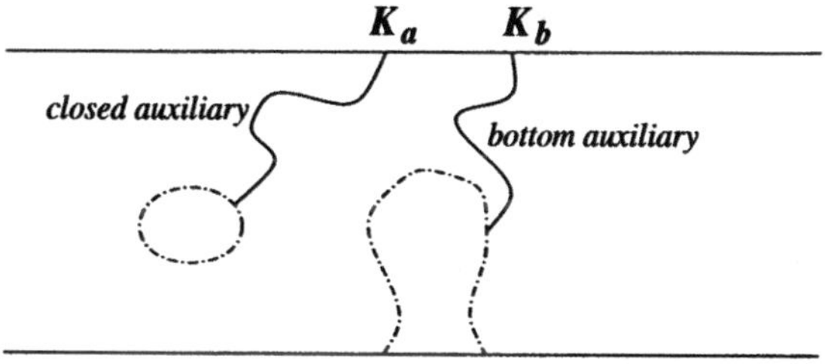

   Note that there are no auxiliary strands for the top or through internal strands.
4. **Orientations:** For planar tangles an orientation of the plane $\mathbb{R}_x \times [-1, 1]$ naturally induces an orientation on all internal and external types of ribbons. This follows easily from the fact that in all of the elementary diagrams from Section 2.1.3 we can impose orientations on the internal and external strands that are compatible with a given plane orientation at the out-going and in-going strands. Consequently, the "upper side" of a ribbon $[-L, L] \times [0, 1]$ will be shown upwards at *both* intervals at the top or bottom where it is attached. Hence, with the exception of the closed components, we may think of the orientation of the external and internal strands as being induced by an orientation of the boundary of the diagram.

   For the auxiliary ribbons we also allow π-twists instead of just 2π-twists. This causes a flip of orientation of the strand relative to a given plane orientation.

2.2.3 Conditions for Admissibility

We now use the notions and definitions developed in the previous sections to give the precise condition for a generic tangle projection to be admissible:

Definition 2.2.2.
An admissible tangle *is a tangle projection in* $\mathbb{R} \times [0,1]$, *with the following constraints:*

- *Its parts can be internal, external, and auxiliary ribbon pieces, as well as coupons.*
- *The projection must be generic, meaning that the tangle must consist of elementary slices, whose possible local pictures are either crossings, maxima and minima, twists, coupons, joints, or boundary connections, as depicted previously.*
- *The* components *must be either external strands (source or target), internal strands (closed, top, bottom, or through), or auxiliary strands (closed or bottom) as described above.*
- *Furthermore, for a diagram as in (2.2.1) with 1-arrows* $[g_{sc}, a/b]$ *and* $[g_{tg}, a/b]$, *there will be required that intervals and joints are occupied exactly once:*
  1. *Each of the* g_{sc} *pairs of internal intervals at the top line is connected to either exactly one top ribbon or one pair of through ribbons.*
  2. *Each of the* g_{tg} *pairs of internal intervals at the top line is connected to either exactly one bottom ribbon or one pair of through ribbons.*
  3. *Each external interval at the top or bottom line of the diagram is connected to exactly one external strand.*
  4. *To each closed internal component and to each bottom internal ribbon is connected exactly one auxiliary ribbon.*

To illustrate the implication of this definition further as well as to keep better track of the ribbon combinatorics let us introduce the following numbers:

$$\begin{array}{ll} \mathsf{C} = \text{\# of closed internal ribbons} & \mathsf{E}_s = \text{\# of source external ribbons} \\ \mathsf{B} = \text{\# of bottom internal ribbons} & \mathsf{E}_t = \text{\# of target external ribbons} \\ \mathsf{T} = \text{\# of top internals ribbons} & \mathsf{P}_t = \text{\# of coupons} \\ \mathsf{H} = \text{\# of pairs of through ribbons} & \mathsf{A} = \text{\# of auxiliary ribbons} \end{array} \tag{2.2.2}$$

The definition of an admissible tangle implies then the equations:

$$\begin{array}{llll} g_{sc} = \mathsf{T} + \mathsf{H} & \mathsf{A} = \mathsf{B} + \mathsf{C} & a = \mathsf{E}_s \\ g_{tg} = \mathsf{B} + \mathsf{H} & & b = \mathsf{E}_t \end{array} \tag{2.2.3}$$

We also assume that the auxiliary intervals that are connecting to auxiliary ribbons are numbered consecutively, i.e., $K_1, K_2, \ldots, K_{\mathsf{A}}$.

2.3 Equivalence Moves of Tangles, and the 2-Arrows in $\mathcal{T}gl$

An admissible tangle determines via surgery a unique homeomorphism class of cobordisms. However, in order to establish an isomorphism between the combinatorial category and the cobordisms category, we have to account for the fact that many admissible tangles may produce the same cobordism. We, therefore, have to introduce equivalence classes on the set of these tangles, such that each class of tangles represents precisely one homeomorphism class of cobordisms.

The equivalences on the set of tangles should be formulated in purely combinatorial terms in order for the presentation to be of any value. We define these equivalences in terms of elementary equivalences that are determined by so called *moves*.

Specifically, a *move*, M, is a prescription about how to substitute certain combinations of local pictures and elementary slices within a given tangle diagram by other local pictures. Examples are the three Reidemeister moves for knots, in which the local pictures are crossings combined by stacking them on top of each other in specific ways. Let us indicate a move by

$$T_1 \xrightarrow{M} T_2 .$$

If for a certain type of moves the number of such local pictures that are changed in that prescription is not dependent on the tangle as a whole then we call it a *local move*. To use again the example of the Reidemeister moves, the second move involves two elementary crossings and the third involves three crossings. Hence, the number of local pictures involved in each move is two or three independently of how the knot looks like globally.

Any given set of moves $M_1, \ldots, M_k$ determines an *equivalence relation* $\mathcal{R}_{M_1,\ldots,M_k}$ on the set of admissible tangle diagrams. It is defined as the smallest equivalence relation, such that $T_1 \xrightarrow{M_j} T_2$ for some j implies that $T_1 \sim_{\mathcal{R}} T_2$ (meaning $(T_1, T_2) \in \mathcal{R}_{M_1,\ldots,M_k}$). More specifically, this implies firstly that the reversal M^- of a move M, such that $T_2 \xrightarrow{M^-} T_1$ is equivalent to $T_1 \xrightarrow{M} T_2$. Hence, we can say that for two tangle diagrams $T_A \sim_{\mathcal{R}} T_B$, if and only if there is a finite sequence of diagrams $T_1, \ldots, T_{n-1}$ and moves $M_{j_1}, \ldots, M_{j_n}$ as well as signs $\varepsilon_1, \ldots, \varepsilon_n \in \{+, -\}$, such that (with the convention $M^+ = M$)

$$T_A \xrightarrow{M_{j1}^{\varepsilon 1}} T_1 \xrightarrow{M_{j2}^{\varepsilon 2}} \cdots \xrightarrow{M_{j_{n-1}}^{\varepsilon_{n-1}}} T_{n-1} \xrightarrow{M_{j_n}^{\varepsilon_n}} T_B \quad .$$

We now give a list of all moves we want to use to eventually define the 2-arrows of our tangle category. They are divided into moves that involve isotopies of the tangles, equivalences for the auxiliary data, and moves related to surgery and the boundaries of cobordisms.

All moves that we define are *local* ones:

2.3.1 Local Moves relating Isotopies and Projections

The "TI"-Moves listed in this paragraph are those arising from choosing different generic projections for the same ribbon tangle in three dimensional space, as well as applying a generic isotopy to each tangle before projecting it. Consequently, all of the following moves apply to every type of ribbon strand.

TI1) *Framing-Flip :* The 2π-twists can be resolved in two ways as flat untwisted ribbon diagrams, each involving two extrema and one crossing, as indicated. Note, the loops to the left and right are related by a rotation of π.
This move replaces the first Reidemeister move for unframed tangles, and applies to all three types of ribbon pieces.

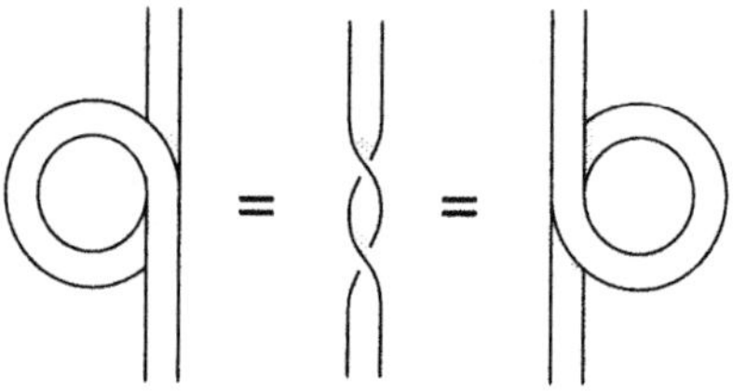

TI2) 2nd *Reidemeister:* This move is analogous to the one for unframed tangles, and involves two strands and two crossings in each diagram. We may assume all framings to be in the plane of projection. It applies to all three ribbon types.

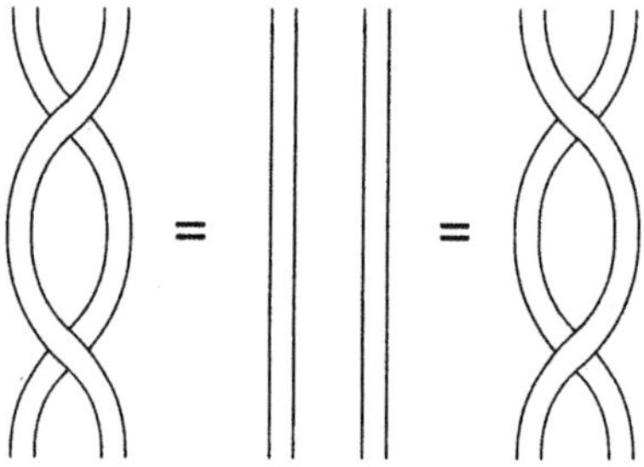

TI3) 3rd *Reidemeister:* Also this move is just as the regular unframed third Reidemeister move with framings in the plane. It involves three strands and three crossings in each diagram, and, again, applies to all combinations of the three ribbon types.

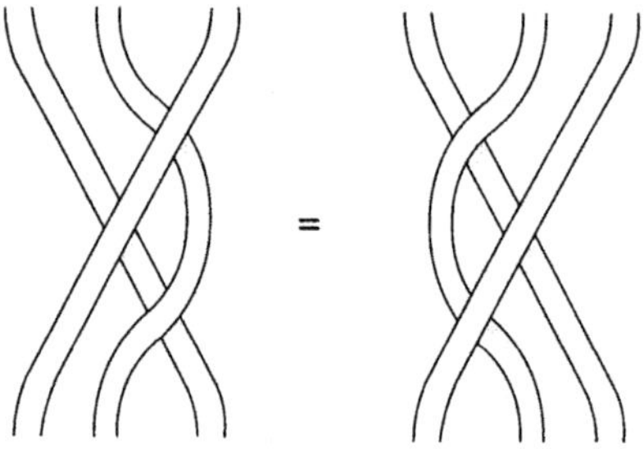

TI4) *Extrema-Cancellation:* This move and the next one have to be introduced besides Reidemeister moves to account for genericity with respect to a preferred vertical direction. Here a local maximum and a minimum that are consecutive on a strand of any type can be cancelled.

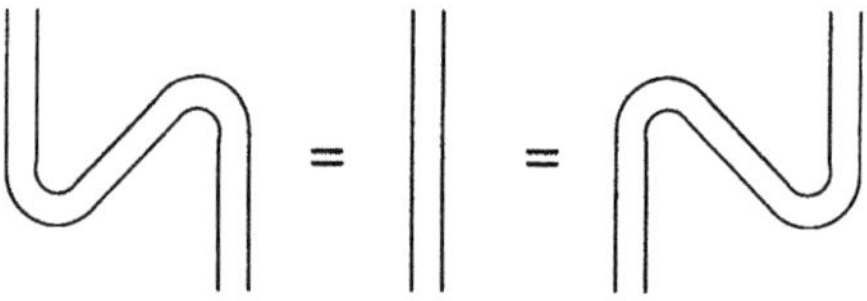

TI5) *"Crossing-Symmetry":* In this move a crossing and an extremum exchange places. The middle strand is slid from left to right. The variations we wish to include but have not depicted are
- Reflection at the horizontal.
- Exchange over- and under-crossings.
- All combinations of ribbon types.

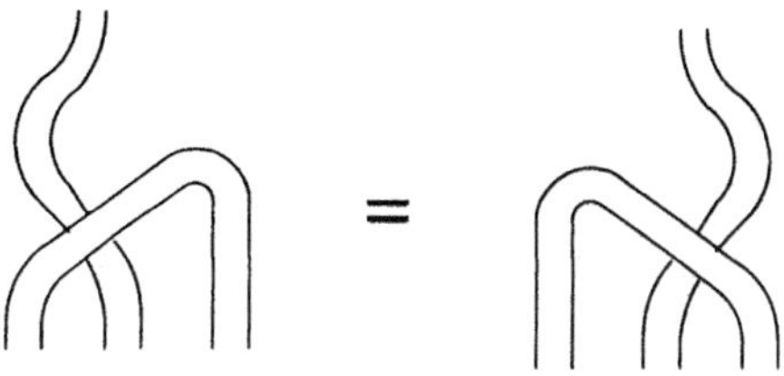

TI6) *Sliding Joints over Extrema:* The following two moves account for isotopies of the joints between internal and auxiliary ribbons. Firstly, a joint may encounter an extremum while it is being slid along an internal ribbon. The version of the resulting move for a maximum is depicted to the right. The corresponding move for a minimum can be easily *deduced* from this one and the move TI4 above.

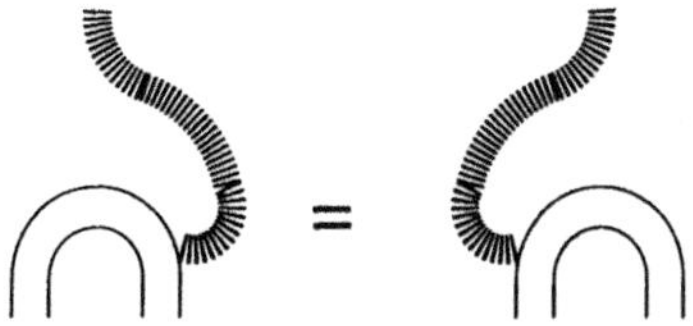

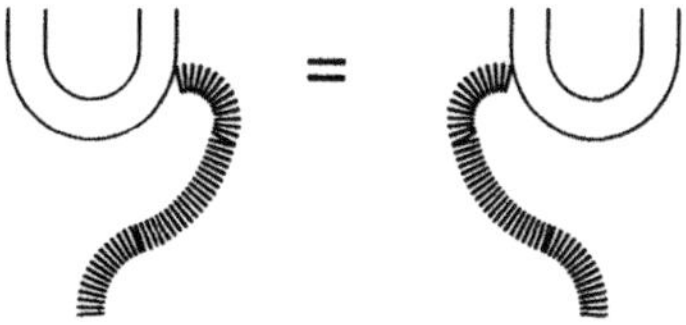

TI7) *Crossing of Joints:* The second elementary critical point of an isotopy of a joint is when it crosses through another strand of an *arbitrary* type as depicted. Here again, it is sufficient to include only one version (for each type) for the move, since all other reflections can be obtained as a combination of this special version and the previous isotopy moves TI1 through TI7.

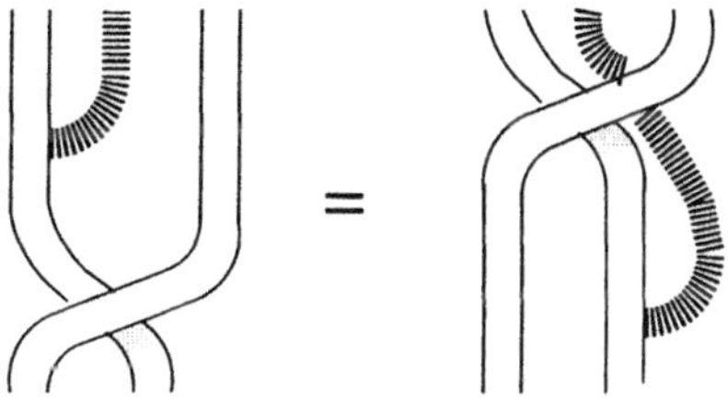

TI8) *Vertical Independence:*
Elementary pictures, that are neighboring in height, and that are connected to different and separated sets of vertical strands can be moved past each other as indicated in the schematic picture.

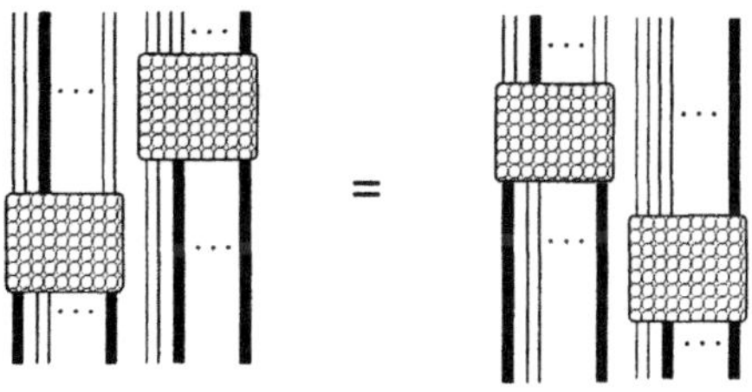

TI9) *Horizontal Independence:* Vertical strands and connected elements may be pushed together or apart (provided no further crossings are generated).

TI10) *Crossings and Cancellations for π-Twists of Unoriented Auxiliary Ribbons:*
The fact that we allow half twists for the auxiliary ribbons entails that we have to introduce the isotopies, where the π-twists are isotoped through a crossing or opposite twists can cancel each other.

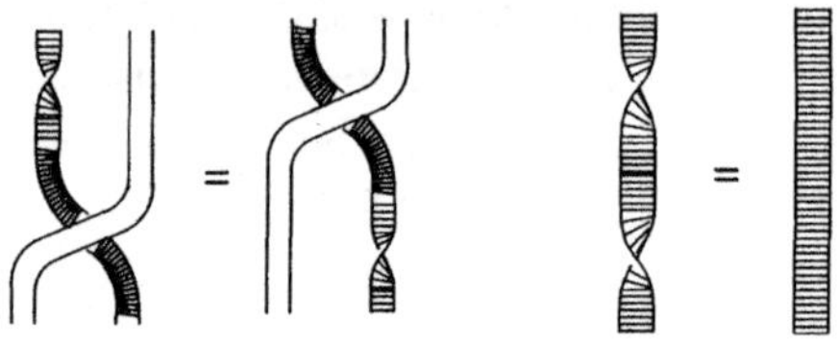

TI11) *Flip over of Joints :*
A joint can be flipped to the other side of an internal ribbon as depicted. This may be thought of as being accomplished by a rotation of the joint around the axis along the internal ribbon. The two opposite π-twists along the internal ribbon that are created in this rotation are, however, omitted.

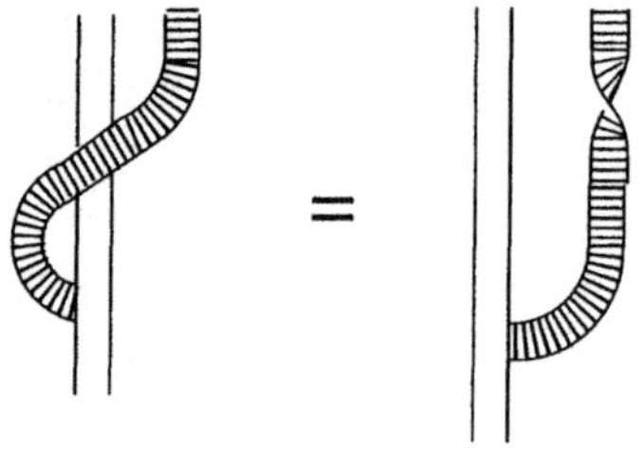

2.3.2 Local Moves for Coupons and Auxiliary Ribbons

The moves in this section concern equivalences for the special elements (decorations), which are not isotopies. For the coupons they imply that the upper and the lower half of a coupon can be isotoped independently. The moves for the auxiliary ribbon will allow us to move from one such ribbon to any other as long as the attachments stay the same.

TD1) *Crossing with Coupons:* A ribbon piece of arbitrary type that runs across the group of parallel strands emerging from a coupon, can be pushed from running over these strands to running underneath the strands.

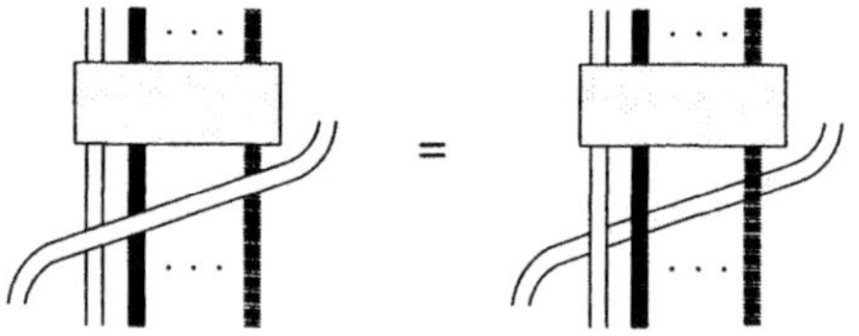

TD2) *2π-Twist at Coupon:* A collective 2π twist can be applied to the group of strands emerging at the top of a coupon. (The twist move of the strand group at the top together with other moves also implies that at the bottom.)

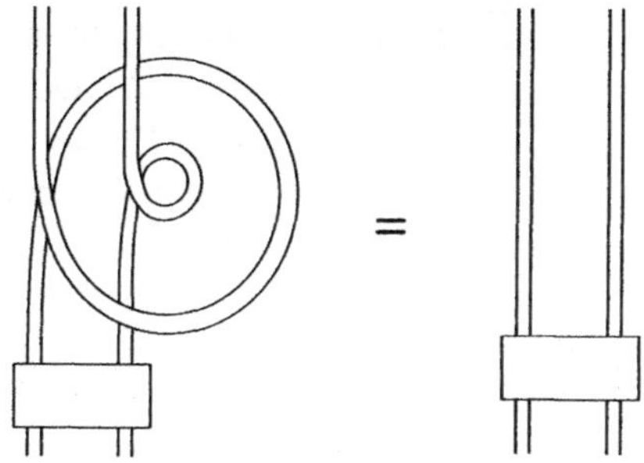

The next three moves concern auxiliary ribbons. For all moves TD3 - TD5 it suffices to consider a configuration only in the vicinity of the top line.

TD3) *Crossings with Auxiliary Ribbon at Top-Line:* An over-crossing of an auxiliary ribbon with any other ribbon can be replaced by an under-crossing and vice versa.

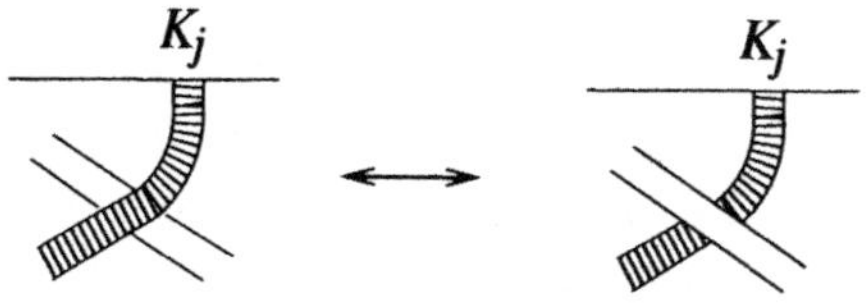

TD4) *Braiding of Splitting-Ribbons:* Right next to the top line we can also switch the attachment intervals for auxiliary ribbons. Since these generate the braid (permutation) group, we can insert any braid and permute the attachment intervals in any way we want.

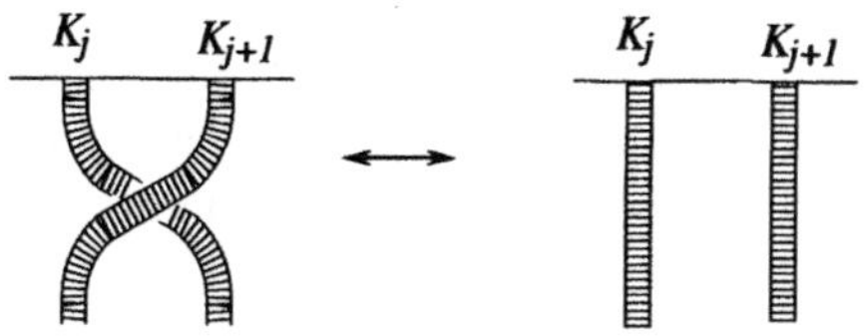

TD5) *π-Twists at Top-Line:* An additional π-twist (and, hence, any twist) can be introduced at an auxiliary strand.

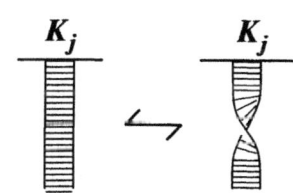

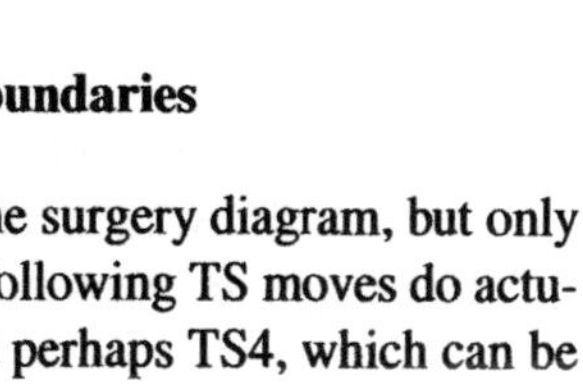

2.3.3 Local Moves for Surgery in Interior and at Boundaries

The previous moves did not describe any changes of the surgery diagram, but only the way it is presented in the plane up to isotopy. The following TS moves do actually change the link in three dimensional space, except perhaps TS4, which can be understood also as an isotopy if we consider a tangle over a sphere.

TS1) *Handle Trading:* The handle trading move replaces an annulus, i.e., a 0-framed unknot bounding a disc D^2, by a coupon. The ribbon strands passing transversally through the disc, in an order determined by their planar projection, are then the strands attached to the coupon at the top and bottom, as depicted on the right.

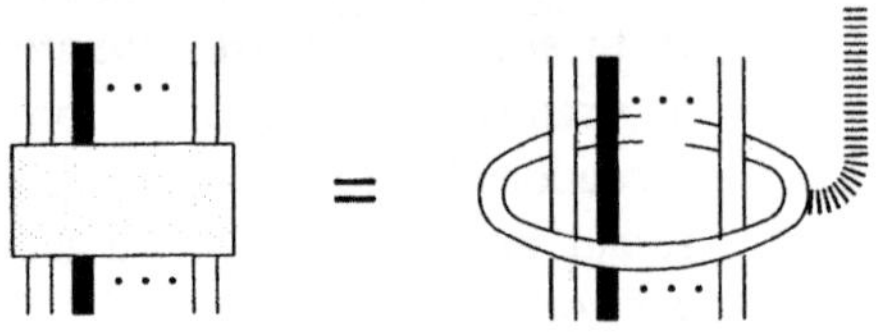

TS2) *1-2-Cancellation:* An *isolated* configuration that consists of one closed internal ribbon, one coupon, and an auxiliary ribbon as depicted can be added to or removed from a diagram. Specifically, the internal ribbon has to be an annulus (i.e., a 0-framed unknot) that passes through the coupon exactly once, and the three elements are inside a disc that contains no other elements of the diagram.

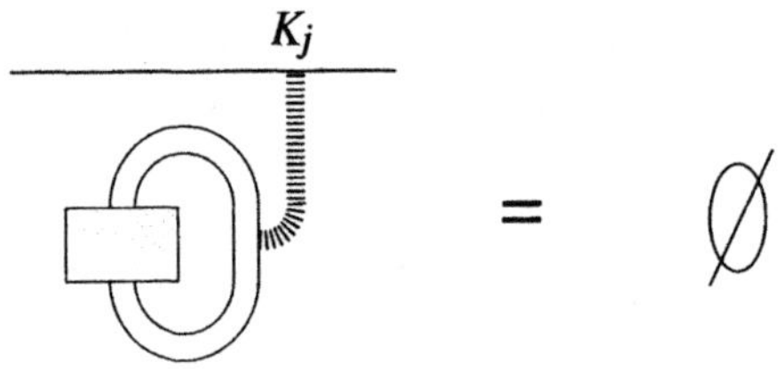

TS3) *σ-Move:* This move accounts for the fact that the boundary of the cobordism is non-simply connected. In this move a pair of through internal ribbon strands is replaced by a short arc top-ribbon, a coupon and an internal (bottom) ribbon with an additional auxiliary ribbon attached. The top arc, coupon and joint all have to lie in a small vicinity of the top line.
In the second version of this move we start from a top ribbon instead of a pair of through ribbons and generate a top and a closed internal ribbons running through the coupon. In the context of other calculi this move turns out to be redundant.

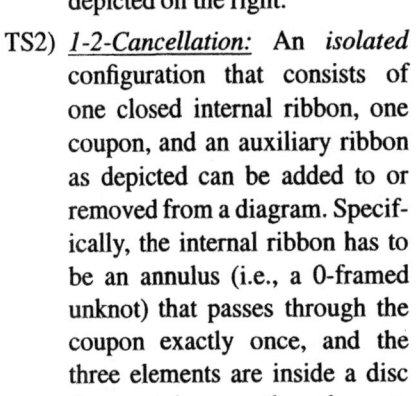

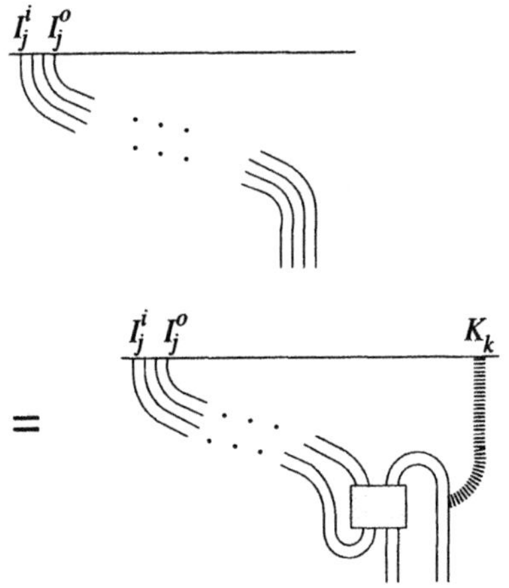

TS4) *τ-Move:* This move involves a small collar of the top-line into the diagram, where all the strands emerging from that top line are still straight and parallel. The only other element that lies in this collar is a strand of arbitrary type that crosses over all other strands. In the τ-Move all the over-crossings are to be replaced by a collective undercrossing of the extra strand.

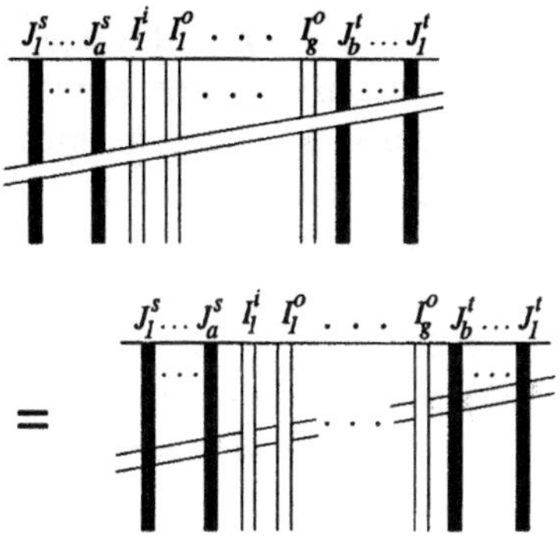

For a presentation of the cobordism category $\widetilde{\mathcal{Cob}}$ with signature extension Ω_4 this is precisely the needed set of moves. Occasionally, we also wish to describe the plain three dimensional cobordisms in the original category $\mathcal{Cob}$. The equivalence of these cobordisms in $\mathcal{Cob}$ that may differ by a signature in $\widetilde{\mathcal{Cob}}$ corresponds to an additional move for the tangle diagrams, which is just Kirby's $\mathcal{O}_1$-Move:

TS5) *Signature-Move:* An isolated configuration, that consists of an unknot with exactly one 2π-twist, or, as depicted, an 8-shape with planar immersion of the ribbon, and the usual auxiliary ribbon, attached to it, can be removed from a tangle diagram.

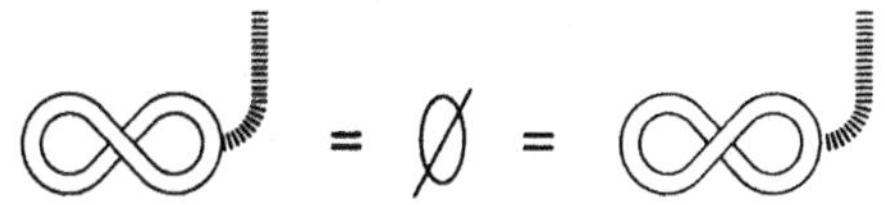

2.3.4 Definition of 2-Arrow sets of $\mathcal{Tgl}$

In order for all these prescriptions to be applicable to the class of tangle diagrams we started out with we need to observe one almost obvious fact, where we also specify the changes in numbers of components a little more precisely:

Lemma 2.3.1. *The moves* TI1 - TI11, TD1 - TD5, *and* TS1 - TS5 *all close within the class of* admissible *tangles as in Definition 2.2.2.*

Proof. The isotopy moves TI and decoration moves TD do not affect the amount of elements and their connectivity at all. The TD4 move changes the order in which two closed or bottom internal ribbons are connected to auxiliary intervals. Also the TS4 move is really only an isotopy and, hence, does not change number or connectivity of ribbons.

For the remaining four moves the number of components changes. We use as in (2.2.2) the numbers C, B, T, H and P. By (2.2.3) the numbers E_s, and E_t do not change during any of the moves, and A is simply the sum of C, and B and, thus, changes as these. The changes are described as follows:

TS1) In the Handle Trade we replace an internal closed ribbon by a coupon. The connectivity of ribbon pieces over the under- and overcrossings with the annulus is the same as that over the coupon edges. In particular, all ribbons attached to the boundaries of the tangle diagram are unchanged.
$\mathsf{C} \mapsto \mathsf{C}+1,\quad \mathsf{P} \mapsto \mathsf{P}-1,\quad \mathsf{T} \mapsto \mathsf{T},\quad \mathsf{B} \mapsto \mathsf{B},\quad \mathsf{H} \mapsto \mathsf{H}.$

TS2) In the 1-2-Cancellation complete components, namely one internal ribbon and one coupon, are removed, and both are separated from the boundary:
$\mathsf{C} \mapsto \mathsf{C}-1,\quad \mathsf{P} \mapsto \mathsf{P}-1,\quad \mathsf{T} \mapsto \mathsf{T},\quad \mathsf{B} \mapsto \mathsf{B},\quad \mathsf{H} \mapsto \mathsf{H}.$

TS3) For the σ-Move it was outlined in the description of the move that we stay within the admissible tangles. If it is applied at a top internal interval pair I^i_j, and I^o_j, to which a top internal ribbon is attached, then this is replaced by a short arc as a top ribbon. In addition, we create one coupon and one closed internal ribbon:
$\mathsf{C} \mapsto \mathsf{C}+1,\quad \mathsf{P} \mapsto \mathsf{P}+1,\quad \mathsf{T} \mapsto \mathsf{T},\quad \mathsf{B} \mapsto \mathsf{B},\quad \mathsf{H} \mapsto \mathsf{H}.$
The second version of the move considers a top internal interval pair I^i_j, and I^o_j, to which a pair of through internal ribbons are attached. In the move they are replaced by again a small arched top ribbon, and we add a coupon. But the additional internal component is now a bottom ribbon.
$\mathsf{C} \mapsto \mathsf{C},\quad \mathsf{P} \mapsto \mathsf{P}+1,\quad \mathsf{T} \mapsto \mathsf{T}+1,\quad \mathsf{B} \mapsto \mathsf{B}+1,\quad \mathsf{H} \mapsto \mathsf{H}-1.$

TS5) As for TS2, this move is simply about removing entire components that are not attached to the boundaries, and, hence, don't change the admissibility of the graph. The only component to be removed here is a closed ribbon and the corresponding auxiliary ribbon:
$\mathsf{C} \mapsto \mathsf{C}-1,\quad \mathsf{P} \mapsto \mathsf{P},\quad \mathsf{T} \mapsto \mathsf{T},\quad \mathsf{B} \mapsto \mathsf{B},\quad \mathsf{H} \mapsto \mathsf{H}.$

Having established a valid equivalence relation on the set of admissible tangles we can now proceed with the next ingredient in the definition of our tangle double category:

Definition 2.3.2. *The* set of 2-arrows *of $\mathcal{T}gl$ for a square as in (2.2.1) is the set of equivalence classes of admissible tangles with respect to that square (as in Definition 2.2.2) with equivalence relations determined by the moves* TI1 - TI11, TD1 - TD5, *and* TS1 - TS4.

We also define 2-arrow sets for another tangle double category, $\mathbf{Tgl}$, with the same 1-arrow and object structure, where we introduce in addition to the above moves also TS5 *as an equivalence. Hence, we have a surjection $\mathcal{T}gl \twoheadrightarrow \mathbf{Tgl}$.*

Before we discuss the composition laws for these 2-arrows, we introduce a few alternative ways of presenting the tangle equivalence classes.

2.4 Tangles in Three-Space

The planar admissible tangles, as defined in Section 2.3.4, can, without difficulty, be visualized as tangles in three dimensional space that have been projected into the plane. Also many of the moves, in particular the TI-Moves, can be identified as corresponding isotopies of tangles in three spaces.

In this section we begin to describe the equivalence classes of the tangles in three space in more detail. Our main result will be to establish a bijection between the equivalence classes of admissible tangles subject to the full set of moves in Section 2.3 and classes of tangles in the thickened sphere $S^2 \times [-1, 1]$ without auxiliary ribbons, and only subject to the five moves, TD1, TD2, TS1, TS2, and TS3.

Common methods of deriving moves for planar projections are transversality or general position arguments. For background and details on the results that we are employing here see for example [Hir76].

2.4.1 Tangles over $\mathbb{R}^2$

We may define the 2-arrows of a double tangle category $\mathcal{T}gl_{\mathbb{R}^2}$ starting from tangles in three spaces that are not yet projected.

The three-space for the tangle will be $\mathbb{R}^2 \times [-1, 1]$. For the plane $\mathbb{R}^2$, we choose a coordinate axis $\mathbb{R}_x \subset \mathbb{R}^2$, i.e., a linear subspace, as well as a linear projection $p_x : \mathbb{R}^2 \twoheadrightarrow \mathbb{R}_x$, which is the identity on $\mathbb{R}_x$. We want to use the same notation for the respective projection $p_x : \mathbb{R}^2 \times [-1, 1] \twoheadrightarrow \mathbb{R}_x \times [-1, 1]$.

With these convention an *embedded (admissible) tangle* $T \in \mathbb{R}^2 \times [-1, 1]$ will be an embedding of ribbons and coupons into $\mathbb{R}^2 \times [-1, 1]$. As in Section 2.2.2 the ribbons are either homeomorphic to $[-L, L] \times [-1, 1]$ or to $S^1 \times [-1, 1]$, and may be one of three types (namely internal, external, or auxiliary). The notion of a connected component is now the ordinary topological one, except that two ribbon pieces ending at and emerging from the same coupon will be considered as parts of the same component, if they are attached to corresponding opposite intervals on the two sides of the coupon. The conditions for an embedded tangle to be admissible are almost literally the same as in Definition 2.2.2. There are only two differences to observe:

For one all ribbons (except the closed internal) are still attached to a *top* or *bottom line* but now these are subspaces $\mathbb{R}_x \times \{\pm 1\} \subset \mathbb{R}^2 \times \{\pm 1\}$ of a top or bottom plane.

Also we now need to explicitly introduce conditions on orientations, since orientability through the projections is no longer automatic. Thus, all embedded internal and external strands as well as the coupons will have fixed orientations, which are compatible at the attachment intervals with the orientation of the bounding lines.

In fact, for ribbons that end in the top or bottom intervals we only need to require *orientability*. The orientations themselves are then determined by a given orientation of the plane $\mathbb{R}_x \times [-1, 1]$ to which the ribbons are tangential at the boundaries of the diagram. For the closed internal ribbons, however, a choice between two possible orientations needs to be made. The following simple observation shows that we can freely switch between these two choices if we pass to equivalence classes of tangles.

Lemma 2.4.1. *The isotopy classes of orientable and oriented tangles in $R^2 \times [-1,1]$ are the same.*

Proof. A closed internal ribbon $S^1 \times [-1,1]$ embedded into three dimensional space can be extended to an embedding of $S^1 \times D^2$ by tubular neighborhood arguments, where D^2 is the two dimensional unit disc and $([-1,1],\{-1,1\},0) \subset (D,\partial D,0)$ is a diameter in that disc. Suppose R_θ is a rotation by an angle θ in the two plane. Then

$$I_\theta : S^1 \times [-1,1] \hookrightarrow S^1 \times D^2 \xrightarrow{R_\theta} S^1 \times D^2 \hookrightarrow R^2 \times [-1,1]$$

defines an isotopy between the original embedding I_0 and the embedding I_π, which has the same ribbon as image as I_0 but with reversed orientation. Hence, as an oriented submanifold with boundary a closed ribbon is ambiently isotopic to itself with reversed orientation. Therefore, the isotopy classes are the same whether we equip every ribbon with an orientation to begin with or not.

The three dimensional version of a 2-arrow for a tangle category is now as in the planar version except that we do not have the finite set of isotopy moves:

Definition 2.4.2. *The objects and 1-arrows of $\mathcal{T}gl_{\mathbb{R}^2}$ are the same as for $\mathcal{T}gl$.*

The set of 2-arrows of $\mathcal{T}gl_{\mathbb{R}^2}$ are equivalence classes of embedded *admissible tangles in $\mathbb{R}^2 \times [-1,1]$, where the generating equivalence moves are the following:*

- *Any ambient isotopy of $\mathbb{R}^2 \times [-1,1]$.*
- *Decoration Moves* TD1 - TD5.
- *Surgery and boundary Moves* TS1 - TS4.

Obviously, we have a well defined map, ex : $\mathcal{T}gl \longrightarrow \mathcal{T}gl_{\mathbb{R}^2}$, on the level of 2-arrow sets, which is obtained by thickening out the tangle projection in $\mathbb{R}_x \times [-1,1] \subset \mathbb{R}^2 \times [-1,1]$, pushing the strands of a crossing away from each other in the direction of the projection p_x. Each of the moves TI1 - TI10 is easily identified with an (ambient) isotopy of the embedded tangle. Also the move TI11 describes essentially an isotopy, which rotates the joint by π along the axis of the internal ribbon. The isotopy itself generates also two opposite π-twists on the internal ribbon. If the internal ribbon is a closed one, these twists can be slid against each other around the remaining part of the ribbon and cancelled. In case we are using an orientation this will be correspondingly switched and after moving the π-twists through is again compatible, e.g., with a plane orientation. A bottom ribbon can be turned into an closed ribbon by use of the TS1 and TS3 Moves and, hence, treated in the same fashion.

Our goal is to show that the map ex is, in fact, a bijection. First, let us introduce the subset of *well positioned* tangles among those that are admissible and embedded into $\mathbb{R}^2 \times [-1,1]$. They are characterized by the positioning of the coupons and joints in three-space.

In order to describe this condition let us first assume that the tangle classes in three space are defined with ribbons to which specific orientations have been assigned. In this case the coupons as well as joints have an orientation compatible with the attached internal or external strands. Thus, to each coupon and joint embedded in three space we can associate a unique normal unit vector $\mathbf{v}_1$. Furthermore, the direction of the strands passing through the coupon and the direction of the internal strand through the joint determine a line $\mathbb{R}\mathbf{v}_2$ in three space, that is a unit vector $\pm\mathbf{v}_2$ unique only up to a sign. Then, a tangle is said to be *well-positioned*, if for all coupons and joints:

- $\mathbf{v}_2$ is in vertical direction, i.e., normal to the planes $\{t\} \times \mathbb{R}^2$ with $t \in [-1, 1]$,
- $\mathbf{v}_1$ is normal to the plane of projection $[-1, 1] \times \mathbb{R}_x$, and observes its orientation.

We then define the 2-arrow sets in $\mathcal{T}gl_{\mathbf{R}^2}^{well-pos}$ as equivalence classes of well-positioned tangles, for which the equivalence moves are the same ones as for $\mathcal{T}gl_{\mathbf{R}^2}$, with the following differences:

- During all isotopies (and other moves) the orientation of coupons and joints has to be fixed.
- The move TI6 for sliding joints over extrema is introduced explicitly.
- Furthermore, we impose the Move TI11 in order to flip a joint over.

It is obvious that by inclusion we have a well defined map of 2-arrow sets:

$$\mathbf{sp} : \mathcal{T}gl_{\mathbf{R}^2}^{well-pos} \longrightarrow \mathcal{T}gl_{\mathbf{R}^2}.$$

The first step in relating projected and embedded tangles is, thus, the following:

Lemma 2.4.3. *The map* **sp** *is bijection on the 2-arrow sets (with orientations).*

Proof. The strategy is to define an inverse, $\mathbf{sp}^{-1} : \mathcal{T}gl_{\mathbf{R}^2} \longrightarrow \mathcal{T}gl_{\mathbf{R}^2}^{well-pos}$, which is defined on the embedded tangles ambiguously up to certain choices, but well defined on equivalence classes. To this end we assign a well positioned tangle to a general embedded tangle as follows.

Around each coupon or joint we choose local coordinates, such that the attached strands go straight and parallel (or perpendicular for the auxiliary ribbon of the joint) from the decoration. The coordinate neighborhood is a ball of radius $a = 2$ and the coupon or joint lies entirely within the ball of radius $a = 1$. Now, for each coupon and joint we can find an element $\underline{g} \in SO(3)$, such that $\underline{g}.\mathbf{v}_1$ is the unit normal $\mathbf{n}_x$ of $[-1, 1] \times \mathbb{R}_x$, and $\underline{g}.\mathbf{v}_2$ is normal to $\{1\} \times \mathbb{R}^2$.

For given $\mathbf{v}_1, \mathbf{v}_2 \in \mathbb{R}^3$ there are always exactly two such elements, say $\underline{g}^+$ and $\underline{g}^-$, related by $\underline{g}^+ = \underline{r} \cdot \underline{g}^-$, where $\underline{r}$ is the rotation by π in the $[-1, 1] \times \mathbb{R}_x$- plane (i.e. $\underline{r}^2 = 1$ and $\underline{r}.\mathbf{n}_x = \mathbf{n}_x$). The general position of a coupon or a joint is precisely given by an element in the quotient space

$$\mathcal{Q} = \frac{SO(3)}{\mathbb{Z}_2(\underline{r})} = \frac{S^3}{\mathbb{Z}_4}.$$

The reason for the second presentation of $\mathcal{Q}$ is that $\mathbb{Z}_2(\underline{r}) \subset SO(3)$ lifts to a group $\mathbb{Z}_4(\tilde{\underline{r}}) \subset SU(2)$ with $\tilde{\underline{r}}^2 = -\mathbb{1}$.

Now, as $SO(3)$ is connected we can find a path $[0,1] \longrightarrow SO(3) : t \mapsto \underline{g}(t)$ with $\underline{g}(0) = \mathbb{1}$ and $\underline{g}(1) = \underline{g}^{\pm}$. From this we can define an isotopy

$$G : [0,1] \times \left(\mathbb{R}^2 \times [-1,1]\right) \longrightarrow \mathbb{R}^2 \times [-1,1]$$

supported only in the ball $D^3_2 \subset \mathbb{R}^2 \times [-1,1]$ around the coupon or joint by the formula:

$$G(s,\mathbf{x}) = \begin{cases} \mathbf{x} & \text{for } 2 \leqslant \|\mathbf{x}\|, \\ \underline{g}(s(2-\|\mathbf{x}\|)).\mathbf{x} & \text{for } 1 \leqslant \|\mathbf{x}\| \leqslant 2, \\ \underline{g}(s).\mathbf{x} & \text{for } \|\mathbf{x}\| \leqslant 1. \end{cases} \tag{2.4.1}$$

We also denote the local homeomorphism

$$G_{\underline{\underline{g}}} : \mathbf{x} \mapsto G(1,\mathbf{x}) \quad \text{for any path} \quad \underline{\underline{g}} : [0,1] \to \mathcal{Q} : t \mapsto \underline{g}(t),$$

where the lifting into $SO(3)$ is uniquely determined by $\underline{g}(0) = \mathbb{1}$. $G_{\underline{\underline{g}}}$ is clearly isotopic to the identity and acts as $\underline{g}(1) = \underline{g}^{\pm}$ on the unit ball D^3_1 and, hence, also on the coupon or joint. By the definition of $\underline{g}^{\pm}$ the image of the respective decoration under $G_{\underline{\underline{g}}}$ is, hence, *well-positioned*. The action of $G_{\underline{\underline{g}}}$ on a coupon is depicted below:

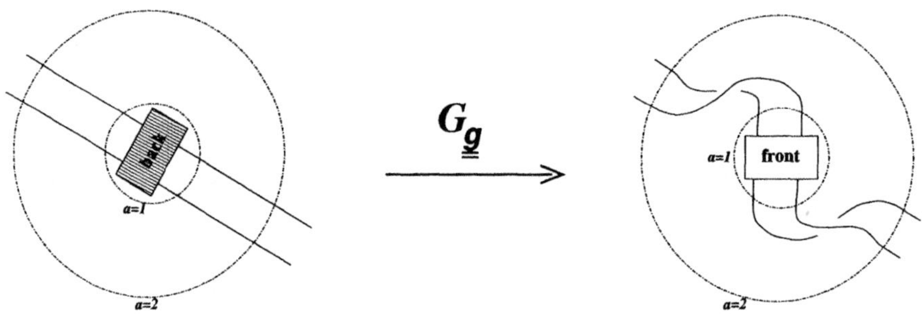

The map $\mathbf{sp}^{-1}$ is now given by, first, selecting a tangle representing a given class in $\mathcal{T}gl_{\mathbb{R}^2}$. Then we choose for each decoration a rotation $\underline{g} \in SO(3)$ that puts the respective coupon or joint into its normal orientation, as well as a path in $SO(3)$ from $\mathbb{1}$ to each $\underline{g}$. The image of $\mathbf{sp}^{-1}$ in $\mathcal{T}gl^{well-pos}_{\mathbb{R}^2}$ is then represented by the well-positioned tangle diagram that we obtain by applying the respective $G_{\underline{\underline{g}}}$ map to each vicinity of a decoration.

It is obvious with this definition that $\mathbf{sp}^{-1} \circ \mathbf{sp} = \mathrm{Id}$ on $\mathcal{T}gl^{well-pos}_{\mathbb{R}^2}$. Since each $G_{\underline{\underline{g}}}$ is via (2.4.1) isotopic to the identity, also the original tangle and the well-positioned one are isotopic in $\mathcal{T}gl_{\mathbb{R}^2}$. Hence, we also have that $\mathbf{sp} \circ \mathbf{sp}^{-1} = \mathrm{Id}$ in $\mathcal{T}gl_{\mathbb{R}^2}$.

It still remains to be shown that $\mathbf{sp}^{-1}$ is actually well defined on the classes of $\mathcal{T}gl_{\mathbb{R}^2}$, and independent of the choices of paths $\underline{\underline{g}} : [0,1] \to \mathcal{Q}$. If we have two homotopic paths $\underline{\underline{p}}$ and $\underline{\underline{q}}$, both starting at $\mathbb{I} \in \mathcal{Q}$, then quite obviously also the corresponding traced out homeomorphisms are *isotopic*, i.e.,

$$\underline{\underline{p}} \sim \underline{\underline{q}} \qquad \Longrightarrow \qquad G_{\underline{\underline{p}}} \sim G_{\underline{\underline{q}}}.$$

Here, the first relation $\sim$ means homotopy of paths $([0,1],\{0\}) \to (\mathcal{Q},\{\mathbb{I}\})$ with start points always constant $\mathbb{I}$ throughout the homotopy. The second relation $\sim$ means isotopy, which is always identity outside of D_2^3, and inside D_1^3 it acts as the family of rotations given by the path of end-points from $\underline{p}(1)$ to $\underline{q}(1)$.

Let $t \mapsto \underline{h}(t)$ be a path starting at the end-point of a path $\underline{\underline{p}}$ in $\mathcal{Q}$, which itself starts at $\mathbb{I}$, i.e., $\underline{h}(0) = \underline{p}(1)$. Up to homotopy we can define a composition $\underline{\underline{h}}*\underline{\underline{p}}$ as in the definition of fundamental groups. Observe then that both $\underline{\underline{h}}^\sharp : t \mapsto \underline{h}(t)\cdot\underline{h}(0)^{-1}$ and $\underline{\underline{h}}^\flat : t \mapsto \underline{h}(0)^{-1}\cdot\underline{h}(t)$ are paths in $\mathcal{Q}$ starting at $\mathbb{I}$. It is not hard to see that the composite path yields the following composition of homeomorphisms:

$$G_{\underline{\underline{h}}*\underline{\underline{p}}} \quad \sim \quad G_{\underline{\underline{h}}^\sharp}\circ G_{\underline{\underline{p}}} \quad \sim \quad G_{\underline{\underline{p}}}\circ G_{\underline{\underline{h}}^\flat}.$$

The isotopies in both cases are constant $\mathbb{I}$ outside D_2^3, *and* constant $\underline{h}(1)$ inside D_1^3. They are constructed by deforming $\underline{\underline{h}}$ and $\underline{\underline{p}}$ into paths that are constant on opposite halves of the interval $[0,1]$. This implies that $G_{\underline{\underline{h}}^{\sharp/\flat}}$ and $G_{\underline{\underline{p}}}$ will be constant on correspondingly different parts of $[1,2]\times S^2 = D_2^3 - D_1^3$.

Now, let us suppose two tangles T_1 and T_2 in $\mathbb{R}^2\times[-1,1]$ to be isotopic to each other, and let T_1^{wp} and T_2^{wp} be the corresponding well positioned tangles. In the (ambient) isotopy moving T_1 to T_2 the center point of each decoration is moved along a path in $\mathbb{R}^2\times[-1,1]$ and at the same time the decoration is rotated around this center point. It is clear that we can separate the given isotopy into two, one from T_1 to a tangle $\hat{T}_2$ and another from $\hat{T}_2$ to T_2. The one from T_1 to $\hat{T}_2$ rotates the decorations around their center points without moving them, and the one from $\hat{T}_2$ to T_2 moves the decorations along the paths without changing their orientations. Since the choice of the well positioned tangles only depends on these orientations, we have that all rotations used to move any tangle along the isotopy from $\hat{T}_2$ to T_2 into a well position are the same for a given decoration. Hence, we can construct an isotopy from $\hat{T}_2^{wp}$ to T_2^{wp} *within* the well-positioned tangles.

It suffices, therefore, to consider only isotopies $\Psi : [0,1] \to \mathcal{D}iff(\mathbb{R}^2\times[-1,1])$ deforming T_1 into T_2 without moving the center points of the decorations. A tangle T_j, with $j = 1,2$, is mapped to T_j^{wp} by $G_{\underline{\underline{p}}^j}$ for some path $\underline{\underline{p}}^j : [0,1] \to \mathcal{Q}$. Hence, $G_{\underline{\underline{p}}^2}\circ\Psi_1\circ G_{\underline{\underline{p}}^1}^{-1} : \mathbb{R}^2\times[-1,1] \to \mathbb{R}^2\times[-1,1]$ is a homeomorphism that maps T_1^{wp} to T_2^{wp}. Moreover, since all of the factors are isotopic to the identity with isotopies that leave center points of decorations fixed, the same is true for the composite. That is, there is $t \mapsto \Phi_t \in \mathcal{D}iff(\mathbb{R}^2\times[-1,1])$ such that $\Phi_0 = id$, $\Phi_t(\mathbf{p}) = \mathbf{p}$ with $\mathbf{p}$ the center of decoration, and $\Phi_1(T_1^{wp}) = T_2^{wp}$. At the point $\mathbf{p}$ we can define a rotation $\underline{\underline{f}} : [0,1] \to SO(3) : t \mapsto D\Phi_t(\mathbf{p})$ that starts at $\mathbb{I}$ and ends at $\mathbb{I}$ or $\underline{r}$. This defines an isotopy $t \mapsto G_t$, where $G_t(\mathbf{x}) = G(t,\mathbf{x})$ is as defined in (2.4.1), that acts close to $\mathbf{p}$

by the same rotations. Hence, the isotopy $t \mapsto G_{\underline{t}}^{-1} \circ \Phi_t$ causes no rotation around p, and can, thus, be "stretched" to an isotopy within the well positioned tangles, from T_1^{wp} to $\tilde{T}_1^{wp} = G_{\underline{f}}^{-1} \circ \Phi_1(T_1^{wp})$.

It remains to be shown that $\tilde{T}_1^{wp}$ and T_2^{wp} are equivalent, that is, two tangles are related by an application of $G_{\underline{f}}$ for any path $\underline{f} : [0,1] \to \mathcal{Q}$ starting or ending in $\mathbb{1} \in \mathcal{Q}$. We know that a homotopy of such paths with fixed endpoints translates to isotopies of $G_{\underline{f}}$ constant at the decoration and with local compact support. It is, therefore, enough to show the equivalence under application of $G_{\underline{f}}$ only for homotopy classes of paths, i.e., for a set of representatives of $\pi_1(\mathcal{Q}) \cong \mathbb{Z}_4$. Also we know that the composition in $\pi_1(\mathcal{Q})$ can be written as compositions of the respective $G_{\underline{f}}$'s so that it really suffices to check only the generators of the fundamental group. Now, the *one* generator of $\pi_1(\mathcal{Q})$ is represented by the rotation $\underline{r} : [0,1] \to SO(3) : t \mapsto \exp(t\pi\Omega_z)$, in the $\mathbb{R}_x \times [-1,1]$ plane, where the matrix Ω_z is $\frac{\pi}{2}$ rotation in $\mathbb{R}_x \times [-1,1]$, which is zero on the normal vector to this plane. Note that $\underline{r}(1)$ is, hence, the rotation by π in the plane.

The action of the generator $G_{\underline{r}}$ is depicted for both types of decorations in the diagrams below. We have applied here additional isotopies on both sides of the move in order to make the effect of $G_{\underline{r}}$ more transparent:

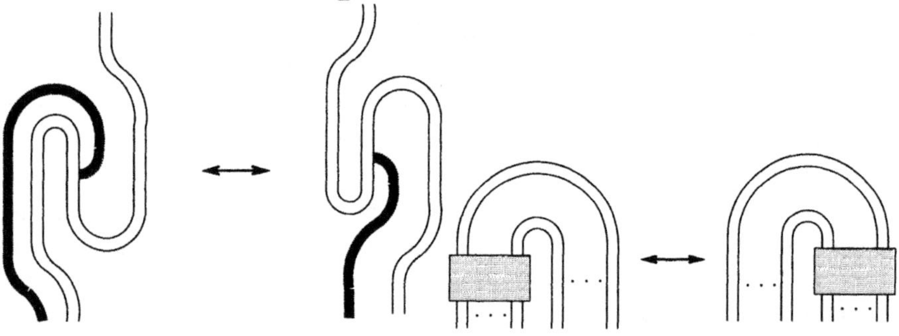

The fact that the coupon move is already implied by the other moves in $\mathcal{T}gl_{\mathbb{R}^2}^{well-pos}$ can be seen by first applying the (reverse) handle trade move TS2 to the coupon on the left side, which yields a diagram without any coupons. The extra annulus can then be moved into the position of the coupon on the other side of the move using only ordinary tangle isotopies. Applying again TS2 in the reverse way, we obtain the coupon in the correct position on the right side.

In the case of the joint we have to apply the additional move TI6 to shift the joint through two extrema. Note, that we also have to consider the equivalence obtained by applying $G_{\underline{r}}$ to the reflected version of the above picture, which follows analogously.

Hence, if two tangle diagrams are isotopic then their well-positioned tangles are equivalent in $\mathcal{T}gl_{\mathbb{R}^2}^{well-pos}$. All other equivalences in $\mathcal{T}gl_{\mathbb{R}^2}$ are also translated in a straightforward fashion into equivalences in $\mathcal{T}gl_{\mathbb{R}^2}^{well-pos}$. Hence, sp^{-1} is well defined, which completes the proof of the assertion of the lemma.

The reduction to well-positioned tangles leads to a bijection of tangles classes also in the case in which we start from tangles that are orientable but not oriented strands. The main difference is now that the normal vector $\mathbf{v}_1$, as defined in the beginning of this section, is defined only up to a sign.

As a result the symmetry group in $SO(3)$ between positionings grows from the $\mathbb{Z}_2(\underline{r})$ subgroup to the abelian group $\mathbb{Z}_2(\underline{r}) \times \mathbb{Z}_2(\underline{s})$ of order four. Here $\underline{s} \in SO(3)$ is a reflection such that $\underline{s}\mathbf{v}_1 = -\mathbf{v}_1$ and $\underline{s} \cdot \underline{r} = \underline{r} \cdot \underline{s}$. Both generators lift in $SU(2)$ to generators $\tilde{\underline{r}}$ and $\tilde{\underline{s}}$ of order four, for which the relations $\tilde{\underline{r}}^2 = \tilde{\underline{s}}^2 = -\mathbb{1}$ and $\tilde{\underline{r}} \cdot \tilde{\underline{s}} \cdot \tilde{\underline{r}} = \tilde{\underline{s}}$ are readily verified. Hence, the abelian group in $SO(3)$ lifts to the quaternion group $Q(8) \subset SU(2)$ of order eight, generated by $\tilde{\underline{r}}$ and $\tilde{\underline{s}}$. Analogously to the oriented case, general positions of decorations or isotopies between well positionings are now given by points or closed paths in $\mathcal{Q}^*$,

$$\mathcal{Q}^* = \frac{SO(3)}{\mathbb{Z}_2(\underline{r}) \times \mathbb{Z}_2(\underline{s})} = \frac{S^3}{Q(8)} \quad \text{with} \quad \pi_1(\mathcal{Q}^*) = Q(8).$$

The additional generator of $\pi_1(\mathcal{Q}^*)$ is given by the one parameter rotation $\underline{\underline{s}} : [0,1] \to SO(3) : t \mapsto \exp(t\pi\Omega_y)$, mapped down by the projection $SO(3) \to \to \mathcal{Q}^*$. Here Ω_y is the skew matrix, which is a rotation of $\frac{\pi}{2}$ of, e.g., the plane $\mathbb{R}^2 \times 1 \subset \mathbb{R}^2 \times [-1,1]$ and whose kernel consists of directions parallel to the line $\{0\} \times [-1,1] \subset \mathbb{R}^2 \times [-1,1]$. In this setting the reduction to well positioned tangles is still a one-to-one correspondence:

Lemma 2.4.4. *The map* **sp** *is a bijection on the 2-arrow sets (without orientations).*

Proof. What remains to be shown is that the isotopy $G_{\underline{\underline{s}}}$, as defined in (2.4.1), and $\underline{\underline{s}}$ as in the previous paragraph, applied to a decoration in $\mathcal{T}\!gl_{\mathbb{R}^2}$ can be reexpressed by the moves in $\mathcal{T}\!gl_{\mathbb{R}^2}^{well-pos}$. In the case of coupons this follows as before by conjugation with the TS1 Move. For joints we observe that $G_{\underline{\underline{s}}}$ is precisely the rotation used to describe TI11. The way the π-twists created by this rotation on the internal strand are related to an orientation switch has been discussed in the paragraphs following Definition 2.4.2.

The next step in our reduction to the planar calculus is to introduce an equivalent category $\mathcal{T}\!gl_{\mathbb{R}^2}^{dec-proj}$. The 2-arrow sets are equivalence classes of tangles with *projectable decorations*. This means that a representative tangle is well positioned and in addition fulfills the following condition:

If $C \subset [-1,1] \times \mathbb{R}^2$ is a decoration then $p_x^{-1}(p_x(C)) = C + \mathbb{R}_y$, where $\mathbb{R}_y = \ker(p_x)$, is disjoint from all other parts of the tangle. In other words, the projection of C into $[-1,1] \times \mathbb{R}_x$ is separated from the projections of all other tangle pieces.

We endow the set of tangles with projectable decorations with all relations from $\mathcal{T}\!gl_{\mathbb{R}^2}^{well-pos}$ and, in addition, with the three-dimensional analog of TI7. The resulting equivalence classes are the 2-arrows of $\mathcal{T}\!gl_{\mathbb{R}^2}^{dec-proj}$. Clearly, the inclusion of the tangles with projectable decorations into the well positioned tangles factors into the equivalence classes, since TI7 can be interpreted as an isotopy in $\mathcal{T}\!gl_{\mathbb{R}^2}^{well-pos}$.

Lemma 2.4.5. *The map* $\mathcal{T}\!gl_{\mathbb{R}^2}^{dec-proj} \to \mathcal{T}\!gl_{\mathbb{R}^2}^{well-pos}$ *is a bijection.*

Proof. The method to show bijection is again by construction of a map from $\mathcal{T}gl^{well-pos}_{\mathbb{R}^2}$ to $\mathcal{T}gl^{dec-proj}_{\mathbb{R}^2}$ that is inverse to the natural one. To this end we consider the lines $\{0_C\} + \mathbb{R}_y$, where $0_C \in C$ is a selected center point of a decoration. The interior $C - \{0_C\}$ has a retraction isotopy to a vicinity around C within the plane in which C lies. This isotopy clearly extends to one from the cylinder over the decoration $(C - \{0_C\}) + \mathbb{R}_y$ to a vicinity. Thus, to any tangle in which all other elements are disjoint from these special lines we can associate an isotopic tangle with projectable decoration, which is unique up to isotopy within $\mathcal{T}gl^{dec-proj}_{\mathbb{R}^2}$.

Any tangle in $[-1,1] \times \mathbb{R}^2$ is obviously isotopic to one that avoids all of the special lines $\{0_C\} + \mathbb{R}_y$, and any isotopy of tangles can be made transverse to these lines. After application of the deformation to the outside of $C + \mathbb{R}_y$ we obtain a projectable tangle, and the tangles shortly before and shortly after the transverse intersection differ by moving a strand from one side of the decoration to the other. For a coupon the resulting move is depicted on the right. This move is already implied by the other moves in $\mathcal{T}gl^{well-pos}_{\mathbb{R}^2}$. As before we show this by applying first TS2, then isotoping the extra strand over the configuration with the annulus, and then reintroducing the coupon by reverse application of TS2.

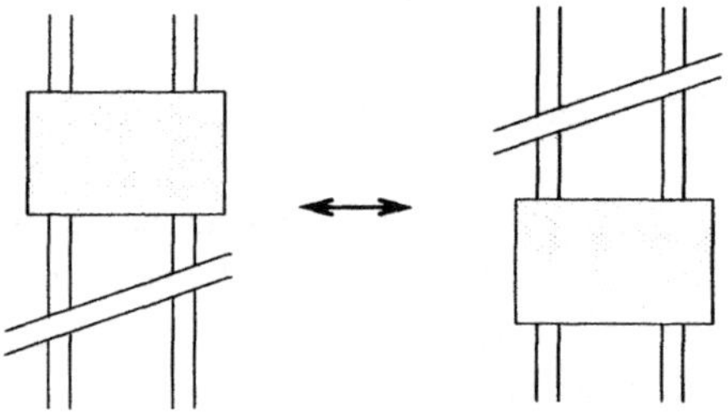

The respective move for joints translates precisely into TI7. We, thus, have a well defined inverse and the lemma is thereby proven.

In the next step of the reduction we put the remaining tangle into general position with respect to the projection map $p_x : \mathbb{R}^2 \times [-1,1] \to \mathbb{R}_x \times [-1,1]$. In particular, we demand that the projection of the ribbons of the tangle are immersions into the plane except for a finite number of isolated positions, where the ribbon is allowed to make a 2π-twist (or π-twist if auxiliary) as in Picture 3 of Section 2.1.3. Moreover, different ribbon parts in the projection may intersect only pairwise and transversally, and the 2π-twists is separated from the intersection.

A 2-arrow set of the category $\mathcal{T}gl^{plan}$ of planar tangles is now defined as the set of equivalence classes of such tangles that are in general position with respect to p_x, where the equivalence relations are those of $\mathcal{T}gl^{dec-proj}_{\mathbb{R}^2}$ and the generalized versions of TI1, TI2, TI3, and TI10, where the diagrams may be rotated in the plane of projection.

As all of the additional moves are interpreted as regular isotopies within $\mathcal{T}gl^{dec-proj}_{\mathbb{R}^2}$ we have a well defined map from equivalence classes of planar maps to respective classes of projectable tangles.

Lemma 2.4.6. *The map $\mathcal{T}gl^{plan}_{\mathbb{R}^2} \to \mathcal{T}gl^{dec-proj}_{\mathbb{R}^2}$ is a bijection.*

Proof. As before, the basic strategy is to find a well defined inverse $\mathcal{T}gl^{dec-proj} \to \mathcal{T}gl^{plan}_{\mathbb{R}^2}$, by expressing isotopies in $\mathcal{T}gl^{dec-proj}$ in terms of the moves in $\mathcal{T}gl^{plan}_{\mathbb{R}^2}$. More precisely, we will show that any projectable tangle in $\mathcal{T}gl^{dec-proj}$ is isotopic to a planar tangle, and, further, that any isotopy between two planar tangles within the set of projectable tangles can be reduced to a series of isotopies within the set of planar tangles interrupted by applications of the additional moves.

Indeed, we first consider an intermediate type of tangle classes, which will replace $\mathcal{T}gl^{dec-proj}$. They are derived in the first three paragraphs of this proof. The classes correspond to ribbon tangles whose planar projections are immersions with the exception of isolated π-twists in the strands. We sometimes call them tangles with planar orientation.

In part *A)* we show that tangles with planar orientations are dense in the set of projectable tangles. Parts *B)* and *C)* discuss two types of critical points of a generic isotopy of projectable tangles. At these points the tangles fail to have planar orientation, and additional moves need to be introduced, similar to the ones for $\mathcal{T}gl^{plan}_{\mathbb{R}^2}$.

The general position arguments involved in the derivation of generic isotopies are technically given by partial jet transversality conditions along special submanifolds in higher dimensional Euclidean spaces. To describe these more precisely we define the following open subset of $\mathbb{R}^6$:

$$\mathcal{X}^0 = \{(\mathbf{b},\mathbf{c}) : \dim\langle \mathbf{b},\mathbf{c}\rangle_{\mathbf{R}} = 2\} \quad \subset \mathbb{R}^3 \times \mathbb{R}^3 = \mathbb{R}^6 .$$

We identify the following subsets of it:

$$\mathcal{X}^1 = \big\{(\mathbf{b},\mathbf{c}) : \dim\langle p_x(\mathbf{b}), p_x(\mathbf{c})\rangle_{\mathbf{R}} = 1\big\},$$
$$\mathcal{X}^2 = \big\{(\mathbf{b},\mathbf{c}) \in \mathcal{X}^1 : p_x(\mathbf{b}) = 0\big\}.$$

Here $\langle \mathbf{x}_1,\dots,\mathbf{x}_k\rangle_{\mathbf{R}}$ denotes the linear subspace spanned by the vectors $\mathbf{x}_1,\dots,\mathbf{x}_k$.

It is clear that $\mathcal{X}^1 \subset \mathbb{R}^3 \times \mathbb{R}^3$ is a submanifold of codimension 1. Further, $\mathcal{X}^2$ is a submanifold of $\mathcal{X}^1$ with codim 1, hence, also a submanifold of $\mathbb{R}^3 \times \mathbb{R}^3$ of codim 2. In fact, $\mathcal{X}^1 \cong T^2 \times \mathbb{R}^3$ and $\mathcal{X}^2 \cong S^1 \times \mathbb{R}^3$, where $S^1 \subset T^2$ is a non separating cycle in the 2-dimensional torus.

A) Generic planar ribbons and π-twists:

A framed tangle piece is given by a pair of paths $t \mapsto \mathbf{x}(t)$ and $t \mapsto \nu(t)$ in $\mathbb{R}^3$, where the first parametrizes the curve in three space that the tangle follows and $\nu(t)$ determines the direction of the framing at the point $\mathbf{x}(t)$. For the curve to be a submanifold and the framing to be well defined we find the condition that for each t the vectors $\frac{d}{dt}\mathbf{x}(t)$ and $\nu(t)$ are linearly independent. In particular, both are always non-zero. More formally, we have a *framed map*:

$$\rho : [t_0,t_1] \longrightarrow \mathcal{X}^0 \subset \mathbb{R}^3 \times \mathbb{R}^3 : t \mapsto \rho(t) = \Big(\frac{d}{dt}\mathbf{x}(t), \nu(t)\Big).$$

By a variant of jet transversality the set of ribbon embeddings, such that the corresponding framed maps $[t_0,t_1] \to \mathcal{X}^0$ are transverse to both $\mathcal{X}^1$ and $\mathcal{X}^2$, is open and dense. Hence, modulo an arbitrarily small isotopy we can always arrange that $\rho \pitchfork \mathcal{X}^1$ and $\rho \pitchfork \mathcal{X}^2$. Counting the codimensions, the latter means simply that ρ never meets $\mathcal{X}^2$, and, moreover, ρ intersects $\mathcal{X}^1$ in a finite number of points with non-zero normal velocity.

Since $im(\rho) \cap \mathcal{X}^2 = \varnothing$, the composite map $t \mapsto \hat{x}(t) = p_x(\mathbf{x}(t)) : [t_0, t_1] \to \mathbb{R}^2$ is an *immersion*. Away from $\mathcal{X}^1$ the vector $\hat{\nu}(t) = p_x(\nu(t)) \in \mathbb{R}^2$ is never tangent to $t \mapsto \hat{x}(t)$. That is, $\hat{\nu}(t)$ always has a non-zero component in the path's normal bundle in $\mathbb{R}^2$. At a special time, $t_i \in [t_0, t_1]$, when ρ intersects $\mathcal{X}^1$, we have that $\hat{\nu}(t)$ switches from one side of $\hat{x}(t)$ to the other. At the intersection time itself we have that $\hat{\nu}(t_i)$ is parallel to $\frac{d}{dt}\hat{x}(t_i)$ or zero. However, since $\rho(t_i) \in \mathcal{X}^0$, we have that $\nu(t_i)$ is *not* parallel to $\frac{d}{dt}\mathbf{x}(t_i)$ so that we can distinguish between over- and undercrossings. In the picture of ribbon tangles this means that locally a ribbon piece is flatly embedded into $\mathbb{R}^2$ for generic times, and at special points $\hat{x}(t_i)$ we encounter a π-twist at the ribbon in one or the other direction. Such a twist is depicted in Picture 3 of Section 2.1.3 (π-twist for auxiliary and 2π-twists for other ribbons).

B) Isotopies of planar ribbons - Cancellation on π-twists:

Let next consider isotopies $R : [s_0, s_1] \times [t_0, t_1] \to \mathcal{X}^0 : (s,t) \mapsto \rho_s(t)$ of framed tangle pieces and make them transverse to the special submanifolds. Specifically, the first condition that we impose for transversality with respect to $\mathcal{X}^1$ is

$$(R, \frac{\partial R}{\partial t}) \pitchfork T\mathcal{X}^1 .$$

The functions with this properties form by jet-transversality a dense and open set within the functions from $[s_0, s_1] \times [t_0, t_1]$ to $T\mathcal{X}^0$, so any given isotopy can be perturbed into one as above. Since $T\mathcal{X}^1 \subset T\mathcal{X}^0$ has codimension 2, the points, at which a transverse $R(s,t)$ intersects $\mathcal{X}^1$, but $\frac{\partial R}{\partial t}(s,t) = \frac{d}{dt}\rho_s(t)$ is tangent to $\mathcal{X}^1$, form a finite discrete subset of $\mathcal{G}$. That is, we have a set $(R, \frac{\partial R}{\partial t})^{-1}(T\mathcal{X}^1) = \{(s_k, t_k) : k = 1, \ldots N\}$. Furthermore, transversality implies that at these points both $\frac{\partial^2 R}{\partial t^2}(s_k, t_k)$ and $\frac{\partial R}{\partial s}(s_k, t_k)$ have non-zero normal components at $\mathcal{X}^1$.

At all other points where $R(s,t)$ intersects $\mathcal{X}^1$ the vector $\frac{\partial R}{\partial t}$ will have non-zero normal component. Thus, our transversality condition, in particular, implies $R \pitchfork \mathcal{X}^1$. This, in turn, means that $\mathcal{G} = R^{-1}(\mathcal{X}^1) \subset [s_0, s_1] \times [t_0, t_1]$ is a one-dimensional submanifold of the st-plane, which clearly contains all the points $(s_k, t_k), k = 1, \ldots, N$.

Since the normal part of $\frac{\partial R}{\partial t}$ is never zero on $\mathcal{G}^0 = \mathcal{G} - \{(s_1, t_1), \ldots, (s_N, t_N)\}$, we can parametrize $\mathcal{G}^0$ using the implicit function theorem. The components $\mathcal{G}^0_\mu$ of $\mathcal{G}^0$ are open paths $s \mapsto t_\mu(s)$, where s runs between two of the special s-coordinates. I.e., $s \in I_\mu = (s_{k^i_\mu}, s_{k^f_\mu})$ and $t_\mu(s) \to t_{k^{i/f}_\mu}$ as $s \to s_{k^{i/f}_\mu}$.

Thus, for $\rho_s(t) = (\frac{d}{dt}\mathbf{x}_s(t), \nu_s(t))$, the points where $\frac{d}{dt}\hat{x}_s(t)$ and $\hat{\nu}_s(t)$ are parallel in the projection into $\mathbb{R}^2$ (given by intersections of ρ_s with $\mathcal{X}^1$) are given by all $\hat{x}_\mu(s) = \hat{x}_s(t_\mu(s))$ for which $s \in I_\mu$. Now, the $\hat{x}_\mu(s)$ give the positions of all of the π-twists along the projected strands for a given parameter s. Hence, the existence of the paths $s \mapsto x_\mu(s)$ obtained from transversality implies that over the interval I_μ the position of the π-twists is moved continuously through the strands.

At the special points (s_k, t_k) we have, by transversality, that for normal component $R^\perp$ (e.g., in a Cartesian submanifold chart) we have $\frac{\partial R^\perp}{\partial s} \neq 0$ so that locally we can parametrize $\mathcal{G}$ in the form $t \mapsto (s(t), t)$ for some function $t \mapsto s(t)$. This is shown by use of implicit functions and the fact that locally $\mathcal{G}$ can be thought of as the zero set of $R^\perp$. Now, as $R^\perp(s(t), t) = 0$ we find by differentiating with respect to t and using that $\frac{\partial R}{\partial s} \neq 0$ but $\frac{\partial R}{\partial s} = 0$ at (s_k, t_k), that $\frac{ds}{dt}(t_k) = 0$. Differentiating again, we find

$$\frac{d^2}{dt^2} s(t_k) = -\frac{\frac{\partial^2}{\partial t^2} R(s_k, t_k)}{\frac{\partial}{\partial s} R(s_k, t_k)} \neq 0.$$

This means that the point (s_k, t_k) is either a non-degenerate local maximum or minimum of the s-coordinate function restricted to $\mathcal{G}$. Hence, e.g., in the case of a maximum, we have that in a small vicinity around (s_k, t_k) there are always two points on $\mathcal{G}$ with given s-coordinate if $s < s_k$ and no points for $s > s_k$. The pairs of points for $s < s_k$ clearly belongs to a pair of paths $s \mapsto t_{\mu_1}(s)$ and $s \mapsto t_{\mu_2}(s)$ parametrizing the respective components of $\mathcal{G}$, and $t_{\mu_i}(s) \to t_k$ as $s \nearrow s_k$ for both $i = 1, 2$. Hence, also the movements $s \mapsto \hat{x}_{\mu_1}(s)$ and $s \mapsto \hat{x}_{\mu_2}(s)$ of the π-twists along the strands converge to an end in the same point $\hat{x}_{s_k}(t_k)$, where they "cancel" each other.

For the normal component $\rho^\perp_{s_k}$ of the critical path we have that $\rho^\perp_{s_k}(t_k) = 0$, $\frac{d}{dt}\rho^\perp_{s_k}(t_k) = 0$, as well as $\frac{d^2}{dt^2}\rho^\perp_{s_k}(t_k) \neq 0$. Thus, the path only touches $\mathcal{X}^1$ and stays on one side of $\mathcal{X}^1$, where its normal component behaves in first order like $\rho^\perp_{s_k}(t) \sim \alpha(t - t_c)^2$ with $\alpha \neq 0$.
The paths $t \mapsto \rho_s(t)$ are depicted for different values of s on the right.

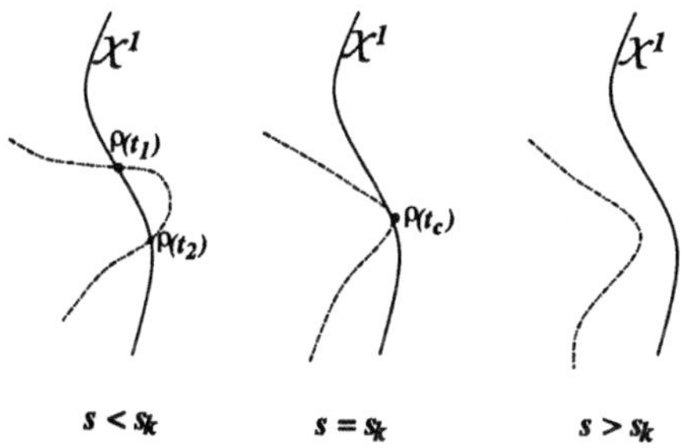

For $s < s_k$ in the maximum-case we have from $\frac{d}{dt}\rho_s(t) = -\frac{\partial R}{\partial s}\frac{ds}{dt}$ so that $\frac{d}{dt}\rho_s(t)$ changes sign as we pass from the point $(s, t_{\mu_1}(s))$ of one branch to the corresponding point $(s, t_{\mu_2}(s))$, since $\frac{\partial R}{\partial s}$ is not zero (hence, does not change sign) close to (s_k, t_k), but $\frac{ds}{dt}$ changes sign at t_k since $\frac{d^2 s}{dt^2} \neq 0$.

This implies that the orientation, in which the vectors $\frac{d}{dt}\hat{x}_s$ and $\hat{\nu}_s$ cross over each other as t runs through $t_{\mu_1}(s)$, is opposite to the crossing orientation at $t_{\mu_2}(s)$. Moreover, since both vectors in $\mathbb{R}^2$ are parallel at (s_k, t_k) and the image of ρ_{s_k} lies in $\mathcal{X}^0$, the vectors $\frac{d}{dt}\mathbf{x}_s(t)$ and $\boldsymbol{\nu}_s(t)$ are well ordered by the normalized component orthogonal to the plane of projection. This ordering does not change close to (s_k, t_k) by continuity, and together with the crossing orientation of $\frac{d}{dt}\hat{x}_s$ over $\hat{\nu}_s$ determines the orientation of the π-twist.

In summary, this means that the two π-twists, with *opposite* orientations that are moved together along a strand, can be canceled against each other. The corresponding picture is given by the right part in the moves TI10 in Section 2.3.1 for auxiliary strands.

C) Isotopies of planar ribbons - Framed first Reidemeister move:

Additionally, let us impose and discuss transversality at $\mathcal{X}^2$. The conditions are:

$$R \pitchfork \mathcal{X}^2 \quad \text{and} \quad (R, \frac{\partial R}{\partial t}) \pitchfork T_{\mathcal{X}^2}\mathcal{X}^1 .$$

Now, $\mathcal{X}^2$ has codimension 2 so that a transverse R meets $\mathcal{X}^2$ only in a bunch of isolated points $R^{-1}(\mathcal{X}^2) = \{(\dot{s}_\alpha, \dot{t}_\alpha)\}_{\alpha=1}^K$. Since $\mathcal{X}^2 \subset \mathcal{X}^1$, this means, in particular, that any of these points is contained in the curve in the st-plane, along which the π-twists occur. That is, $(\dot{s}_\alpha, \dot{t}_\alpha) \in \mathcal{G} = R^{-1}(\mathcal{X}^1$ for all α.

Clearly, $T_{\mathcal{X}^2}\mathcal{X}^1$ has codimension 3 so that the last condition really means $im(R, \frac{\partial R}{\partial t}) \cap T_{\mathcal{X}^2}\mathcal{X}^1 = \varnothing$, and, hence, $\frac{\partial R^\perp}{\partial t}(\dot{s}_\alpha, \dot{t}_\alpha) \neq 0$ for all α. This means, in particular, that these points are all different from the special points (s_k, t_k) from above, which are mapped to $T\mathcal{X}^1$ by $(R, \frac{\partial R}{\partial t})$. Hence, in a vicinity of each $(\dot{s}_\alpha, \dot{t}_\alpha)$ we can parametrize $\mathcal{G}$ by $s \mapsto (s, t(s))$. This implies that, locally, for $s < \dot{s}_\alpha$ as well as for $s > \dot{s}_\alpha$ the paths $t \mapsto \rho_s(t)$ will intersect $\mathcal{X}^1$ exactly once at times close to $\dot{t}_\alpha$. Thus, the local ribbon diagram shortly before and shortly after the passage of the deformation parameter through $\dot{s}_\alpha$ will include one π-twist, which will, however, change its orientation.

Next, let us investigate the more explicit consequences of the transversality conditions at a particular point $(\dot{s}_\alpha, \dot{t}_\alpha)$. To this end we may assume $\dot{s}_\alpha = 0 = \dot{t}_\alpha$ after shifting variables. The requirement $R(0,0) \in \mathcal{X}^2$ then translates to the conditions on vectors in $\mathbb{R}^2$: $\frac{d}{dt}\hat{x}_0(0) = 0$ and $\hat{\nu}_0(0) \neq 0$.

The condition $\frac{\partial R}{\partial t}(0,0) \in T_{R(0)}\mathcal{X}^1$ then means for this case that $\frac{d^2}{dt^2}\hat{x}_0(0)$ is parallel to $\hat{\nu}_0(0)$. Thus, transversality implies that these two vectors are linearly independent.

Moreover, the condition $R \pitchfork \mathcal{X}^2$ implies that $\frac{\partial R}{\partial s}$ and $\frac{\partial R}{\partial t}$ complement the vector space $T_{R(0)}\mathcal{X}^2 \cong \mathbb{R}^4 \subset \mathbb{R}^6$. In coordinate form this is a condition on vectors in the plane of projection: $\frac{d^2}{dt^2}\hat{x}_0(0)$ and $\frac{\partial d}{\partial s dt}\hat{x}_0(0)$ span $\mathbb{R}^2$.

Now, as long as we do not pass over $\frac{d}{dt}\hat{x}_s(t)$ we may apply small deformations to the framing vector. Since in lowest order $\frac{d}{dt}\hat{x}_s(t)$ will be close to $\frac{d^2}{dt^2}\hat{x}_0(0)$, we can use the above condition to make the framing vector parallel to $\frac{\partial d}{\partial s dt}\hat{x}_0(0)$, not only at $(s,t) = (0,0)$ but also in a vicinity of the origin. Also we can always apply a linear transformation to move any two linearly independent vectors to the unit coordinate vectors $\pm\mathbf{e}_x$ and $\mathbf{e}_y$, depending on the relative orientations of $\frac{d^2}{dt^2}\hat{x}_0(0)$ and $\frac{\partial d}{\partial s dt}\hat{x}_0(0)$.

After some rescaling of vectors and parameters we may summarize the situation as follows:

$$\frac{d}{dt}\hat{x}_0(0) = 0, \quad \frac{d^2}{dt^2}\hat{x}_0(0) = \pm 2\mathbf{e}_x, \quad \frac{\partial d}{\partial s dt}\hat{x}_0(0) = \mathbf{e}_y, \quad \text{and} \quad \nu_s(t) = \pm\mathbf{e}_y,$$

in a vicinity of $(0,0)$. Among the four indicated cases (two for each sign) we consider in the following only the positive signs. The other cases follow via reflections.

Note, that from the above formulae we can see explicitly that the strand direction at the π-twist $s \mapsto \frac{d}{dt}\hat{x}_s(t(s))$ flips its direction, whereas $s \mapsto \hat{\nu}_s(t(s))$ maintains its direction. Hence, the π-twist does change its orientation as s passes through 0.

We may further apply an s-dependent isotopy $\psi_s : \mathbb{R}^2 \to \mathbb{R}^2 : \mathbf{y} \mapsto \mathbf{y} - \hat{x}_s(0)$ so that for the isotoped path $\hat{x}'_s(t) = \psi_s(\hat{x}_s(t))$ we have $\hat{x}'_s(0) = 0$. We shall, thus, assume this condition on $\hat{x}_s(t)$ to also hold. Hence, in a power expansion of $\hat{x}_s(t)$ there will be no pure powers of s occurring. From the above we find that to second order we have:

$$\hat{x}_s(t) = t^2\mathbf{e}_x + st\mathbf{e}_y + O_2(s,t),$$

where $\frac{O_2(s,t)}{t^2+s^2} \to 0$ as $(s,t) \to 0$. In order to find a useful model for this move we have to add higher order terms in order to fix two points on the strand for all s-parameters so that we can consider the move as a local move inside a diagram that remains unchanged outside a vicinity of $\hat{x}_0(0) = 0$. More precisely, we wish to add third and forth order terms, such that, as before, $\hat{x}_s(0) = 0$ and, further, $\hat{x}_s(-1) = \hat{a}$ and $\hat{x}_s(1) = \hat{b}$, where $\hat{a}, \hat{b} \in \mathbb{R}^2$ will be distinct and independent of s.

If we write $\hat{x}_s(t) = x(s,t)\mathbf{e}_x + y(s,t)\mathbf{e}_y$, the easiest functions that fulfill these conditions (and are even and odd in t for the x or y variable, respectively) are the following:

$$x(s,t) = t^2 \quad \text{and} \quad y(s,t) = s(t - t^3) - t^3,$$

so that $\hat{a} = \mathbf{e}_x + \mathbf{e}_y$ and $\hat{b} = \mathbf{e}_x - \mathbf{e}_y$. The paths $t \mapsto \hat{x}_s(t)$ for $t \in [-1,1]$ are depicted below for different values of s.

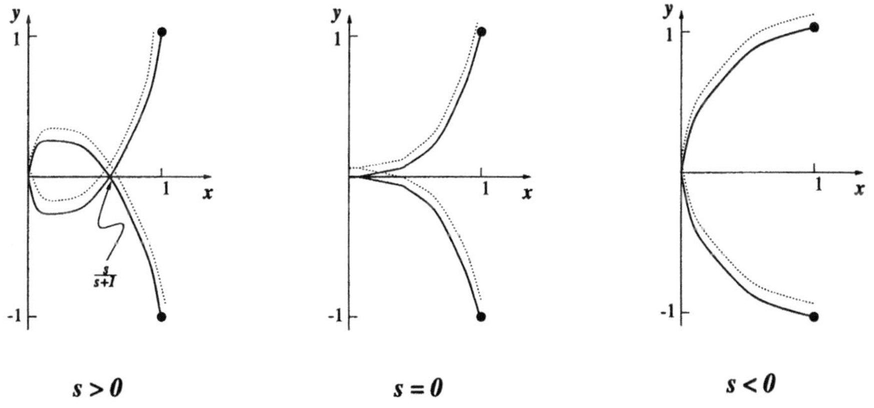

The dotted lines are the push-off's of each curve along the framing in the case of $\nu_s(t) = \mathbf{e}_y$, which then indicates the position of the corresponding ribbons. Below the sequence of paths is redrawn as a sequence of ribbon immersions, including twists, over- and undercrossings. Particularly, the crossings between solid and dotted lines in the above graphs become π-twist, and, as explained, these twists change orientation throughout the move:

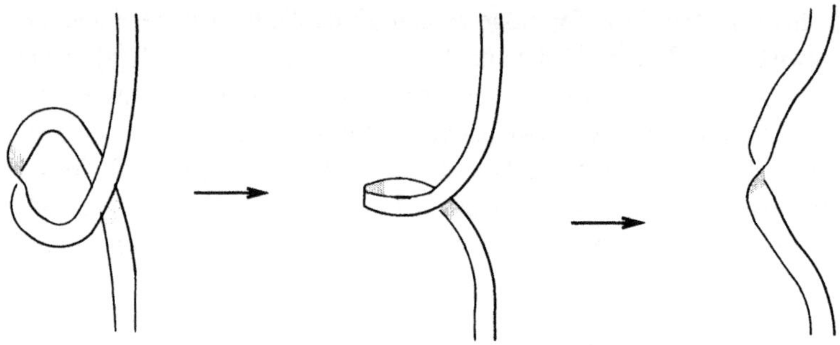

Recall that this picture includes three further versions that are obtained by mirror reflections in the plane as well as the switching of over- and undercrossings.

Summarizing the findings from parts *A)*, *B)*, and *C)*, we find that the 2-arrows from $\mathcal{T}gl_{\mathbb{R}^2}^{dec-proj}$ are in one-to-one correspondence with equivalence classes of tangles with planar orientation (i.e., ribbons with planar immersion and π-twists) if we add the following two moves:

- Isotopies and cancellations of π-twists.
- The framed versions of the first Reidemeister move as depicted above.

The remaining parts of the proof are now mostly of combinatorial nature.

D) Combination into 2π-twists, and the TI1-Move:

The orientation of a ribbon can be either equal or opposite to the orientation of the plane it is projected into. The relative orientation between ribbon and plane obviously changes through every π-twist so that any component in an admissible tangle has an even number of π-twists. Furthermore, a given orientation of a component also determines uniquely how these π-twists can be either paired up into 2π-twists, by combining consecutive π-twists with the same helicity, or cancelled against each other, if the corresponding neighboring π-twists are opposite to each other. Specifically, two consecutive twists belong together if the planar ribbon piece in between them has orientation different from the one chosen for the plane of projection. A switch in orientation yields the only other way of combining the π-twists. In the discussion of the TI11-Move such orientation flips have been related to π-twists being moved through a component.

Given these recombinations of twists we may consider now the smaller set of tangles, where we only allow 2π-twists along the internal and external strands.

The moves for these tangles are isotopies and cancellations now for the 2π-twists, and the corresponding version for the Reidemeister move from the previous section. The latter is obtained by adding the partner π-twist of the one in the move to all the pictures, and then moving the π-twists together. The resulting move is precisely the TI1-Move from Section 2.3.1.

Observe that no move for the cancellation of 2π-twists is listed in this section. The reason is simply that we can use the TI1-Move to obtain a planar immersed ribbon as depicted on the right. It is easy to see that this ribbon can be made into a straight vertical ribbon by an isotopy entirely within the planar immersed ribbons (without twists).
Hence, the TI1-Move is the only move that needs to be added as we pass from tangles from $\mathcal{T}gl_{\mathbb{R}^2}^{dec-proj}$ to those with planar ribbon immersions and 2π-twists.

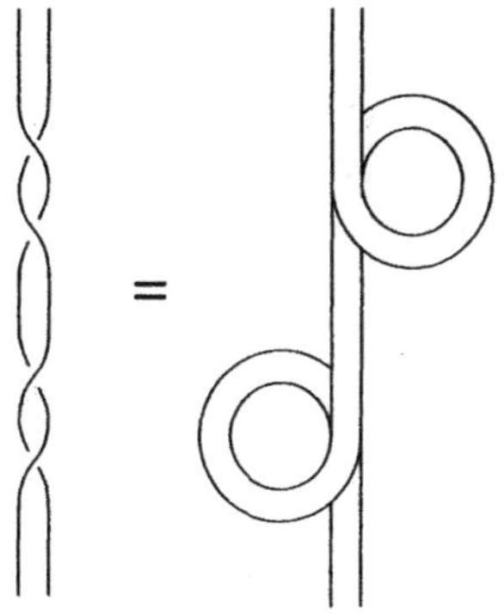

E) Clean intersections, and reduction to planar tangles:

The reduction to planar tangles and the second and third Reidemeister Moves as the resulting additional equivalences are obtained as usual. A derivation using transversality proceeds roughly as follows. For a manifold N we denote $N^k = N \times N \times \ldots \times N(k - \text{times})$. A subvariety in N^k is given by the set of tuples, for which two or more entries coincide. We denote this variety as $\Delta^k(N) = \{(x_1, \ldots, x_k) \in N : x_i = x_j \text{ for some } i \neq j\}$. It has a natural stratification into submanifolds depending on the partition of the set of indices for which entries coincide. Of course $\Delta^2(N) \subset N^2$ is the usual diagonal and, hence, itself a submanifold $\cong N$. Furthermore, $\Delta^3(N) = \Delta^3_1(N) \cup \Delta^3_2(N)$. Here, $\Delta^3_1(N)$ consists of all triples for which exactly two entries are the same and the third is different. $\Delta^3_2(N) = \{(x,x,x) : x \in N\}$ is the set of triples for which all three entries are the same. Moreover, for a differentiable function $f : M \to N$ we denote

$$f^{(k)} : M^k - \Delta^k(M) \longrightarrow N^k : (x_1, \ldots, x_k) \mapsto (f(x_1), \ldots, f(x_k)).$$

By an application of Baire's property (see [Hir76], Sections 2.4 and 3.2) one can show that the set of functions $f \in C^\infty(M,N)$, for which $f^{(k)}$ is transverse to any of the strata of $\Delta^k(N)$, is dense in $C^\infty(M,N)$.

For ordinary tangles we have that M is a compact 1-manifold, i.e., a union of closed intervals and circles, $N = \mathbb{R}_x \times [-1,1]$ is the 2-dimensional plane of projection, and the relevant transversality conditions are:

$$f^{(2)} \pitchfork \Delta(N) \qquad \text{and} \qquad f^{(3)} \pitchfork \Delta^3_2(N)\,.$$

In the first condition $\Delta(N)$ has codimension 2, which is the dimension of the domain of $f^{(2)}$. Hence, intersections will occur at a finite number of isolated points in $(M \times M) - \Delta(M)$, which are precisely the self intersections of the immersed 1-manifold M in the plane N. Thus, also the crossing parts of strands at these self intersections are transverse.

Furthermore, we have that $\Delta^3_2(N)$ has codimension 4, whereas the domain of $f^{(3)}$ is of dimension 3. Hence, transversality means that $f^{(3)} \cap \Delta^3_2(N) = \varnothing$, that is, there are no triple self intersections of strands. In summary, any tangle can be perturbed so that it is an embedding expect for isolated, transverse double crossings as required for tangles in $\mathcal{T}gl^{plan}$.

Of course, the additional feature of framings can be included in this discussion, if we extend the above functions to include a framing vector as used in the previous parts, and find corresponding submanifolds in $(\mathcal{X}^0)^{(k)}$. Given the codimensions of the intersections of the diagonals with the $\mathcal{X}^1$ and $\mathcal{X}^2$ manifolds, we can arrange that the crossings are also separated from the twists.

However, since we already isolated the 2π-twists in the previous parts, we may think of them as special points or markings on the 1-dimensional manifold M. Clearly, we can always move a special point away from self intersection points. Thus, for any tangle with planar orientation and 2π-twists there is an arbitrarily small isotopy that moves it into a planar tangle.

F) The other Reidemeister Moves:

For an isotopy $F : M \times [0,1] \to N : (x,s) \mapsto f_s(x)$ we may define similarly

$$F^{(k)} : \left(M^k - \Delta^k(M)\right) \times [0,1] \to N^k : (x_1,\dots,x_k,s) \mapsto (f_s(x_1),\dots,f_s(x_k)).$$

and require by two variations of the transversality theorems implied above analogous conditions:

$$F^{(2)} \pitchfork \Delta(N) \quad \text{and} \quad F^{(3)} \pitchfork \Delta^3_2(N) \quad \text{and} \quad F^{(4)} \pitchfork \Delta^4_3(N).$$

Here $\Delta^4_3(N) = \{(x,x,x,x) : x \in N\}$, which has codimension 6. Since the domain of $F^{(4)}$ only has dimension 5, its image does not meet $\Delta^4_3(N)$. This means there are no quadruple points occurring during isotopies.

The first condition means that the double points form a one-dimensional submanifold in $\left(M \times M - \Delta^2(M)\right) \times [0,1]$. As in part *B)* we may use a refined jet transversality condition, which guarantees that the s-coordinate function is regular on the manifold except for isolated points where it has non-degenerate maxima or minima.

This means that with the exception of a finite number of s-values, the double crossing are simply moving along paths in the plane, and at the special s-values we have cancellations of two opposite crossings that collide. This type of cancellation is then easily identified as the second Reidemeister Move. By further dimension counting arguments we can arrange that the cancellation points are separated from triple intersection points or the special twist-points in M. As a result, we obtain precisely the TI2-Moves.

Finally, $\Delta^3_2(N)$ has codimension 4, which is also the dimension of the domain of $F^{(3)}$ so that $F^{(3)}$ meets $\Delta^3_2(N)$ only in a finite number of isolated points. Hence, during an isotopy isolated triple points may occur at discrete times. Also, we can make $F^{(3)}$ transverse to $\Delta^3_1(N)$, which is of codimension 2 and indicates all pure double points. The set of parameters in $M^3 \times [0,1]$, where this occurs is, thus, a two dimensional submanifold. Away from the points where the second Reidemeister Move applies it is, e.g., of the form $\{(t,\tau_1(s),\tau_2(s),s) : t \in M, s \in [0,1]\} - \Delta^3(M) \times [0,1]$, where $s \mapsto (\tau_1(s),\tau_2(s))$ is a regular path in $M \times M - \Delta^2(M)$. Two other versions follow from permutations of the entries.

A point in $\Delta^3_2(N)$ is in the closure of three such local components of $\Delta^3_1(N)$, which can be parametrized in the above way, since we can separate triple points from crossing-cancellation points. It follows that a triple point occurring in a transverse isotopy is always the transverse intersection of three paths of double points. Further analyzing the transversality conditions and the geometry of over and undercrossings, we are left precisely with the third Reidemeister Move. Picturing again the 2π-twists as special markings on M, we may assume that they are separated from any triple points by an isotopy. Hence, we obtain the TI3-Move as the relevant equivalence.

The TI2 and TI3 Moves, thus, guarantee that isotopic plain tangles (without framing or twists) are equivalent to each other.

Although in an isotopy a 2π-twist marking can be removed from the cancellation and triple points at discrete times, it can not be separated anymore from a crossing or double point. All we can do is to make the passage of a twist through a crossing a transverse one as indicated in the picture on the right.

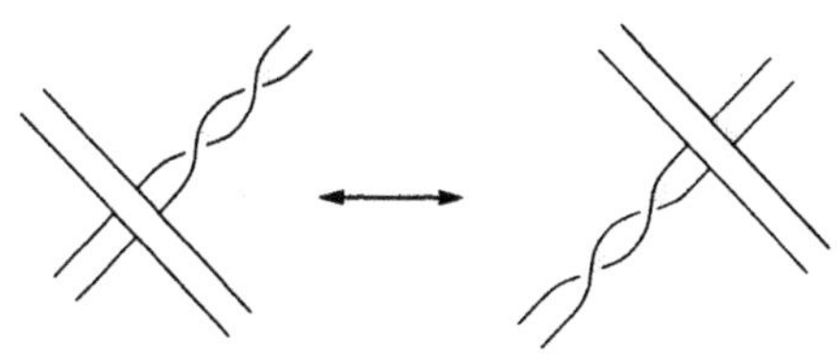

This move, however, is already implied by the other moves for the same reason from Part *D)* that the twist-cancellation is redundant. More precisely, we can substitute the 2π-twists on both sides of the move by planar loops using the TI1-Moves. These diagrams are then equivalent by an isotopy within the class of tangles without twists.

Only for the auxiliary ribbons, where we also consider π-twists instead of only 2π-twists this trick can not be applied, and we have to introduce TI10 explicitly.

Summarily, we have found in Parts *A)* through *F)* that any projectable tangle can be perturbed into a planar tangle, and any isotopy within the projectable tangles can be replaced by an isotopy within the planar tangles and applications of the moves TI1, TI2, TI3, and TI10. This, therefore, concludes the proof of Lemma 2.4.6.

Finally, we wish to replace the planar tangles by those used to define the original tangle category $\mathcal{T}gl$. The only new restriction on the tangles is that all of the strands are drawn parallelly to the vertical direction in the plane $\mathbb{R}_x \times [-1, 1]$ (i.e., parallelly to the segment $\{0\} \times [-1, 1]$) except for the crossings, local maxima and local minima. This condition can be relaxed to requiring that the strands should never be horizontal (i.e., never parallel to the line $\mathbb{R}_x \times \{0\}$), except for the non-degenerate maxima and minima.

The remaining moves to be introduced are TI4, TI5, TI8, and TI9. Here, the TI9-Move really only expresses the ambiguity of the position of a vertical strand that is obtained by straightening a monotonous strand as in the relaxed definition of admissible tangles. We proceed as before by showing that an isotopy within the class of planar tangles can be replaced by isotopies within the admissible tangles from $\mathcal{T}gl$ together with applications of the additional moves. This will establish the missing link in our chain of bijections:

Lemma 2.4.7. *The natural map $\mathcal{T}gl \to \mathcal{T}gl^{plan}_{\mathbb{R}^2}$ is a bijection.*

Proof. For the path function $f : M \to \mathbb{R}_x \times [-1,1]$ the first relevant jet-transversality condition is

$$\frac{df}{dt} \pitchfork \mathbb{R}_x \times \{0\}.$$

This precisely implies the condition that the vertical coordinate has only isolated and non-degenerate maxima and minima, and is elsewise regular. That is, except for the extrema the strands, which are in the image of f, can be locally parametrized as $]0,\varepsilon[\to \mathbb{R}_x \times [-1,1] : z \mapsto (x(z), z)$.

Similarly to the 2π-twists, the extremal points can be thought of as special markings on the 1-manifold M, and, hence, we may also require them to be separated from the 2π-twists as well as the crossing points.

For an isotopy $F : M \times [0,1] \to \mathbb{R}_x \times [-1,1] : (t,s) \mapsto f_s(t)$ we may require

$$\left(\frac{\partial F}{\partial t}, \frac{\partial^2 F}{\partial t^2}\right) \pitchfork \left(\mathbb{R}_x \times \{0\}\right) \times \left(\mathbb{R}_x \times \{0\}\right) \quad \text{and} \quad \frac{\partial F}{\partial t} \pitchfork \mathbb{R}_x \times \{0\}.$$

The second condition implies that the vertical component of $\frac{\partial F}{\partial t}$ vanishes along a 1-dimensional submanifold $\mathcal{E}$ in the two-dimensional "ts-plane" $M \times [0,1]$. The first transversality also implies that the intersection set, where both the vertical components of $\frac{\partial F}{\partial t}$ and $\frac{\partial^2 F}{\partial t^2}$ vanish, consists of a finite number of isolated points on that 1-manifold in the ts-plane. They are easily identified as the local maxima or minima of the s-coordinate restricted to $\mathcal{E}$.

Analogously to the proofs in previous paragraphs, these points are, thus, identified as cancellations of a local maximum with a local minimum. Such a special cancellation point in the ts-plane can, certainly, be separated from the paths representing the movement of 2π-twists and crossings. Hence, we obtain the TI4-Move.

During an isotopy away from the cancellations the markings of crossing points, twist points and extrema move along paths in the ts-plane. Generally, they can not be made disjoint so that intersections remain where the admissibility condition is violated. Nevertheless, we can make these intersections transverse to each other and allow only double points of paths. Each such passage then yields an additional equivalence move. With three types of paths there are, thus, six types of intersections. We have already discussed collisions of the same type in the extrema cancellation, twist cancellations, and Reidemeister moves. Also we already dealt with the move, in which crossings meet twists. Thus, there remain two types:

A) Crossings and Extrema: At a generic crossing point not both strands can be horizontal. Hence, the path of a marking for an extremum will intersect only one part of the paths of the crossing points for a given parameter s. A transverse intersection of a crossing and an extremum on one of the strands, thus, amounts to moving the crossing point from one side of the extremum to the other. This is exactly what is being done in the TI5-Move.

B) Twists and Extrema: A transverse intersection of a path of a marking for a 2π-twist, and the marking for an extremum can be seen as an isotopy moving a 2π-twist from one side of a maximum (or minimum) to the other as illustrated in the picture to the right.

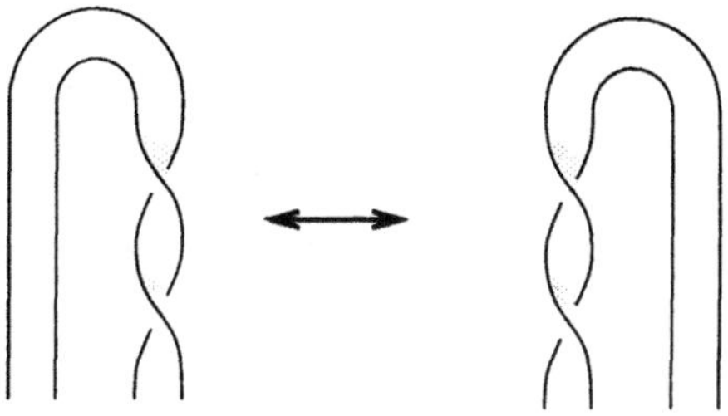

This move, however, is already implied by the TI1 through TI5 Moves. To see this one proceeds exactly as in Part *D)* or *F)* of the proof of Lemma 2.4.6, and replaces the 2π-twist by loops, which can be isotoped as before.

In case we deal only with π-twists the extrema slide needs to be included explicitly, see TI10.

Finally, admissibility implies also that the special ingredients of a tangle diagram occur at different heights. That is, the tuple of vertical coordinates of K elements has to lie in $[-1,1]^K - \Delta^K([-1,1])$, with notation as introduced in Part *D)* of the proof of Lemma 2.4.6. Now, $\Delta_1^K([-1,1])$, the set of tuples for which exactly two entries but no others coincide, has codimension 1, and all other strata of $\Delta^K([-1,1])$ have higher codimensions. Hence, if an isotopy of the vertical coordinates is transverse to all of the strata of $\Delta^K([-1,1])$, there will be never more than two vertical coordinates of elements coinciding. Moreover, such a coincidence will occur at a finite number of isolated times, and in that instance of the isotopy one element will be moved to the other side of another element of the tangle diagram. This is precisely the definition of the TI8-Move.

This completes the proof of Lemma 2.4.7.

Putting together Lemmas 2.4.3, 2.4.5, 2.4.6, and 2.4.7 we obtain the following sequence of bijections on the 2-arrow sets of tangle categories, each of which is induced by the natural inclusion of classes of tangles:

$$\mathcal{T}gl \longrightarrow \mathcal{T}gl^{plan}_{\mathbb{R}^2} \longrightarrow \mathcal{T}gl^{dec-proj}_{\mathbb{R}^2} \longrightarrow \mathcal{T}gl^{well-pos}_{\mathbb{R}^2} \longrightarrow \mathcal{T}gl_{\mathbb{R}^2}.$$

We summarize this result in the following theorem.

Theorem 2.4.8. *The natural inclusion of admissible tangles as framed tangles in* $\mathbb{R}^2 \times [-1,1]$ *induces a bijection on the 2-arrow sets of the respective tangle categories:*

$$\mathcal{T}gl \xrightarrow{\sim} \mathcal{T}gl_{\mathbb{R}^2}.$$

2.4.2 Tangles over S^2

Among the remaining moves in $\mathcal{T}gl_{\mathbb{R}^2}$ the equivalence TI4 really expresses an isotopy, but of a tangle over S^2 instead of over $\mathbb{R}^2$. In order to relate presentations over the sphere to those over the plane we introduce a homeomorphism of the following form. It can be obtained, e.g., from a stereographic projection:

$$\varphi : (S^2 - \{\infty\}, S^1_b - \{\infty\}) \xrightarrow{\cong} (\mathbb{R}^2, \mathbb{R}_x). \tag{2.4.2}$$

Here, we have selected a special point, $\infty \in S^2$, (point at infinity) on the sphere as well as a special equator, S^1, on the same sphere, which runs through the point at infinity. That is, we have $\infty \in S^1 \subset S^2$. The homeomorphism, φ, between the punctured sphere and the Euclidean plane restricts to a homeomorphism between the punctured equator $S^1 - \{\infty\}$ and $\mathbb{R}_x$, the special coordinate axis of $\mathbb{R}^2$. Hence, from the intervals on $\mathbb{R}_x$, as introduced in Section 2.1.1, we obtain a corresponding sequence of intervals on the equator S^1. Let us denote the natural homeomorphism

$$\varphi : (S^2 \times [-1,1]) - L \longrightarrow \mathbb{R}^2 \times [-1,1],$$

extended to the cylinders over the punctured sphere and plane by the same letter. We denote here by $L = \{\infty\} \times [-1,1]$ the special line segment "at infinity" in $S^2 \times [-1,1]$.

Given the intervals on the equator we may give a definition of an *admissible tangle in* $S^1 \times [-1,1]$ in exactly the same way as we defined admissible tangles in $\mathbb{R}^2 \times [-1,1]$ or $\mathbb{R}_x \times [-1,1]$ in Section 2.2. An obvious example of an admissible tangle in $S^2 \times [-1,1]$ is given by $T_s = \varphi^{-1}(T_R)$, where T_R is an admissible tangle in $R^2 \times [-1,1]$.

Conversely, it is clear that any admissible tangle T_S over S^2 is represented by an admissible tangle over R^2 if and only if

$$T_S \cap L = \varnothing,$$

simply because T_S then lies in the domain of φ. An arbitrary tangle T_S can be made disjoint from the line segment at infinity by almost any small perturbation into "general position".

More precisely, if we denote by $f : M^1 \hookrightarrow S^2 \times [-1,1]$ the embedding of the one dimensional manifold M^1, i.e., the union of intervals and circles, then the condition that the tangle T_S without framing (or thickness) is disjoint from L is equivalent to the transversality condition $f \pitchfork L$. Hence, standard transversality theorems guarantee that we can always achieve that T_S is of the form $\varphi^{-1}(T_R)$ by an arbitrarily small isotopy of f.

Now, for an isotopy of tangles over the sphere, $F : M^1 \times [0,1] \to S^2 \times [-1,1] : (t,s) \mapsto f_s(t)$, transversality $F \pitchfork L$ implies that only for a finite number of points $\{(t_j, s_j)\}_{j=1}^N$ we have $f_{s_j}(t_j) \in L$. Moreover, at these points the partial derivatives of F together with the directions span $\mathbb{R}^3$ so that, after application of a suitable continuous family of linear transformations, we may assume that $\frac{\partial F}{\partial s}|_{(t_j,s_j)}$ is perpendicular to the plane of projection $T_\infty S^1 \times [-1,1]$ and $\frac{\partial F}{\partial t}|_{(t_j,s_j)}$ is parallel to $T_\infty S^1 \times \{0\}$, that is $\frac{\partial(\varphi \circ F)}{\partial t}$ is parallel to $\mathbb{R}_x \times \{0\}$ in $\mathbb{R}^2 \times [-1,1]$ as (t,s) approaches (t_j, s_j). Furthermore, the isotopy can be modified easily so that an intersection point $f_{s_j}(t_j) \in L = \{\infty\} \times [-1,1]$ lies arbitrarily close to the top endpoint $\{\infty\} \times \{-1\} \in \partial L$ of the line segment at infinity.

Having restricted to isotopies of tangles in $S^2 \times [-1,1]$ for which the intersections of strands with the line L have the special positions as described above, we may now determine how such a crossing translates into a move for a tangle over $\mathbb{R}^2$.

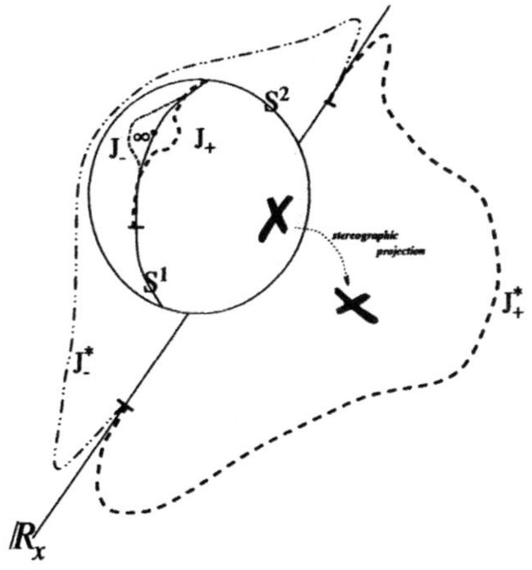

To this end we consider a curve segment $J_+ \subset S^2$, which runs along the equator $S^1 \subset S^2$ but is slightly pushed off towards one side of S^1 closely to the point ∞. Also we introduce a corresponding line segment J_-, which is detoured along the opposite side of S^1. The image of J_+ under the stereographic projection $\varphi : S^2 - \{\infty\} \to \mathbb{R}^2$ is, hence, a curve $J_+^* = \varphi(J_+) \subset \mathbb{R}^2$ in the Euclidean plane that starts and ends tangentially in the coordinate axis $\mathbb{R}_x$ and encompasses an arbitrarily large area in the positive half plane on one side of R_x. Similarly, $J_-^* = \varphi(J_-)$ is a curve that surrounds a large area on the opposite half plane of $\mathbb{R}^2$.

The special positioning of the transverse intersection of a strand with L allows us now to interpret this part of an isotopy equivalently as a substitution of an interval piece in the tangle of the form $J_+ \times \{1-\varepsilon\} \subset S^2 \times [-1,1]$ by a corresponding piece $J_- \times \{1-\varepsilon\} \subset S^2 \times [-1,1]$. In the stereographic projection this means, consequently, that a strand $J_+^* \times \{1-\varepsilon\} \subset \mathbb{R}^2 \times [-1,1]$ that runs in front of the rest of the tangle, relative to the plane of projection, and which is close to the top boundary, is replaced by a strand $J_+^* \times \{1-\varepsilon\} \subset \mathbb{R}^2 \times [-1,1]$, which runs behind all other parts of the tangle.

Hence, in the projection into the plane $\mathbb{R}_x \times [-1,1]$ this translates into a move where a strand running in front and close to the top of all other strands is replaced by a strand that runs behind everything else. This describes precisely the TS4-Move.

Since all equivalences of tangles over S^2 can be expressed by equivalence relations over $S^2 - \{\infty\}$ together with instances of an isotopy where a strand passes through L in a given way, we obtain exactly the equivalence classes for tangles over $\mathbb{R}^2$ if we add TS4 to the moves that are inherited from the tangles over S^2.

Hence, if we define the two-arrow set of $\mathcal{T}gl_{S^2}$ to be the set of equivalence classes of admissible tangles in $S^2 \times [-1,1]$ subject to the equivalence moves

- all of the TD Moves,
- the moves TS1, TS2, and TS3,
- ambient isotopies,

then they are in a natural one-to-one correspondence with the two-arrow set of the tangle category over $\mathbb{R}^2$. More precisely, we can summarize our observations as follows.

Lemma 2.4.9. *The natural map $\mathcal{T}gl_{S^2} \longrightarrow \mathcal{T}gl_{\mathbb{R}^2}$ induced by the homeomorphism φ^{-1} acting on tangles is a bijection.*

2.4.3 Removing Auxiliary Tangles

The auxiliary tangle pieces do indeed contain no relevant topological information but serve merely to keep track of some combinatorial rules. Let us, thus, introduce double categories $\mathcal{T}gl^{\mathbf{nx}}$, $\mathcal{T}gl^{\mathbf{nx}}_{R^2}$, $\mathcal{T}gl^{\mathbf{nx}}_{S^2}$, etc., and $\mathcal{T}gl^{\mathbf{s};*}$, $\mathcal{T}gl^{\mathbf{s};*}_{S^2}$, $\mathcal{T}gl^{\mathbf{s};*}_{R^2}$, etc., whose two-arrow sets are generated by tangles without any auxiliary strands.

For all of these types of categories we omit in Definition 2.2.2 condition *4)* of the fourth item (since $\mathsf{A} = 0$), and also the moves TI6, TI7, TI10, TI11, TD3, TD4 and TD5 from Section 2.3 become obsolete. It is obvious from equation 2.2.3 that the admissible tangles in the ordinary $\mathcal{T}gl_{\#}$ categories have only *top* or *through* internal ribbons, i.e., $\mathsf{B} = \mathsf{C} = 0$. We want to consider two ways in which one can define a map from the 2-arrow set of an ordinary $\mathcal{T}gl_{\#}$-type category to a 2-arrow set generated by tangles without auxiliary strands. We confine our discussion to the cases of $\mathcal{T}gl^{\mathbf{nx}}_{S^2}$ and $\mathcal{T}gl^{\mathbf{s};*}_{R^2}$.

Omission Map: As already explained above, the classes of 2-arrows of $\mathcal{T}gl^{\mathbf{nx}}_{S^2}$ are generated simply by tangles without auxiliary strands but for which all other types of strands can occur and fulfill the usual admissibility conditions. The equivalences are precisely those from $\mathcal{T}gl_{S^2}$ that do not involve the auxiliary tangles. We may define a map that takes an ordinary admissible tangle as in Definition 2.2.2 and simply removes all auxiliary tangles so that we obtain an admissible tangle as it is used for $\mathcal{T}gl^{\mathbf{nx}}_{S^2}$. The image of the moves TD3, TD4, and TD5 for admissible tangles in $\mathcal{T}gl^{\mathbf{nx}}_{S^2}$ are then either obvious isotopies or simply void. Hence, the omission of auxiliary ribbons factors into a well defined map denoted as follows:

$$\mathsf{O} : \mathcal{T}gl_{S^2} \longrightarrow \mathcal{T}gl^{\mathbf{nx}}_{S^2}. \tag{2.4.3}$$

In fact, the following is not much harder to show:

Lemma 2.4.10. *The map* O *is a bijection on the 2-arrow sets.*

Proof. As before we attempt to construct an inverse map O^{-1}. On the level of tangles we, thus, have to assign to a tangle that is admissible for $\mathcal{T}gl^{\mathbf{nx}}_{S^2}$ one which has all the necessary auxiliary strands in order to be admissible in the sense of Definition 2.2.2. We do this by simply adding any arbitrary strand to each bottom and close internal ribbon in the diagram that joins these ribbons with the intervals on the top-equator $S^1 \times \{1\}$ as prescribed in Definition 2.2.2. In order to show that an unspecified assignment such as this leads to a well defined map O^{-1} on the equivalence classes, which is then, naturally, the desired inverse, we need to verify that any two choices of auxiliary ribbons lead to tangles that lie within the same equivalence class of $\mathcal{T}gl_{S^2}$.

To this end we observe that TD3 can be used to turn an overcrossing of an auxiliary strand with any other strand anywhere in a tangle diagram into an undercrossing. Specifically, we can apply an isotopy, which moves a given crossing along the auxiliary strand all the way up to the top line. Here, TD3 can be applied, after which we perform the inverse of the previous isotopy only with the crossing reversed. An auxiliary ribbon can, thus, be isotoped through any other ribbons and thereby we can unknot and untangle them in any given way from themselves and the rest of the tangle diagram. Furthermore, we can change the number of twists of an auxiliary ribbon using TD5. As a result, we are able to move any arbitrary auxiliary strand to any other provided they start at the same tangle component and end in the same auxiliary interval at the top line. Now, the intervals in which the strands end can be permuted in any way by applying the Moves TD4, each of which clearly induces the action of a generator of the symmetric group on the set of intervals. Finally, if we are considering orientations on tangle components we also want to be able to change the side of a ribbon, that is the boundary component, to which an auxiliary ribbon is attached. Here, we can use Move TI11, where the π-twist on the auxiliary ribbon can be eliminated by using Move TD5 again.

In summary, any configuration of auxiliary strands can be changed into any other configuration by application of the Moves TD3, TD4, and TD5. Hence, $\mathbf{O}^{-1}$ is well defined and $\mathbf{O}$ a bijection.

Again we may put the established one-to-one relations together to obtain the following bijection between the 2-arrow sets of the original category of planar tangles and the tangles over S^2 without auxiliary strands:

$$\mathcal{T}gl \;\stackrel{\sim}{\longleftrightarrow}\; \mathcal{T}gl^{\mathtt{nx}}_{S^2}. \tag{2.4.4}$$

In this equivalence the category $\mathcal{T}gl$ is combinatorially more convenient but involves all nineteen moves from Section 2.3, whereas the description of $\mathcal{T}gl^{\mathtt{nx}}_{S^2}$ is already close to the actual topological surgery presentation and is subject only to five moves, namely TD1, TD2, TS1, TS2, and TS3, besides the ambient isotopies.

Splitting Map: The second method of removing auxiliary ribbons unveils their purpose in the construction of topological quantum field theories in later chapters. A configuration of auxiliary ribbons represents a prescription for how to cut each of the bottom and closed internal ribbons and attach the resulting ends to the intervals at the top line. The target tangle category for the *splitting map* that constructs this correspondence is different from the naïve one used for the omission map. We use the following notations for the map and category:

$$\mathsf{S} : \mathcal{T}gl_{S^2} \longrightarrow \mathcal{T}gl^{\mathtt{s};*}_{S^2}.$$

The tangles we consider admissible for the 2-arrow sets of $\mathcal{T}gl^{\mathtt{s};*}_{S^2} = \bigcup_{\mathsf{A}=0}^{\infty} \mathcal{T}gl^{\mathtt{s};\mathsf{A}}_{S^2}$ are those which contain no closed or bottom internal ribbons and, hence, as tangles of either $\mathcal{T}gl_{S^2}$ or $\mathcal{T}gl^{\mathtt{nx}}_{S^2}$ also contain no auxiliary ribbons. Furthermore, in the subset $\mathcal{T}gl^{\mathtt{s};\mathsf{A}}_{S^2}$, labeled by an integer $\mathsf{A} \geqslant 0$, precisely A of the top internal ribbons or through ribbon pairs end in pairs of closed intervals, K_j^- and K_j^+, on the top line with $j = 1, \ldots, \mathsf{A}$, instead of pairs of internal intervals $I_j^{i/o}$. They are characterized by the conditions $K_j^- \cap K_j^+ = \varnothing$ and $K_j^- \cup K_j^+ \subset K_j$ where K_j is an auxiliary interval as described in Section 2.1.1.

The 2-arrows of $\mathcal{T}gl^{\mathtt{s};*}_{S^2}$ are again equivalence classes of admissible tangles, and the equivalence moves used to define the classes of $\mathcal{T}gl^{\mathtt{s};\mathsf{A}}_{S^2}$ encompass all those of

$\mathcal{T}gl_{S^2}$ or $\mathcal{T}gl^{\mathrm{nx}}_{S^2}$. However, we have to introduce an additional set of equivalence moves defined as follows.

TD3* Move : A strand of any type that crosses over the pair of internal ribbons emerging from a pair of intervals, $K_j^{-/+}$, parallel and right next to the top line can be changed to an undercrossing of both strands. Hence, it is basically a refinement of Move TS4. The local situation is illustrated on the right.

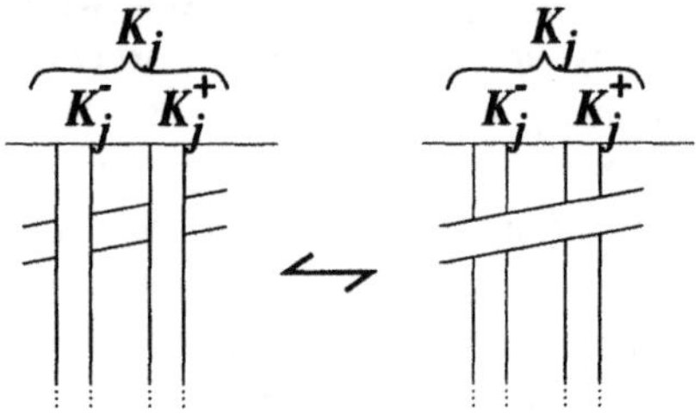

TD3* Move

TD4* Move : Two neighboring pairs of strands emerging from interval pairs $K_j^{-/+}$ and $K_{j+1}^{-/+}$ can be crossed over each other as indicated in the local equivalence of diagrams given on the right.

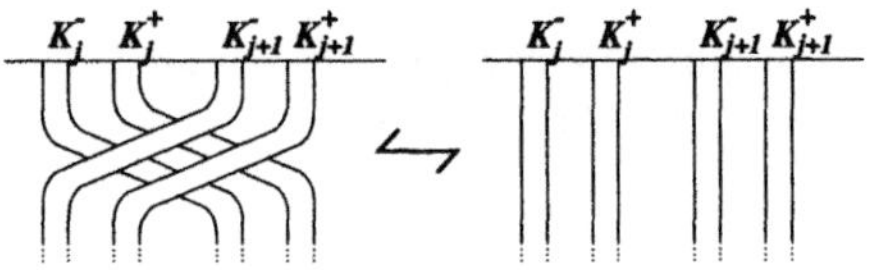

TD4* Move

TD5* Move : At a pair of strands emerging from a pair of intervals $K_j^{-/+}$ we can introduce a braid with an additional 2π-twist as depicted. Note that the twist can be replaced by a planar loop using Move TI1.

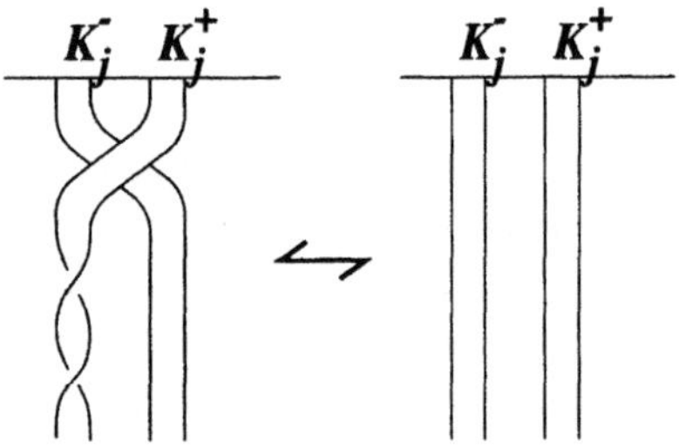

TD5* Move

In addition to these moves we also have to find substitutes for the Moves TS1, TS2, and TS3 from Section 2.3.3 that contain neither auxiliary ribbons nor closed or bottom internal ribbons. They will be called TS1*, TS2*, and TS3*. The equivalences they imply are all local moves in a vicinity of an interval pair $K_j^{-/+}$ at the top line. Note that in all cases the interval pair $K_j^{-/+}$ disappears on one side of the equivalence. It is tacitly assumed that the numbering of the following interval pairs is then decremented by one.

TS1* Move : Here we have a top internal ribbon at the K-intervals in the form of an arc or loop, and a collection of parallel strands passing through it. This top ribbon can be removed if we insert a coupon on the collection of the passing strands.

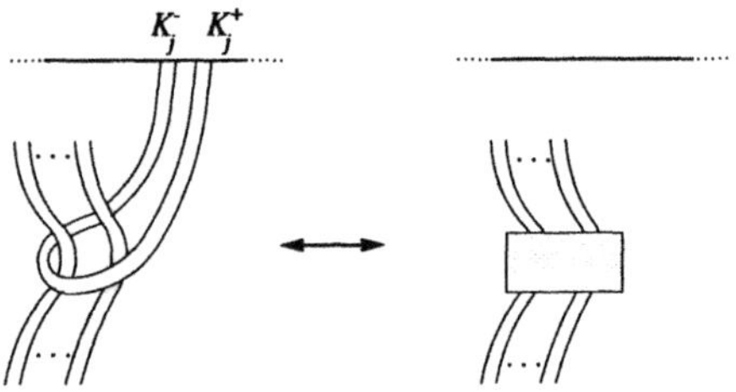

<ins>TS2* Move:</ins> A top ribbon that is attached in a simple arc to a pair of K-intervals and which runs once through a coupon can be deleted together with that coupon if both ribbon and coupon are isolated from the rest of the diagram.

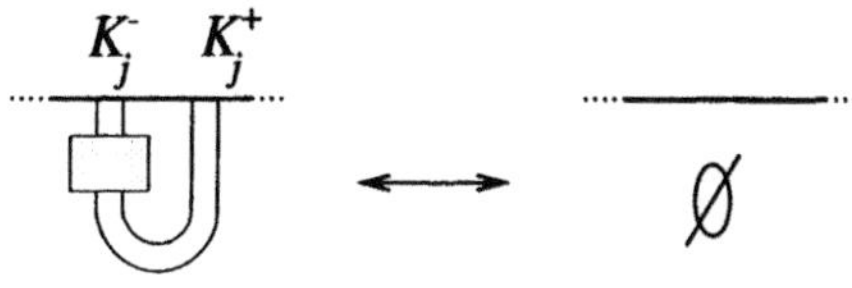

<ins>TS3* Move:</ins> A top ribbon that is attached in a simple arc to a pair of K-intervals and which runs once through a coupon can be deleted together with that coupon if both, ribbon and coupon, are isolated from the rest of the diagram.

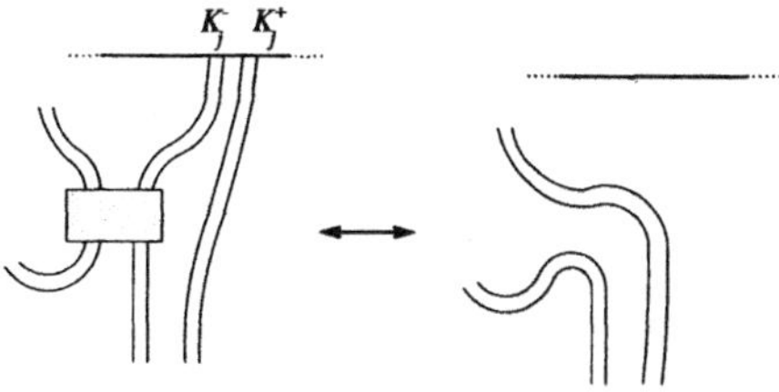

We define the splitting map first on arbitrary admissible tangles in the sense of Definition 2.2.2 with $\mathsf{A} = \mathsf{B} + \mathsf{C}$ auxiliary ribbons, where B and C are the numbers of bottom and closed internal ribbons respectively.

The definition of the map $\mathbf{S}$ starts with an operation on representing admissible tangles from $\mathcal{T}gl_{S^2}$ in the sense of Definition 2.2.2. For a given tangle each auxiliary tangle is cut along its center line in two ribbons (parallel to each other and the original auxiliary ribbon). Both of these strands are considered internal ribbons. At the top-boundary the splitting will result in two attaching intervals, K_j^+ and K_j^-, instead of the one K_j as depicted in Figure 2.1.*A* below. Figure 2.1.*B* shows how internal strands are merged together at a joint of an auxiliary strand and an internal strand. The pictures for the action of the splitting map at extrema and crossings are given in Figures 2.1.*C* and *D*. Finally, Figure 2.1.*E* illustrates the prescription to replace a π-twist in an auxiliary ribbon by a braid with 2π-twist on one internal ribbon. Observe that the diagram of internal ribbons contains only the admissible orientable pieces for the corresponding planar tangles so that the orientations are compatible for a given projection. For an opposite π-twist the corresponding inverse of the braid is chosen.

It is obvious from this description that an admissible tangle with $\mathsf{A} = \mathsf{B} + \mathsf{C}$ auxiliary ribbons, where B and C are the numbers of bottom and closed internal ribbons respectively, will, thus, be transformed to a tangle admissible for $\mathcal{T}gl^{\mathbf{s};*}_{S^2}$, which has $\mathsf{T}^* = \mathsf{T} + \mathsf{C}$ top internal ribbons and $\mathsf{H}^* = \mathsf{H} + \mathsf{B}$ pairs of through internal ribbons. The other ribbon types, namely the bottom and closed ones, are absent after application of the splitting operation, i.e., $\mathsf{B}^* = \mathsf{C}^* = 0$. Hence, the resulting tangle is in fact admissible for $\mathcal{T}gl^{\mathbf{s};*}_{S^2}$.

Lemma 2.4.11. *The splitting operation on the auxiliary strands factors into equivalence classes. The resulting splitting map,* $\mathbf{S} : \mathcal{T}gl_{S^2} \to \mathcal{T}gl^{\mathbf{s};*}_{S^2}$, *is a* bijection.

Proof. To show that $\mathbf{S}$ maps equivalence classes to equivalence classes let us start with isotopy classes. All isotopies involving an auxiliary strand other than the TI1 Move are easily translated into isotopies of two parallel strands. It is also sufficient

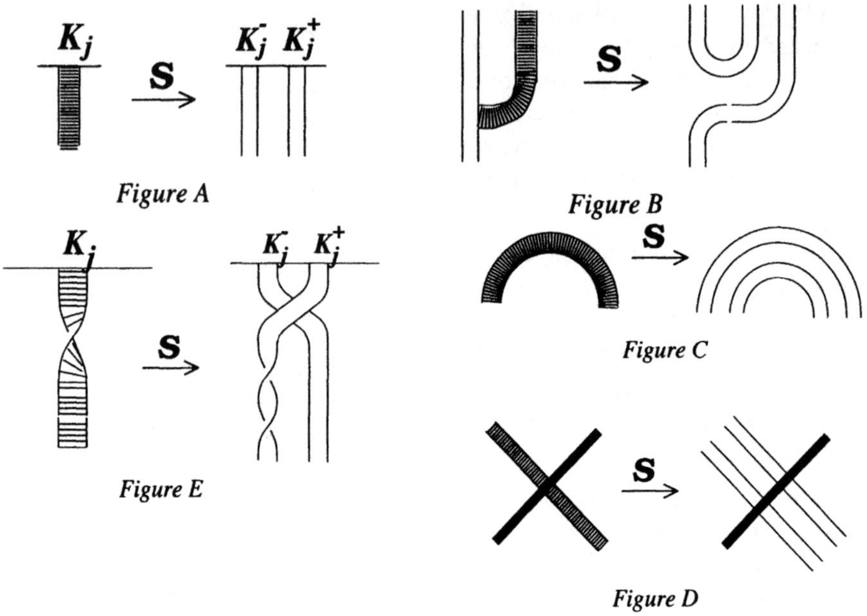

Fig. 2.1. The splitting map S

to define the splitting of a π-twist only at the top-line since any such twist can be isotoped all the way up. The substitution given for a π-twist is also compatible with the TI1 Move, since two of the braids of internals strands given in Figure E above are easily identified as a collective 2π-twist on the two parallel strands, to which we can then apply TI1 Moves individually.

The other equivalences for $\mathcal{T}gl_{S^2}$, namely TD1, TD2, TD3, TD4, TD5, TS1, TS2, and TS3, are easily seen to translate into the Moves TD1, TD2, TD3*, TD4*, TD5*, TS1*, TS2*, and TS3* for $\mathcal{T}gl^{\mathbf{s};*}_{S^2}$ under the splitting operation. Hence, the map S is well defined.

The construction of an inverse S^{-1} is straightforward. As depicted on the right hand side we simply connect the pairs of internal strands at an interval pair $K_j^{-/+}$ to each other and insert an auxiliary ribbon between the new internal ribbon and the larger interval K_j as shown.

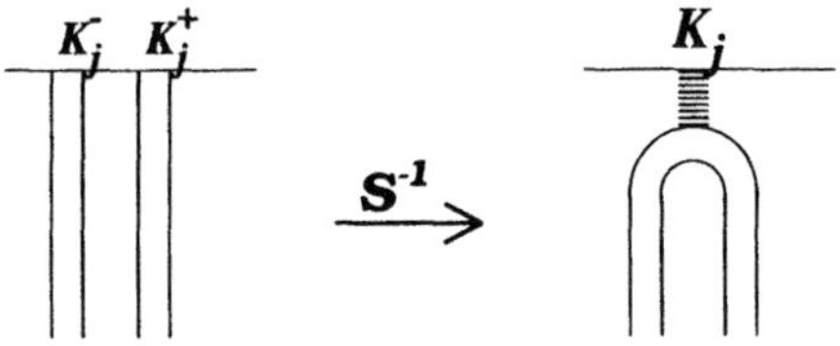

The reasons that S^{-1} is well defined are completely analogous to those that ensured the existence of S. The fact that $\mathrm{S} \circ \mathrm{S}^{-1} = id$ is immediately clear from the diagrams at the top line. We only need to use some obvious isotopies along the auxiliary ribbons in order to see that the opposite composition also yields the identity.

A Composite Correspondence:

In previous sections we have shown that the tangle classes of S^2 are the same as the original classes of planar tangles. From the first paragraph in this section we know that for the tangles over S^2 we may omit the auxiliary ribbons and still obtain the same category. Moreover, the second paragraph allows us to consider planar tangles without auxiliary ribbons as well as no bottom and closed internal ribbons using the splitting map. Combining all of these maps we find that there is a one-to-one correspondence between two tangle categories that both contain no auxiliary ribbons, one with tangles over S^2 and the other with tangles in the plane $\mathbb{R}_x \times [-1,1]$:

Corollary 2.4.12. *The composite of* $\mathbf{O}$, $\mathbf{S}^{-1}$, *and the relations between tangles in* $S^2 \times [-1,1]$ *and* $\mathbb{R}_x \times [-1,1]$ *yield a natural bijection between equivalence classes of tangles as follows:*

$$\mathcal{T}gl^{\mathbf{nx}}_{S^2} \quad \overset{\sim}{\longleftrightarrow} \quad \mathcal{T}gl^{\mathbf{s};*}.$$

The 2-arrow sets of both categories are also in bijection with those of $\mathcal{T}gl$.

Note that the tangles for the category over S^2 are subject to only *five* moves aside from ambient isotopies, namely,

TD1, TD2, TS1, TS2, and TS3.

The moves for the planar category are

TI1-TI11, TD1, TD2, TD3*, TD4*, TD5*, TS1*, TS2*, TS3*, and TS4.

In applications a tangle from $\mathcal{T}gl^{\mathbf{nx}}_{S^2}$ will represent a surgery presentation of a cobordism. We will discuss a modified presentation in the next section, in which also the Moves TD1 and TD2 will disappear. The remaining surgery moves TS1, TS2, and TS3 will be interpreted in the proof of our main theorem on tangle presentations of cobordisms in the following chapter as handle trade, cancellation and as a "σ-Move" for representing handle decompositions.

The planar category involves a lot more moves. However, it contains only top and through internal ribbons and external ribbons, but no bottom or closed internal ribbons as well as no auxiliary ribbons. This form will allow us to construct the TQFT functor in a systematic way.

2.5 Alternative Calculi and Further Equivalences

The tangle diagrams used to generate $\mathcal{T}gl^{\mathbf{nx}}_{S^2}$ are almost what we consider a surgery diagram for three manifolds, and the moves used in the definition of the 2-arrow sets of $\mathcal{T}gl^{\mathbf{nx}}_{S^2}$ give rise to what is often called a surgery calculus. A well known surgery calculus is Kirby's calculus of links, see [Kir78]. Its purpose is to establish a bijective correspondence between equivalence classes of links, which are tangles without boundary that consist only of closed interior ribbons, and homeomorphism classes of three manifolds without boundaries.

The description of $\mathcal{Tgl}^{\mathbf{nx}}_{S^2}$ is deduced from the so called "Bridged Link" calculus, derived in [Ker99]. The Bridged Link calculus is nonetheless equivalent to a generalization of the original Kirby Calculus, as shown in [Ker98a]. Another variant of the Kirby Calculus is the Fenn Rourke picture, which results from peculiar combinatorial reduction and has technical advantages in particular applications.

In this section we will either show or review the equivalences between the mentioned calculi in the context of admissible tangles. The relevant bijective relations can be inferred entirely on the level of combinatorial tangle classes with no reference, yet, to three manifolds.

We will also discuss how we can go from tangle diagrams over S^2 back to tangles over $\mathbb{R}^2$ *without* introducing the TS4 Move, but instead by eliminating one of the external strands. This reduction will be an essential tool in the construction of the horizontal compositions for the tangle double categories.

2.5.1 From Coupons to Bridged Links

The coupons used in our previous descriptions of admissible tangles can be substituted by pairs of balls inside $S^2 \times [-1, 1]$. Following the calculus of "Bridged Links" as in [Ker99] we can build a tangle category, $\mathcal{Tgl}^{\mathbf{BL}}_{S^2}$, starting with these ingredients and an even smaller set of moves. Let us summarize the definition of $\mathcal{Tgl}^{\mathbf{BL}}_{S^2}$ more precisely.

As before, an admissible tangle for $\mathcal{Tgl}^{\mathbf{BL}}_{S^2}$ contains external and internal ribbons inside $S^2 \times [-1, 1]$ that can end in the internal and external intervals at the boundary $S^2 \times \{\pm 1\}$. We can also have pairs of surgery balls embedded in $S^2 \times [-1, 1]$ in which ribbons can end. For any pair of balls (B, B'), with $B \cong D^3 \cong B'$, we also have an orientation reversing diffeomorphism between their bounding spheres, i.e., a map $\phi^B : \partial B \overset{\cong}{\rightarrow} -\partial B'$.

The condition for attaching ribbons to the balls is that if some type of ribbon ends in an interval $p \subset \partial B$ another ribbon has to emerge at the image of this interval $p' = \phi^B(p) \subset \partial B'$ under the identification map. The two ribbon pieces ending in p and p' are thus considered part of the same component of the tangle. With this notion of components we can then use the original Definition 2.2.2 of admissible tangles. However, now the tangles contain neither coupons nor auxiliary strands.

Among the moves are first of all, of course, isotopies. Here we also include isotopies of the attaching intervals as long as the isotopy on a sphere $\partial B'$ is the composite of the isotopy on the partner sphere ∂B and ϕ^B. Moreover, we also allow isotopies of the identification map ϕ^B itself. Since the oriented mapping class group of the sphere is trivial we can therefore move from one ϕ^B to any other.

Besides the isotopies we introduce the following three moves:

TS1$^\spadesuit$ Move: Two partner balls, B and B', are right next to each other, such that the map ϕ^B coincides with the mirror reflection at the plane in the middle between the balls. Ribbons are entering at intervals $p_1, p_2, \ldots$ at ∂B and emerge at corresponding intervals $p_1', p_2', \ldots$ at $\partial B'$. For this configuration the pair of balls can be replaced by an internal closed ribbon in the form of an annulus A. The ribbons that were passing through the balls are now passing through the disc bounded by this annulus as depicted.

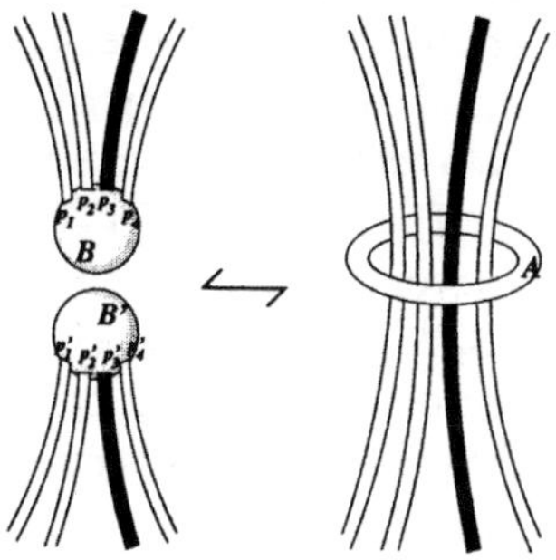

TS2$^\spadesuit$ Move: Consider a pair of balls with only one internal ribbon passing through them. Let the pair of balls and the closed internal ribbon be separate from the rest of the diagram. Such a configuration can be eliminated all together from a tangle diagram.

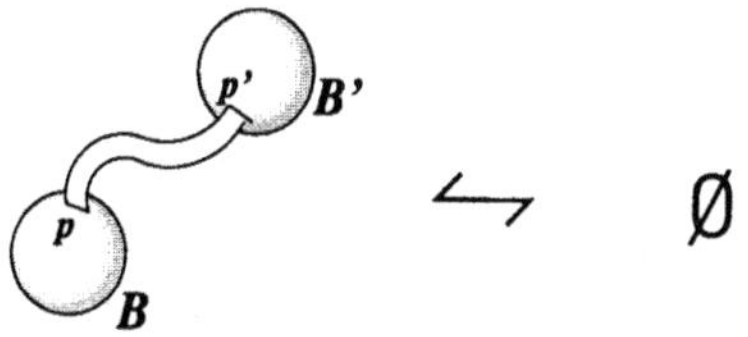

TS3$^\spadesuit$ Move: On the right we have a top internal ribbon passing through a pair of balls, B and B', at intervals p and p' such that the entire component consists of two small strips from the top line to these intervals. Furthermore, another bottom internal ribbon is passing through the same pair of balls at intervals q and q', but no other ribbons run through B and B'. We can eliminate the top component and the pair of balls, and connect, as depicted, the end intervals q and q' of the second bottom ribbon to the top line so that it becomes a pair of through internal ribbons.

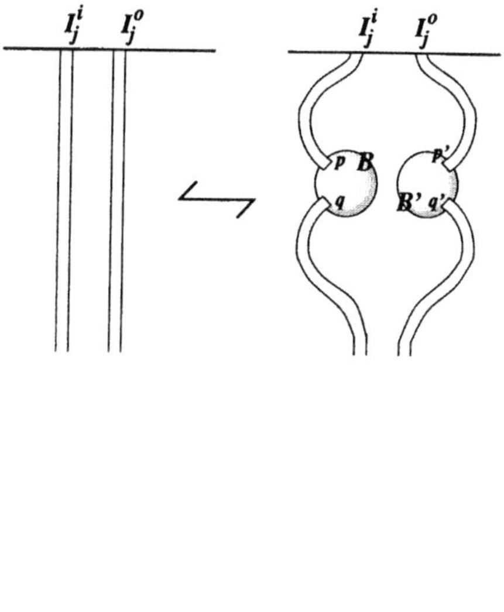

Remark 2.5.1. In another version of the TS3$^\spadesuit$ Move we may assume that the strands on the right belong to a top ribbon instead of a pair of through ribbons. This is, however, redundant since we can move the surgery balls around this connected top ribbon next to each other, and then apply a cancellation move.

In order to see that the definition of $\mathcal{T}gl^{\mathbf{BL}}_{S^2}$ is equivalent to all of the other definitions of tangle categories, in particular $\mathcal{T}gl^{\mathbf{nx}}_{S^2}$, we introduce an intermediate tangle category $\mathcal{T}gl^{\mathbf{rec}}_{S^2}$. It contains the same elements and conditions for admissibility as $\mathcal{T}gl^{\mathbf{BL}}_{S^2}$, but, in addition, we require that for every pair of balls (B, B') there is a *recombination ribbon* denoted as r^B. This is a usual ribbon piece $\cong [-L, L] \times [0, 1]$, embedded into $\mathbb{R}^2 \times [-1,1]$, which is disjoint from the tangle diagram except at the endpoints. There we require $\{L\} \times [0,1] \subset \partial B$ and $\{-L\} \times [0,1] \subset \partial B'$, such that the identification map ϕ^B between the spheres not only maps the two end interval onto each other but, moreover, is the *identity* on the interval $[0,1]$.

The moves in $\mathcal{T}gl^{\mathbf{rec}}_{S^2}$ consist, firstly, of the modified moves in $\mathcal{T}gl^{\mathbf{BL}}_{S^2}$, where a recombination ribbon is introduced between the balls in the pictures for TS1♠, TS2♠, and TS3♠ which is a flat, planar piece in the plane of projection of each illustration. We shall maintain the same notation for these moves. Besides these three modifications of moves in $\mathcal{T}gl^{\mathbf{BL}}_{S^2}$ there are two additional moves that enter the definition of $\mathcal{T}gl^{\mathbf{rec}}_{S^2}$. They both concern the recombination ribbons and are depicted below as Moves TD1♠ and TD2♠.

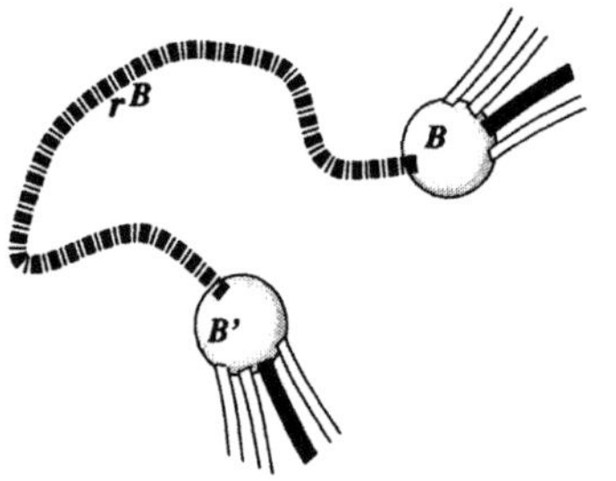

TD1♠: 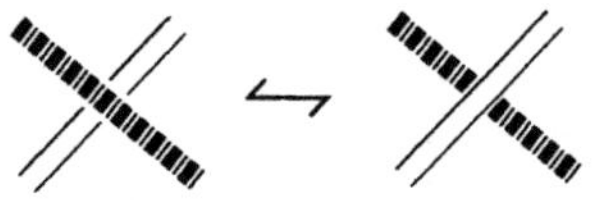

TD2♠: 

The two additional moves guarantee that we can change every recombination ribbon r^B between two given balls to any other such ribbon. The crossing move TD1♠ allows us to change the path of the center of the ribbon in any given way, since the paths are in a simply connected three space and every homotopy can be deformed into a differential isotopy interrupted by transverse intersections between strands. The Move TD2♠ allows us to change the framing by any integer. Notice that we have an orientation condition at the end points of the r^B's so that π-twists would lead to an inadmissible recombination ribbon.

Since the choice of these additional ribbons is, thus, arbitrary the following is immediately implied:

Lemma 2.5.2. *There is a natural bijection between 2-arrow sets of the categories,*

$$\mathcal{T}gl^{\mathbf{rec}}_{S^2} \longrightarrow \mathcal{T}gl^{\mathbf{BL}}_{S^2},$$

which is induced by the omission of the recombination ribbons.

Once the recombination ribbons are included, we can find maps that relate tangle classes with coupons to those with surgery balls:

Lemma 2.5.3. *There is a natural bijection between the 2-arrow sets of the categories,*

$$\mathcal{R} : \mathcal{Tgl}^{\mathbf{nx}}_{S^2} \longrightarrow \mathcal{Tgl}^{\mathbf{rec}}_{S^2},$$

which is induced by replacing coupons by pairs of balls.

Proof. The definition of $\mathcal{R}$ on a representative tangle is quite straightforward. A coupon is simply replaced by a pair of balls with a straight recombination ribbon between them. As in the illustration below, the in and out going strands are entering the balls along the equators that are obtained as the intersection of the plane through the coupon and the two spheres. The extra ribbon also lies in the plane of the coupon.

The fact that $\mathcal{R}$ factors into tangle classes is immediate since the moves TS1, TS2, TS3, TD1, and TD2 in $\mathcal{Tgl}^{\mathbf{nx}}_{S^2}$ are, with the given positionings, readily implied by the moves TS1♠, TS2♠, TS3♠, TD1♠, and TD2♠ in $\mathcal{Tgl}^{\mathbf{rec}}_{S^2}$.

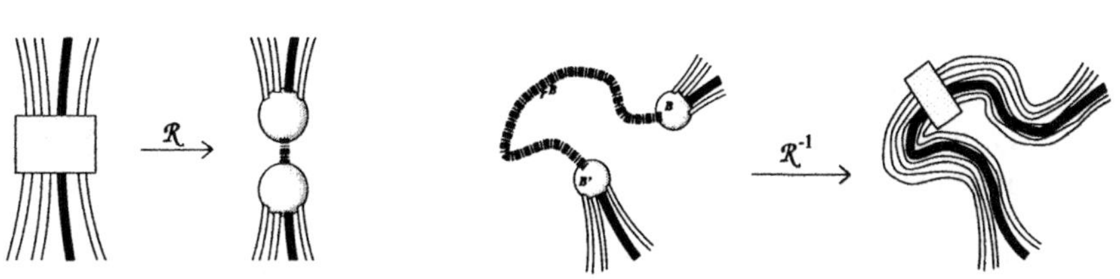

An inverse map $\mathcal{R}^{-1}$ is defined as follows. The strands attached to the spheres are aligned by an isotopy along an equator (or rather a pair thereof) that also contains the attachment point for the recombination ribbon r^B. As depicted above we can, then, replace the ribbon r^B by parallel strands that continue the incoming and outgoing strands at the balls. More precisely, the equators to which the incoming and outgoing strands are attached bound discs D^2 inside the balls. We can find a homeomorphism $D^2 \cup r^B \cup D^2 \cong [0,1] \times [-L, L]$ such that the identification map ϕ^B between the spheres maps $[0,1] \times \{-L\}$ as the natural identity to $[0,1] \times \{L\}$. The strands are, thus, attached along disjoint intervals $F_j \subset [0,1] \times \{\pm L\}$. The parallel strands are, therefore, generated by replacing $D^2 \cup r^B \cup D^2$ by $(\dot{\cup}_j F_j) \times [0, L]$. Finally, a coupon is introduced right across these parallel strands by slightly expanding a rectangle $[0,1] \times [-\varepsilon, \varepsilon] \subset [0,1] \times [-L, L]$

Now, in order to show that $\mathcal{R}^{-1}$ is actually well defined on equivalence classes let us first check that the ambiguities in the above construction are taken care of by equivalences in $\mathcal{Tgl}_{S^2}$. To begin with, the homeomorphism $D^2 \cup r^B \cup D^2 \cong [0,1] \times [-L, L]$ (with fixed attachment intervals) is not unique but all such homeomorphisms are isotopic to each other. An isotopy between two different homeomorphisms can be lifted to an ambient isotopy in three space and, hence, to an isotopy for a tangle from $\mathcal{Tgl}_{S^2}$. Moreover, two ways of moving the attaching intervals into position along equator on the sphere may differ by a braiding of the strands in a vicinity of the sphere, $S^2 - I \cong int(D^2)$, with the end interval I of the recombination ribbon removed, and a corresponding opposite braiding for the outgoing strands on the other sphere. After applying the recombination $\mathcal{R}^{-1}$ we will, thus, have a braid right on top of the coupon as well as its opposite at the bottom of the coupon. But if we apply the TS1 Move in $\mathcal{T}gl_{S^2}$ to this coupon we easily see that the braids can be pushed through the annulus from the TS1 Move and cancelled against each other. Reversing the TS1 Move we, thus, obtain an equivalent tangle with no braids at the coupon. Changes in choosing the identification homeomorphism ϕ^B between the two spheres are dealt with in the exact same way.

It remains to show that if two tangles are equivalent in $\mathcal{T}gl^{\mathbf{rec}}_{S^2}$ their images under $\mathcal{R}^{-1}$ are also equivalent in $\mathcal{T}gl^{\mathbf{nx}}_{S^2}$. Isotopies in $\mathcal{T}gl^{\mathbf{rec}}_{S^2}$ also include deformations of the recombination ribbons but those can be expressed as collective isotopies of the parallel strands from the substitution. The Moves TD1$^\spadesuit$, and TD2$^\spadesuit$ can be expressed by the Moves TD1 and TD2 in $\mathcal{T}gl_{S^2}$, if we also use the fact that we are free to choose where to place the coupon along the parallel strands or recombination ribbon. For TD1$^\spadesuit$ we put it right at the crossing, and for TD2$^\spadesuit$ we place the coupon outside but right after the 2π-twist.

The recombination of tangles in the moves TS1$^\spadesuit$, TS2$^\spadesuit$ and TS3$^\spadesuit$ leads exactly to the pictures for the Moves TS1, TS2 and TS3.

Finally, it is obvious that the map $\mathcal{R}^{-1}$ that we have, thus, constructed is the inverse to the previous map $\mathcal{R}$ on sets of tangle classes.

In summary, we have the following fundamental bijection between 2-arrow sets of tangle categories, one given by planar pictures with twenty moves and the other by three dimensional surgery data with only three moves:

$$\mathcal{T}gl \quad \overset{\sim}{\longleftrightarrow} \quad \mathcal{T}gl^{\mathbf{BL}}_{S^2}. \tag{2.5.1}$$

This correspondence can be used to find a few further equivalence relations in $\mathcal{T}gl$, which will be very useful in later computations with tangles. Let us discuss the following three:

• *β-Move ($\mathcal{O}_3$-Move):* Let a coupon C have exactly one internal strand passing through it, which is a part of a closed ribbon component R of the tangle diagram. Then this tangle is equivalent to the one where both the coupon and the closed internal component R have been removed from the diagram.

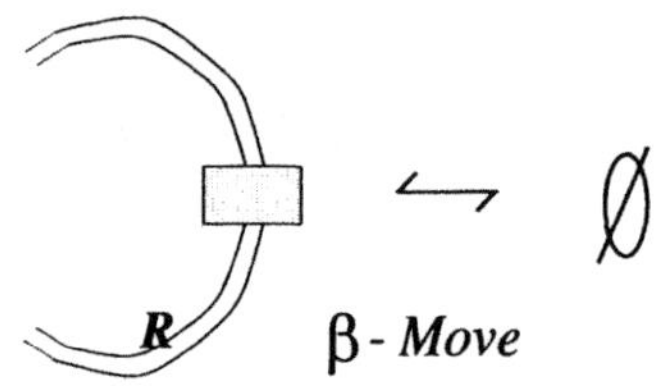

If we substitute the coupon with an annulus via the TS1-Move we obtain precisely what is called the $\mathcal{O}_3$-move in [Ker98a] and β-Move in [FR79]. It is easily derived as an equivalence in the category $\mathcal{T}gl^{\mathbf{BL}}_{S^2}$, where the coupon is replaced by a pair of balls. The resulting configuration is, clearly, contractible to an isolated cancellation picture such as in the TS2$^\spadesuit$-Move.

• <u>General Cancellation:</u> Here we have a closed internal ribbon A, which enters and exits a coupon C exactly once, which may have also other ribbons passing through. As indicated in the diagram on the right, the ribbon A looks like an isolated annulus except for the piece running through C.
The cancellation move is given by removing (or adding) both the annulus A and the coupon C.

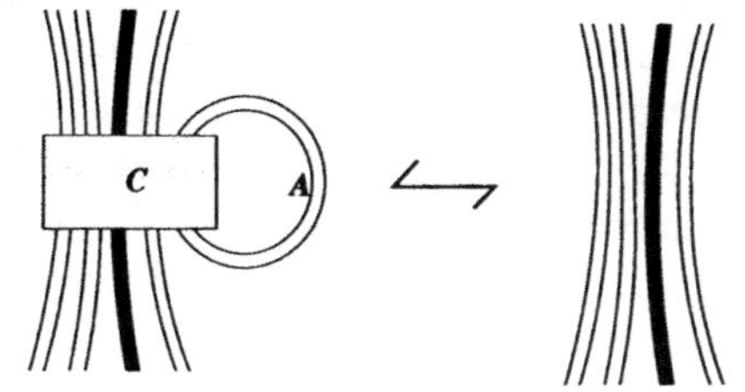

The fact that this is an equivalence in $\mathcal{T}gl^{\mathbf{BL}}_{S^2}$ is explicitly shown in [Ker98a] or [Ker99]. The short diagrammatic proof starts by replacing the coupon C by another annulus A^* using the TS1-Move. Now, A surrounds only A^* and we can use the TS1-Move again to replace A by a coupon that is placed in a piece of A^* only. The β-Move from above then allows us to remove A^* together with the extra coupon.

• <u>Connecting Annulus:</u> This move considers *two separate* internal ribbons R_1 and R_2, which run through the same coupon C. We assume that there are no other ribbons running through C.

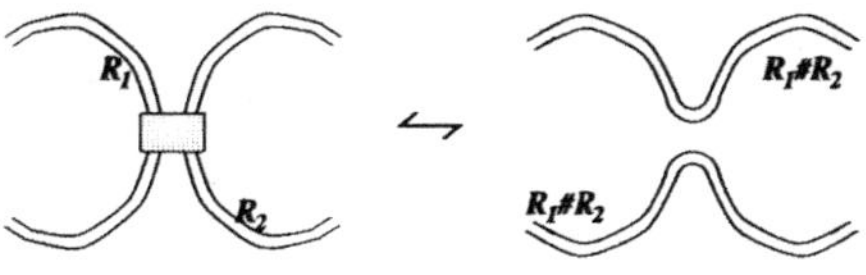

This configuration is equivalent to the one where R_1 and R_2 are replaced by their connected sum $R_1\#R_2$. The connecting operation is performed in the plane of the coupon. See the picture on the right. C itself is removed from the diagram.

There are several proofs of this equivalence. One way is to replace the coupon by two balls as in $\mathcal{T}gl^{\mathbf{BL}}_{S^2}$, move one of them along all of R_2 until it reaches its partner ball on the same strand and then cancel them both using the generalized cancellation. Another proof replaces the coupon by an annulus via the TS1-Move and then applies a 2-handle slide as in the next section, followed by a β-Move.

As they are, in both proofs we really assume that R_1 and R_2 are closed internal ribbons. They can be easily generalized though to the cases where R_1 and R_2 are top, bottom, or through internal ribbons. To this end one applies the TS3-Move, or rather its ribbon version described in Section 2.5.2 below, at the ends of the ribbons. This turns R_1 and R_2 into closed ribbons. After applying the previous arguments for the Connecting-Annulus-Move the TS3-Move is applied backwards to the resulting configuration so that $R_1\#R_2$ has the same boundary attachment intervals as R_1 and R_2 before the move. Thus, the only global condition to observe in this move is that R_1 and R_2 belong to different ribbon components.

2.5.2 Kirby and Fenn Rourke Moves

Although the present versions of setting up equivalence classes of tangles will be predominately used to present cobordisms and construct TQFT functors, it is interesting to also introduce the versions that directly generalize the original calculi of Kirby [Kir78] and Fenn Rourke [FR79]. They give presentations of closed 3-manifolds in terms of equivalence classes of links in S^3. In this section we shall describe categories $\mathcal{T}gl^{\mathbf{Ki}}_{S2}$ and $\mathcal{T}gl^{\mathbf{FR}}_{S2}$, which are equivalent to the category $\mathcal{T}gl^{\mathbf{BL}}_{S2}$ and specialize to the link calculi for closed 3-manifolds in corresponding situations.

Specifically, the results from [Kir78] and [FR79] pertain only to the special case of trivial 1-arrows with $g_1 = g_2 = a = b = 0$, when there are no boundaries. Hence, we have to add boundary moves to the definitions of $\mathcal{T}gl^{\mathbf{Ki}}_{S2}$ and $\mathcal{T}gl^{\mathbf{FR}}_{S2}$. Also, they do not contain decorations such as coupons or surgery balls, and at least [Kir78] uses non local equivalence moves, and one needs to prove that both type of calculi yield the same equivalence classes. Finally, both [Kir78] and [FR79] consider only 3-manifolds themselves without the additional framing or signature structure that we include in our cobordism categories. This means that some equivalences from [Kir78] and [FR79] need to be relaxed in the definitions of $\mathcal{T}gl^{\mathbf{Ki}}_{S2}$ and $\mathcal{T}gl^{\mathbf{FR}}_{S2}$.

A) The Category $\mathcal{T}gl^{\mathbf{Ki}}_{S2}$: The notion of a set of admissible tangles in $\mathcal{T}gl^{\mathbf{Ki}}_{S2}$ is precisely the same as for $\mathcal{T}gl^{\mathbf{BL}}_{S2}$ or $\mathcal{T}gl^{\mathbf{nx}}_{S2}$ except that we do not allow surgery balls or coupons in the tangle but only ribbons.

The equivalences TS1, TS2, and TS3 are replaced by the following three moves, which generate all equivalences in $\mathcal{T}gl^{\mathbf{Ki}}_{S2}$. In each case it is not very difficult to see that they are already equivalences in either $\mathcal{T}gl^{\mathbf{BL}}_{S2}$ or $\mathcal{T}gl^{\mathbf{nx}}_{S2}$:

1) *Hopf-Link Move:* A Hopf link consists of two unknots with linking number 1 as depicted. In this move a Hopf link in which one of the two components has 0-framing can be removed or added.

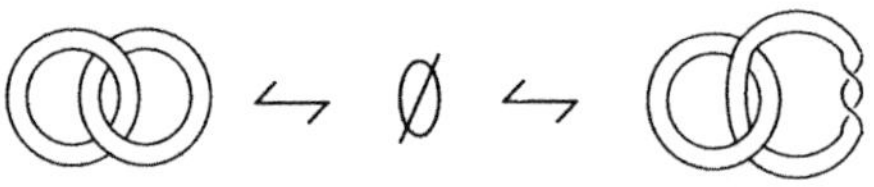

In fact, we can assume that the other component has framing either 0 or 1, since the framing can be changed by 2 by slides over the other 0-framed annulus. The fact that this is an equivalence in $\mathcal{T}gl^{\mathbf{nx}}_{S2}$ is immediate from the TS2-Move.

2) *2-Handle Slide ($\mathcal{O}_2$-Move):* For the 2-handle slide we start with a ribbon R_1 and a distinct closed ribbon R_2. An auxiliary strip, s, is fit between them so that one end of s ends on the boundary of R_1, and the other on the boundary of R_2. We then slice R_2 down the middle into two parallel ribbons R_a and R_b.

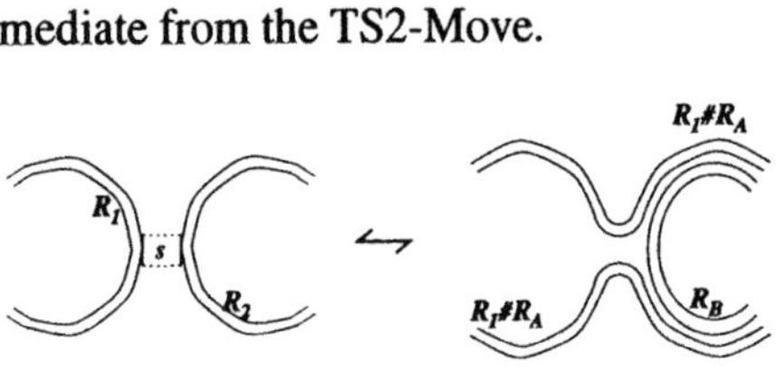

As in the diagram on the right we then also cut the strip s from one point on the boundary of R_1 to another point on the boundary of R_A and then continue this cut in both directions to also cut R_1 and R_A at this point. As a result, we obtain one component $R_1\#R_A$, which is basically the connected sum of the two original ribbons, and, in addition, the component R_B, which is another pushed off copy of R_2.

The proof that the 2-handle slide can be obtained from the defining moves in $\mathcal{T}gl^{\mathbf{BL}}_{S^2}$ is given in [Ker98a]. The basic idea is to introduce a coupon C with an annulus A in the place of the auxiliary strip s via the general cancellation move from the previous section for $\mathcal{T}gl^{\mathbf{nx}}_{S^2}$. In $\mathcal{T}gl^{\mathbf{BL}}_{S^2}$ the coupon is replaced by a pair of balls, one of which we can drag along R_2. The annulus A then becomes stretched into the ribbon R_B, and R_1 is extended along R_2 to $R_1\#R_A$. At the same time R_2 is shrunk to a strip, which is, then, used for an opposite cancellation to eliminate any balls or coupons again via cancellation. For details see [Ker98a].

3) *Ribbon-TS3 (σ-) Move:* For a pair of internal through ribbons we can always insert a link configuration of an additional top ribbon and an additional internal annulus as depicted on the right. Instead of through ribbons we have a bottom type ribbon.
This is easily recognized as the TS3-Move with the coupon replaced via the TS1-Move.

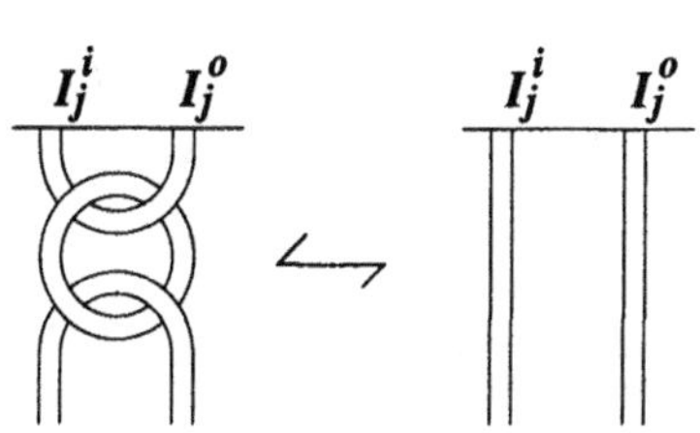

Remark 2.5.4. Besides this version of the Ribbon-TS3 Move we can also consider the one where we have a top ribbon, C, instead of a pair of through ribbons to start with. The move, as depicted, would again introduce a top ribbon T, and a closed annulus A, but the top ribbon would be turned into a closed internal ribbon R. This move, however, is redundant, since we can slide T over R, which yields the original top ribbon C and is no longer linked to A. The pair of ribbons A and R are, thus, as in the β-Move configuration, and can, therefore, be removed.

Since all three equivalences of $\mathcal{T}gl^{\mathbf{Ki}}_{S^2}$ are also equivalences in the other categories we have a well-defined, natural map $\mathcal{J} : \mathcal{T}gl^{\mathbf{Ki}}_{S^2} \to \mathcal{T}gl^{\mathbf{BL/nx}}_{S^2}$, which takes a representative tangle from $\mathcal{T}gl^{\mathbf{Ki}}_{S^2}$ and maps it to its class in $\mathcal{T}gl^{\mathbf{BL/nx}}_{S^2}$.

We also can define a map $\mathcal{K} : \mathcal{T}gl^{\mathbf{nx}}_{S^2} \longrightarrow \mathcal{T}gl^{\mathbf{Ki}}_{S^2}$, on the representative tangles, which replaces every coupon in a tangle diagram by an annulus using the TS1-Move.

In order for $\mathcal{K}$ to be well defined we need to check that all equivalences in $\mathcal{T}gl^{\mathbf{nx}}_{S^2}$ can be expressed by a combination of the three moves of $\mathcal{T}gl^{\mathbf{Ki}}_{S^2}$. The fact that this is true for TS1, TS2, and TS3 is quite obvious. The Moves TD1 and TD2 are only slightly more subtle:

After replacing the coupon on both sides of the TD1-Move by an annulus A the resulting move is found to be simply a 2-handle slide of the extra strand running across over A.

For the TD2-Move we remark that a 2π-twist on a collection of parallel strands can be created by 2-handle-sliding them over an unknot A^1 with framing 1 or -1. See below the Fenn-Rourke Move for details. Moreover, the coupon in the TD2-Move is replaced under $\mathcal{K}$ by a 0-framed annulus A^0. Hence, in $\mathcal{T}gl^{\mathbf{Ki}}_{S^2}$ the TD2-

Move means that we can add or remove a 1-framed unknot A^1 around a collection of strands, if these are also surrounded by a 0-framed annulus A^0. But this is easily seen to be true by 2-handle-sliding A^1 over A^0.

Now, $\mathcal{K}$ and $\mathcal{J}$ are obviously inverses of each other on the equivalence classes so that we obtain another bijective correspondence of tangle classes.

Lemma 2.5.5 ([Ker98a]). *The natural map*

$$\mathcal{J} : \mathcal{T}gl_{S^2}^{\mathbf{Ki}} \longrightarrow \mathcal{T}gl_{S^2}^{\mathbf{BL/nx}}$$

defines a bijection on the equivalence classes of tangles.

The proof in [Ker98a] is different from our outline only in so far that it constructs the inverse on $\mathcal{T}gl_{S^2}^{\mathbf{BL}}$ instead of on $\mathcal{T}gl_{S^2}^{\mathbf{nx}}$. But by Lemmas 2.5.3 and 2.5.2 we know these two are also the same. The correspondence from Lemma 2.5.5 also identifies coupons with what is often referred to as a "dotted circle" in Kirby's language [Kir89].

B) The Category $\mathcal{T}gl_{S^2}^{\mathbf{FR}}$: Shortly after Kirby introduced his calculus [Kir78] of links Fenn and Rourke [FR79] singled out a smaller set of equivalence moves that still generates the same equivalence classes of links and, hence, also gives presentations of three manifolds. Specifically, the general 2-handle slide or $\mathcal{O}_2$-Move is replaced by a special one, in which the closed ribbon over which we slide is a 1-framed unknot A^1. If we apply this special 2-handle slide to all strands running through the unknot A^1 the resulting equivalence is what is called the κ-Move in [FR79].

The Fenn-Rourke Calculus generalizes to the tangle category situation in the same straightforward way as the Kirby Calculus. The admissible tangles of $\mathcal{T}gl_{S^2}^{\mathbf{FR}}$ are precisely the same those for $\mathcal{T}gl_{S^2}^{\mathbf{Ki}}$. Only the equivalence moves are given by the following three:

1) *Signature Cancellation:* Here the two separated and isolated unknots, one with framing number $+1$, the other with framing number -1, can be cancelled against each other.

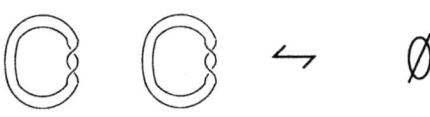

If one of the unknots is slid over the other this configuration turns into a Hopf link with a ± 1-framed component and a 0-framed component, and, hence, this is also an equivalence in the Kirby Calculus.

2) *Fenn-Rourke Move (κ-Move):* Starting point of this move is a collection of parallel strands that are passing through a ring A^1 with framing +1 or -1. The unknot A^1 can be separated from the other ribbons if at the same time a full 2π-twist is applied to the collection of strands in *opposite* direction of the A^1-framing.

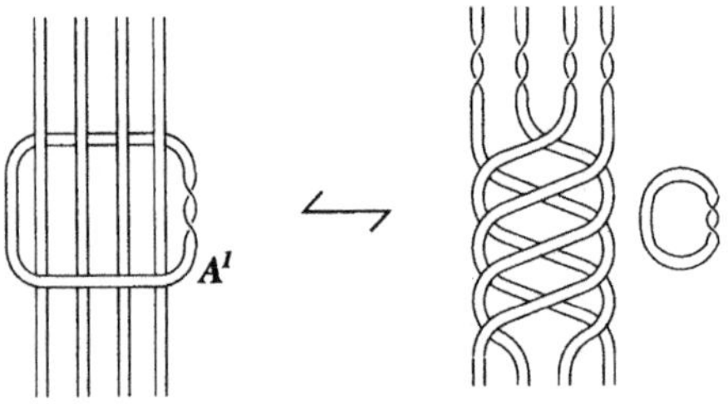

The case for one framing is depicted on the right. In addition, also the case with opposite framings has to be introduced as an equivalence. Note, that unlike in the original case the isolated 1-framed unknot is not discarded.

3) Ribbon-TS3 (σ-Move): This is exactly the same move as in the definition of $\mathcal{T}gl^{\mathbf{Ki}}_{S^2}$.

For each of the moves above all equivalences in $\mathcal{T}gl^{\mathbf{FR}}_{S^2}$ are clearly also equivalences in $\mathcal{T}gl^{\mathbf{Ki}}_{S^2}$. It follows that the identity on representing admissible tangles factors into a map $\mathcal{I} : \mathcal{T}gl^{\mathbf{FR}}_{S^2} \to \mathcal{T}gl^{\mathbf{Ki}}_{S^2}$. Fenn Rourke have proven for links in S^3 that any 2-handle slide can be obtained as a combination of κ-Moves. The proof is literally the same for tangles in $S^2 \times [-1, 1]$. Hence, we obtain the following.

Lemma 2.5.6 ([FR79]). *The natural map*

$$\mathcal{I} : \mathcal{T}gl^{\mathbf{FR}}_{S^2} \longrightarrow \mathcal{T}gl^{\mathbf{Ki}}_{S^2}$$

defines a bijection on the equivalence classes of tangles.

Obviously, this implies that $\mathcal{T}gl^{\mathbf{FR}}_{S^2}$ is also in a natural bijective correspondence with every other tangle category we have discussed in this chapter.

2.6 Compositions and $\mathcal{T}gl$ as a Double Category

Until now we have only defined the 2-arrows of $\mathcal{T}gl$ as nothing but sets, namely sets of equivalence classes of tangles. We have already described the easy composition rules for the 1-arrows of $\mathcal{T}gl$, which were the same as for the 2-cobordisms. However, we still need to explain the two composition operations for 2-arrows in $\mathcal{T}gl$.

In this section we shall introduce both the rules for vertical as well as horizontal compositions of admissible tangles. We verify that they factor into the equivalence classes so that we have two binary operations on compatible elements of the 2-arrow set of $\mathcal{T}gl$. Finally, we prove that these operations satisfy the axioms of a double category as defined in the introduction and Appendix B.1.

2.6.1 Vertical Compositions

We start with two tangles T_u and T_l that are admissible for $\mathcal{T}gl$. Furthermore, the target horizontal 1-arrow of T_u should be the same as the source horizontal 1-arrow of T_l.

Specifically, let us assume that the square, in the sense of the diagram (2.2.1) of Section 2.2.1, of T_u has horizontal 1-arrows $[g_{sc}, a/b]$ and $[g_{int}, a/b]$ and vertical 1-arrows α_u and β_u. Correspondingly, let T_l have horizontal 1-arrows $[g_{int}, a/b]$ and $[g_{tg}, a/b]$ and vertical 1-arrows α_l and β_l.

The composite tangle $T_l \circ_v T_u$ has then horizontal 1-arrows $[g_{sc}, a/b]$ and $[g_{tg}, a/b]$ and vertical 1-arrows $\alpha_l \circ \alpha_u$ and $\beta_l \circ \beta_u$. Schematically, we can write this as a composition of squares similar as for cobordisms as follows:

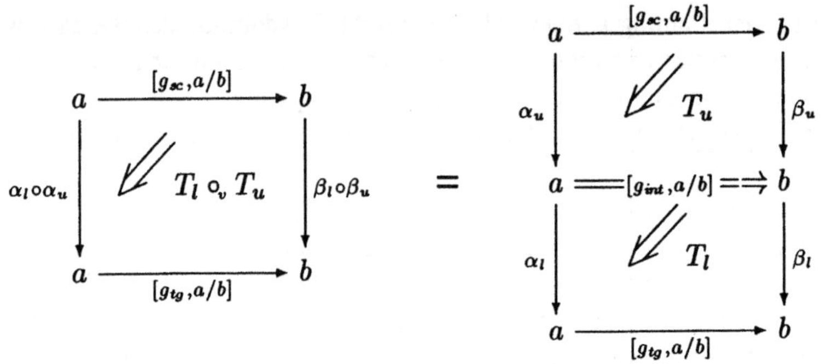

(2.6.1)

For two representing tangles T_u and T_l with these numbers we notice that the succession of internal and external intervals at the bottom-line of T_u and at the top-line of T_l, as described in Section 2.1.1, are exactly the same. That is, from left to right we have a initial external intervals, then $2g_{int}$ internal intervals (i.e., g_{int} pairs of such), and then b final external intervals.

Hence, we can place T_u on top of T_l such that external and internal intervals of the two bounding lines match exactly in succession. The internal and external strands that come together at these intervals are, thus, connected. The resulting tangle is shown in the diagram in Fig. 2.2.

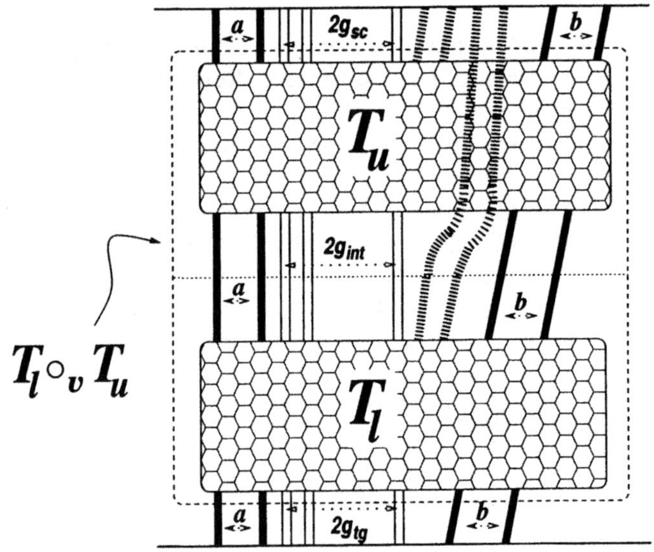

Fig. 2.2. The vertical composition

In addition to the internal and external strands we have to find a rule for the auxiliary strands. The ones from T_u are connected at the top-line at the intervals $K_1, \ldots, K_{\mathsf{A}_u}$ (see Section 2.1.1) as already in T_u. The A_l auxiliary strands from T_l, however, will be extended in parallel over T_u and connected in their order to the intervals $K_{A_u+1}, ..., K_{A_u+A_l}$ at the top-line.

Now, this is only a prescription of how to create a new tangle out of two given ones. We still need to make sure that it does in fact define a composition of elements in $\mathcal{T}gl$.

Lemma 2.6.1. *The composition $(T_l, T_u) \mapsto T_l \circ_v T_u$ of tangles, as described above, closes within the admissible tangles for $\mathcal{T}gl$.*

Furthermore, it factors into the equivalence classes of tangles of $\mathcal{T}gl$, and, hence, defines a composition on $\mathcal{T}gl$.

Proof. First, let us prove that the resulting tangle is admissible again. The fact that the composition of the a initial external tangles leads to a combined permutation of $\alpha_l \circ \alpha_u$ is clear, and similar to the permutations for the final 1-arrow composition.

The possible combinations that arise when the g_{int} pairs of internal strands of each tangle are connected to each other at the intermediate line can be easily identified: if a pair in T_u belongs to a bottom ribbon, and is combined with a pair that belongs to a top-ribbon in T_l then their combination in $T_l \circ_v T_u$ clearly yields a closed internal ribbon. Furthermore, a bottom-pair from T_u combines with a through-pair from T_l into a bottom-pair for $T_l \circ_v T_u$, a through-pair in T_u and a top-pair in T_l give a top-pair, and two through-pairs are connected to another through-pair. This implies, in particular, that the new top and through ribbons are attached to no auxiliary ribbons and the new closed and bottom ribbons have exactly one auxiliary ribbon each.

The counts of the ribbon types can be summarized by using the notation as in (2.2.2) from Section 2.2.3. For example, we denote by C_u the number of closed internal ribbons of T_u, by C_l the number of closed ones in T_u, and by C the total number in $T_l \circ_v T_u$. If N is the number of newly created closed internal ribbons as a result of the composition we obtain the following identities:

$$\mathsf{C} = \mathsf{C}_u + \mathsf{C}_l + N, \quad \mathsf{B} = \mathsf{B}_u + \mathsf{B}_l - N, \quad \mathsf{T} = \mathsf{T}_u + \mathsf{T}_l - N,$$
$$\mathsf{H} = \mathsf{H}_u + \mathsf{H}_l + N - g_{int}, \quad \mathsf{A} = \mathsf{A}_u + \mathsf{A}_l.$$

The last two, for through ribbons and auxiliary ribbons, follow from the other ones by virtue of (2.2.3). In particular, the fact that the number A of auxiliary ribbons is precisely the sum of the already present auxiliary ribbons implies that we do not have to add any more ribbons of this type. The tangle as given in the above picture, therefore, is already admissible.

It remains to prove that the composition is well defined on the equivalence classes in $\mathcal{T}gl$. That is, we need to show that if $T_l' \sim T_l$ via the moves then also $T_l' \circ_v T_u \sim T_l \circ_v T_u$. For all moves that can be localized in the interior of T_l this is obvious because for the composite we can use simply the same equivalence move. Thus, we only need to consider the TD3, TD4, TD5, TS3, and TS4-Moves.

For the TD-Moves the crossings or twists can be easily pushed up (or down) along the parallel extensions of the auxiliary ribbons of T_l.

If we apply the TS3-Move to T_l, the result in $T_l \circ_v T_u$ will be that the strands of the internal pair of T_l, ending at the top-line, are connected to each other by an arc.

The same happens for the pair of strands of T_u ending at the bottom line. Moreover, the arcs pass through a common coupon. This configuration is, however, easily recognized as a case of the *Connecting Annulus Move* at the end of Section 2.5.1. Applying this we recover the picture where the pairs of internal ribbons are connected to each other directly, which is the original one for $T_l \circ_v T_u$ before the application of the TS3-Move to T_l. A priori, the Connecting Annulus Move only applies when the involved ribbons are closed internal ribbons. This, however, is easily generalized if we conjugate the move with TS3-Moves at the top and bottom of the diagram for $T_l \circ_v T_u$.

The effect of the TS4-Move on T_l in $T_l \circ_v T_u$ is to turn an overcrossing of a strand over all of the parallel strands in the middle section into an undercrossing. Such a strand can be easily moved to either the upper or lower boundary line of $T_l \circ_v T_u$, where we can apply the TS4-Move to this composite tangle.

For the vertical composition rules it is clear that for any horizontal 1-arrow $[g, a/b]$ we can find an identity tangle $id_{[g,a/b]}$, such that for any tangle $T : [g_{sc}, a/b] \to [g_{tg}, a/b]$ we have $T \circ_v id_{[g_{sc},a/b]} = id_{[g_{tg},a/b]} \circ_v T$.
A representing tangle is simply given by parallel, straight vertical external and through strands for every interval as depicted on the right.

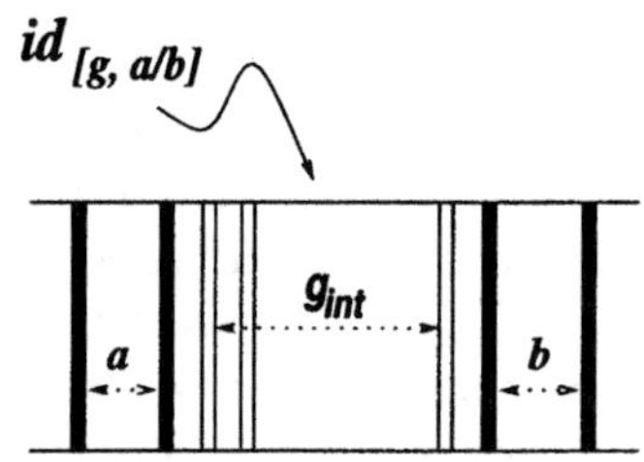

Already with only the vertical composition $\mathcal{T}gl$ forms an ordinary category, whose objects are the horizontal 1-arrows and whose morphisms are the 2-arrow sets given by the equivalence classes of tangles. This category is, naturally, a quotient of a special subcategory of the naïve category of isotopy classes of ribbons tangles in a slice of three space (see for example [FY92], [JS91], and [RT90]).

Furthermore, $\mathcal{T}gl$ thought of as an ordinary category under the vertical composition $\circ_v$ decomposes into subcategories $\mathcal{T}gl(a, b)$ for which the number a of initial external strands and the number b of final external strands are fixed. We summarize the decomposition in the following identity:

$$(\mathcal{T}gl, \circ_v) = \bigcup_{(a,b)\in\mathbb{N}^0\times\mathbb{N}^0} (\mathcal{T}gl(a, b), \circ_v). \tag{2.6.2}$$

2.6.2 Horizontal Compositions

The definition of the horizontal composition proceeds in a similar way as for the vertical one. First, we introduce a composition rule for admissible tangles and then prove that it factors into the equivalence classes.

The composition rule for the admissible tangles will, however, go slightly beyond simply putting the tangles next to each other. Due to the small 1-arrow sets that we have chosen here, which already in the situation of cobordisms in Section 1.2 required the natural transformation in form of the cobordism α, we also define the horizontal composition of tangles in two steps. For two given admissible tangles T_l

and T_r, such that the target vertical 1-arrow, $\beta \in S_b$, of T_l coincides with the source vertical 1-arrow of T_r, we first define a tangle $T_r \triangle_h T_l$. This tangle is, however, not quite admissible in the sense of Section 2.2.3 in that the internal intervals of some of the pairs at the top and bottom line are not next to each other. The order of the internal intervals is a permutation of the standard one proposed in Section 2.1.1.

In order to correct the order of the intervals, special braids, $\boldsymbol{\Upsilon}$ and $\boldsymbol{\Upsilon}^{-1}$, are added at the top and bottom line in the second step of the construction. These are chosen to be exactly the presentations of the braids from (1.1.7), where they appear as elements of the mapping class group of the sewn surface.

For a larger class of 1-arrows and, thus, admissible tangles the class of the composite $T_r \triangle_h T_l$ only depends on the equivalence classes of the factors, and the special braids are clearly also well defined as classes. The two steps of the construction of the horizontal composition can therefore be summarized in the formula:

$$T_r \circ_h T_l \quad = \quad \boldsymbol{\Upsilon} \circ_v \left(T_r \triangle_h T_l\right) \circ_v \boldsymbol{\Upsilon}^{-1}. \tag{2.6.3}$$

Schematically, the structure of the composition is illustrated in the following diagram of 1-arrows and 2-arrows. The composite $T_r \triangle_h T_l$ may be thought of as the two boxes in the middle of the right diagram:

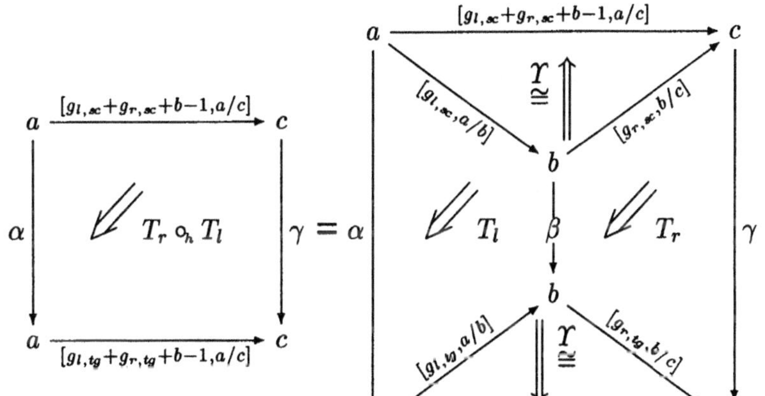

(2.6.4)

We start with the definition of $T_r \triangle_h T_l$ on the level of representing tangles. For simplicity, let us omit the auxiliary tangles in our discussion. They can be easily dealt with in a similar way as in the case of the vertical composition.

The first step is to juxtapose the two tangles into one diagram. This means that the external and internal ribbons of T_l and T_r will be on the same top line or bottom line respectively. Let us denote the source external intervals of T_l by ${}^lJ^s_j$ with $j = 1, \dots, a$ and the target external intervals by ${}^lJ^t_k$ with $k = 1, \dots, b$. Similarly, the external intervals of T_r are ${}^rJ^s_k$ with $k = 1, \dots, b$ and ${}^rJ^t_k$ with $k = 1, \dots, c$. In the same way, the internal intervals of T_l at the *top line* are given by ${}^lI^i_1, {}^lI^o_1, {}^lI^i_2, \dots, {}^lI^i_{g_{l,sc}}, {}^lI^o_{g_{l,sc}}$, and those of T_r are ${}^rI^i_1, {}^rI^o_1, \dots, {}^rI^o_{g_{r,sc}}$.

Thus, we obtain the following sequence of intervals at the top line of the juxtaposed tangles

$$ {}^lJ^s_1 <^l J^s_2 < \ldots <^l J^s_a <^l I^i_1 <^l I^o_1 < \cdots <^l I^i_{g_{l,sc}} <^l I^o_{g_{l,sc}} < \qquad \text{(Int-1)}$$

$$ <^l J^t_b < \cdots <^l J^t_2 <^l J^t_1 <^r J^s_1 <^r J^s_2 < \cdots <^r J^s_b < \qquad \text{(Int-2)}$$

$$ <^r I^i_1 <^r I^o_1 < \cdots <^r I^i_{g_{r,sc}} <^r I^o_{g_{r,sc}} <^r J^t_c < \cdots <^r J^t_2 <^r J^t_1 \qquad \text{(Int-3)}$$

Next, we disconnect the two strands ending in the neighboring intervals ${}^lJ^t_1$ and ${}^rJ^s_1$ from the top line and attach them to each other. An unambiguous way of reconnecting these strands can be given by inserting between the two strands ending at the top line a strip that is parallel and bordering to the top line, such that it runs between the endpoints of the intervals ${}^lJ^t_1$ and ${}^rJ^s_1$. Hence, the two short sides of the strip will lie on the edges of the strands ending in ${}^lJ^t_1$ and ${}^rJ^s_1$. This is where the two strands and the strip are glued together, and the combined strand is then detached from the top line. We denote this resulting strand by C_{sc}.

In the same way the corresponding strands are connected to each other at the bottom line. Analogously, we denote the glued strand by C_{tg}. Note, that with this operation the succession of the intervals at the top and bottom line is changed, since ${}^lJ^t_1$ and ${}^rJ^s_1$ are missing. Specifically, in the above sequence the second line has to be replaced by

$$ \ldots {}^lJ^t_b < \cdots <^l J^t_3 <^l J^t_2 <^r J^s_2 <^r J^s_3 < \cdots <^r J^s_b \ldots \qquad \text{(Int*-2)}$$

Furthermore, we redeclare the typed strands ending in these intermediate external intervals as well as the components to which C_{sc} and C_{tg} belong as being *internal* instead of external.

Finally, an internal annulus A is placed around the interior part of the tangle T_l. Of course, by the TS1-Move, we may, equivalently, place a coupon on either the group of the strands at the top end of T_l or the group at the bottom end of T_l.

The result of these procedures that construct a representative of $T_r \triangle_h T_l$ is summarized in the tangle picture in Fig. 2.3.

$T_r \triangle_h T_l$ =

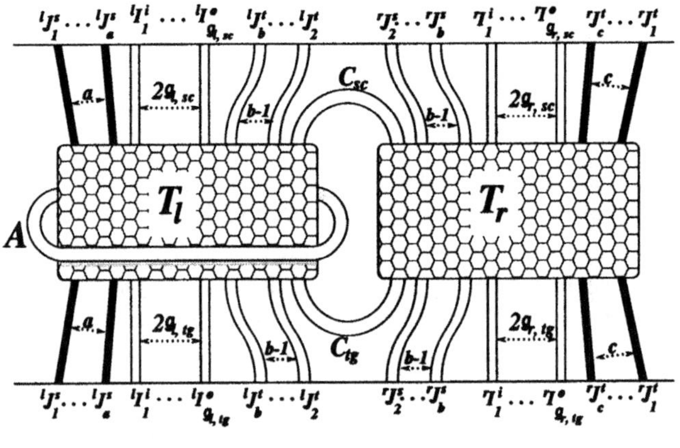

Fig. 2.3. The essential part of horizontal composition of tangles

Before we continue with the definition let us analyze further the connectivity of the tangle components in the tangle for $T_r \triangle_h T_l$.

For the intermediate (formerly external) intervals listed in (Int*-2) we note that they occur in pairs $\mathbf{p}_j = \{{}^lJ^t_j, {}^rJ^s_j\}$ with $j = 2, \ldots, b$. Recall that the condition for composing two tangles horizontally is that the vertical 1-arrows are the same, that is, the external strands induce the same permutation β. This means that the strand in T_l starting in ${}^lJ^t_j$ at the top line will end in ${}^lJ^t_{\beta(j)}$ at the bottom line, and at the same time the external strand in T_r starting in ${}^rJ^s_j$ at the top line will end in ${}^rJ^s_{\beta(j)}$ at the bottom line.

Hence, in the tangle for $T_r \triangle_h T_l$ the intervals of the pair $\mathbf{p}_j$ (with $j \geqslant 2$) will be connected to the intervals of the pair $\mathbf{p}_{\beta(j)}$ if $\beta(j) \neq 1$. The, now internal, strands can, thus, be considered a through pair, with the only difference that the intervals in $\mathbf{p}_j$ and $\mathbf{p}_{\beta(j)}$ are not adjacent.

If $j_1 = \beta^{-1}(1) \geqslant 2$, then the external strands of T_l and T_r, starting, respectively, in the intervals ${}^lJ^t_{j_1}$ and ${}^rJ^s_{j_1}$ at the top line, will end in ${}^lJ^t_1$ and ${}^rJ^s_1$. It follows that in the tangle for $T_r \triangle_h T_l$ the intervals of the pair $\mathbf{p}_{j_1}$ are connected to the piece C_{tg}, and, hence, form a top component in the sense of Section 2.2.2, again with the modification that the intervals are not adjacent.

Analogously, if $\beta(1) \geqslant 2$, the strands ending in the intervals of $\mathbf{p}_{\beta(1)}$ at the bottom line are connected to the piece C_{sc}, and, thus, form a bottom component in the tangle for $T_r \triangle_h T_l$ for this pair of intervals. In both cases the stated prerequisites that C_{tg} or C_{sc} is connected to the top or bottom line to form a top or bottom internal component respectively is equivalent to $\beta(1) \neq 1$.

For the case of $\beta(1) = 1$ the intervals ${}^lJ^t_{j_1}$ and ${}^rJ^s_{j_1}$ at the top lines of T_l and T_r are connected to the same intervals at the bottom lines. The gluing operation described for these strands will, thus, result in one closed internal component that contains both C_{sc} and C_{tg}. Since the permutation β leaves the set $\{2, \ldots, b\}$ invariant, all of the other intermediate external strands become through strands. Thus, there are no additional top or bottom components and, together with the annulus A, we have two additional closed internal components.

If we denote the number of closed, top, bottom and through components of T_l, T_r, and $T_r \triangle_h T_l$ in the same way as in the previous chapter, we can again summarize the results of the horizontal operation by a count of these numbers. The relations are listed below for the two cases $\beta(1) \neq 1$ and $\beta(1) = 1$.

For $\beta(1) \neq 1$:	For $\beta(1) = 1$:
$\mathsf{C} = \mathsf{C}_l + \mathsf{C}_r + 1,$	$\mathsf{C} = \mathsf{C}_l + \mathsf{C}_r + 2,$
$\mathsf{T} = \mathsf{T}_l + \mathsf{T}_r + 1,$	$\mathsf{T} = \mathsf{T}_l + \mathsf{T}_r,$
$\mathsf{B} = \mathsf{B}_l + \mathsf{B}_r + 1,$	$\mathsf{B} = \mathsf{B}_l + \mathsf{B}_r,$
$\mathsf{H} = \mathsf{H}_l + \mathsf{H}_r + b - 2.$	$\mathsf{H} = \mathsf{H}_l + \mathsf{H}_r + b - 1.$

In order to determine whether $T_r \triangle_h T_l$ can be well defined on equivalence classes of tangles we have to compare tangles for $T_r \triangle_h T_l$ and $T'_r \triangle_h T'_l$ where $T_r \sim T'_r$ and $T_l \sim T'_l$ in $\mathcal{T}gl$. However, as we already mentioned in the beginning of this section, it is clear that the composite $T_r \triangle_h T_l$ is not quite an admissible tangle because the intervals of the pairs $\mathbf{p}_j$ are not neighboring. Hence, it is a priori not clear which notion of equivalence we wish to use in order to relate $T_r \triangle_h T_l$ and $T'_r \triangle_h T'_l$.

In Section 7.1 we will relax the neighboring condition so that also $T_r \triangle_h T_l$ is admissible in a larger tangle category for which the set of equivalence moves is similarly generalized. For our purposes in this section it will suffice that we consider provisionally a smaller set of moves for the composed tangle with the interval structure of $T_r \triangle_h T_l$. Specifically, we want to consider the following *restricted set of moves* for the combined diagrams:

- All moves for $\mathcal{T}gl$ that do not involve the boundaries of the tangle diagram (i.e., TI1-TI11, TD1-TD5, TS1, and TS2).
- The TS3-Move for the $(g_{r,sc} + g_{l,sc})$ internal pairs of intervals $\{{}^lI^i_1, {}^lI^o_1\}, \ldots,$ $\{{}^lI^i_{g_{r,sc}}, {}^lI^o_{g_{l,sc}}\}, \{{}^rI^i_1, {}^lI^o_1\}, \ldots, \{{}^lI^i_{g_{r,sc}}, {}^rI^o_{g_{r,sc}}\}$ at the top line, as well as the corresponding $(g_{r,tg} + g_{l,tg})$ pairs of intervals at the bottom line.
- The TS4-Move for a strand that runs over *all* of the $a + 2(g_{r,sc} + g_{l,sc} + b - 1) + c$ strands emerging at the top line in the combined diagram.

It is useful to observe first that in its equivalence class in this sense the diagram for the tangle for $T_r \triangle_h T_l$ is really symmetrical despite the extra annulus A.

Lemma 2.6.2. *The tangle for $T_r \triangle_h T_l$, as depicted above, is equivalent in $\mathcal{T}gl$ to the one where the annulus A is placed around T_r instead of T_l.*

Proof. This is a straightforward application of the TS4-Move from the above restricted list of moves to a part of the ribbon A.

Lemma 2.6.3. *Let us suppose that $T_r \sim T'_r$ and $T_l \sim T'_l$ in $\mathcal{T}gl$. Then $T_r \triangle_h T_l \sim T'_r \triangle_h T'_l$ by the restricted set of moves above so that the operation on tangles factors into a binary operation $\triangle$ on their respective equivalence classes.*

Proof. It suffices to consider equivalences $T_r \sim T'_r$ and $T_l \sim T'_l$ that are given by one of the elementary moves from Section 2.3. All of these are interior moves away from the boundary and are, thus, included in the first group of the reduced set of moves in the above list. The only part of the operation that yields the tangle for $T_r \triangle_h T_l$ that affects also the interior of the individual tangles T_r and T_l is the redeclaration of parts of the external strands as internal strands. It is, however, easy so see that for all of the moves we can replace any external strand on the right and left side of the equivalence and still obtain a valid equivalence move. (Note, however, that the converse type replacement may obviously lead to invalid or meaningless moves.)

Thus, we only need to be concerned about equivalences $T_r \sim T'_r$ and $T_l \sim T'_l$ that are given by the TS3 and TS4 Move. A TS3 Move for either T_r or T_l leads quite obviously to equivalent horizontal composites by the second part of the above restricted set of moves.

We suppose now that T_l and T'_l differ by a TS4-Move. A strand that runs over the entire group of $a + 2g_{l,sc} + b$ strands emerging at the top of T_l can be slid over A by a 2-handle slide, which is implied by the restricted set of moves in the same way as for $\mathcal{T}gl$. This move results in the strand running now behind the group of strands of T_l. Hence, the picture we obtain is really $T_r \triangle_h T_l'$, which is, thus, equivalent to $T_r \triangle_h T_l$. Alternatively, we could have also used the form of $T_r \triangle_h T_l$ where instead of the annulus A a coupon is inserted at the top end of T_l. The TS4-Move for a strand is then simply realized by a TD1-Move of this strand with the extra coupon.

For a TS4-Move on T_r the argument is precisely the same once we have applied Lemma 2.6.2 and placed the annulus or coupon on this part of the composition.

It remains to reorder the intermediate intervals from (Int*-2) so that the intervals of a pair $\mathbf{p}_j$ become neighboring. This is achieved by adjoining corresponding braids, $\boldsymbol{\Upsilon}$ and $\boldsymbol{\Upsilon}^{-1}$, at the top and bottom of the diagram, which permute only these intervals.

Obviously, such braids always exist but we want to make a special selection here that is compatible with the choices of braids in (1.1.7). We consider only plat ribbon braids, i.e., braid diagrams with framings in the plane of projection. Moreover, the braid should only include the $2(b-1)$ intermediate, new internal strands. Thus, a representative can be chosen from the braid group in $2(b-1)$ strands as follows:

$$B_{2(b-1)} = \left\langle \sigma_j, {}_{j=1,\ldots,2b-3} \;\middle|\; \sigma_{i+1} \cdot \sigma_i \cdot \sigma_{i+1} = \sigma_i \cdot \sigma_{i+1} \cdot \sigma_i \right\rangle .$$

In this group we define the elements $U_{i,k}$ given by the following products of generators of $B_{2(b-1)}$ for $1 < i < k < 2b-2$:

$$U_{i,k} = (\sigma_k \cdot \sigma_{k-1} \cdot \ldots \cdot \sigma_{i+1} \cdot \sigma_i)(\sigma_{k+1} \cdot \sigma_k \cdot \ldots \cdot \sigma_{i+2} \cdot \sigma_{i+1}) \quad \in B_{2(b-1)} .$$

This braid moves the intervals in i-th and $i+1$-st position to those in k-th and $k+1$-st position and all intervals from positions $i+2$ until $k+1$ two notches down. See below for a picture of the diagram for $U_{i,k}$:

$U_{i,k} =$ 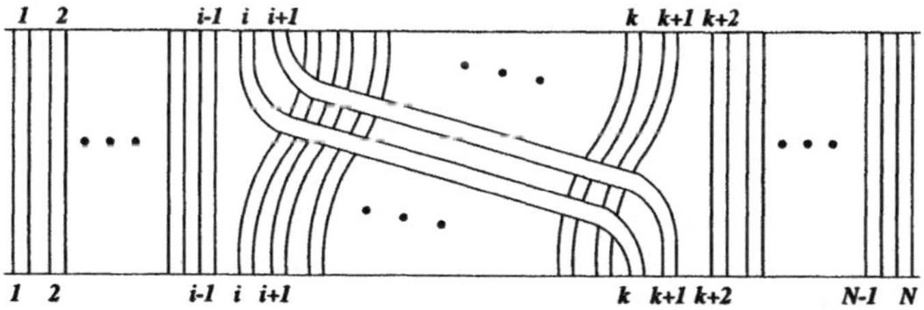

Now the braid $U_{b-1,2b-3}$ applied to the $2(b-1)$ intermediate intervals moves the intervals in $\mathbf{p}_2$, which occupy positions $b-1$ and b, into the last two positions but does not change the order of the remaining intervals. Hence, if we label start and end interval of a strand in the diagram for $U_{b-1,2b-3}$ by the same interval, the order of the intermediate part (Int*-2) at the top line of this diagram will, thus, be changed to the following at the bottom line:

$${}^lJ^t_b < \cdots <^l J^t_3 <^r J^s_3 < \cdots <^r J^s_b <^l J^t_2 <^r J^s_2 .$$

Next, if we apply to this ordering of intervals the braid $U_{b-2,2b-5}$, it involves only the first $2(b-2)$ strands, that is, all except for ${}^lJ^t_2$ and ${}^rJ^s_2$. The pair $\{{}^lJ^t_3, {}^rJ^s_3\}$ is permuted to the end of these strands in positions $2b-5$ and $2b-4$ so that ${}^lJ^t_4$ and ${}^rJ^s_4$ are now neighboring. The application of such braids can be repeated until the intervals of all pairs $\mathbf{p}_j$ are neighboring in the order

$$ {}^lJ^t_b <{}^r J^s_b <{}^l J^t_{b-1} <{}^r J^s_{b-1} < \cdots <{}^l J^t_3 <{}^r J^s_3 <{}^l J^t_2 <{}^r J^s_2 . $$

The braid Υ is, hence, given on the intermediate intervals by the corresponding composite of the special braids $U_{i,k}$, and it is identity for the $a + 2g_{l,sc(tg)}$ external and internal intervals before this group as well as for the $2g_{r,sc(tg)} + c$ external and internal intervals after this group. We summarize the definition of Υ in the following formula:

$$ \Upsilon \;=\; \mathbb{1}_{a+2g_l} \times \bar{U}_b \times \mathbb{1}_{2g_r+c}, \qquad (2.6.5) $$

$$ \text{where} \qquad \bar{U}_b = U_{2,3} \cdot U_{3,5} \cdot \ldots \cdot U_{j,2j-1} \cdot \ldots \cdot U_{b-1,2b-3}. \qquad (2.6.6) $$

In the above analysis of the connectivity structure of the tangle we had already observed that all tangle components are of the types described in Section 2.2.2 if the $\mathbf{p}_j$ are admitted as interval pairs bounding a component. Having the missing neighboring condition now corrected by the conjugation with Υ the combined tangle now fits the Definition 2.2.2 with $b-1$ additional in- or out- pairs $({}^{int}I^i_j, {}^{int}I^o_j) = ({}^lJ^t_j, {}^rJ^s_j)$ of internal intervals. All together we, thus, have $g_{l,sc} + g_{r,sc} + b - 1$ internal pairs at the top line, and $g_{l,tg} + g_{r,tg} + b - 1$ pairs at the bottom line.

Lemma 2.6.4. *The tangle for $T_l \circ_h T_r$ obtained from (2.6.3) using the above constructions for Υ and $\triangle_h$ is an admissible tangle for $\mathcal{Tgl}$ and for the square given on the left hand side of (2.6.4).*

Observe that each of the moves of the restricted set of equivalences for $T_l \triangle_h T_r$ translates into a move of $\mathcal{Tgl}$ for $T_l \circ_h T_r$ in this construction. As a matter of fact, the only equivalence moves that we need to add to those coming from the restricted set in order to obtain all equivalences of $\mathcal{Tgl}$ are the TS3-Moves for the new internal interval pairs $({}^{int}I^i_j, {}^{int}I^o_j)$ with $j = 2, \ldots, b$.

We also know from Lemma 2.6.3 that the class of $T_l \triangle_h T_r$ only depends on the classes of T_l and T_r, and that the vertical composition $\circ_v$ that we use to adjoin Υ also only depends on the equivalence classes of tangles. Combining these facts we conclude that $\circ_h$ also factors into the classes of tangles.

Lemma 2.6.5.

The product $\circ_h$ is a well defined binary operation on the 2-arrow sets of $\mathcal{Tgl}$.

2.6.3 Double Category Structure of $\mathcal{Tgl}$

The two binary operations $\circ_v$ and $\circ_h$ defined in Sections 2.6.1 and 2.6.2 do, in fact, make $\mathcal{Tgl}$ into a double category in the sense of Ehresmann [Ehr63a] and as explained in the introduction and Appendix B.1. The relations that we need to verify are the associativity of each product by itself, as well as the *interchange law* relating the two products. We start with the easiest condition.

Lemma 2.6.6. *The multiplication $\circ_v$ is associative.*
Hence, $(\mathcal{T}gl, \circ_v)$ forms a (vertical) category, which decomposes as in (2.6.2).

Proof. From the definitions in Section 2.6.1 it is obvious that the composition is already associative on the level of representing tangles.

The corresponding statement for the horizontal composition requires only little more attention:

Lemma 2.6.7. *The multiplications $\circ_h$ and $\triangle_h$ are associative.*
In particular, $(\mathcal{T}gl, \circ_h)$ forms a (horizontal) category.

Proof. We prove associativity only for $\circ_h$, where also the special braids $\bar{U}_b \in B_{2(b-1)}$ from (2.6.6) enter the definition of the tangles. The proof for $\triangle_h$ follows simply by omitting these braids.
Let us have three 2-arrows T_1, T_2, and T_3, such that the structure of objects and matching 1-arrows is as in the composite diagram of squares below.

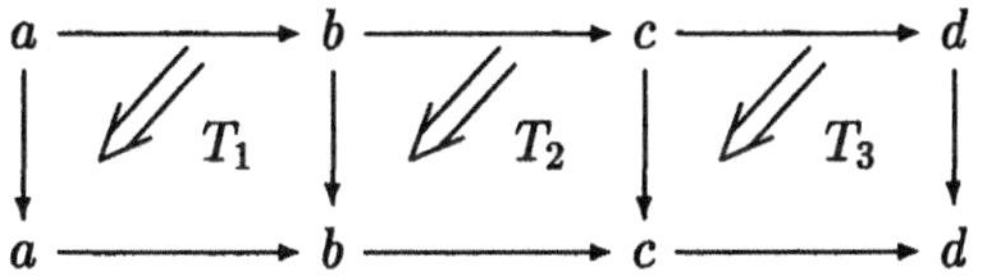

$$
\begin{array}{ccccccc}
a & \longrightarrow & b & \longrightarrow & c & \longrightarrow & d \\
\Big\downarrow \;\; \nearrow T_1 & & \Big\downarrow \;\; \nearrow T_2 & & \Big\downarrow \;\; \nearrow T_3 & & \Big\downarrow \\
a & \longrightarrow & b & \longrightarrow & c & \longrightarrow & d
\end{array}
\qquad (2.6.7)
$$

We have to show that the tangle constructed for $T_3 \circ_h (T_2 \circ_h T_1)$ is equivalent in $\mathcal{T}gl$ to the one we obtain for $(T_3 \circ_h T_2) \circ_h T_1$. Both are drawn in the following picture. In these diagrams the framings of all depicted tangle components are understood to be in the plane of projection.

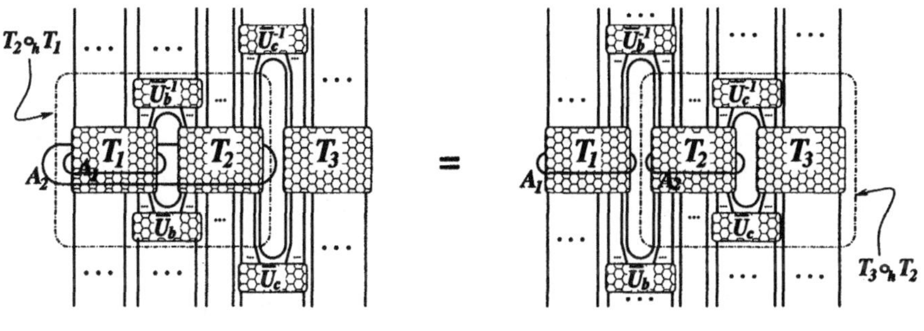

On the left side the tangle representing the composite $(T_2 \circ_h T_1)$ is surrounded by a dashed line and the annulus used in its construction is A_1. In order to obtain

$T_3 \circ_h (T_2 \circ_h T_1)$ another annulus A_2 is placed around the tangle for $(T_2 \circ_h T_1)$ and T_3 is connected to the right.

The product $(T_3 \circ_h T_2)$ is similarly the part of the diagram on the right side, which lies within the dashed box. In there an annulus A_2 is placed only around T_2. The composite $(T_3 \circ_h T_2) \circ_h T_1$ is constructed from this by connecting T_1 to it on the left and placing an annulus A_1 around it.

Also we have to adjoin the braids Υ for each of the compositions. Its form in (2.6.5) shows that we only need to insert braids $\bar{U}_{b/c}^{\pm 1}$ only at the groups of intermediate strands that are of external type in the individual tangles and become internal in the composites.

Since for the two composites these groups of strands are disjoint, the two Υ braids commute with each other. In the diagram above this manifests itself in the fact that the blocs representing the braids $\bar{U}_{b/c}^{\pm 1}$ can be moved past each other using simply the TI8-Moves and, hence, the braid conjugations are the same for the two orders of composition.

Finally, we still need to match the different configurations of annuli A_1 and A_2 in order to show that the two composite tangles are equivalent. This is easily done by using the 2-handle slide or $\mathcal{O}_2$-Move that was introduced in Section 2.5.2, which is by Lemma 2.5.5 also valid as an equivalence in $\mathcal{T}gl$. Specifically, if we slide the A_2-component over the A_1-component in the left diagram for $T_3 \circ_h (T_2 \circ_h T_1)$ we obtain precisely the annuli in the right diagram for $(T_3 \circ_h T_2) \circ_h T_1$.

The only remaining axiom for a double category is verified next.

Lemma 2.6.8. *The* interchange law *for double categories holds between the horizontal products $\circ_h$ or $\triangle_h$ and the vertical product $\circ_v$.*

Proof. We shall prove the assertion only for the product $\triangle_h$. For $\circ_h$ we need to include the braids $\bar{U}_b^{\pm 1}$ for the horizontal compositions. They do, however, cancel out in the relevant vertical compositions.

Let us, thus, assume that we have four admissible tangles $T_{i,j}$ with $i, j = 1, 2$. Their 1-arrow and object structure are such that they fit together in a composite of squares as in the following diagram.

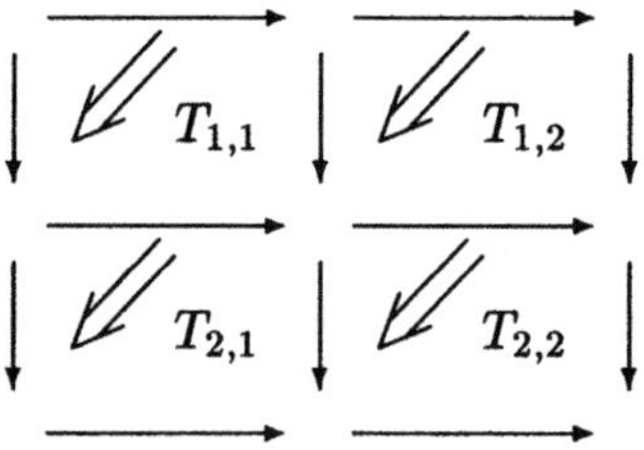

(2.6.8)

The interchange law requires the following equation, which has to hold modulo the equivalence moves in $\mathcal{T}gl$:

$$(T_{2,2} \triangle_h T_{2,1}) \circ_v (T_{1,2} \triangle_h T_{1,1}) \quad = \quad (T_{2,2} \circ_v T_{1,2}) \triangle_h (T_{2,1} \circ_v T_{1,1}).$$

In the sequence of diagrams below we describe how we can apply equivalence moves to the tangle constructed in the first order of composition until we arrive at the tangle obtained for the second composite form. All strands (apart from those replaced by the blocs for the $T_{i,j}$-factors) are understood to have framings in the plane of projection. That is, they are plat if considered as ribbons in the plane.

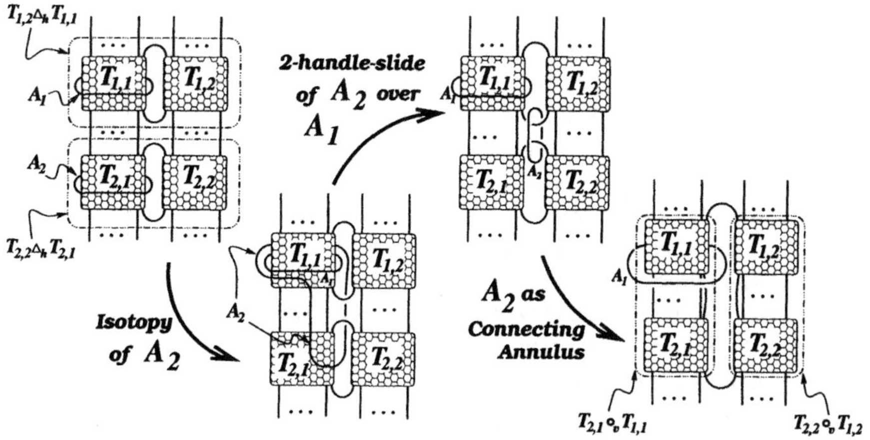

In the first diagram for $(T_{2,2}\Delta_h T_{2,1}) \circ_v (T_{1,2}\Delta_h T_{1,1})$ the subdiagrams for the factors $(T_{2,2}\Delta_h T_{2,1})$ and $(T_{1,2}\Delta_h T_{1,1})$ are surrounded by a dashed line. The annuli used to construct these two horizontal compositions are A_2 and A_1, respectively.

The first step consists of an isotopy in which the greater part of the annulus A_2 surrounding $T_{2,1}$ is pushed upwards over the bloc representing $T_{1,1}$ so that A_2 runs parallel to A_1 around this part. Next, we apply the $\mathcal{O}_2$-Move again, as described in Section 2.5.2, and slide A_2 over A_1, with an auxiliary strip chosen as short as possible between the two annuli, and which is plat in the plane of projection.

As a result the annulus A_2 will no longer surround the tangles $T_{2,1}$ or $T_{1,1}$, but instead will bound a disc through which only the lower connecting strand C^2_{tg} of the product $(T_{2,2}\Delta_h T_{2,1})$ and the upper connecting strand C^1_{sc} of the product $(T_{1,2}\Delta_h T_{1,1})$ runs. After we replace this annulus A_2 by coupon using a TS1-Move, this yields precisely the situation of the Connecting Annulus Move stated at the end of Section 2.5.1.

The result of this move is that the annulus A_2 disappears and the strands C^2_{tg} and C^1_{sc} are connected to each other. Thus, the strands of $T_{2,1}$ and $T_{1,1}$ are now all connected to each other as for a vertical composition. The same is true for the components $T_{2,2}$ and $T_{1,2}$ so that we obtain the subdiagrams for $(T_{2,2} \circ_v T_{1,2})$ and $(T_{2,1} \circ_v T_{1,1})$ surrounded in the last diagram by dashed lines. The remaining A_1 plays now the role of the annulus needed to construct the final horizontal composition between the $(T_{2,2} \circ_v T_{1,2})$ and $(T_{2,1} \circ_v T_{1,1})$ parts. We arrive at the picture for the composition order $(T_{2,2} \circ_v T_{1,2})\Delta_h(T_{2,1} \circ_v T_{1,1})$, and have, thus, proven the interchange law for equivalence classes of tangles.

Our constructions and observations in Section 2.6 combine to the following central result of this chapter.

Theorem 2.6.9. $(\mathcal{Tgl}, \circ_v, \circ_h)$ *is a double category.*

2.7 Special Cases and Applications

2.7.1 Isolated Strands Category for $\beta(1) = 1$

Consider a horizontal composition of two tangles T_l and T_r as in (2.6.4). Already in the counting of ribbon types we distinguished the cases $\beta(1) = 1$ and $\beta(1) \neq 1$ for the intermediate vertical 1-arrow $\beta \in S_b$.

A simple example of a tangle T_l with $\beta(1) = 1$ is one in which the external strand running from J_1^t at the top line to the corresponding interval J_1^t at the bottom line is a straight vertical strip, X, that is isolated from the rest of the tangle. This means that all other strands of the tangle are completely to the left of the external strip X. The purpose of this section is to show that all tangles can be reduced to this special case in a coherent way. More precisely, we will identify the classes of tangles with $\beta(1) = 1$ with classes of tangles for which the first external strand is isolated in the described manner.

To this end let us describe the relevant maps on subspaces of admissible tangles more formally. The first, $\mathbf{Inc}^+$, is described in a straightforward way. On the space of admissible tangles with source and target objects a and $b-1$ and source and target vertical 1-arrows $\alpha \in S_a$ and $\beta^\# \in S_{b-1}$ it adds an unknotted external strand to the right of all other components of the tangle. The result is a tangle with target object b and target vertical 1-arrow $\beta = \beta^\# \times 1 \in S_b$. Note that every $\beta \in S_b$ with $\beta(1) = 1$ is of this form, where $\beta^\#$ is understood as a permutation on the integers $\{2, \ldots, b\}$. The notation for the set of admissible tangles for a given square of objects and 1-arrows is as in the following description of the map $\mathbf{Inc}^+$.

$$\mathbf{Inc}^+ : \mathrm{Adm}\begin{bmatrix} & a \rightarrow b-1 & \\ \alpha\downarrow & & \downarrow\beta^\# \\ & a \rightarrow b-1 & \end{bmatrix} \longrightarrow \mathrm{Adm}\begin{bmatrix} & a \rightarrow b & \\ \alpha\downarrow & & \downarrow\beta^\#\times 1 \\ & a \rightarrow b & \end{bmatrix} \tag{2.7.1}$$

The map can also be illustrated in a bloc tangle diagram. Below we label the target external intervals by $J_b^t, J_{b-1}^t, \ldots, J_2^t$, with the last being the rightmost.

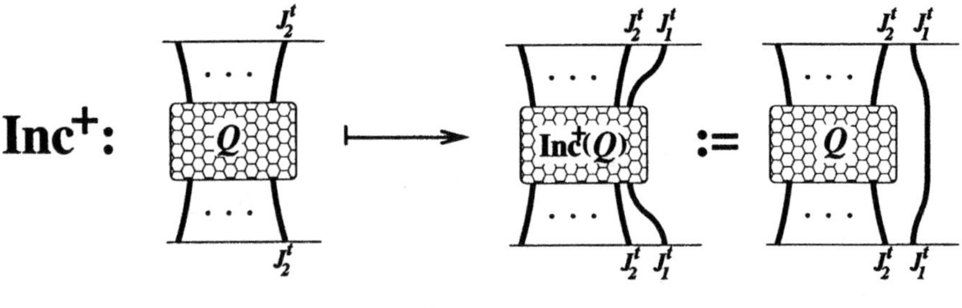

The objective is to show that $\mathbf{Inc}^+$ factors into a map on the equivalence classes of the tangles in (2.7.1), and, as such, is a bijection. Obviously, we cannot hope to find an inverse for **Inc**$^+$ on the level of tangles, but we may attempt to find a map **Isl**$^+$ that assigns to a tangle with $\beta(1) = 1$ a tangle with isolated last strand, such that it inverts **Inc**$^+$ up to equivalence. The structure of **Isl**$^+$ is, in the same notation as before, given as follows:

$$\mathbf{Isl}^+ : \mathrm{Adm}\begin{bmatrix} & a \to b & \\ \alpha\downarrow & & \downarrow\beta^{\#}\times 1 \\ & a \to b & \end{bmatrix} \longrightarrow \mathrm{Adm}\begin{bmatrix} & a \to b-1 & \\ \alpha\downarrow & & \downarrow\beta^{\#} \\ & a \to b-1 & \end{bmatrix}. \qquad (2.7.2)$$

The definition of **Isl**$^+$ is given by diagrams (v) and (vi) in the illustration below, if we delete from both pictures the strand labeled X. The sequence of the six diagrams also serves to prove that $\mathbf{Inc}^+(\mathbf{Isl}^+(T))$, given by diagrams (v) and (vi), is indeed equivalent in $\mathcal{T}gl$ to the original tangle T:

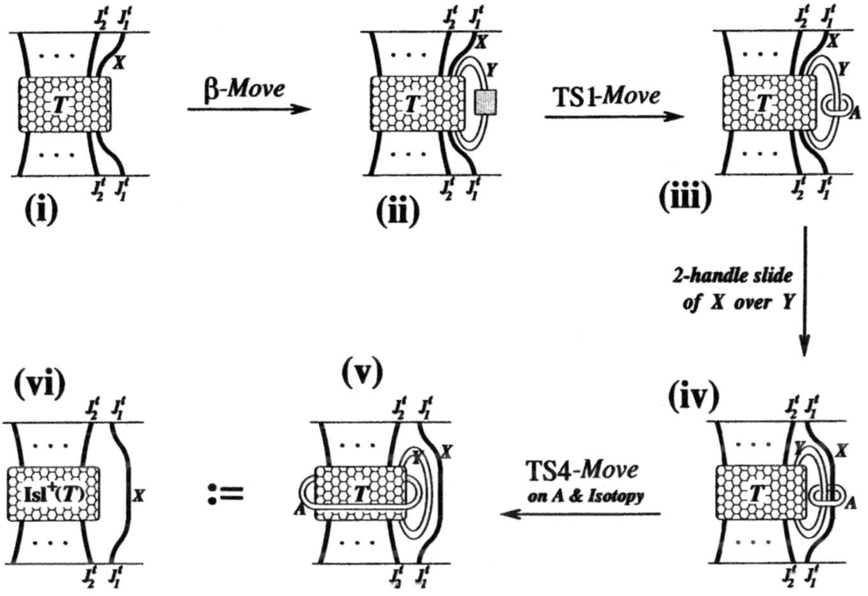

In diagram (i) we know because of $\beta(1) = 1$ that the first target external strand X runs from the interval J_1^t at the top line to the interval J_1^t at the bottom line. A β-Move is given by introducing an arbitrary ribbon Y, which has an elsewise isolated coupon placed on it. In diagram (ii) we choose Y to be parallel to X inside of the tangle T using the ribbon width or framing to determine the local directions of the push off. If Y enters at the top of T to the right of X it will also emerge to the right of X at the bottom of T due to the orientability constraints on X. The coupon on Y is then replaced by a small annulus A around Y using the TS1-Move so that we are left with diagram (iii).

Next, we perform a 2-handle slide of the external strand X over Y in the obvious way. The result is the diagram in (iv) where X runs now straight from the top to the bottom and links only to the annulus A. In the bloc of T in (iv) we have, thus, substituted the strand X from (i) completely by Y.

Finally, we can stretch a front piece of the annulus A along either the top or bottom line and apply the TS4-Move to push it behind the entire group of lines. Instead of the two vertical strands X and Y the resulting annulus A will now surround the bloc T. In particular, we have, thus, managed to isolate the external strand X, and produced the desired picture for $\mathbf{Isl}^+(T)$.

We can also consider the opposite composition, namely $\mathbf{Isl}^+(\mathbf{Inc}^+(Q))$. Here, the first external strand of $\mathbf{Inc}^+(Q)$ is already isolated so that its replacement by the internal strand Y results in an annulus $\widehat{Y}$, which surrounds only the annulus A. The resulting diagram is as depicted below.

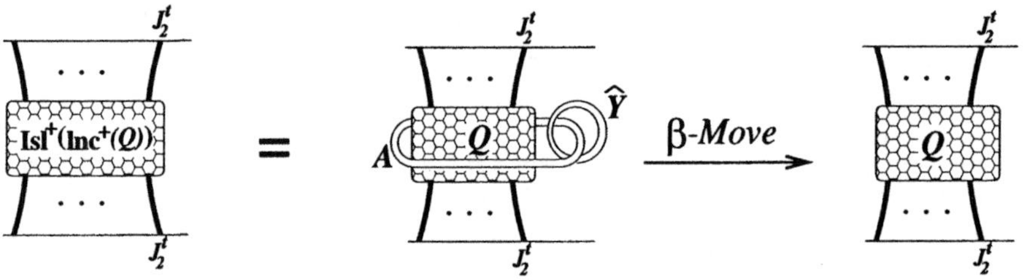

If we replace the annulus $\widehat{Y}$ by a coupon via a TS1-Move, we can further apply a β-Move by which both the ribbon A and the coupon for $\widehat{Y}$ are eliminated. We recover the original tangle, proving that $\mathbf{Isl}^+(T)$ and $\mathbf{Inc}^+(Q)$ are two-sided inverses of each other up to the given equivalences. Let us summarize this and the specific moves used for the equivalences next.

$$\mathbf{Inc}^+\big(\mathbf{Isl}^+(T)\big) \sim_{\mathrm{TS1,TS4},\beta,\mathcal{O}_2} T, \qquad \mathbf{Isl}^+\big(\mathbf{Inc}^+(Q)\big) \sim_{\mathrm{TS1},\beta} Q. \tag{2.7.3}$$

In the picture of tangles of S^2 we can imagine that an isolated external strand is moved to the line $L = \{\infty\} \times [-1,1]$ at infinity in $S^2 \times [-1,1]$. The remainder $\mathbf{Isl}^+(T)$ of the tangle is, thus, naturally understood as a tangle in the complement $S^2 \times [-1,1] - L \cong \mathbb{R}^2 \times [-1,1]$. The transition from tangles in $\mathbb{R}^2 \times [-1,1]$ to tangles in $S^2 \times [-1,1]$ is reflected in the introduction of the TS4-Move. It is, thus, nearby to assume that the equivalences we really want to consider for $\mathbf{Isl}^+(T)$ should exclude this TS4-Move in order to make the two maps identifications.

For the precise statement let us introduce special tangle categories $\mathcal{T}\!\mathit{gl}^{\beta(1)=1}$ and $\mathcal{T}\!\mathit{gl}^{tg1\to\infty}$. The first is a full subcategory $\mathcal{T}\!\mathit{gl}^{\beta(1)=1} \subset \mathcal{T}\!\mathit{gl}$ and consists of all 2-arrows for which the target vertical 1-arrow fulfills $\beta(1) = 1$ or, equivalently, $\beta = \beta^\# \times 1$. Also, for $\mathcal{T}\!\mathit{gl}^{tg1\to\infty}$ the target vertical 1-arrow will always fulfill $\beta(1) = 1$. However, the set of admissible tangles for a given square is given by the corresponding set of admissible tangles with one less external strand. More precisely, for an object b of $\mathcal{T}\!\mathit{gl}^{tg1\to\infty}$ the object of the tangle as a tangle in $\mathcal{T}\!\mathit{gl}$ will be $b-1$ as given in the following definition:

$$\mathrm{Adm}_{\mathcal{T}gl^{tg1\to\infty}} \begin{bmatrix} a \to b \\ \alpha\downarrow \qquad \downarrow\beta^\#\times 1 \\ a \to b \end{bmatrix} \quad = \quad \mathrm{Adm}_{\mathcal{T}gl} \begin{bmatrix} a \to b-1 \\ \alpha\downarrow \qquad \downarrow\beta^\# \\ a \to b-1 \end{bmatrix}. \tag{2.7.4}$$

The equivalence moves on the set of admissible tangles is now given as

$$\{\text{Moves of } \mathcal{T}gl^{tg1\to\infty}\} = \{\text{Moves of } \mathcal{T}gl\} - \{\text{TS4} - \text{Move}\}$$

With these notions of equivalence for the domain and range spaces of the two maps of tangles we may now formulate the following lemma.

Lemma 2.7.1. $\mathbf{Isl}^+$ *and* $\mathbf{Inc}^+$ *factor into the equivalence classes of tangles so that they become well defined maps between the 2-arrow sets of* $\mathcal{T}gl^{\beta(1)=1}$ *and* $\mathcal{T}gl^{tg1\to\infty}$ *for a given square.*

Proof. For the first map the lemma asserts that if two tangles T and T' with $\beta(1) = 1$ are equivalent in $\mathcal{T}gl$ then $\mathbf{Isl}^+(T)$ has to be equivalent to $\mathbf{Isl}^+(T')$ in $\mathcal{T}gl^{tg1\to\infty}$. Let T and T' be related by an internal move, i.e., any move other than TS3 or TS4. The tangle bloc in diagram (v) is basically unchanged, and all internal moves are valid in $\mathcal{T}gl^{tg1\to\infty}$. Therefore, $\mathbf{Isl}^+(T)$ and $\mathbf{Isl}^+(T')$ are also related by the corresponding internal move. The only way that the bloc T in (v) is changed from the original tangle is that the type of the first strand changes from external to internal. However, it is clear that all moves that are valid for external strands are also valid for internal ones. Of the two remaining boundary Moves TS3 also follows by a quick inspection of the diagram in (v). The diagram of each TS3-Move on T can be moved without obstruction to the boundary of $\mathbf{Isl}^+(T)$, where it is interpreted as a TS3-Move in $\mathcal{T}gl^{tg1\to\infty}$.

Hence, we only need to show that if a tangle T' is obtained from T by a TS4-Move, where a strand S running over and across the top group of strands is pushed behind the entire group, then $\mathbf{Isl}^+(T)$ can be related to $\mathbf{Isl}^+(T)$ *without* the use of any TS4-Move as those are not available in $\mathcal{T}gl^{tg1\to\infty}$. In order to show this let us consider a variation of the defining picture in (v). We move the annulus A to the top of T until it surround the group of strands emerging from T. By a TS1-Move we can, thus, replace the annulus by a coupon that lies on the group. The TS4-Move in T can now be realized by pushing the strand S through this extra coupon at the top using the TD1-Move valid in $\mathcal{T}gl^{tg1\to\infty}$.

The corresponding proof for $\mathbf{Inc}^+$ is trivial since in this direction we are adding instead of subtracting equivalence moves.

Observe also that the equivalence moves indicated in equations (2.7.3) are valid in the respective categories. Hence, the factored maps from Lemma 2.7.1 are indeed inverses of each other.

It is quite obvious from the definition of the admissible tangles for $\mathcal{T}gl^{\beta(1)=1}$ and $\mathcal{T}gl^{tg1\to\infty}$ that these sets of tangle classes are closed under the vertical composition $\circ_v$, in which representing tangles are stacked on top of each other and the permutations are multiplied in vertical direction. Hence, these two sets indeed form vertical categories, whose objects are the horizontal 1-arrows of $\mathcal{T}gl$. Both $\mathcal{T}gl^{\beta(1)=1}$ and $\mathcal{T}gl^{tg1\to\infty}$ are also right ideals for the horizontal composition $\circ_h$, but they clearly do not close under $\circ_v$.

Combining these observations we arrive at the following correspondence of categories.

Proposition 2.7.2. *We have the following isomorphisms of (vertical) categories:*

$$\mathbf{Isl}^+ : \left(\mathcal{Tgl}^{\beta(1)=1}, \circ_v\right) \xleftarrow{\sim} \left(\mathcal{Tgl}^{tg1\to\infty}, \circ_v\right) : \mathbf{Inc}^+ .$$

There is an obvious variation of this proposition, in which we isolate the left-most source external strand in the case where the source vertical 1-arrow does not permute 1. In this case we have analogous maps $\mathbf{Isl}^-$ and $\mathbf{Inc}^-$, which establish an isomorphism between similar vertical categories $\mathcal{Tgl}^{\alpha(1)=1}$ and $\mathcal{Tgl}^{sc1\to\infty}$. The defining diagrams are simply mirror reflections of the ones above.

Further generalizations can be found by realizing that the isolation procedure used to define $\mathbf{Isl}^-$ can indeed be applied to *any* strand in the diagram. However, in our context we are mainly interested in other descriptions of the horizontal composition, for which the outer most external strands play a special role.

2.7.2 IXB-Decomposition and Specializations of Compositions

As an application of the strand isolation techniques from the previous section we shall give in this section alternate ways of describing horizontal compositions.

Lemma 2.7.3. *For* $\beta(1) = 1$ *the composition* $T_2 \triangle_h T_1$ *is, equivalently, given by the simple juxtaposition* $\left[\mathbf{Isl}^+(T_1)\middle|\mathbf{Isl}^-(T_2)\right]$ *of the tangles with isolated strands as in Proposition 2.7.2.*

Proof. The proof follows directly by applying the definition of the horizontal composition from Figure 2.3 to the equivalent presentations of T_2 and T_1 as described in Section 2.7.1. The diagram below shows that the juxtaposition can be obtained by a simple application of the β-Move, as introduced at the end of Section 2.5.1:

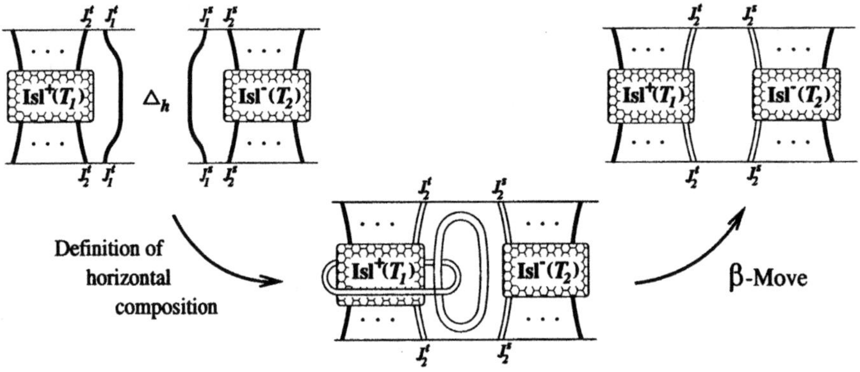

Among the most basic horizontal compositions for which $\beta(1) \neq 1$ is the simple braid of two target external strands, one straight source external strand and no internal strands. The tangle is depicted on the right side in the diagram below and denoted $\mathbf{c}^+$. Another tangle $\mathbf{c}^-$ is given by switching the source and target strands. In the diagram below the horizontal composite $\mathbf{m} = \mathbf{c}^- \circ_h \mathbf{c}^+$ is worked out explicitly:

$\mathbf{c}^{+}$ $\mathbf{c}^{-}$ $\mathbf{m}$

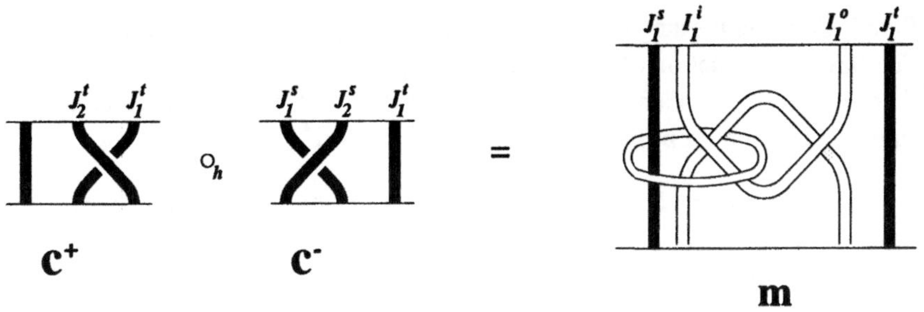

The horizontal arrows of $\mathbf{c}^{+}$ are clearly both $[0, 1/2]$ and the target permutation is just the non-trivial element $(1, 2) \in S_2$. We summarize the structure of the horizontal composition as depicted in the following diagram:

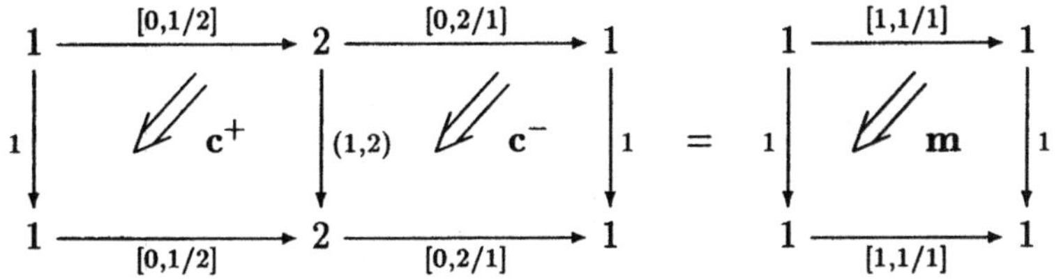

$$
\begin{array}{ccccccc}
1 \xrightarrow{\;[0,1/2]\;} 2 \xrightarrow{\;[0,2/1]\;} 1 & & 1 \xrightarrow{\;[1,1/1]\;} 1 \\
\end{array}
$$

$$
\begin{array}{ccc}
1 \Big\downarrow \quad \mathbf{c}^{+} \quad \Big\downarrow (1,2) \quad \mathbf{c}^{-} \quad \Big\downarrow 1 & = & 1 \Big\downarrow \quad \mathbf{m} \quad \Big\downarrow 1 \qquad (2.7.5) \\
1 \xrightarrow{\;[0,1/2]\;} 2 \xrightarrow{\;[0,2/1]\;} 1 & & 1 \xrightarrow{\;[1,1/1]\;} 1
\end{array}
$$

Finally, for $k = 2, \dots, b$ let us define special braids, $\mathbf{b}_k^{\pm}$, with horizontal arrows $[0, 1/b]$, and target vertical arrow the cyclic permutation $(2, 3, \dots, k)^{-1} \in S_b$, as depicted below:

$\mathbf{b}_k^{+}$ $\mathbf{b}_k^{-}$

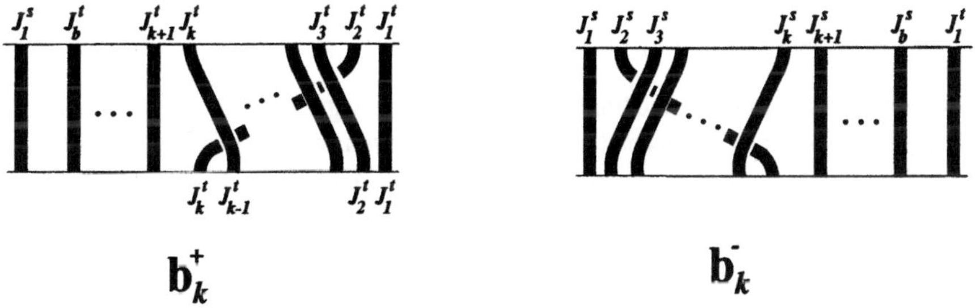

Given these special braid elements we can now state what we consider an "IBX-decomposition" of a tangle with respect to its target permutation.

Proposition 2.7.4. *For any tangle T in $\mathcal{T}gl$ with target vertical arrow $\beta \in S_b$ there is a tangle Q_T^{+} in $\mathcal{T}gl^{tg1 \to \infty}$ such that we have the following equivalent presentation for T:*

$$T \cong (\mathbf{b}_{\beta(1)}^{+} \circ_h \mathbf{I}_{[g, a/1]}) \circ_v (\mathbf{c}^{+} \circ_h \mathbf{I}_{[g, a/(b-1)]}) \circ_v \mathrm{Inc}^{+}(Q_T^{+}).$$

An analogous statement exists for Q_T^{-}, in which horizontal target and sources are exchanged.

Proof. Consider the tangle $T^* = (\mathbf{c}^+ \circ_h \mathbb{I}_{[g,a/(b-1)]})^{-1} \circ_v (\mathbf{b}_{\beta(1)} \circ_h \mathbb{I}_{[g,a/1]})^{-1} \circ_v T$. The target permutation of T^* is, thus, given by $\beta^* = (1,2) \circ (2,3,\ldots,\beta(1)) \circ \beta$ so that $\beta^*(1) = 1$ and, hence, $T^* \in \mathcal{T}gl^{\beta(1)=1}$. We can, therefore, apply the equivalence from Proposition 2.7.2 and set $Q_T^+ = \mathbf{Isl}^+(T^*)$.

An interesting application of the IXB-decomposition is an alternate way of describing the horizontal composition. After writing the product of two tangles T_2 and T_1 as in Proposition 2.7.4 we can use the interchange law, Lemma 2.7.3 and (2.7.5), and obtain:

$$T_2 \triangle_h T_1 = \Big((\mathbb{I}_{[g,1/c]} \circ_h \mathbf{b}^-_{\beta(1)}) \circ_v (\mathbb{I}_{[g,(b-1)/c]} \circ_h \mathbf{c}^-) \circ_v \mathbf{Inc}^-(Q^-_{T_2}) \Big) \circ_h$$
$$\circ_h \Big((\mathbf{b}^+_{\beta(1)} \circ_h \mathbb{I}_{[g,a/1]}) \circ_v (\mathbf{c}^+ \circ_h \mathbb{I}_{[g,a/(b-1)]}) \circ_v \mathbf{Inc}^+(Q^+_{T_1}) \Big)$$

$$= \Big(\mathbb{I}_{[g,1/c]} \circ_h \Big[\hat{\mathbf{b}}^+_{\beta(1)} \Big| \hat{\mathbf{b}}^-_{\beta(1)} \Big] \circ_h \mathbb{I}_{[g,a/1]} \Big) \circ_v$$
$$\circ_v \Big(\mathbb{I}_{[g,(b-1)/c]} \circ_h \mathbf{m} \circ_h \mathbb{I}_{[g,a/(b-1)]} \Big) \circ_v \Big[Q^+_{T_1} \Big| Q^-_{T_2} \Big]. \qquad (2.7.6)$$

Here $[A, B]$ denotes the ordinary juxtaposition of two tangles, where the intermediate external strands are reinterpreted as internal ones, and $\hat{\mathbf{b}}^{\pm}_{\beta(1)}$ stands for the braid, which is obtained by removing the outermost straight strand between intervals $J_1^{s/t}$. The horizontal compositions with the identities are simply given by substituting the single source or target external strand by the missing set of parallel internal and external strands required by the horizontal 1-arrow.

Remark 2.7.5. Note that the final formula in (2.7.6) can be used as a definition of the horizontal composition instead of the diagram given in Figure 2.3 in Section 2.6.2 as all ingredients are independently well defined. This definition was in fact the first version used in our construction of topological quantum field theories, which was somewhat more convenient in the case of $\beta(1) = 1$ but cumbersome in the more general case. The reader is invited to show that the definition suggested by (2.7.6) implies the one given in Figure 2.3.

3. Isomorphism between Tangle and Cobordism Double Categories

In the previous two chapters we introduced two double categories. One is the double category of framed relative cobordisms $\widetilde{\mathcal{Cob}}$ from Chapter 1, and the other is the double category of equivalence classes of admissible tangles $\mathcal{Tgl}$ from Chapter 2. The purpose of this chapter is to show that they are isomorphic as double categories. This provides us with a systematic combinatorial presentation of 3-manifolds with corners that respects the two gluing operations. The main result of the first three chapters of this book is summarized in the following presentation theorem.

Theorem 3.0.6. *There is an isomorphism double functor of double categories,*

$$\mathfrak{Surg} : \mathcal{Tgl} \quad \xrightarrow{\cong} \quad \widetilde{\mathcal{Cob}},$$

with the following properties:

(A) *It is identity on the objects and* 1*-arrows.*
(B) *In the case of closed manifolds,* $\mathfrak{Surg}$ *is constructed as surgery along (bridged) link diagrams.*
(C) *It strictly respects the horizontal and vertical composition,* $\circ_h$ *and* $\circ_v$*.*

Let us give a few more explanations on the stated properties of the functor in Theorem 3.0.6:

(A) is nearby as the sets of objects and 1-arrows of both categories are already identical by construction. Recall, that the objects are given by non-negative integers, and 1-arrows $[g, a/b] : a \to b$ for given objects are also labeled by non-negative integers $g \geqslant 0$. Less abstractly, we may say the $\mathfrak{Surg}$ assigns to each circle S^1 in the boundaries an external interval $J_k^{s/t}$ and to each surface handle an interval pair $\{I_k^i, I_k^o\}$ on the real line.

In the case of a closed 3-manifold M, as in *(B)*, we have neither boundaries nor corners so that all objects are 0 and all vertical 1-arrows are of the form $[0, 0/0]$. (We assume that any S^2 in the boundary is filled up with a 3-ball.) The square associated to M in double category picture is thus the trivial one:

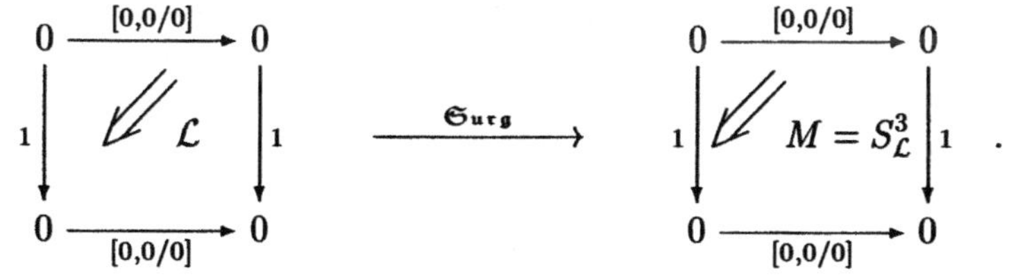

For a tangle with this source-target square there are no intervals, external or internal, on the top or bottom lines. Hence, we obtain a diagram that is contained entirely in the interior of the strip, and contains no external strands, top, bottom or through internal strands. We can, thus, have a diagram of only closed internal ribbons and coupons in S^3. By replacing the coupons either by ball pairs or 0-framed annuli we obtain a surgery diagram $\mathcal{L} \subset S^3$ and its image under $\mathfrak{Surg}$ is the 3-manifold $M = S^3_{\mathcal{L}}$ obtained by surgery along the link components as will be explained in more detail below. Our construction, thus, generalizes the known surgery presentations, such as those in [Kir78], [FR79], or [Ker99].

The property in *(C)* is merely reemphasizing the obvious, since the functoriality is already implied in the notion of a double functor, as defined in Section B.1. The statement of the theorem additionally implies that this is an isomorphism functor, which implies a bijection between corresponding 2-arrow sets.

Theorem 3.0.6 allows us to reduce the construction of a TQFT functor defined on the topological double category to the construction of a functor on a combinatorial category, which is explicitly given in terms of generators and relations. The possibility of two independent gluing operations also leads to a variety of other applications, which we will discuss elsewhere. They include, for example, systematic ways of constructing surgery diagrams for fibered knot complements or other manifolds with boundaries and fibrations.

The proof of Theorem 3.0.6 that we give here, will follow in many parts the presentation of ordinary cobordisms given in [Ker99], with only trivial objects. The inclusion of the S^1-corners, however, requires a number of generalizations of the arguments given there. Although the verification of the vertical composition law is still fairly similar to the closed case in [Ker99], the proof of the horizontal functoriality deals with a completely new situation. The particular steps of the proof are organized as follows.

$\mathcal{T}gl \longrightarrow \widetilde{\mathcal{C}ob}$: Departing from a tangle in $S^2 \times [-1,1]$ we construct a bridged link in a standard connected cobordism $\mathcal{H}_{g_{sc}}^{g_{tg}}$ in Sect. 3.3.1. From such a bridged link we construct a relative cobordism and a 4-manifold (an object of $\widetilde{\mathcal{C}ob}$) in Sect. 3.2.3 according to bridged link calculus of Sect. 1.5.2.

$\widetilde{\mathcal{C}ob} \longrightarrow \mathcal{T}gl$: Departing from an object of $\widetilde{\mathcal{C}ob}$ (a 3-cobordism and a 4-manifold) we construct a bridged link in $\mathcal{H}_{g_{sc}}^{g_{tg}}$ in Lemma 3.3.3. From such a link we construct a tangle diagram in $S^2 \times [-1,1]$ in Sect. 3.3.2.

The fact that the obtained maps between sets of equivalence classes are inverse to each other is proved in Lemma 3.3.1, Proposition 3.3.2, Lemma 3.3.3 and Lemma 3.3.4.

Summary of Content:

In Section 3.1 we provide additional tools to manipulate handle decompositions besides those already given in Section 1.5. In particular, we describe the procedure of handle trading in general dimensions and its particular geometries in dimensions 4 and 3. This is used to reduce the handle decompositions of bounding 4-manifolds to ones which have only index 1 and 2 handles. A closely related variant of the handle trade move is the β-Move, given in the coupon version in Section 2.5.1.

We show that any β-Move corresponds to an index-2 surgery in the interior of the bounding 4-manifold, and vice versa.

A second technical tool is introduced in Section 3.2, where we discuss the natural stratification of the space of smooth functions on a bounding 4-manifold W, as studied in generality, for example, in [Cer70]. The dense stratum consists of Morse functions with distinct critical values and critical manifolds in general position so that it describes a nice handle decomposition. We recall the standard arguments to show that the space of functions with only index 1 and 2 critical points is path connected. The new feature of our discussion is the inclusion of the external strand data on the M_o-boundary part of W, which leads to an additional stratification of our function spaces. Combining this with results from the previous section we obtain the relevant surgery calculi on the standard handlebodies.

In Section 3.3 we finally construct the map on the 2-arrow sets that assigns to each class of admissible tangles a homeomorphism class of extended relative cobordism. An inverse map is constructed as well. Construction of the latter starts with an assignment of concrete tangles to 3-manifolds with corners using standard link pictures in standard handlebodies and surgery along these links. A boundary move is introduced that pushes an arbitrary link diagram on a handlebody into a standard one and, thus, produces an admissible tangle. We construct also a map from the set of tangles of bridged links in $S^2 \times [-1, 1]$ to bridged link diagrams in handlebodies. We show that the assignment factors into the equivalence classes and, hence, projects to respective 2-arrows sets. Moreover, using the surgery calculi from the previous sections we prove that the resulting map $\mathfrak{Surg}$ on the two arrow sets is bijective.

The final piece in the proof of Theorem 3.0.6, namely, that the map $\mathfrak{Surg}$ respects both of the compositions, is given in Section 3.4. The vertical functoriality of $\mathfrak{Surg}$ is verified very similarly to the proof for ordinary cobordisms as in [Ker99]. However, in order to show horizontal functoriality we heavily rely on the discussion of the handlebody and surgery geometry from Section 1.6. Our choice of partial link diagram in standard handlebody (Sect. 3.3.1), which is a part of the assigning of a bridged link to a tangle, is explained by Lemma 3.4.2, where the structure of vertical identity morphisms is studied.

3.1 Trading and Eliminating Handles

In the handle decomposition of a 4-manifold we can trade a j-handle, whose attaching sphere S^{j-1} bounds a j-disc in a level 3-manifold with consistent framing, for a $3 - j$-handle. This changes the homeomorphism class of the 4-manifold, but not its signature or boundary.

We describe this handle trading procedure for general dimensions in Section 3.1.2, showing also that trading a j-handle for an $d - 1 - j$-handle in a decomposition corresponds to an index-$j + 1$-surgery on the d-manifold. The handle trade move for dimension $d = 4$ and index $j = 1$ is described explicitly in the framework of bridged links.

A general index-2 surgery on a 4-manifold is identified with β-Moves in Section 3.1.3. This is used to show that handle decompositions of two 4-manifolds with equal boundary and signature are related by a series of such β-Moves. Since the β-Moves are implied by the standard link calculi this allows us to substitute freely the 4-manifolds from which we obtain surgery presentations of 3-manifolds.

3.1.1 Handle Decompositions, Connectivity, and Elimination of top Handles

It is a basic fact that any compact, differentiable manifold admits a Morse function, see [Mil69] or [Hir76], which is well behaved at the boundaries. This fact together with Lemma 1.5.1 shows that any $n+1$-cobordism $N : V_0 \to V_1$ can be written in the form

$$N \cong (V_0 \times [0,1]) \cup \mathrm{e}_0^n \cup \ldots \cup \mathrm{e}_0^n \cup \mathrm{e}_1^n \cup \ldots \cup \mathrm{e}_{n-1}^n \cup \mathrm{e}_n^n \cup \ldots \cup \mathrm{e}_n^n. \quad (3.1.1)$$

As stated in Lemma 1.5.3 we can assume that these attachments are in the order of the indices. I.e., the 0-handles are attached first, then the 1-handles, and so forth, until the n-handles are attached last.

One simple question in this context is how the connectivity changes as we add on more and more handles to $V_0 \times [0,1]$. Handles with index $j \geqslant 2$ are attached along thickened S^{j-1}-spheres in the boundaries, hence, along connected pieces. The connectivity therefore does not change under these handle attachments. More formally, we have that

$$\pi_0(i) : \pi_0(V) \xrightarrow{\quad\cong\quad} \pi_0(V \cup \mathrm{e}_j^n) \qquad j \geqslant 2,$$

is a bijection. Here $i : V \hookrightarrow V \cup \mathrm{e}_j^n$ is the natural inclusion.

Attaching 0-handles e_0^n means adding a ball D^n, and, thus, increasing the number of components by one.

Finally, the only way we can decrease the number of components is by adding a 1-handle, e_1^n, such that the two attaching discs $S^0 \times D^{n-1}$ lie in different components. Thus the total number of connected components is determined by the difference in the number of 0-handles and "connecting" 1-handles in the decomposition. A useful application of this observation is the following.

Lemma 3.1.1. *Any handle decomposition of a* connected *cobordism N as in* (3.1.1) *can be reduced to one without 0-handles e_0^n and n-handles e_n^n by performing Smale cancellations (see Lemma 1.5.4).*

Proof. In order for the entire handle decomposition to be connected every 0-handle has to be connected to other components by the subsequent handle attachments. The only attachments that will perform this task are 1-handles. Thus to each 0-handle e_0^n there has to be a 1-handle, which at one end is attached to this e_0^n and at the other end to either another 0-handle or $V_0 \times [0,1]$. Hence, the geometric intersection number of the 1-handle and 0-handle is one and we can apply Lemma 1.5.4 to cancel them. In this way the 0-handles are inductively removed.

For the n-handles observe that we can replace a Morse function f by $1-f$ so that an index-j singularity turns into an index-$n-j$ singularity. This also yields an opposite handle decomposition, in which we start from $V_1 \times [0,1]$ and the n-handles in the original decomposition become 0-handles that are adjoined first. Repeating the previous argument in the reverse decomposition shows now that every n-handle in the original decomposition can be cancelled against an $n-1$-handle.

For a connected, differentiable 4-dimensional cobordism, W, the observations of this section can be summarized in the fact that there is always a decomposition of the form

$$W = V \times [0,1] \cup e_1^4 \cup \ldots \cup e_1^4 \cup e_2^4 \cup \ldots e_2^4 \cup e_3^4 \cup \ldots \cup e_3^4. \qquad (3.1.2)$$

3.1.2 Handle Trading and Elimination of 3-Handles

Another move, that is central in the study of surgery calculi, is the handle trading move. In this operation a j-handle of a handle decomposition of an n-dimensional manifold, N, is exchanged for an $n-j-1$-handle if the critical manifolds are contractible. This changes the n-manifold N itself but its boundary ∂N is unchanged under handle trading. Another description of this operation is given in [Wal60] under the name of *modifications*. In the following lemma we give the more precise prerequisites and show that a handle trading is equivalent to an n-dimensional surgery on the interior of N, thus corresponding to an elementary $n+1$-dimensional cobordism between N and its *modified* version with the same boundary.

Lemma 3.1.2. *Suppose N is a differentiable n-manifold and $N' = N \cup e_j^n$ is a handle attachment, such that the embedded attaching manifold $S^{j-1} \times D^{n-j} = S^{j-1} \times [0,\varepsilon] \times D^{n-j-1} \subset \partial N$ extends to an embedding $D^j \times D^{n-j-1} \subset \partial N$, where $S^{j-1} \times [0,\varepsilon]$ appears as a collar of the D^j factor.*

Then we can perform an index-$j+1$-surgery, $\mathcal{G}$, in the interior of N' in a vicinity of the handle attachment, such that the surgered manifold

$$N'_{\mathcal{G}} \cong N \cup e^n_{n-1-j}.$$

The attachments of the $n-j-1$-handle has properties analogous to those required for the original j-handle attachment with an extension of the embedding $D^j \times [0,\varepsilon] \times S^{n-j-2} \subset D^j \times D^{n-j-1} \subset \partial N$ into the same *bi-disc as for the j-handle.*

In particular, we have that

$$\partial(N \cup e_j^n) = \partial N' = (\partial N)\#(S^j \times S^{n-j-1}) = \partial(N'_{\mathcal{G}}) = \partial(N \cup e^n_{n-1-j}).$$

Proof. In order to describe the attachment of the j-handle in better detail let us represent this $e_j^n = D^j \times D^{n-j}$ in a convenient way. First, let us write $D^{n-j} = [0,\varepsilon] \times D^{n-j-1}$ as in the prerequisites of the lemma. Also, we represent the other disc factor as a cone $D^j = CS^{j-1} = S^{j-1} \times [a,b] \big/_{S^{j-1}\times\{a\}}$ so that we obtain in summary

$$\begin{aligned} \mathrm{e}^n_j &= D^j \times [0,\varepsilon] \times D^{n-j-1} \\ &= S^{j-1} \times [a,b] \times [0,\varepsilon] \times D^{n-j-1} \big/_{S^{j-1}\times\{b\}\times\{y\}\sim\{c\}\times\{b\}\times\{y\}}. \end{aligned}$$

The j-handle is, hence, attached to a disc $D^{n-1}_{\$} = D^j \times D^{n-j-1}$ along the collar product $S^{j-1} \times [0,\varepsilon] \times D^{n-j-1}$. We may thicken this disc now into the interior of N with a collar $D^n_{\$} = D^j \times [p,q] \times D^{n-j-1}$ such that $D^{n-1}_{\$} = \partial N \cap D^n_{\$} = D^j \times \{p\} \times D^{n-j-1}$. Also, the D^j is now presented as a cone over an S^{j-1} so that

$$\begin{aligned} D^n_{\$} &= D^j \times [p,q] \times D^{n-j-1} \\ &= S^{j-1} \times [0,1] \times [p,q] \times D^{n-j-1} \big/_{S^{j-1}\times\{1\}\times\{y\}\sim\{c\}\times\{1\}\times\{y\}}. \end{aligned}$$

The union $D^n_{\$} \cup \mathrm{e}^n_j$ is now obtained by identifying $\partial D^j \times [0,\varepsilon] \times D^{n-j-1} \subset \partial \mathrm{e}^n_j$ with $collar(D^j) \times \{p\} \times D^{n-j-1} \subset \partial D^n_{\$}$. The factor D^{n-j-1} appears in all spaces and can, thus, be omitted to simplify the discussion of the identification. In the parametrization it, therefore, suffices to find the identification space $S^{j-1} \times \{a\} \times [0,\varepsilon] \subset S^{j-1} \times [a,b] \times [0,\varepsilon]/_\sim$ with $S^{j-1} \times [0,\varepsilon] \times \{p\} \subset S^{j-1} \times [0,1] \times [p,q]/_\sim$.

The identification space is, hence, given by products of S^{j-1}-spheres with 2-dimensional squares, which are glued together along line segments in their boundary. Furthermore, the S^{j-1}-factor is collapsed to a point at each point along other line segments in the boundaries of the squares. The squares and the relevant edges are depicted in the top portion of Figure 3.1.

After gluing them together we obtain an area with two line segments, along which the $j-1$-sphere is pointwise collapsed. Deforming this area again into a square $[0,1] \times [g,h]$, such that the special line segments are opposite edges, we obtain for the union

$$\begin{aligned} \mathrm{e}^n_j \cup D^n_{\$} &= S^{j-1} \times [0,1] \times [g,h] \times D^{n-j-1} \big/_{S^{j-1}\times\{0/1\}\times\{y\}\sim\{c\}\times\{0/1\}\times\{y\}} \\ &= \Sigma S^{j-1} \times [g,h] \times D^{n-j-1} = S^j \times D^{n-j}_T, \end{aligned}$$

where $\Sigma S^{j-1} = S^j$ denotes, as usual, the double cone.

Since $\overline{N - D^n_{\$}} \cong N$ and $D^n_{\$} \cap \overline{N - D^n_{\$}} \cong D^{n-1}$, we find that

$$N \cup \mathrm{e}^n_j = (N \cup_{D^{n-1}} D^n_{\$}) \cup \mathrm{e}^n_j = N \cup_{D^{n-1}} (D^n_{\$} \cup \mathrm{e}^n_j) \cong N \cup_{D^{n-1}} (S^j \times D^{n-j}_T). \tag{3.1.3}$$

The index-$j+1$ surgery on the n-manifold $N \cup \mathrm{e}^n_j$ can now be described by the embedding of the attaching manifold $\mathcal{G} : S^j \times D^{n-j} \hookrightarrow S^j \times D^{n-j}_T \subset N \cup_{D^{n-1}} (S^j \times D^{n-j}_T)$, which is identity on the S^j-factor and a constant embedding $D^{n-j} \hookrightarrow \overset{\circ}{D}{}^{n-j}_T$ on the second factor so that $D^{n-j}_T - D^{n-j} \cong [0,\delta) \times S^{n-j-1}$. We, thus, find that

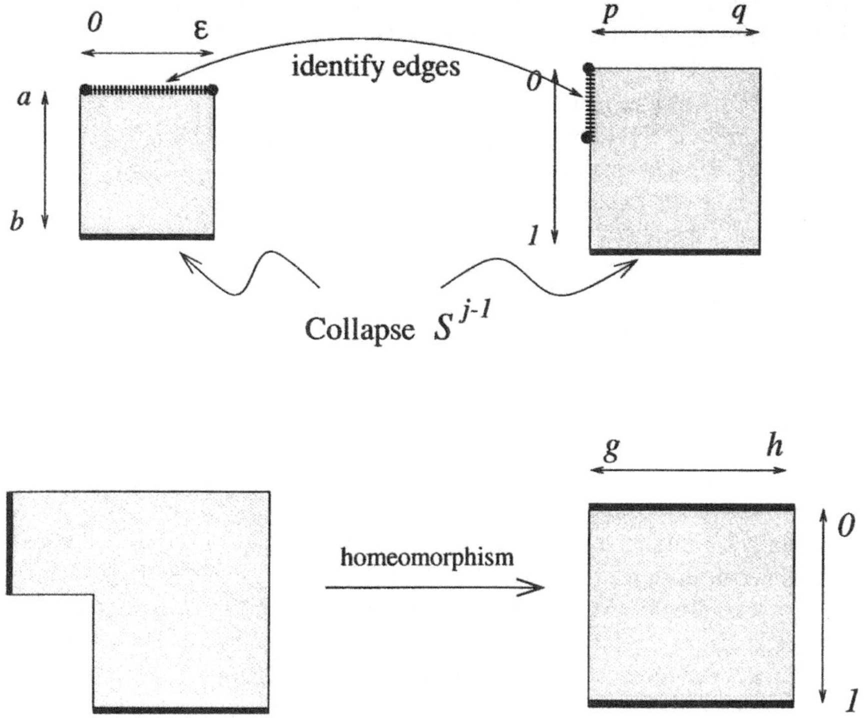

Fig. 3.1. Attachments and Collapsing along Square Factors

$$\overline{N \cup e^n_j - im(\mathcal{G})} = N \cup_{D^{n-1}} (S^j \times [0,\delta] \times S^{n-j-1}).$$

The second step in the index-$j+1$ surgery is to reglue a $D^{j+1} \times S^{n-j-1}$ along the $S^j \times \{\delta\} \times S^{n-j-1}$ boundary piece. Reinterpreting $S^j \times [0,\delta]$ as a collar of D^{j+1} we then obtain that

$$\begin{aligned}(N \cup e^n_j)_{\mathcal{G}} &= \overline{N \cup e^n_j - im(\mathcal{G})} \cup D^{j+1} \times S^{n-j-1} \\ &= N \cup_{D^{n-1}} ((S^j \times [0,\delta] \cup D^{j+1}) \times S^{n-j-1}) \\ &= N \cup_{D^{n-1}} (D^{j+1} \times S^{n-j-1}) \cong N \cup e^n_{n-j-1}.\end{aligned}$$

The last isomorphism is precisely the identification (3.1.3) with j replaced by $n-j-1$. This also implies that the attachment of e^n_{n-j-1} has the same property as that of the original e^n_j attachment.

Since the index-$j+1$ surgery was entirely inside the n-manifold, the boundary, obviously, does not change. Also from the gluing formula (3.1.3) the connected sum expression for the boundary is evident.

For our purposes we mainly consider the case of dimension $n = 4$ and indices $j = 1$ and $j = 2$. The necessary configuration for trading a $j = 1$-handle for a handle of index $n-j-1 = 4-1-1 = 2$ is obtained from a bi-disc $D^2 \times D^1 \subset \partial W$ in the 3-dimensional boundary as prescribed in Lemma 3.1.2.

The full cylinder given by such a product of a 2-disc and an interval is depicted in the diagram on the right hand side. The attaching data for the 1-handle is given by $D^3 \times S^0 = D^2 \times [0, \varepsilon] \times S^0 \subset D^2 \times D^1$, where $[0, \varepsilon] \times S^0 \subset D^1$ is a closed collar with $\{0\} \times S^0 = \partial D^1$, i.e., vicinities of the endpoints of the interval. In the diagram they appear as the thickenings of the discs at the top and bottom of the cylinder.

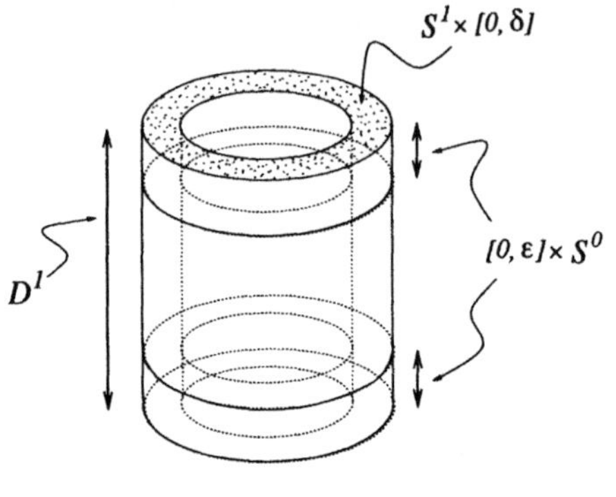

The attachment of the corresponding 2-handle is given by choosing a closed collar $S^1 \times [0, \delta] \subset D^2$ so that $S^1 \times D^2 \cong S^1 \times [0, \delta] \times D^1 \subset D^2 \times D^1$ is its attaching manifold. In the diagram this torus is given as the thickened cylinder that is bounded by annuli at the top and bottom and by $S^1 \times D^1$-pieces on the sides.

In a general connected handle decomposition, with handles ordered as in Lemma 1.5.3, this case of the handle trading move will occur on a manifold to which 1-handles have already been added and further 2- and 3-handles are added after the two traded handles. Using 1-handle slides as in Figure 1.12 of Section 1.5.2 the cylinder $D^2 \times D^1$ can, however, be isotoped into a position where it is disjoint from all other 1-handles. Furthermore, for our purposes we shall disregard 3-handle attachments so that the surgery moves in question involve besides the attaching data for the involved 1-handle and 2-handle further 2-handles. In general position they will be disjoint from the attaching manifold of the special 2-handle and pass transversally through the ascending attaching spheres for the 1-handle.

In the surgery picture of the bounding 3-manifold the handle trading move, thus, appears as depicted in Figure 3.2. The annulus R indicates the index-2 surgery along the torus $S^1 \times [0, \delta] \times D^1 \subset D^2 \times D^1$, and the two balls D and D' indicate the index-1-surgery along the piece $D^2 \times [0, \varepsilon] \times S^0 \subset D^2 \times D^1$ in the above diagram. The strands $a, b, c, \ldots$ indicate the additional 2-handles that are added later on.

For a direct proof that this move does not change the homeomorphism type of the surgered 3-manifold see also [Ker99].

Remark 3.1.3. Handle trading does not change the signature.

More precisely, suppose the 4-manifold W' is obtained from W by trading a handle e^4_j for a handle e^4_{3-j}. Then $\mathrm{sign(W)} = \mathrm{sign(W')}$.

This is a simple consequence of the fact stated in Lemma 3.1.2 that W can be obtained from W' by an index-$j+1$ surgery. This implies, in particular, that W and W' are cobordant by a 5-cobordism, and, hence, their signatures have to coincide.

Another useful application of the modification procedure in four dimensions is the elimination of the index-3 handles in a decomposition of a manifold as in (3.1.2).

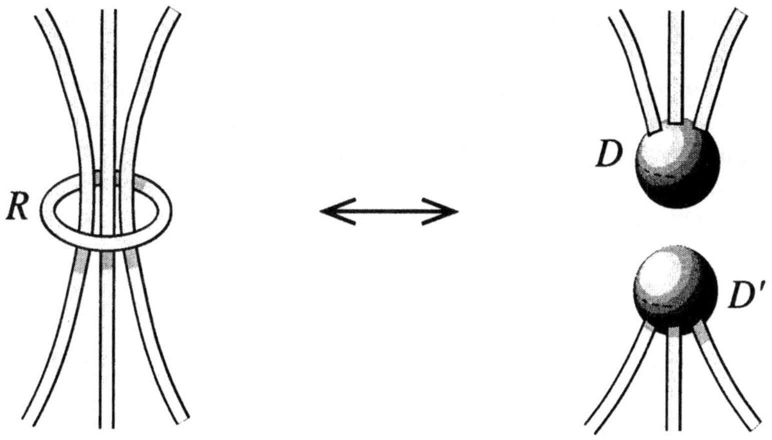

Fig. 3.2. Handle Trading (Modification)

Lemma 3.1.4. *Let W be a 4-cobordism between 3-manifolds V and V^* (meaning $\partial W = V \cup_{\partial V} V^*$) with a handle decomposition* $W = V \times [0,1] \cup \mathrm{e}_1^4 \cup \ldots \cup \mathrm{e}_1^4 \cup \mathrm{e}_2^4 \cup \ldots \cup \mathrm{e}_2^4 \cup \mathrm{e}_3^4 \cup \ldots \cup \mathrm{e}_3^4$.

If V^* is also connected, then all of the 3-handles can be replaced by 1-handles. This means we obtain a cobordism $W_T : -V \to V^*$ between the same 3-manifolds, with the same signature $\mathrm{sign(W)} = \mathrm{sign(W_T)}$, *and a handle decomposition of the form*

$$W_T = V \times [0,1] \cup \mathrm{e}_1^4 \cup \ldots \cup \mathrm{e}_1^4 \cup \mathrm{e}_2^4 \cup \ldots \cup \mathrm{e}_2^4.$$

Proof. The operation needed here is a handle trade for the opposite cobordism $-W : -V^* \to V$. As explained in Section 3.1, this corresponds to replacing a Morse function f by $1 - f$, for which we obtain a handle decomposition $-W = V^* \times [0,1] \cup \mathrm{e}_1^4 \cup \ldots \cup \mathrm{e}_1^4 \cup \mathrm{e}_2^4 \cup \ldots \mathrm{e}_2^4 \cup \mathrm{e}_3^4 \cup \ldots \cup \mathrm{e}_3^4$. A 3-handle on the original decomposition becomes a 1-handle in this decomposition. Since the 1-handles are attached to a connected manifold V^*, we can move the attachment balls together and trade the 1-handles for 2-handles as described in the previous paragraphs. As a result we obtain a cobordism $-W_T : -V^* \to V$ with decomposition $-W_T = V^* \times [0,1] \cup \mathrm{e}_2^4 \cup \ldots \cup \mathrm{e}_2^4 \cup \mathrm{e}_3^4 \cup \ldots \cup \mathrm{e}_3^4$. The corresponding decomposition for the direct cobordism $W : -V \to V^*$ is, thus, of the prescribed form. The equality of signatures follows from Remark 3.1.3.

3.1.3 4-dim Surgery and the β-Moves

In Lemma 3.1.2 the handle trade move has been identified as a special index-$j+1$ surgery on an n-dimensional manifold. This specializes in the case of $j = 1$, i.e., trading an index-1 handle for an index-2 handle in dimension $n = 4$, to surgery along a framed path $\mathcal{G} : S^1 \times D^3 \hookrightarrow W$, where W is the respective 4-manifold.

In this section we discuss the presentation of an index-2 surgery on a 4-manifold for general paths $\mathcal{G}$.

The first observation will be that such a path can always be put into a convenient position in an intermediate level manifold. To be more precise, we consider a 4-manifold W with a handle decomposition as in (3.1.2), with order of handles as in Lemma 1.5.3. For this let us denote the 4-manifold obtained by only attaching the 1-handles but omitting the 2- and 3-handles by

$$\check{W} \;=\; V \times [0,1] \cup \mathrm{e}_1^4 \cup \ldots \cup \mathrm{e}_1^4$$

so that $$W = \check{W} \cup \mathrm{e}_2^4 \cup \ldots \cup \mathrm{e}_2^4 \cup \mathrm{e}_3^4 \cup \ldots \cup \mathrm{e}_3^4.$$

Also, let us denote the intermediate 1-surgered 3-manifold by $\check{V} = V_{\mathcal{G}_1}$, where $\mathcal{G}_1$ denotes the surgery subdiagram that consists only of the index-1 data. We, thus, have

$$\partial\check{W} = V \cup_{\partial V} \check{V} \qquad \text{and} \qquad \check{V} \subset W.$$

The following lemma has already been given in a slightly more special case in [Ker99]. It states that any path in W can be isotoped into a path in the intermediate level manifold $\check{V}$.

Lemma 3.1.5. *Any embedding $S^1 \hookrightarrow W$ is isotopic to an embedding $S^1 \hookrightarrow \check{V} \subset W$.*

Proof. The index-1 handles, that enter the cobordism $\check{W} : V \to \check{V}$, may be interpreted as index-3 handles if we view this cobordism in opposite direction. This means we can construct $-\check{W}$ by attaching 3-handles to $\check{V} \times [0,1]$. Combining this with the other handles we, thus, have a homeomorphic presentation of the total 4-manifold W as

$$W = \mathrm{e}_3^4 \cup \ldots \cup \mathrm{e}_3^4 \cup (\check{V} \times [0,1]) \cup \mathrm{e}_2^4 \cup \ldots \cup \mathrm{e}_2^4 \cup \mathrm{e}_3^4 \cup \ldots \cup \mathrm{e}_3^4, \qquad (3.1.4)$$

where the first set of 3-handles is attached along the $\check{V} \times 0$ part of the boundary and the second set of 3-handles to the $\check{V} \times 1$ part, and $\check{V} \subset W$ is, for example, identified with $\check{V} \times \frac{1}{2}$. Recall that a j-handle, $\mathrm{e}_j^4 = D^j \times D^{4-j}$, in 4 dimensions is attached along an embedded piece $S^{j-1} \times D^{4-j}$ in the boundary. In a j-handle we can identify the $4-j$-dimensional disc $d_{4/j} = \{0\} \times D^{4-j} \subset \mathrm{e}_j^4$. It is disjoint from the boundary piece along which e_j^4 is attached, and we easily see that $\mathrm{e}_j^4 - d_{4/j} \cong [0,\delta) \times S^{j-1} \times D^{4-j}$ with attaching along the $\{0\} \times S^{j-1} \times D^{4-j}$ part. Hence, attaching $\mathrm{e}_j^4 - d_{4/j}$ to a manifold amounts to no more than thickening the boundary collar locally along the respective $S^{j-1} \times D^{4-j}$ so that we have a homeomorphism, that is isotopic to the identity and coincides with the identity away from the handle and boundary:

$$\overline{W \cup (\mathrm{e}_j^4 - d_{4/j})} \;\;\cong\;\; W.$$

In particular, if we remove the discs $d_{4/j}$ from every handle of a given decomposition we obtain such a homeomorphism on the original manifold. In the presentation of the 4-manifold in (3.1.4) we can consider the manifold

$$
\begin{aligned}
W^{!!} &= \overline{W - (d_{4/3} \cup \ldots \cup d_{4/3} \cup d_{4/2} \cup \ldots \cup d_{4/2})} \\
&= \overline{(e_3^4 - d_{4/3}) \cup \ldots \cup (e_3^4 - d_{4/3}) \cup (\check{V} \times [0,1]) \cup (e_2^4 - d_{4/3}) \cup \ldots \cup (e_2^4 - d_{4/3}) \cup (e_3^4 - d_{4/3}) \cup \ldots \cup (e_3^4 - d_{4/3})} \\
&\cong (\check{V} \times [0,1]),
\end{aligned}
$$

where we have removed the 1- and 2-discs $d_{4/3}$ and $d_{4/2}$. Now, since their codimensions in the 4-manifold are 3 and 2, respectively, any path $S^1 \hookrightarrow W$ can be isotoped away from these discs so that it lies entirely inside $W^{!!}$. By use of the homeomorphism constructed above it can be further pushed into the piece $\check{V} \times [0, 1]$.

It is now a standard result that any path $S^1 \to \check{V}$ in a 3-manifold can be made into an embedding by an arbitrarily small deformation, see [Hir76]. For the composite $S^1 \hookrightarrow \check{V} \times [0, 1] \to \check{V}$, where the first map is the previous embedding and the second the canonical projection, we may choose this deformation, such that it lies entirely within a tubular neighborhood of the embedded path in $\check{V} \times [0, 1]$. As we are dealing with compact manifolds the space of embeddings is open in the natural topologies, [Hir76], and we can, thus, make the deformation in $\check{V}$ small enough so that the resulting one in $\check{V} \times [0, 1]$ is in fact an isotopy. After application of the latter we, thus, have that $S^1 \hookrightarrow \check{V} \times [0, 1] \to \check{V}$ is an embedding. By continuously collapsing the interval $[0, 1]$ to its midpoint $\frac{1}{2}$ we, thus, obtain an isotopy of the path into $\check{V} \times \frac{1}{2}$.

Already in [Kir78] and [FR79] a special move between surgery links was introduced, which we shall call here the β-move. It assumes a configuration in which an arbitrary ribbon R is surrounded at one point by another surgery ribbon in the form of an annulus A, i.e., a trivially framed unknot. We assume no other ribbons pass through this annulus and R passes through A exactly once. In this case the two ribbons A and R can be omitted from the surgery diagram without changing the homeomorphism class of the surgered 3-manifold. The move is indicated in Figure 3.3.

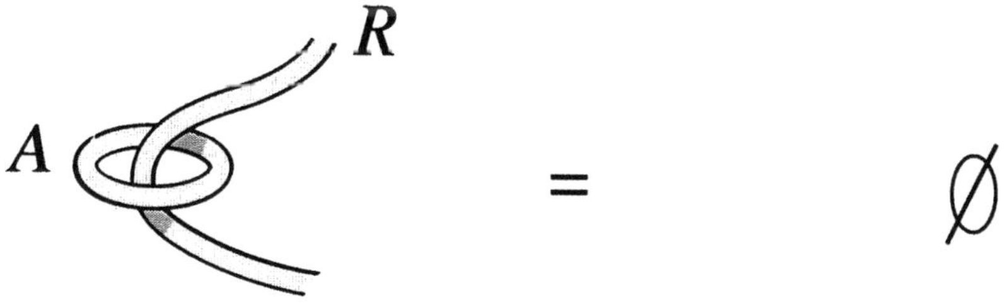

Fig. 3.3. β-Move

It is easy to see that the β-move is already implied by the cancellation move given by a 1-2-cancellation as described in Figure 1.15 of Section 1.5.2 and a handle trade move as given in Figure 3.2 in Section 3.1.2. The argument can, for example, be found in [Ker98a] and, for convenience, is summarized in Figure 3.4. Here trading the annulus A for an index-1 surgery yields a diagram, in which R is opened up with the pair of surgery balls at its ends. It can, thus, be contracted by an isotopy until R is only a short local strip as in the picture for the 1-2-cancellation.

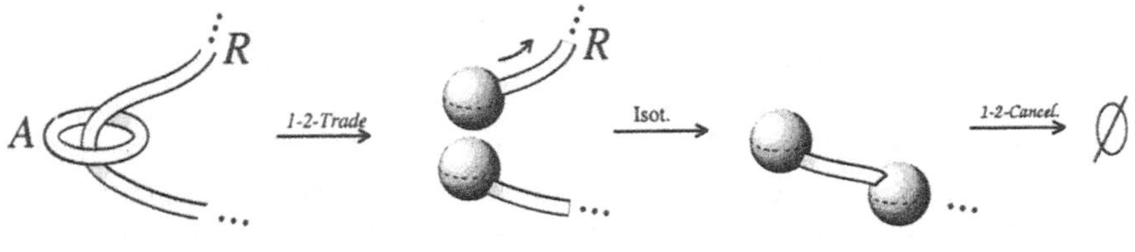

Fig. 3.4. Proof of β-Move

The fact that the β-Move does not change the homeomorphism class of the 3-manifold can also be explained by identifying it as a surgery on the interior of the bounding 4-manifold. This is the content of the following lemma.

Lemma 3.1.6. *Suppose $R \subset \check{V}$ is an embedded circle. Let $Y = (\check{V} \times [0,1])_R$ be the 4-manifold obtained by doing an index-2 surgery along $R \times \frac{1}{2} \subset \check{V} \times [0,1]$ with an arbitrary framing. Then*

$$Y \cong \check{V} \times [0,1] \cup e_2^4 \cup e_2^4,$$

where the two 2-handles are attached at $\check{V} \times \{1\}$ along a configuration as given on the left side of Figure 3.3 of the β-Move. In particular, Y and $\check{V} \times [0,1]$ have the same boundaries.

Proof. The normal bundle $\nu_4(R)$ of the curve $R \times \frac{1}{2} \subset \check{V} \times [0,1]$ naturally splits into the normal bundle $\nu_3(R) \subset T\check{V}$ of R in $\check{V}$ and the trivial line bundle $\varepsilon_1(R) = R \times \mathbb{R}$ given by the direction of the interval factor $[0,1]$ so that we have $\nu_4(R) = \nu_3(R) \oplus \varepsilon_1(R)$. Hence, any framing $\alpha : \nu_3(R) \xrightarrow{\sim} \varepsilon_2(R)$ of R in $\check{V}$ can be extended to a framing of R in $\check{V} \times [0,1]$, which is of the block form

$$\alpha \oplus id : \nu_4(R) = \nu_3(R) \oplus \varepsilon_1(R) \xrightarrow{\sim} \varepsilon_3(R) = \varepsilon_2(R) \oplus \varepsilon_1(R). \qquad (3.1.5)$$

Any other framing can be obtained from the given one by applying a bundle isomorphism to $\varepsilon_3(R)$, which, for some choice of metric, is the same as giving a path $R = S^1 \to SO(3)$. Isotopy classes of framings are, thus, determined by elements in $[S^1, SO(3)]$. Since $\pi_1(SO(3)) = \mathbb{Z}/2$ is abelian the latter contains only one non-trivial element. A representative of such a path can be given by any rotation $\theta \to \exp(\theta\Omega)$ in a direction $ker(\Omega)$, where Ω is a 3×3-skew matrix with $||\Omega|| = 1$. We choose Ω, such that the direction of rotation coincides with the $\varepsilon_1(R)$ summand of $\varepsilon_3(R)$.

As a result the bundle isomorphism of $\varepsilon_3(R)$ is also of block form and, hence, in any isotopy class of framings there is one which has a block form as in (3.1.5).

For a tubular neighborhood associated to such a framing, given by an embedding $\xi : S^1 \times D^3 \hookrightarrow \check{V} \times [0,1]$, we may, thus, assume that it restricts to a tubular neighborhood embedding $S^1 \times D^2_{**} \hookrightarrow \check{V} \times \{\frac{1}{2}\}$. Here $D^2_{**} \subset D^3$ is a fixed 2-disc that separates the 3-ball into an upper and lower half $D^3_+ \cup D^3_- = D^3$ such that $S^1 \times D^3_+ = \xi^{-1}(\check{V} \times [\frac{1}{2}, 1])$ and $S^1 \times D^3_- = \xi^{-1}(\check{V} \times [0, \frac{1}{2}])$, and both $D^3_\pm$ are closed three balls with $D^3_+ \cap D^3_- = D^2_{**}$.

In addition, we shall denote $D^2_+ = \partial D^3_+ - D^2_{**}$ the opposite disc in the boundary of the three ball D^3_+ so that $\partial D^3_+ = D^2_+ \cup_{S^1_{**}} D^2_{**}$, where $S^1_{**} = \partial D^2_{**} = \partial D^3_+ = D^2_{**} \cap D^2_+$. The disc D^2_- is defined analogously. The arrangement of these components is depicted schematically in the diagram on the right hand side.

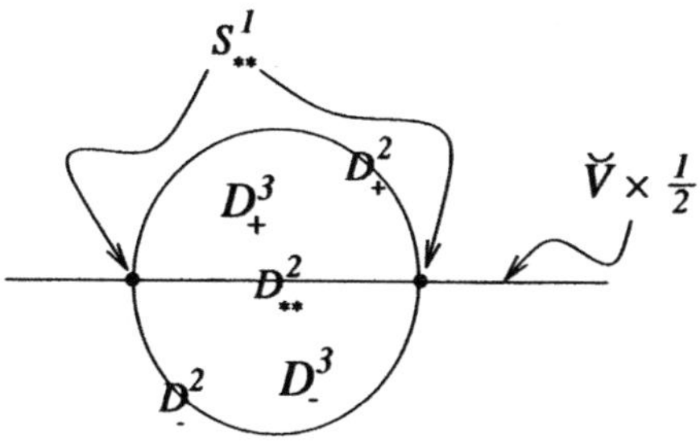

The manifold $\check{V} \times [0,1]$, thus, naturally splits into the four pieces $\check{V} \times [0, \frac{1}{2}] - \xi(S^1 \times D^3_-), \xi(S^1 \times D^3_-), \xi(S^1 \times D^3_+)$, and $\check{V} \times [\frac{1}{2}, 1] - \xi(S^1 \times D^3_+)$. In the process of surgery the second and third part is removed and replaced by $D^2 \times S^2$ with $S^2 = D^2_+ \cup_{S^1_{**}} D^2_-$. This yields a decomposition of the surgered manifold as follows:

$$
\begin{aligned}
&\big(\check{V} \times [0,1] - \xi(S^1 \times D^3)\big) \cup_{S^1 \times S^2} D^2 \times S^2 \\
&\quad = \big(\check{V} \times [0, \tfrac{1}{2}] - \xi(S^1 \times D^3_-)\big) \cup_{S^1 \times D^2_-} D^2 \times D^2_- \cup_{D^2 \times S^1_{**}} D^2 \times D^2_+ \cup \\
&\qquad\qquad\qquad\qquad \cup \big(\check{V} \times [\tfrac{1}{2}, 1] - \xi(S^1 \times D^3_+)\big) \\
&\quad \cong \check{V} \times [0,1] \cup_{R \times D^2_-} D^2 \times D^2_- \cup_{D^2 \times S^1_{**}} D^2 \times D^2_+ .
\end{aligned}
$$

In the last homeomorphism we use that $\check{V} \times [0, \frac{1}{2}] - \xi(S^1 \times D^3_-) \cong \check{V} \times [0,1]$, since we can describe the removal of the piece $S^1 \times D^3_-$ also by collapsing it into $S^1 \times D^2_-$. By the same argument we have that $\check{V} \times [\frac{1}{2}, 1] - \xi(S^1 \times D^3_+) \cong \check{V} \times [0,1]$, but gluing this to a 4-manifold with boundary piece $\cong \check{V}$, as in our situation, does not change the homeomorphism class, since it simply thickens the collar. Hence, we can omit it.

The remaining two pieces are easily identified as 2-handle attachments. The first one is along the path R with tubular neighborhood $R \times D^2_-$, and the second is attached to $\check{V} \times [0,1] \cup_{R \times D^2_-} D^2 \times D^2_-$ along the curve $S^1_{**} \times 0 = \partial D^2_- \times 0 \subset D^2_- \times D^2$ with framing determined by the D^2-coordinates. Specifically, we can choose a point $p \in \partial D^2_-$ and isotop this curve to $S^1_{**} \times \{p\}$. Hence, it becomes a meridian on the torus $R \times D^2_-$ so that after pushing it off further by a bit it surrounds the torus as the annulus A does in Figure 3.3. The relevant framing is determined, for example, by a fixed direction in the D^2-factor, which may be precisely a tangent vector of R at this point or, after a $\frac{\pi}{2}$-twist, the annulus framing as indicated in Figure 3.3. Hence, the attaching diagram of the two 2-handles is precisely the one prescribed by the β-Move.

Note, that the framing of the ribbon R in $\check{V}$ is only defined modulo 2 since the relative framing numbers in the 4-manifold $\check{V} \times [0,1]$ are $\mathbb{Z}/2$-valued. In the three dimensional surgery picture this manifests itself in the fact that sliding R over A changes the framing precisely by a 4π-twist.

The relevance of index-2 surgeries on 4-manifolds explains itself in the following lemma.

Lemma 3.1.7. *Let W and W' be two differentiable, compact, oriented and connected 4-manifolds with equal signatures and boundaries* ($\mathrm{sign}(W) = \mathrm{sign}(W')$ *and* $\partial W = \partial W'$*).*

Then there is a 4-manifold W'', which can be obtained by a series of index-2 surgeries on W, or, at the same time, as the result of another series of index-2 surgeries on W'.

Proof. Since $\partial W = \partial W'$ we can consider the manifold $W^0 = W' \cup_{\partial W} -W$ with $\mathrm{sign}(W^0) = \mathrm{sign}(W') - \mathrm{sign}(W) = 0$. This means, in particular, that W^0 is the boundary of a 5-dimensional, differentiable manifold, and, hence, that W and W' are cobordant.

In a handle decomposition for this 5-dimensional manifold we can assume by connectivity that there are no index-0 or index-5 handles using Lemma 3.1.1. Furthermore, if a 1-handle e^5_1 is attached along a connected 4-manifold we can find a path $\cong D^1$ between the corresponding attachment points as the prerequisite of Lemma 3.1.2 with $n = 5$ and $j = 1$. The 1-handle can, thus, be traded for a handle of index $5 - 1 - 1 = 3$, which changes the 5-manifold but not its boundary W^0.

Hence, we know that we can cobord W to W' using only handle attachments of index 2, 3, and 4. Looking at this in opposite direction we have a cobordism from W' to W with corresponding handles of indices 3, 2, and 1. Since W' is also assumed to be connected, we can again trade the 1-handles for 3-handles, which appear in the original cobordism from W to W' as handles of index $2 = 5 - 3$.

In summary, we have found a 5-manifold of the form $Q = W \times [0,1] \cup e^5_2 \cup \ldots \cup e^5_2 \cup e^5_3 \cup \ldots \cup e^5_3$, with $\partial Q = W^0$. We then consider the intermediate level manifold of this cobordism $Q : W \to W'' \to W'$, where $W \to W''$ is given by the part $W \times [0,1] \cup e^5_2 \cup \ldots \cup e^5_2$ with all the 2-handles of Q included but the 3-handles omitted, and $W'' \to W'$ is $W'' \times [0,1] \cup e^5_3 \cup \ldots \cup e^5_3$ obtained by attaching all the 3-handles. In the opposite direction the latter cobordism is from W' to W'' with the handle decomposition $W \times [0,1] \cup e^5_2 \cup \ldots \cup e^5_2$ of handles of indices $2 = 5 - 3$.

Translating handle attachments into surgery we, thus, obtain the assertion that W'' is given by index-2 surgery on either W or W'.

3.2 Stratified Function Spaces and External Strands on W

In the previous section we introduced special surgery operations that suffice to change any 4-manifold into another provided their boundaries and signatures coincide. In this section we shall consider a fixed 4-manifold and discuss how any two handle decompositions as in Lemma 3.1.4 can be moved into each other. The crucial technical tool that we employ here is the use of stratified function spaces and their connectivities, as studied in the work of Cerf [Cer70], see also [HW73]. A function in a codimension-0 stratum is a Morse function with additional properties so that it defines a handle decomposition. In [Kir78] this was already used to derive Kirby's calculus of links in S^3. The moves are 2-handle slides ($\mathcal{O}_2$-Moves) and the insertion of an isolated ± 1 framed unknot ($\mathcal{O}_1$-Moves).

Kirby's calculus needs to be extended if we consider surgery on manifolds that are not simply connected. The simplest example that shows that the original Kirby Moves are not sufficient is given by a full torus T. We consider the framed, two-component link $\mathcal{L} = A \sqcup B \subset T$, as depicted below. Here A is obtained by pushing in the short meridian of ∂T with framing parallel to the boundary, and B runs along the core of T also with framing parallel to ∂T. The surgered manifold $T_{\mathcal{L}}$ does not change under the modification move, where we can replace the annulus A by a pair of surgery balls. These in turn can be cancelled via the connecting strand B. Hence $T_{\mathcal{L}} \cong T$.

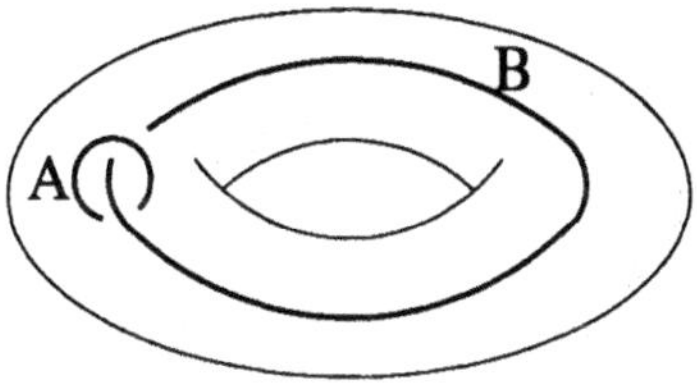

Let us denote by W the 4-manifold obtained by adding two 2-handles along $\mathcal{L} \times \{1\}$ in $T \times [0,1]$. Clearly we have that $i_* : H_1(T \times 0) \to H_1(W)$ is the trivial map since the 2-handle attached along B kills the homology generator of $H_1(T)$ along the core of T. Now the $\mathcal{O}_2$-Moves do not change W at all. Also connecting $\mathbb{CP}^2$ to a 4-manifold W in the interior, which is the content of the $\mathcal{O}_1$-Move, does not change the kernel of $H_1(\partial W) \to H_1(W)$. So $i^* = 0$ for all surgery presentations of T that can be obtained from $\mathcal{L} \subset T$ via application of the original two Kirby Moves. In particular, we cannot obtain the trivial surgery presentation of T on T given by the empty link, since for this $i_* : H_1(T \times 0) \to H_1(T \times [0,1])$ is injective. Thus, in this example two $\mathcal{O}_{1,2}$-inequivalent links produce via surgery homeomorphic 3-manifolds.

In [Ker99] we find a generalization of Kirby's calculus which applies to non-simply connected manifolds. To this end we will consider the corresponding stratified spaces of functions on the 4-manifolds as introduced in Section 1.6.2. Here we generalize the situations further from [Kir78] and [Ker99] by considering also the external strands in the representation. Other extensions of Kirby's calculus to non-simply connected manifolds are given in [FR79] and [Rob97].

The ordinary stratification of a function space according to [Cer70] is introduced in Section 3.2.1 for the case of the 4-manifolds W appearing in the construction of $\widetilde{\mathcal{COB}}$. The basic ingredients of a graphic, in the sense of [Cer70], of a path of functions, as well as their rules for manipulation are reviewed. We obtain a connected stratified space in which the codimension-0 stratum consists of Morse functions with distinct, ordered, critical values of index either 1 or 2, and critical manifolds in general position. In addition, we have five types of codimension-1 strata, which contain cancellation (birth/death) points, passages of critical manifolds through each other, and crossings of values. This stratification is refined in Section 3.2.1, where we require the ascending critical manifolds of a generic function to be transverse to the markings of external strands in the boundary of W. We obtain two further 1-strata that are illustrated in some detail. Section 3.2.3 then summarizes the surgery calculi from [Kir78], [FR79], and [Ker99] and describes how they generalize to our situation. The calculi are obtained by eliminating redundant moves from the list of equivalence moves that is obtained from both the passages through the 1-strata of functions representing handle decompositions and the handle trade manipulations of 4-manifolds.

3.2.1 Natural Stratification of Functions on W

Consider a relative cobordism $M : \Sigma_{g_{sc},a/b} \to \Sigma_{g_{tg},a/b}$ as in the definition of $\mathcal{C}ob$ in Section 1.2 and its closure $M_o = \phi(M) : \Sigma_{g_{sc}} \to \Sigma_{g_{tg}}$ to an ordinary cobordism as defined in (1.6.8). Let us denote also by

$$\tau = \sqcup_{j=1}^{a+b} \tau_j \subset M_o \qquad \text{with} \quad \tau_j \cong D^2 \times [0,1]$$

the union of the full cylinders that are filled. This implies that M is obtained from M_o by $M = \overline{M_o - \tau}$.

Furthermore, consider the closure $\langle M_o \rangle$ as in (1.6.6), as well as a compact, differentiable 4-manifold W bounding it so that $\partial W = \langle M_o \rangle$.

Let $\mathcal{F}$ be the space of smooth functions

$$f : W \longrightarrow [f_0, f_1]$$

for which the restriction onto the boundary fulfills the following properties:

- $f^{-1}(f_0) = \mathcal{H}_{g_{sc}}^{g_{tg}} \subset \partial W$,
- $f^{-1}(f_1) = M_o \subset \partial W$,
- the restriction of f onto the third boundary piece

$$f^{|} : (-\Sigma_{g_{sc}} \sqcup \Sigma_{g_{tg}}) \times [f_0, f_1] \longrightarrow [f_0, f_1]$$

is the canonical projection.

In the following we shall describe a dense subspace of $\mathcal{F}$ which is even smaller than the ones used in [Cer70] but is still connected.

First, let us specify a stratified subspace of the following form:

$$\mathcal{F}^C = \mathcal{F}^0 \cup \mathcal{F}^1_\alpha \cup \mathcal{F}^1_\beta \subset \mathcal{F}. \tag{3.2.1}$$

The codimension-0 stratum $\mathcal{F}^0$ consists of all Morse functions on W, which obey the properties of $\mathcal{F}$, and for which all critical values are distinct. The codimension-1 stratum $\mathcal{F}^1_\alpha$ consists of functions with distinct critical values, which are non-degenerate except for one. This special critical point is a death or birth point and is given for some choice of local coordinates by the form in (1.5.3) with $t = 0$. The other codimension-1 stratum $\mathcal{F}^1_\beta$ consists of Morse functions, for which exactly two critical values coincide but all others are distinct.

As already mentioned at the end of the proof of Lemma 1.5.4 it can be deduced from results in [Hir76] or [Cer70] that the space $\mathcal{F}^C$ is path connected, and that any path can be perturbed into one which is transverse to the codimension-1 stratum $\mathcal{F}^1_\alpha \cup \mathcal{F}^1_\beta$. In particular, the path meets the codimension-1 stratum only at a finite number of discrete times. For a path segment $s \mapsto f_s$ within $\mathcal{F}^0$ it follows from the implicit function theorem that there are paths $s \mapsto a_\nu(s)$ in W such that $\{a_1(s), \ldots, a_M(s)\}$ are the critical points of f_s. Let us also denote by $c_\nu(s) = f_s(a_\nu(s))$ the paths of the critical values in $[f_0, f_1]$.

The path of functions $s \mapsto f_s$ is now conveniently illustrated by a *graphic*, which is the combination of the graphs of the functions $s \mapsto c_\nu(s)$, labeled by the respective index of the Morse singularity of f_s at $a_k(s)$. For a path segment in the codimension-0 stratum we, thus, obtain a set of smooth and disjoint curves as depicted in the left graphic of Figure 3.5.

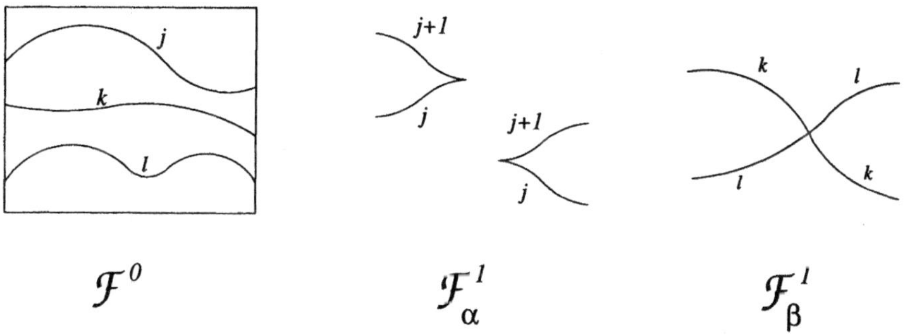

Fig. 3.5. Graphics of Strata in $\mathcal{F}^C$

At a point s^*, where the path passes through the codimension-1 stratum $\mathcal{F}^1_\alpha$ either two graphs $c_\nu(s)$ and $c_\mu(s)$ converge and end in a cusp point $c_k(s^*) = c_l(s^*)$ or they originate in such a point. For $c_\nu(s) > c_\mu(s)$ we must then have for the indices $index(a_\nu) = index(a_\mu) + 1$. In appropriate coordinates the passages through such a point is described by the one-parameter family of functions in (1.5.3), and the graphics for both the birth and the death case with $j = index(a_\nu)$ are shown in the middle of Figure 3.5. Finally, a passage through the codimension-1 stratum $\mathcal{F}^1_\beta$ is described by a crossing of the graphs as on the right in the same figure.

The graphic of any generic path in $\mathcal{F}^C$ can, thus, be given as a combination of the elements in Figure 3.5. Let us briefly review here also the possible deformations of such generic paths into each other. They can be chosen to pass transversally through codimension-2 strata, such as triple values, second order degenerate birth-death points, or double values with one cancellation point. The detailed list of possible moves between paths can, for example, be taken from [HW73] or [Cer70], and is summarized in the following list:

- *Independent Trajectories:* If the ascending and descending manifolds of two critical points $a_\nu(s)$ and $a_\mu(s)$ are disjoint for every s in a given interval, then their paths can be deformed independently from each other.
- *Triangle Lemma:* The order of three successive intersection points with indices i_1, i_2, and i_3, can be interchanged as illustrated, if $i_1 = i_3 \leqslant 3$, or $inf(i_1, i_3) \leqslant i_2 - 1$, or $i_1 = i_2 = i_3 \leqslant 2$.

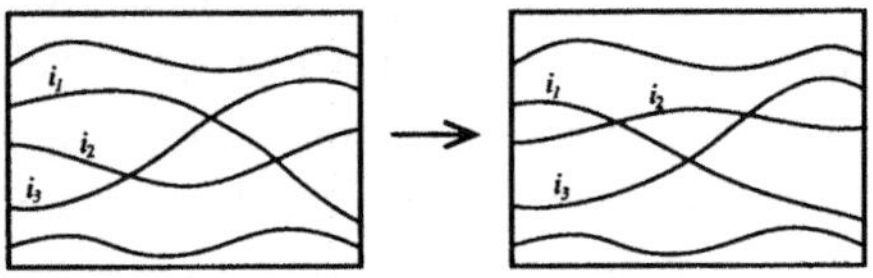

- *Beak Lemma:* The vicinity of a birth (death) point can be pushed above or below other graphs for $i < 4$ and $i > 0$ respectively.

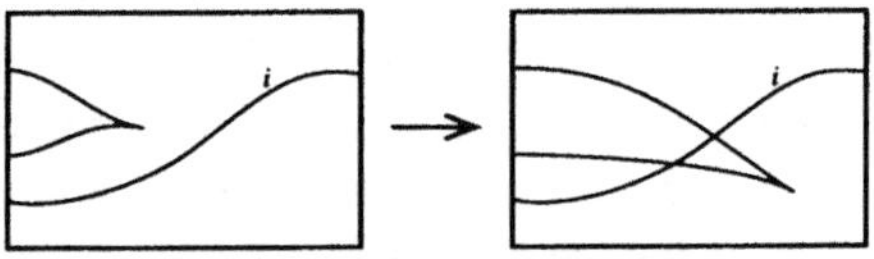

- *Dovetail Lemma:* Adjacent birth and death points can be cancelled against each other.

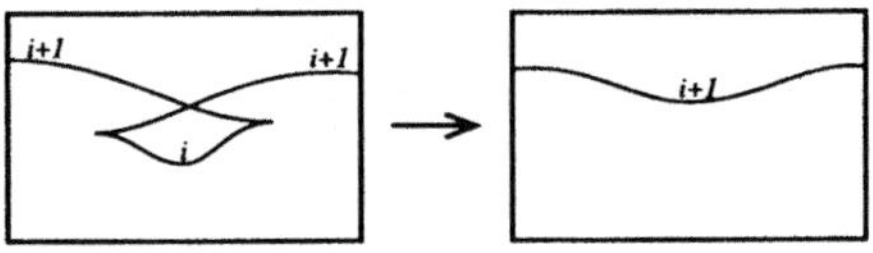

In order to define the critical manifolds mentioned in the Lemma of Independent Trajectories we will tacitly assume in the following some fixed metric on W. For example, the descending critical manifold is obtained by continuing the subspace of the instable directions of the critical Morse singularity via the gradient flow of the function on W.

A typical application for the principle of Independent Trajectories is given by two graphs of c_ν and c_μ such that over a segment we have $c_\nu > c_\mu$, but for the indices $j_{\nu/\mu} = index(a_{\nu/\mu})$ we have that $j_\nu < j_\mu$. In an intermediate 3-manifold the dimensions of the descending manifold of a_ν is $j_\nu - 1$ and the ascending intersects the level manifold in a $3 - j_\mu$-dimensional submanifold. Thus, generically, a one-parameter family of these manifolds will not intersect so that they can be deformed independently of each other.

In the case, where $j_\nu = j_\mu$, and the one-parameter deformation of the critical manifolds is transverse, they will intersect in one point for finite number of discrete parameters s. In a graphic we shall indicate such an intersection point, as usual, by a vertical double arrow, as in the left part of Figure 3.6.

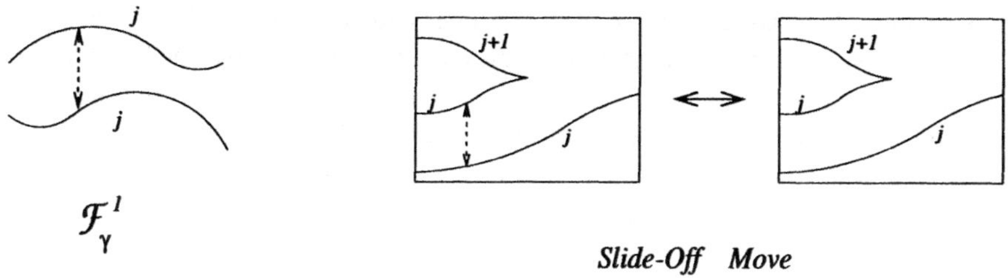

Fig. 3.6. Intersecting critical manifolds of same index

If such an intersection occurs right next to a birth or death point it can be removed, or *slid off*, as indicated on the right of Figure 3.6. This has been remarked in [Kir78] and is implied by results in [Cer70].

More generally, we assume from now on that for a path in $\mathcal{F}^C$ the critical manifolds of the functions at the end points intersect transversally and that the one-parameter deformations of the critical manifolds are transverse. Hence, if we assume that the space $\mathcal{F}^0$ consists of functions, for which all ascending and descending critical manifolds are transverse, we need to adjoin a third codimension-1 stratum $\mathcal{F}^1_\gamma$. In $\mathcal{F}^1_\gamma$ the intersection can become degenerate in lowest order, as, for example, the point of intersection for the manifolds of equal index. This means that really the space $\mathcal{F}^C = \mathcal{F}^0 \cup \mathcal{F}^1_\alpha \cup \mathcal{F}^1_\beta \cup \mathcal{F}^1_\gamma$ is connected.

In Lemma 3.1.4 we have seen that a 4-manifold W, bounding $\langle M_o \rangle$, can always be chosen such that it is built up only of handles of index 1 and 2. Hence, there is also a Morse function on W, which has only index-1 and index-2 singularities. The natural question, which we shall address next, is whether for a fixed 4-manifold of this type a stratified function space with singularities of only these indices is connected.

To this end we introduce several function subspaces with corresponding restrictions on the possible indices of the singularities.

Let ${}_{[1,2]}\mathcal{F}^0$ be the space of Morse functions, which have singularities of only index 1 and 2 with distinct values, and for which all critical manifolds are transverse to each other. In addition, we assume that all critical values of index 2 are larger than all critical values of index 1. This corresponds to the genericity property asserted in Lemma 1.5.3. The only birth or death points allowed in the corresponding subspace ${}_{[1,2]}\mathcal{F}^1_\alpha$ are between an index-1 and an index-2 singularity. The definition of ${}_{[1,2]}\mathcal{F}^1_\beta$ is analogously given by restricting the indices and their ordering so that we may have two trajectories both of index 2 or two trajectories both of index 1 crossing each other. However, the ordering condition excludes crossings of different indices.

For ${}_{[1,2]}\mathcal{F}^1_\gamma$ we are left with three possibilities of the degenerate intersection of critical manifolds. We denote by ${}_{[1,2]}\mathcal{F}^1_{2-2}$ and ${}_{[1,2]}\mathcal{F}^1_{1-1}$ the spaces of functions for which two critical manifolds of equal index, namely 1 or 2, intersect transversally in one point. It is implied by transversality that any path of functions can be deformed to such that these pairs of critical manifolds meet only at a finite number of times in a single point. The descending manifold of an index-2 singularity and the ascending of an index-1 singularity in an intermediate 3-manifold are a curve and a 2-dimensional surface, respectively, and meet transversally in a point. Hence, as a codimension-1 stratum we have ${}_{[1,2]}\mathcal{F}^1_{1-2}$, where the curve meets the surface tangentially. The transverse passage through ${}_{[1,2]}\mathcal{F}^1_{1-2}$, thus, implies the cancellation of two intersection points or their creation.

The following result is already implied by the arguments in [Kir78] and the proof, we shall outline here for convenience, is taken from [Ker99].

Lemma 3.2.1. *The space*

$$
{}_{[1,2]}\mathcal{F}^C = {}_{[1,2]}\mathcal{F}^0 \cup {}_{[1,2]}\mathcal{F}^1_\alpha \cup {}_{[1,2]}\mathcal{F}^1_\beta \cup {}_{[1,2]}\mathcal{F}^1_{1-1} \cup {}_{[1,2]}\mathcal{F}^1_{2-2} \cup {}_{[1,2]}\mathcal{F}^1_{2-1}
$$

is path-connected. Moreover, the set of paths that pass transversally through the codimension-1 strata at a finite number of points is dense.

Proof. The goal is to consider a general path of functions in $\mathcal{F}^C$ that starts and ends with functions from ${}_{[1,2]}\mathcal{F}^0$ and to show that such a path can be modified by the use the elementary lemmas stated above into a path entirely within ${}_{[1,2]}\mathcal{F}^C$.

The ordering condition, that index-1 critical values are always smaller than index-2 critical values, is readily realized by the use of the principle of independent trajectories. Suppose an index-1 critical value is larger than an index-2 critical value. The descending manifold of the former and the ascending manifold of the latter will be points and curves in an intermediate 3-manifold with codimension 2. Hence, also a one-parameter family of such configurations can be perturbed so, that these manifolds remain disjoint. Any index-1 path segment can, thus, be pushed below an index-2 segment, and also birth and death points can be moved below all other index-2 trajectories and above all index-1 trajectories by means of the Beak Lemma.

A trajectory of a critical point of index-0 can occur only if it is created by a 0-1-birth point at some times and ended by a 0-1-death point as in $\mathcal{F}^1_\alpha$ at another time. The procedure to remove such a trajectory from a graphic has been described in [Kir78] (Figure 3.9 through Figure 3.11) using only triangle, beak, and dovetail lemma, as well as the independent trajectory principle and the slide-off move. In particular, the proof makes no reference to the manifold the functions are defined on and, thus, applies literally also to our more general situation. The index-4 trajectories are treated in exactly the same way.

In order to remove the index-3 trajectories we start by arranging them in a nested pattern as indicated in the first picture of Figure 3.7.

The procedure to arrive at this graphic is as for the index 0 and 4 trajectories in [Kir78]. It uses mostly the fact that the beaks at the end can be moved into any position. Also we employ the move from Figure 3.6 in order to slide off the vertical arrows between index-3 trajectories at the birth and death points at the ends. Thus the critical manifolds of the index-3 trajectories are mutually disjoint and can, therefore, be moved independently.

As depicted in the second picture of Figure 3.6 we next create additional birth and death points at the innermost trajectory by application of the Dovetail Lemma

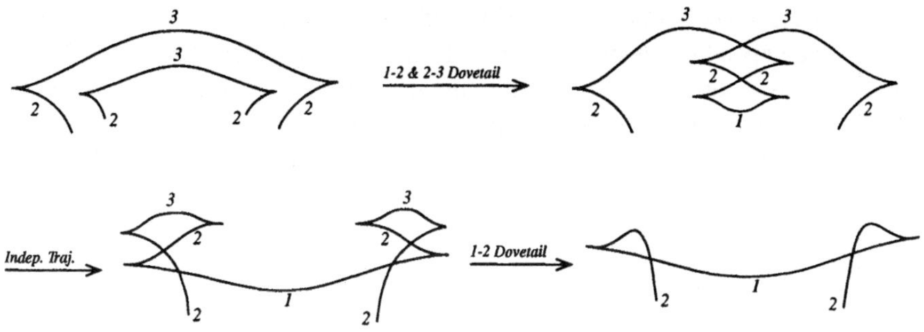

Fig. 3.7. Intersecting critical manifolds of same index

twice. The small additional trajectory in the double tail can be chosen independently so that we can use the Beak Lemma to pull the 3-2-cancellation points past each other, and then spread them apart as in the third picture of Figure 3.6. A final application of the Dovetail Lemma leaves us with a graphic with an index-1 instead of an index-3 trajectory.

This process is repeated until all index-3 parts are eliminated. A similar argument can be obtained from Figure 3.11 in [Kir78] by turning every graphic there upside down and replacing every index j by $3-j$.

The restriction to ${}_{[1,2]}\mathcal{F}^1_{\alpha/\beta}$ is now obvious and the transversality arguments that allow us to restrict to the spaces ${}_{[1,2]}\mathcal{F}^1_{i-j}$, with $i, j = 1, 2$, have been given in the preceding paragraph. This completes the proof.

3.2.2 Stratification in Presence of External Strands

The stratification in Lemma 3.2.1 is sufficient to describe equivalences for surgery on general 3-manifolds, such as the handlebodies $\mathcal{H}^{g_{tg}}_{g_{sc}}$, yielding ordinary 3-cobordisms M_o. However, for the presentation of relative cobordisms we also need to consider the tubular pieces $\boldsymbol{\tau} \subset M_o$ that are removed to obtain M.

The additional restriction that we, therefore, impose on the functions on W is that, for given metric, the ascending manifolds of their critical points are transverse to the strand $\boldsymbol{\tau} = \tau_1 \sqcup \ldots \sqcup \tau_{a+b}$. The subspace of functions in ${}_{[1,2]}\mathcal{F}^0$ for which this is the case is denoted by ${}_X\mathcal{F}^0$.

In order to illustrate this additional transversality condition let us recall that the ascending manifolds of index-2 singularities meet M_o in 1-dimensional closed curves, $\mathcal{L}^{opp}_\nu$, since they are by assumption disjoint. There are no singularities of higher index so that the gradient flow is regular. The ascending manifolds of the index-1 singularities are spheres $\mathcal{P}^{opp}_\mu \cong S^2$ in each level set until they meet a critical point of index 2. There the surface will be punctured at the previous intersection point(s) with the respective index-2 descending manifold, and the created hole is bounded by the respective curve $\mathcal{L}^{opp}$.

A typical generic configuration of ascending manifolds is depicted in Figure 3.8. Note, that, depending on the intersection numbers of the descending index-2 manifolds and the ascending index-1 manifolds, an $\mathcal{L}^{opp}$-curve in M_o can bound several $\mathcal{P}^{opp}$, and every $\mathcal{P}^{opp}$ can have several $\mathcal{L}^{opp}$ in its boundary. In Figure 3.8, for example, $\mathcal{L}_2^{opp}$ bounds both $\mathcal{P}_1^{opp}$ and $\mathcal{P}_2^{opp}$, and $\mathcal{P}_1^{opp}$ contains also $\mathcal{L}_1^{opp}$ in its boundary.

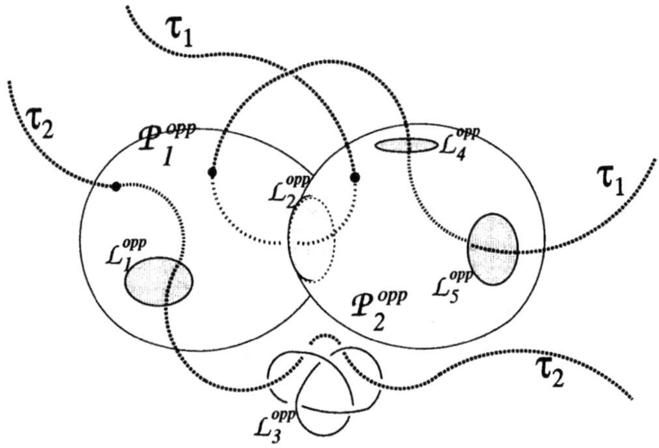

Fig. 3.8. External strands $\pitchfork \mathcal{L}^{opp}$ and $\pitchfork \mathcal{P}^{opp}$

Transversality, $\tau \pitchfork \mathcal{A}^{opp}$, where $\mathcal{A}^{opp} = \mathcal{L}_1^{opp} \cup \ldots \cup \mathcal{P}_M^{opp}$ is the combined ascending manifold complex, implies now that the strands τ_j are disjoint from the curves $\mathcal{L}^{opp}$ in M_o and that each τ_j intersects a given $\mathcal{P}_\mu^{opp}$ in a finite number of points. Here, we disregard for simplicity the thickness of the τ_j, and consider them as curves rather than tubular neighborhoods thereof.

We introduce the codimension-1 function spaces ${}_X\mathcal{F}^1_\alpha$, ${}_X\mathcal{F}^1_\beta$, ${}_X\mathcal{F}^1_{1-1}$, ${}_X\mathcal{F}^1_{2-2}$, and ${}_X\mathcal{F}^1_{2-1}$ that are given by the ones from Lemma 3.2.1 with the transversality condition for the ascending critical manifolds in $\mathcal{A}^{opp} \subset M_o$. This space is, in general, no longer connected, since the ascending manifold cannot be isotoped freely anymore. Still a given isotopy of $\mathcal{A}^{opp}$ can be made transverse to τ by small perturbations. Since all isotopies can extended to ambient ones we may consider, equivalently, isotopies of the strands τ_j with fixed $\mathcal{A}^{opp}$.

For a generic isotopy the strand τ_j is also allowed to be tangential to the interior of a surface piece $\mathring{\mathcal{P}}_\nu \subset \mathcal{A}$. More precisely, the component normal to $\mathcal{P}_\nu$ of the second order derivative of the curve will be non-zero. Let us denote by ${}_X\mathcal{F}^1_{\mathcal{P}}$ the space of all functions in ${}_{[1,2]}\mathcal{F}^0$ for which one such tangential point occurs but all other parts of τ are transverse to $\mathcal{A}^{opp}$. A transverse passage of a one-parameter family of functions through the codimension-1 stratum ${}_X\mathcal{F}^1_{\mathcal{P}}$ is depicted in the upper part of Figure 3.9. On the right side the curve τ_j is a small loop between its two intersection points with $\mathcal{P}_k^{opp}$, and after the move it is locally disjoint from $\mathcal{P}_k^{opp}$ as shown on the upper right side of Figure 3.9.

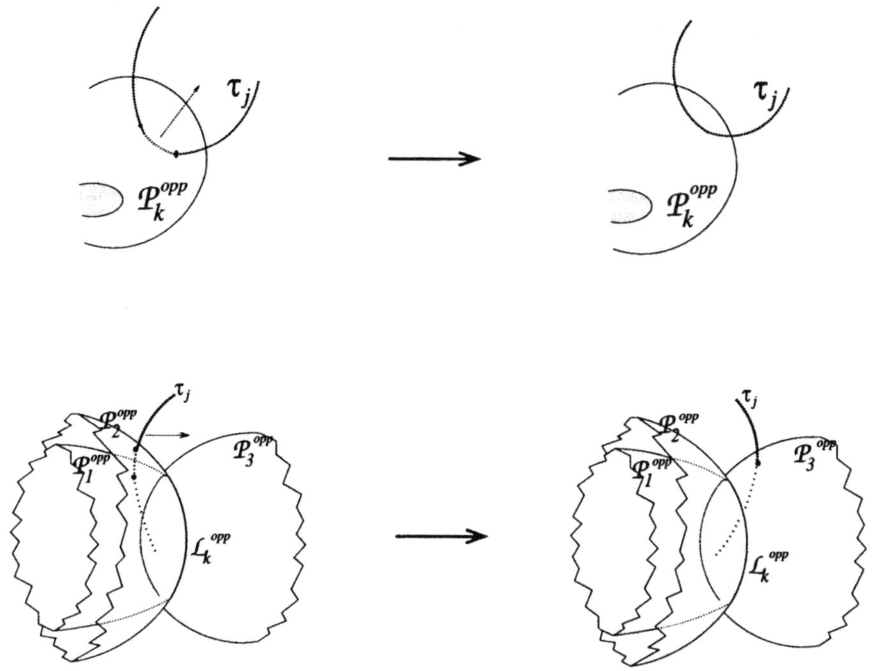

Fig. 3.9. Transverse Moves of Externals Strands through opposite surgery data

To make the space of isotopies between $\mathcal{A}^{opp}$ and τ connected we also have to allow the possibility of a strand τ_j passing transversally through a curve $\mathcal{L}_k^{opp}$ at a finite number of distinct points and times. The subspace of ${}_{[1,2]}\mathcal{F}^0$ of functions, for which $\mathcal{A}^{opp}$ is transverse to τ except for one transverse intersection point of one τ_j with one $\mathcal{L}_k^{opp}$ will be denoted, analogously, by ${}_X\mathcal{F}^1_{\mathcal{L}}$. A passage of a path of functions through this stratum is depicted in the lower part of Figure 3.9. Note, that $\mathcal{L}_k^{opp}$ may be bounding several surface pieces in $\mathcal{A}^{opp}$ so that τ_j will locally intersect one group of these pieces before the move and the complementary part after the move. As an example, we have that τ_j runs through $\mathcal{P}_1^{opp}$ and $\mathcal{P}_2^{opp}$ on the left lower side of Figure 3.9 and only through $\mathcal{P}_3^{opp}$ on the right side.

The moves applied by the addition of the strata ${}_X\mathcal{F}^1_{\mathcal{L}}$ and ${}_X\mathcal{F}^1_{\mathcal{P}}$ make the space of embeddings of $\mathcal{A}^{opp}$ that are transverse to τ connected again. Clearly, any passage through the other codimension-1 strata can be separated from these moves for a generic path of functions. Our findings can, thus, be summarized in the following result.

Proposition 3.2.2. *The space of functions on W*

$$ {}_X\mathcal{F}^C = {}_X\mathcal{F}^0 \cup {}_X\mathcal{F}^1_{\alpha} \cup {}_X\mathcal{F}^1_{\beta} \cup {}_X\mathcal{F}^1_{1-1} \cup {}_X\mathcal{F}^1_{2-2} \cup {}_X\mathcal{F}^1_{2-1} \cup {}_X\mathcal{F}^1_{\mathcal{L}} \cup {}_X\mathcal{F}^1_{\mathcal{P}} $$

is path connected and any paths can be made to pass transversally through the codimension-1 strata at a finite number of points.

3.2.3 Construction of Cobordisms and Surgery Calculi

The intersection of the descending manifolds of a Morse function $f \in {}_X\mathcal{F}^0$ with the handlebody $\mathcal{H}^{g_{tg}}_{g_{sc}} \subset \partial W$ yields a complex $\mathcal{A}$ of framed 1- and 0-manifolds. As explained in Lemma 1.5.1 and [Mil69] or [Hir76] this defines a handle decomposition of W into 1-handles and 2-handles as expressed in Lemma 3.1.4 with $V = \mathcal{H}^{g_{tg}}_{g_{sc}}$.

When we thicken the 1-manifolds in $\mathcal{A} \subset \mathcal{H}^{g_{tg}}_{g_{sc}}$ to ribbons and the 0-manifolds to spheres in order to describe the attachings we obtain a "bridged link" as depicted in Figure 1.10 of Section 1.5.2. $M_o \subset \partial W$ is, thus, obtained by surgery along this bridged link. The bridged link $\mathcal{A}$ is transported to the opposite manifold complex $\mathcal{A}^{opp}$ in M_o by the gradient flow of the Morse function. The boundary $\partial(\mathcal{P}^{opp}_\nu \times [-\varepsilon, \varepsilon]) \cong -\mathcal{P}^{opp}_\nu \sqcup \mathcal{P}^{opp}_\nu$ of the ascending sphere of an index-1 singularity in M_o is mapped by this to the boundary of the neighborhood of the corresponding surgery balls.

Moreover, the torus bounding a tubular neighborhood of a curve $\mathcal{L}^{opp}_\mu \subset M_o$ is mapped to a torus bounding the tubular neighborhood of a corresponding curve $\mathcal{L}_\mu \subset \mathcal{H}^{g_{tg}}_{g_{sc}}$ if $\mathcal{L}^{opp}_\mu$ is disjoint from any surfaces in $\mathcal{A}^{opp}$ but exchanging the long and short meridians. If a curve $\mathcal{L}^{opp}_\mu$ is a common boundary piece in the surfaces $\mathcal{P}^{opp}_1, \ldots, \mathcal{P}^{opp}_a$ then the corresponding surgery ribbon $\mathcal{L}_\mu$ in $\mathcal{H}^{g_{tg}}_{g_{sc}}$ passes through the respective pairs of surgery balls.

Now, the gradient flow of a function in ${}_X\mathcal{F}^0$ also maps the strands τ_j in M_o to respective strands t_j in $\mathcal{H}^{g_{tg}}_{g_{sc}}$. As the τ_j are disjoint from the $\mathcal{L}^{opp}_\mu$ in $\mathcal{A}^{opp}$, the t_j will be disjoint from the ribbons $\mathcal{L}_\mu$ in the bridged link $\mathcal{A}$. Furthermore, if a strand τ_j passes through a surface $\mathcal{P}^{opp}_\mu$ the strand t_j will exit and enter through the corresponding pair of surgery balls in $\mathcal{H}^{g_{tg}}_{g_{sc}}$.

Thus, the strands t_j in $\mathcal{H}^{g_{tg}}_{g_{sc}}$ play precisely the same rôle as the external strands in the generalization of bridged links described in Section 2.5.1. Let us denote $\mathbf{t} = t_1 \sqcup \ldots \sqcup t_{a+b}$, where we now consider external strands thickened to tubular neighborhoods $\cong D^2 \times ([0, p_1] \sqcup [p_1, p_2] \sqcup \ldots \sqcup [p_{n-1}, p_n])$, where p_k are the points at which t_j ends in one surgery sphere and continues at the corresponding partner surgery sphere.

The handle decomposition on W given by the Morse function also results in a surgery presentation $M_o \cong \left(\mathcal{H}^{g_{tg}}_{g_{sc}}\right)_{\mathcal{A}}$. The presentation of the 3-cobordism $M_o \subset \partial W$ extends to a presentation

$$M = \overline{M_o - \tau} \cong \overline{\left(\mathcal{H}^{g_{tg}}_{g_{sc}} - \mathbf{t}\right)_{\mathcal{A}}} \tag{3.2.2}$$

of the relative cobordism M up to homeomorphisms. The surgery along the index-1 spheres in $\mathcal{A}$ is performed as before, except that now spheres with holes are connected to each other after the 3-balls have been removed so that the tubular boundary pieces combine to closed tori after all index 1-surgeries are performed.

In summary, a bridged link, here given by a pair $(\mathcal{A}, \mathbf{t})$ of an ordinary bridged link $\mathcal{A}$ in $\mathcal{H}^{g_{tg}}_{g_{sc}}$ together with external strands $\mathbf{t}$, defines, up to a homeomorphism, a relative cobordism $M_{\mathcal{A},\mathbf{t}}$, as well as a corresponding 4-manifold $W_{\mathcal{A},\mathbf{t}}$. Since we only consider homeomorphism classes of cobordisms in the double category $\widetilde{\mathcal{C}ob}$, we have a well defined assignment

$$(\mathcal{A}, \mathbf{t}) \longrightarrow \widetilde{M_{\mathcal{A},\mathbf{t}}} = \left[M_{\mathcal{A},\mathbf{t}}, \operatorname{sign}(W_{\mathcal{A},\mathbf{t}})\right]$$

from bridged links with external strands in $\mathcal{H}^{g_{tg}}_{g_{sc}}$ to 2-arrows of $\widetilde{\mathcal{C}ob}$, given as in the description (1.6.10), (1.6.11) as a pair of an ordinary cobordisms and an integer, namely, the signature of the bounding 4-manifold. Combining these observations with the results in Lemma 3.1.6, Lemma 3.1.7, and Proposition 3.2.2 we obtain the following statement on equivalences of surgery data of this kind on the standard handlebodies.

Proposition 3.2.3. *Let the 2-arrows $\widetilde{M_{\mathcal{A}^1,\mathbf{t}^1}}$ and $\widetilde{M_{\mathcal{A}^2,\mathbf{t}^2}}$ in $\widetilde{\mathcal{C}ob}$, obtained from bridged links with $a + b$ external strands $(\mathcal{A}^j, \mathbf{t}^j)$ on $\mathcal{H}^{g_{tg}}_{g_{sc}}$, be equal. Then we can obtain $\mathcal{A}^2, \mathbf{t}^2$ from $\mathcal{A}^1, \mathbf{t}^1$ by applying a finite number of operations on the link, which can be any of the following:*

- *Isotopies.*
- *1-Handle Slides in $\mathcal{A}$ as described in Figure 1.12.*
- *2-Handle Slides in $\mathcal{A}$ as described in Figure 1.14.*
- *1-2-Slides in $\mathcal{A}$, pulling a ribbon loop through a surgery sphere as, for example, the strand a through ϕ in Figure 1.11.*
- *2-Handle Slides of an external strand t_j over a ribbon in $\mathcal{A}$. Specifically, in Figure 1.14 the ribbon labeled Hi is the external strand and Lo is a surgery ribbon in $\mathcal{A}$.*
- *1-2-Slides of a small loop of an external strand t_j through a pair of surgery spheres in $\mathcal{A}$. The strand t_j can be, thus, the strand a in Figure 1.11 being pulled through ϕ.*
- *Cancellations as in Figure 1.15, where some of the strands $a, \ldots$ (but not C_1) running through the surgery spheres may be external strands.*
- *The β-Move as described in Figure 3.3.*

Proof. The connected 4-manifolds $W_{\mathcal{A}^1,\mathbf{t}^1}$ and $W_{\mathcal{A}^2,\mathbf{t}^2}$ have by definition the same signature and boundary so that by Lemma 3.1.6 there is a 4-manifold W_3, which can be obtained from either $W_{\mathcal{A}^1,\mathbf{t}^1}$ and $W_{\mathcal{A}^2,\mathbf{t}^2}$ by 4-dimensional index-2 surgery. Furthermore, Lemma 3.1.6 states that this surgery is equivalent to adding a pair of 2-handles $e^4_2 \cup e^4_2$ as prescribed by the β-Move. They can be put into general position with respect to the original handle decompositions $(\mathcal{A}^j, \mathbf{t}^j)$ and, thus, be added to render new attaching diagrams $(\mathcal{A}^{1-3}, \mathbf{t}^{1-3})$ and $(\mathcal{A}^{2-3}, \mathbf{t}^{2-3})$, which both describe the same 4-manifold W_3.

Considering now the stratified space ${}_X\mathcal{F}^C$ of functions on W_3 allows us to prove that the diagrams $(\mathcal{A}^{1-3}, \mathbf{t}^{1-3})$ and $(\mathcal{A}^{2-3}, \mathbf{t}^{2-3})$ can be moved into each other by means of the remaining equivalences. Each of these diagrams corresponds to a generic handle decomposition and, hence, to Morse functions $f_1, f_2 \in {}_X\mathcal{F}^0$ on W_3. By Proposition 3.2.2 there is a path between f_1 and f_2, which passes a finite number of times transversally through the listed codimension-1 strata.

While the path segments in ${}_X\mathcal{F}^0$ are identified as isotopies of the attachment diagrams, a passage through each of the codimension-1 strata can be identified with one of the moves in the proposition. Specifically, going through ${}_X\mathcal{F}^1_\alpha$ corresponds to a Smale Cancellation as described in Lemma 1.5.4 and (1.5.3). A passage through ${}_X\mathcal{F}^1_\beta$ does not affect the isotopy class of the attachment diagram since all critical manifolds of the same index are disjoint so that the implied reordering of the attachings of handles are independent. Recall that crossings of trajectories of different indices have been excluded already in the definition of ${}_{[1,2]}\mathcal{F}^C$.

The intersection of two index-2 critical manifolds as in a passage through ${}_X\mathcal{F}^1_{2-2}$ has been analyzed in Figure 1.13, where it is identified with a 2-handle slide modulo isotopy. Similar inspection of the relative position of critical manifolds for the passages through ${}_X\mathcal{F}^1_{1-1}$, and ${}_X\mathcal{F}^1_{2-1}$ shows that they correspond to 1-Handle Slides and the 1-2-Slides respectively.

Since the moves described in Figure 3.9 can be obtained by replacing the index-2 descending critical manifold of functions passing through ${}_X\mathcal{F}^1_{2-2}$ or ${}_X\mathcal{F}^1_{2-1}$ by strands τ_j, we see that passages through ${}_X\mathcal{F}^1_{\mathcal{L}}$ or ${}_X\mathcal{F}^1_{\mathcal{P}}$ translate to the slides of external strands over index-1 and index-2 surgery data.

The above list of moves is, in fact, highly redundant. There are several ways to select a smaller list of moves for which the assertion in Proposition 3.2.3 still holds. The two we shall quote here are the calculus of bridged links from [Ker99] as well as the respective variant of Kirby's calculus derived from [Kir78]. The inclusion of external strands does not change the proofs, which can be taken literally from the original references.

Theorem 3.2.4. *Two bridged link diagrams $\mathcal{A}^1, \mathbf{t}^1$ and $\mathcal{A}^2, \mathbf{t}^2$ on $\mathcal{H}^{g_{tg}}_{g_{sc}}$ with external strands yield the same 2-arrows $\widetilde{M_{\mathcal{A}^1,\mathbf{t}^1}} = \widetilde{M_{\mathcal{A}^2,\mathbf{t}^2}}$ in $\widetilde{Cob}$ if an only if one bridged link diagram can be moved to the other by a series of operations, which may be any of the following:*

- *Isotopies of the bridged link diagrams.*
- *Handle Trade Move with external strands as depicted for TS1♠ in Section 2.5.1.*
- *Isolated 1-2-Cancellation Move as in TS2♠ of Section 2.5.1.*

The elimination of the handle slide moves as well as the β-Move can be taken literally from the proof of Proposition 6 in [Ker99] if the possibility of external strands is included. For the version corresponding to Kirby's calculus of links we restrict the possible handle decompositions of the 4-manifolds to those of index 2. Hence,

$$W = \mathcal{H}^{g_{tg}}_{g_{sc}} \times [0,1] \cup \mathrm{e}^4_2 \cup \ldots \cup \mathrm{e}^4_2,$$

so that there are no index-1 surgery balls in the diagram $\mathcal{A}$ representing the attachment data. Hence, $\mathcal{A}$ is an ordinary framed link in $\mathcal{H}^{g_{tg}}_{g_{sc}}$ with external strands.

Theorem 3.2.5. *Two link diagrams $\mathcal{A}^1, \mathbf{t}^1$ and $\mathcal{A}^2, \mathbf{t}^2$ (without index-1 surgery balls) on $\mathcal{H}^{g_{tg}}_{g_{sc}}$ with external strands yield the same 2-arrows $\widetilde{M_{\mathcal{A}^1,\mathbf{t}^1}} = \widetilde{M_{\mathcal{A}^2,\mathbf{t}^2}}$*

in $\widetilde{Cob}$ if an only if one bridged link diagram can be moved to the other by a series of operations, which may be any of the following:

- *Isotopies of the link diagrams.*
- *2-Handle Slide of either a surgery or an external ribbon over a surgery ribbon from $\mathcal{A}$.*
- *β-Move, as in Figure 3.3.*

The equivalence of these two calculi on a formal level has been proven in [Ker98a], thus, providing a proof of Theorem 3.2.5 given Theorem 3.2.4. A variant of the extended Kirby calculus in Theorem 3.2.5 has already been described in [Kir78] and [FR79].

The set of all possible β-Moves can be narrowed down further by allowing only a finite number of long ribbons $R_1 \ldots R_{g_{sc}+g_{tg}+2}$ for the configuration from Figure 3.3. Here, the first $(g_{sc} + g_{tg})$ ribbons R_j will go along paths representing each generator of $\pi_1(\mathcal{H}_{g_{sc}}^{g_{tg}})$. Any ribbon R^* may, thus, be slid over these generators according to its word presentation in the fundamental group of the handlebody until it runs inside a sphere in $\mathcal{H}_{g_{sc}}^{g_{tg}}$. If A^* is an annulus around R^* in an arbitrary β-Move we can slide any other ribbon over A^* so that the location of this annulus acts as a gap in R^*. R^* can, thus, be unlinked from all of the other link components, such as the annuli of the generating β-Moves, as well as unknotted. As a result A^* and R^* form now a Hopf link, where A^* has trivial and R^* has arbitrary framing. Sliding R^* over A^* has the effect of changing the framing number by 2 so that we can make the framing of R^* either 1 or 0. These two possibilities are the remaining two choices of representing ribbons and correspond to the Hopf-link-Move as described in Section 2.5.2.A.

3.3 From Tangle Classes to Cobordism Classes

As a first step in proving the existence of the double isomorphism functor in Theorem 3.0.6 we will construct a bijective map between the 2-arrow sets of the tangle and cobordism double category. The objects and 1-arrows are already naturally isomorphic. Specifically, we consider a map

$$\mathfrak{Surg}_2 : 2 - Arrows\{\mathcal{T}gl^{\mathrm{BL}}_{S^2}\} \longrightarrow 2 - Arrows\{\widetilde{Cob}\} \qquad (3.3.1)$$

on the version of the tangle category that is defined with bridged links over S^2 as in Section 2.5.1. We proceed by first defining an assignment of concrete representing tangles in $S^2 \times [-1, 1]$ to a surgery diagram with external strands on handlebodies. Following, we shall prove that it factors into the respective equivalences classes and that the resulting map is a bijection.

3.3.1 From Tangles over S^2 to Bridged Links on $\mathcal{H}$

In order to extend a tangle diagram over a sphere to a surgery diagram in $\mathcal{H}_{g_{sc}}^{g_{tg}} \cong \mathcal{H}^+_{g_{tg}} \# \mathcal{H}^-_{g_{sc}}$ let us define partial link diagrams on the standard handlebodies $\mathcal{H}^{\pm}_g$

with a 3-ball removed. By a partial links in, e.g., $\overline{\mathcal{H}^+_{g_{tg}} - D^3}$ we mean a link, whose strands are allowed to end in the boundary $\partial(\overline{\mathcal{H}^+_{g_{tg}} - D^3}) \cong \Sigma_{g_{tg}} \sqcup S^2_+$. For the spheres $S^2_\pm$ that bound the removed 3-balls in the handlebodies $\mathcal{H}^\pm_g$ we assume that they have a singled out equator $S^1_b \subset S^2_\pm$, on which the intervals $J^{s/t}_j$ and $I^{i/o}_k$ are marked in the order as in (2.1.2) of Section 2.1.1. The relation to the tangle picture over $\mathbb{R}^2$ is described in Section 2.4.2.

First, we describe a partial link for $\mathcal{H}^+_{g_{tg}}$ with the following components:

$$\mathcal{B}^+_{g_t,a/b} = \varepsilon^s_1 \sqcup \ldots \sqcup \varepsilon^s_a \sqcup \beta_1 \sqcup \ldots \sqcup \beta_g \sqcup \varepsilon^t_1 \sqcup \ldots \sqcup \varepsilon^t_b \quad \subset \overline{\mathcal{H}^+_{g_{tg}} - D^3}. \quad (3.3.2)$$

They are arranged as depicted in Figure 3.10. Specifically, the β_j are (internal)

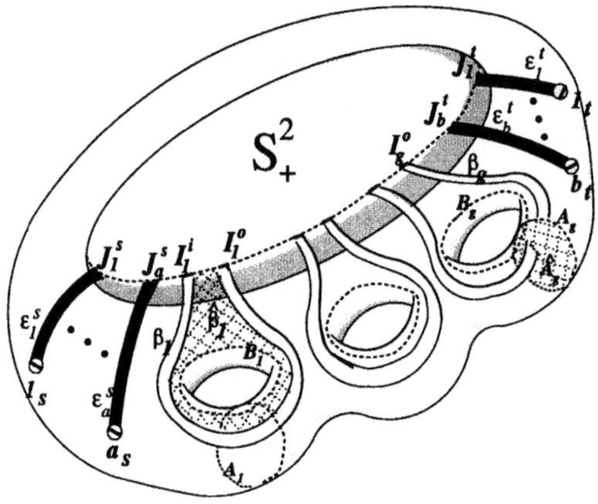

Fig. 3.10. Standard Diagram in $\mathcal{H}^+_g$

surgery ribbons starting and ending in the internal intervals on S^2_+ so that $I^i_j \sqcup I^o_j \subset \partial\beta_j$. The ribbons β_j can be extended to an annulus $\hat{\beta}_j$, which bounds the elementary curve $B_j \subset \Sigma_g$. Here B_j is a homology generator for $\mathcal{H}^+_g$ and it bounds a disc $\widehat{B_j}$ in $\mathcal{H}^-_g$.

The $a+b$ ribbons $\varepsilon^{s/t}_j$ run from an external interval $J^{s/t}_j$ on S^2_+ to the diameter of a j-th hole on the standard surface $\Sigma_{g,a/b}$.

Similarly, we define a standard diagram in $\mathcal{H}^-_{g_{sc}}$ as in Figure 3.11. The components of the partial link are as follows:

$$\mathcal{B}^-_{g_{sc},a/b} = \delta^s_1 \sqcup \ldots \sqcup \delta^s_a \sqcup \alpha_1 \sqcup \ldots \sqcup \alpha_g \sqcup \gamma_1 \sqcup \ldots \sqcup \gamma_g \sqcup \delta^t_1 \sqcup \ldots \sqcup \delta^t_b$$
$$\subset \overline{\mathcal{H}^-_{g_{sc}} - D^3} \cong \overline{D^3 - \mathcal{H}^+_{g_{sc}}}. \quad (3.3.3)$$

Here the α_j are closed internal surgery ribbons that can be obtained by pushing off a neighborhood of the curves $A_j \subset \Sigma_g$ on the standard bounding surface. Any A_j that

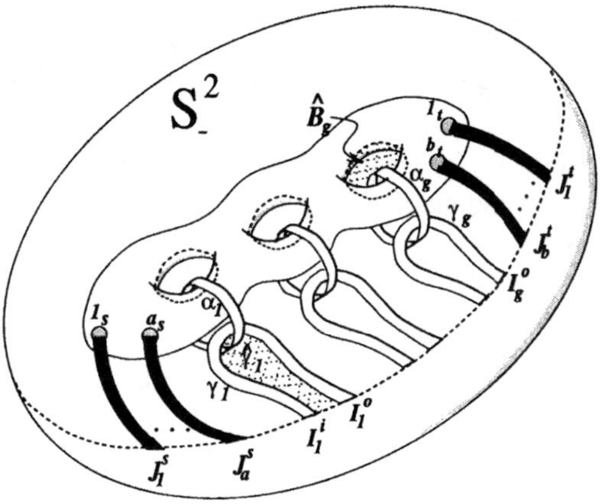

Fig. 3.11. Surgery with 1-handle

bounds a disc $\widehat{A_j}$ in $\mathcal{H}_g^+$, is a homology generator of $\mathcal{H}_g^-$, and intersects $\widehat{B_j}$ without intersecting any of the other curves. The surgery ribbons γ_j link with the α_j and start and end in the intervals I_j^i and I_j^o on S^2_-. Finally, the $\delta_j^{s/t}$ are external strands that run as the $\varepsilon_j^{s/t}$ between the diameters of the holes on the standard surface and the external intervals on S^2_-.

Suppose now that we have a tangle T of bridged links in $S^2 \times [-1,1]$ that represents a 2-arrow of $\mathcal{T}gl_{S^2}^{\mathbf{BL}}$ as described in Section 2.5.1. From this we can now construct a bridged link diagram $\mathcal{B}_T = (\mathcal{A}_T, \mathbf{t}_T)$, in $\mathcal{H}_{g_{sc}}^{g_{tg}}$ by gluing it together with the standard diagrams as follows:

$$\begin{aligned} \mathcal{B}_T \;=\; \mathcal{B}^+_{g_{tg},a/b} \bigcup_{I^{i/o},J^{s/t}} T \bigcup_{I^{i/o},J^{s/t}} \mathcal{B}^-_{g_{sc},a/b} \subset \\ \subset (\mathcal{H}^+_{g_{tg}} - D^3) \bigcup_{S^2\times\{+1\}=S^2_+} S^2\times[-1,1] \bigcup_{S^2\times\{-1\}=S^2_-} (\mathcal{H}^-_{g_{sc}} - D^3) \cong \mathcal{H}^{g_{tg}}_{g_{sc}}. \end{aligned} \tag{3.3.4}$$

In particular, we glue here the $S^2\times[-1,1]$ to $\overline{\mathcal{H}^+_{g_{tg}} - D^3}$ along the S^2-boundary components so that the equators and interval markings on these spheres coincide. As a result, the ribbon piece β_j in $\overline{\mathcal{H}^+_{g_{tg}} - D^3}$ will be attached to the component of the tangle T in $S^2\times[-1,1]$ that starts and ends in the intervals $I_j^{i/o}$ in the top equator in $S^2 \times \{1\}$. If the latter is a top ribbon component, as specified in Section 2.2.2, we, thus, obtain a closed ribbon in $(\mathcal{H}^+_{g_{tg}} - D^3) \cup_{S^2\times\{+1\}=S^2_+} S^2\times[-1,1]$. In case the components in T attached to $I_j^{i/o}$ are through ribbons, the combined component will start and end in $S^2 \times \{-1\}$.

Similarly, the $a+b$ external pieces $\varepsilon_j^{s/t}$ in $\overline{\mathcal{H}^+_{g_{tg}} - D^3}$ are joined to the $a+b$ external strands in T along the common external intervals $J_j^{s/t}$ in $S^2 \times \{+1\} = S^2_+$. As a result we obtain $a+b$ external ribbon components, where each one starts in a diameter of a hole in $\Sigma_{g_{tg},a/b}$ and ends in the corresponding external interval in $S^2 \times \{+1\} = S^2_+$, depending on the permutation given by the vertical 1-arrow of T.

We continue to glue also $\overline{\mathcal{H}^-_{g_{sc}} - D^3}$ to $(\mathcal{H}^+_{g_{tg}} - D^3) \cup_{S^2\times\{+1\}=S^2_+} S^2 \times [-1,1]$ along $S^2 \times \{-1\}$. The remaining internal surgery ribbons that are not closed but end in this sphere will, thus, be joined with the γ_j ribbons in $\overline{\mathcal{H}^-_{g_{sc}} - D^3}$ along the $I_j^{i/o}$ intervals to form closed ribbons. Furthermore, the external ribbons ending in the $J_j^{s/t}$ intervals in $S^2 \times \{-1\}$ are extended by the ribbons segments $\delta_j^{s/t}$. Hence, every external ribbon in the combined diagram runs from a diameter of a hole in $\Sigma_{g_{tg},a/b}$ to a corresponding hole in the opposite surface $\Sigma_{g_{sc},a/b}$, where the holes are permuted as prescribed by the permutations of the vertical 1-arrows of T.

Finally, the connected sum definition of $\mathcal{H}^{g_{tg}}_{g_{sc}}$ in (1.6.5) is the same as writing $(\mathcal{H}^+_{g_{tg}} - D^3) \bigcup_{S^2_+=S^2_-} (\mathcal{H}^-_{g_{sc}} - D^3)$. It is obvious that gluing a collar $S^2 \times [-1,1]$ in between the identification yields a naturally homeomorphic manifold as suggested in (3.3.4). Thus, in summary, the gluing of manifolds and bridged link diagrams as in (3.3.4) produces a well defined bridged link diagram on $\mathcal{H}^{g_{tg}}_{g_{sc}}$ with the correct 1-arrow data.

This completes our definition of the assignment $T \mapsto \mathcal{B}_T$ of admissible bridged link tangles in $S^2 \times [-1,1]$ to bridged link surgery diagrams on a corresponding handlebody.

3.3.2 The Boundary Move and Factorization into Classes

Clearly, a generic bridged link diagram $\mathcal{B}$ in $\mathcal{H}^{g_{tg}}_{g_{sc}}$ is usually not of the form $\mathcal{B}_T$ for some tangle over $S^2 \times [-1,1]$, since the partial diagrams in the $\overline{\mathcal{H}^{\pm}_g - D^3}$-pieces are not of the form $\mathcal{B}^{\pm}_{g,a/b}$. In this section we shall define a procedure that assigns to a generic bridged link $\mathcal{B}$ a tangle $T_{\mathcal{B}}$, unique up to isotopy, such that $\mathcal{B}$ and $\mathcal{B}_{T_{\mathcal{B}}}$ are equivalent surgery diagrams. This operation was already described in [Ker99] in the absence of external strands.

In order to describe this operation for $\mathcal{H}^-_{g_{sc}} - D^3$ we first need to restrict ourselves to bridged links that are transverse to the discs $\widehat{B_j} \subset \mathcal{H}^-_{g_{sc}} - D^3$ bounded by the curves $B_j \subset \Sigma_{g_{sc}}$. Specifically, this means that all surgery balls are disjoint from all of the $\widehat{B_j}$ and that we may assume that all strands pass perpendicularly through the discs. A typical picture of the link diagram in a vicinity of $\widehat{B_j}$ is shown on the left of Figure 3.12.

In the first step we perform an un-cancellation at the disc $\widehat{B_j}$ so that the strands that had been passing through $\widehat{B_j}$ run now through a pair of surgery spheres connected by an additional strand α_j. We move the two surgery balls around the handle along the A_j curve until they meet on the other side as depicted in the middle part of

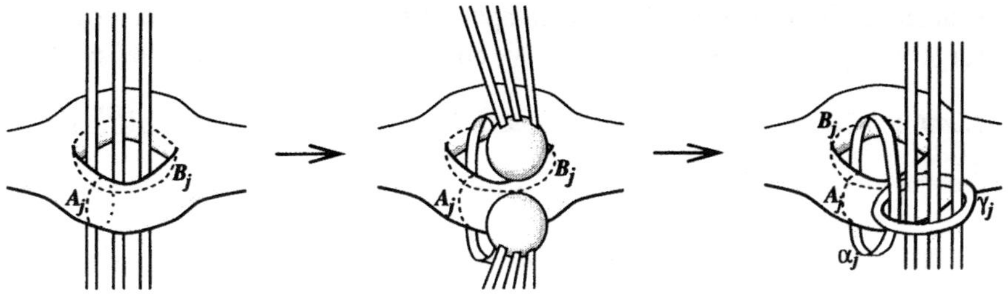

Fig. 3.12. Surgery with 1-handle

Figure 3.12. Finally, we apply a handle trade move as in Figure 3.2 of Section 3.1.2. Thus, the strands that were originally intersecting $\widehat{B}_j$ are now passing through an annulus γ_j as on the right of Figure 3.12.

Note that the strands in a collar vicinity U of the surface with added discs, $\Sigma_{g_{sc},a/b} \cup \widehat{B}_1 \cup \ldots \cup \widehat{B}_g$, are already in the standard configuration as prescribed in Figure 3.11. We assume here that the α_j lie entirely inside of U, that the respective pieces of the γ_j annuli are the only other internal surgery ribbons in U, and that the external strands, emerging from $\Sigma_{g_{sc},a/b}$ run in straight segments inside of U.

Since the addition of thickenings of the discs $\widehat{B}_j$ corresponds to adding 3-dimensional 2-handles, the component of ∂U that lies in the interior of $\overline{\mathcal{H}_{g_{sc}}^- - D^3}$ is a 2-sphere. In particular, $\overline{\mathcal{H}_{g_{sc}}^- - D^3} - U \cong S^2 \times [0,\delta]$. We can construct an isotopy between the following two homeomorphisms of $\overline{\mathcal{H}_{g_{sc}}^- - D^3} \cup_{S^2} S^2 \times [-1,1]$ to itself. The first is the identity, and the second homeomorphism ψ maps U onto $\overline{\mathcal{H}_{g_{sc}}^- - D^3}$, and squeezes the complement $\cong S^2 \times [0,\delta)$ into a collar of the $S^2 \times [-1,1]$ part.

Defining the isotopy in normal direction of the intermediate S^2-coordinates we see that the image of ψ of the link configuration in U as in the local picture of Figure 3.12 in the entire handlebody yields the standard configuration Figure 3.11. The analog of this procedure can be defined on a generic link diagram in the handlebody $\overline{\mathcal{H}_{g_{tg}}^+ - D^3}$. Here the un-cancellation process starts at the discs $\widehat{A}_j$, introducing a ribbon β_j and a pair of surgery spheres. These are led around the B_j-curves. The annulus of the subsequent handle trade move is then pushed entirely into $S^2 \times [-1,1]$ by the corresponding isotopy. The combined operation on both handlebodies, thus, yields a bridged link diagram, which is in standard position in the $\overline{\mathcal{H}_g^\pm - D^3}$-pieces and, hence, of the form $\mathcal{B}_T$, where T is an admissible bridged link tangle over S^2 given by restriction to the $S^2 \times [-1,1]$-piece. We denote the resulting assignment as follows:

$$\Psi : \left\{\text{Diagrams on } \mathcal{H}_{g_{sc}}^{g_{tg}} \text{ with } \pitchfork A_j, B_j\right\} \longrightarrow \left\{\text{Admissible Tangles over } S^2\right\} : \tag{3.3.5}$$

$$\mathcal{B} \mapsto \Psi(\mathcal{B}) = T_{\mathcal{B}} \qquad \text{where} \qquad \mathcal{B}_{T_{\mathcal{B}}} \cong \mathcal{B}.$$

The equivalence in (3.3.5) is obvious from the fact that Ψ has been defined only via equivalence moves of bridges link diagrams for framed 3-manifolds, as enumerated in Section 3.2.3. If we compose the assignments between tangles and surgery diagrams in the opposite way we obtain an equivalence as follows:

Lemma 3.3.1. *For an admissible tangle T for the double category $\mathcal{T}gl^{\mathbf{BL}}_{S^2}$ the tangle*

$$T^* = \Psi(\mathcal{B}_T) = T_{\mathcal{B}_T}$$

is obtained by application of the Ribbon-TS3-Move from Section 2.5.2 at every interval pair $I_j^{i/o}$ at the top and bottom line.

Proof. The surgery diagram $\mathcal{B}_T$, that we start out with in the construction of Ψ, is already of standard type. Hence, the only strand in Figure 3.12, passing through B_j, is the original α_j. Applying the subsequent operations we also obtain ribbons α'_j and γ'_j that are chain linked together. The map ψ pushes the configuration of α_j, γ'_j, and part of α'_j into $S^2 \times [-1,1]$. Close to $S^2 \times \{-1\}$ the resulting link diagram has a form that is then easily identified with the one for the Ribbon-TS3-Move from Section 2.5.2. The push-out by ψ on the external strands has, up to isotopy, no effect since they are already positioned parallel to the direction of the deformation. The description of T^* close to $S^2 \times \{+1\}$ is analogous.

Note, that the Ribbon-TS3-Move can also be given as a combination of the TS3-Move from Section 2.3.3 or the TS3♠-Move from Section 2.5.1, combined with a handle trade move. Since the TS3-Moves are part of the definition of $\mathcal{T}gl^{\mathbf{BL}}_{S^2}$ this result allows us now to show that the assignment of tangles to surgery data from the previous section yields a well defined map between 2-arrow sets.

Proposition 3.3.2. *The assignment*

$$T \mapsto \mathcal{B}_T = \Big(\mathcal{A}_T, \mathfrak{t}_T\Big) \mapsto \tilde{M}_T = \Big(\overline{(\mathcal{H}^{g_{tg}}_{g_{sc}} - \mathfrak{t}_T)_{\mathcal{A}_T}}, \mathrm{sign}(W_{\mathcal{A}_T})\Big),$$

with composites as defined in (3.3.4) *and* (3.2.2)*, factors into a map $\mathfrak{Surg}_2$ between 2-arrow sets as proposed in* (3.3.1).

Proof. For well-definedness we have to ensure that if two tangles, T and T', are equivalent in $\mathcal{T}gl^{\mathbf{BL}}_{S^2}$ then also the manifolds $\tilde{M}_T$ and $\tilde{M}_{T'}$, are homeomorphic and the signatures coincide. Clearly, we need to consider only elementary moves for the T and T' as given in Section 2.5.1. An isotopy of a tangle T extends naturally to an isotopy of $\mathcal{B}_T$, which does not change the homeomorphism class of either the represented 4-manifold or its bounding 3-manifold.

The TS1♠ and TS2♠-Moves are examples of local surgery moves as described in Sections 1.5.2 and 3.1.2. Both moves preserve the homeomorphism class of the surgically presented 3-manifold. The cancellation move leaves also the homeomorphism class of the bounding 4-manifold invariant and changes only its handle decomposition. The handle trade move corresponds to an index-2 surgery on the 4-manifold so that it preserves at least its signature. See Proposition 3.2.3.

The TS3♠-Move is a combination of the TS2♠-Move and the Ribbon-TS3-Move from Section 2.5.2. If two tangles, T and T', are related by a Ribbon-TS3-Move, then $\mathcal{B}_T$ can be moved into $\mathcal{B}_T$ by applying the combination of an uncancellation, isotopy, and a handle trade move, used to define Ψ as in Figure 3.12, to only the handle corresponding to the interval pair of the Ribbon-TS3-Move in consideration. Being a combination of moves in Proposition 3.2.3 the boundary move neither changes the homeomorphism class of the 3-manifold or the signature of the 4-manifold.

3.3.3 Bijectivity of 2-Arrows Sets

In this section we prove that the map $\mathfrak{Surg}_2$ proposed in (3.3.1) and constructed in (3.3.2) is a bijection, which is implied by the notion of the isomorphism double functor that we wish to construct. We start with the easier proof of surjectivity and then show that the map is also into.

Lemma 3.3.3. *The map $\mathfrak{Surg}_2$ between 2-arrow sets from Proposition 3.3.2 is surjective.*

Proof. We need to show that any pair of a relative cobordism, M, and an integer, $\sigma \in \mathbb{Z}$, can be obtained by a surgery and cut out procedure as in (3.2.2). Surjectivity of $\mathcal{S}g$ from (1.6.11), as asserted in Lemma 1.6.1, shows that there is a triple $(W, M_o, \mathbf{b})$, representing a 2-arrow in $\widetilde{\mathcal{COB}}$. The 4-manifold is, thus, a relative cobordism between $\mathcal{H}_{g_{sc}}^{g_{tg}}$ and M_o and can be presented by attaching handles to $\mathcal{H}_{g_{sc}}^{g_{tg}} \times [0, 1]$.

As described in Section 3.1.1 we can eliminate all handles of index 0 and 4 by cancellation, using the fact that W is connected. The handle trade move allows us to eliminate, furthermore, all 3-handles from the decomposition as described in Section 3.1.2. This changes the homeomorphism class of the 4-manifold but not its signature σ.

Finally, we put all handles into general position with respect to each other and with respect to the tubular pieces $\mathfrak{b} \subset M_o$. In other words, we use the fact that the function space ${}_X\mathcal{F}^0$ from Section 3.2.2 is dense. The 4-manifold W and the 3-cobordism are, thus, described by a bridged link diagram $\mathcal{B} = (\mathcal{A}, \mathfrak{t})$ on $\mathcal{H}_{g_{sc}}^{g_{tg}}$ with external strands.

To this link diagram we apply the operation $\Psi : \mathcal{B} \mapsto T_{\tilde{M}} = T_{\mathcal{B}}$. The 4-manifold W'' represented by $\mathcal{B}_{T_\mathcal{B}}$ has the same boundary $\langle M_o \rangle$ and signature σ as the one given by $\mathcal{B}$, although they may be not homeomorphic. Hence, $(W'', M_o, \mathbf{b})$ still represents the same given element $[M, \sigma]$ in $\widetilde{\mathcal{C}ob}$.

Since it is also obtained by assigning $T_{\tilde{M}}$ to the respective surgery diagram and surgered manifolds, as in Proposition 3.3.2, we, thus, have $\tilde{M}_{T_{\tilde{M}}} = [M, \sigma]$ and, hence, have found a preimage.

Bijectivity of $\mathfrak{Surg}_2$ going from the 2-arrow sets of $\mathcal{T}gl^{\mathbf{BL}}_{S^2}$ to the 2-arrow sets of $\widetilde{\mathcal{C}ob}$ follows from the next lemma.

Lemma 3.3.4. *The map $\mathfrak{Surg}_2$ from Proposition 3.3.2 is injective.*

Proof. To prove injectivity we need to show that if for two admissible tangles, T and T', the surgery diagrams, $\mathcal{B}_T$ and $\mathcal{B}_{T'}$, yield homeomorphic framed, relative 3-cobordisms, $\tilde{M}_T$ and $\tilde{M}_{T'}$, then the tangles are equivalent in $\mathcal{T}gl^{\mathbf{BL}}_{S^2}$. From Theorem 3.2.4 we see that in this case $\mathcal{B}_T$ can be moved to $\mathcal{B}_{T'}$ by a sequence of applications of isotopies, handle trade moves, and isolated 1-2-cancellations.

The aim of the proof is to generate from this sequence of surgery diagrams a sequence of equivalences of tangles for $\mathcal{T}gl^{\mathbf{BL}}_{S^2}$, which correspond to surgery diagrams that are of standard form in the handlebody parts. To this end we can apply the map Ψ from (3.3.5) continuously to every surgery diagram in the chain of equivalences and, thus, obtain a list of tangle diagrams. Note, however, that Ψ is defined only for configurations that are transverse to the discs $\widehat{A_j}$ and $\widehat{B_j}$. All handle trade moves and cancellation moves can easily be separated from these discs, and any generic configuration can also be made transverse. Yet an isotopy can usually not be deformed to such that each of its configurations is transverse to the discs. Still we can choose the isotopy itself to be transverse. This means that at a finite number of times we allow either a surgery ball to pass transversally through one of the discs or a ribbon become tangential to a disc, while a small loop is being pulled through, analogous to the situation in the top part of Figure 3.9.

Hence, we may assume that the two surgery diagrams $\mathcal{B}_T$ and $\mathcal{B}_{T'}$ are related by a sequence of equivalences, $\mathcal{B}_T \longrightarrow \mathcal{B}_1 \longrightarrow \mathcal{B}_2 \longrightarrow \ldots \longrightarrow \mathcal{B}_N \longrightarrow \mathcal{B}_{T'}$, in which arrow may be one of the following:

(1) Isotopies, in which each configuration is transverse to the discs $\widehat{A_j}$ and $\widehat{B_j}$.
(2) A loop of a ribbon being pulled transversally through a disc $\widehat{A_j}$ or $\widehat{B_j}$.
(3) A surgery ball being pulled transversally through a disc $\widehat{A_j}$ or $\widehat{B_j}$.
(4) Handle Trade Moves as in Theorem 3.2.4, that are separated from the discs $\widehat{A_j}$ and $\widehat{B_j}$.
(5) 1-2 Cancellation Moves as in Theorem 3.2.4, that are separated from the discs $\widehat{A_j}$ and $\widehat{B_j}$.

In between these moves the surgery diagrams $\mathcal{B}_j$ have strands transverse to the discs so that we can apply the operation Ψ. We obtain a sequence of tangles as in the following diagram:

$$\begin{array}{ccccccccccccc} & & \mathcal{B}_T & \longrightarrow & \mathcal{B}_1 & \longrightarrow & \mathcal{B}_2 & \longrightarrow & \ldots & \longrightarrow & \mathcal{B}_N & \longrightarrow & \mathcal{B}_{T'} & & \\ & & \downarrow{\scriptstyle\Psi} & & \downarrow{\scriptstyle\Psi} & & \downarrow{\scriptstyle\Psi} & & & & \downarrow{\scriptstyle\Psi} & & \downarrow{\scriptstyle\Psi} & & \\ T & \longrightarrow & T_{\mathcal{B}_T} & \longrightarrow & T_{\mathcal{B}_1} & \longrightarrow & T_{\mathcal{B}_2} & \longrightarrow & \ldots & \longrightarrow & T_{\mathcal{B}_N} & \longrightarrow & T_{\mathcal{B}_{T'}} & \longrightarrow & T' \end{array}$$

If two diagrams $\mathcal{B}_j$ and $\mathcal{B}_{j+1}$ are related by an isotopy as in (1), the part of the isotopy outside the vicinity U from the construction of Ψ in Section 3.3.2 simply becomes an isotopy of the tangle in $S^2 \times [-1, 1]$. Since we may push any surgery diagram and isotopy slightly off the surface Σ in the boundary of the handlebody, we can assume that the diagrams intersect U only in the vicinities of the discs $\widehat{A_j}$ and $\widehat{B_j}$.

It is clear from Figure 3.12 that any strand running through a disc $\widehat{B_j}$ will run in the corresponding position through the disc bounded by the annulus in the configuration of the Ribbon-TS3-Move as depicted in Section 2.5.2. The image under Ψ of an isotopy of strands near a disc $\widehat{B_j}$ is now easily identified as an isotopy near this annulus.

Likewise, the images of the moves described in (2) and (3) can be identified as pulling a loop or a surgery ball through this annulus, which is obviously realized by a simple isotopy in $S^2 \times [-1,1]$ as well. Hence, if diagrams $\mathcal{B}_j$ and $\mathcal{B}_{j+1}$ are isotopic by either (1), (2), or (3), then the tangles $T_{\mathcal{B}_j}$ and $T_{\mathcal{B}_{j+1}}$ are isotopic in $S^2 \times [-1,1]$.

The surgery moves in (4) and (5) are chosen so, that they are disjoint from the vicinity U. Hence, their images under Ψ are handle trade moves and cancellations in $S^2 \times [-1,1]$, respectively. Thus, if $\mathcal{B}_{k+1}$ is obtained from $\mathcal{B}_k$ by one of these moves, the arrow from $T_{\mathcal{B}_k}$ to $T_{\mathcal{B}_{k+1}}$ describes indeed an equivalence in $\mathcal{T}gl^{\mathbf{BL}}_{S^2}$.

From Lemma 3.3.1 we know that also the arrows $T \to T_{\mathcal{B}_T}$ and $T_{\mathcal{B}_{T'}} \to T'$ describe equivalences. We have, thus, found that the bottom row in the above diagram describes a chain of moves of tangles in $\mathcal{T}gl^{\mathbf{BL}}_{S^2}$ between T and T' so that both tangles represent the same 2-arrow in $\mathcal{T}gl^{\mathbf{BL}}_{S^2}$. The assignment of tangles classes to cobordism classes is, thus, injective.

3.4 Verification of Compositions

Vertical and horizontal compositions between 2-arrow sets with coinciding 1-arrows were defined for the double category $\widetilde{\mathcal{C}ob}$ in Section 1.6.3 and for the double category $\mathcal{T}gl^{\mathbf{BL}}_{S^2}$ in Section 2.6. In order for $\mathfrak{Surg}_2$ to give rise to a double functor between these categories it must respect the two compositions. Specifically, the map on the 2-arrow sets must be functorial in vertical direction, meaning $\mathfrak{Surg}_2(T_l \circ_v T_u) = \mathfrak{Surg}_2(T_l) \circ_v \mathfrak{Surg}_2(T_u)$, as well as functorial in horizontal direction, which translates, similarly, into $\mathfrak{Surg}_2(T_2 \circ_h T_1) = \mathfrak{Surg}_2(T_2) \circ_h \mathfrak{Surg}_2(T_1)$. Again we start with the easier case, namely, the vertical compositions.

Lemma 3.4.1. *The map $\mathfrak{Surg}_2$ is vertically functorial.*

Proof. Let two tangles T_l and T_u be given such that the target horizontal 1-arrow $[g_{int}, a/b]$ of T_u coincides with the source horizontal 1-arrow of T_l. We obtain a bridged link diagram $\mathcal{B}_{T_u}$ on $\mathcal{H}^{g_{int}}_{g_{sc}} = (\mathcal{H}^-_{g_{sc}} - \mathring{D}^3) \cup S^2 \times [-1,1] \cup (\mathcal{H}^+_{g_{int}} - \mathring{D}^3)$ as well as the diagram $\mathcal{B}_{T_l}$ on $\mathcal{H}^{g_{tg}}_{g_{int}} = (\mathcal{H}^-_{g_{int}} - \mathring{D}^3) \cup S^2 \times [-1,1] \cup (\mathcal{H}^+_{g_{tg}} - \mathring{D}^3)$. To these we associate the 4-manifolds $W_{\mathcal{B}_{T_{u/l}}}$ by attaching 4-dimensional 1-handles and 2-handles e^4_1 and e^4_2 to the thickened handlebodies $\mathcal{H}^{g_{int}/g_{tg}}_{g_{sc}/g_{int}} \times [f_0, f_1]$.

The main step in the vertical composition is to glue these two 4-manifolds together over the common boundary pieces $\Sigma_{g_{int}} \times [f_0, f_1]$, as defined in (1.6.12).

Since the handles are attached to different parts of the boundaries of the thickened handlebody, namely to $\mathcal{H}^{g_{int}/g_{tg}}_{g_{sc}/g_{int}} \times f_1$, it does not matter whether we attach first the handles to each handlebody and then perform the vertical gluing or whether we first put the thickened handlebodies together and then attach both sets of handles to the composite. The exchange of the attaching operations is more formally expressed as the following equation. Here we make use of the fact that the gluing over the handlebody products with $[f_0, f_1]$ over the surface product with the same interval is clearly the same as first gluing the handlebodies over the surface and then taking the product with $[f_0, f_1]$:

$$
\begin{aligned}
W_{\mathcal{B}_{T_l}} \circ_v W_{\mathcal{B}_{T_u}} &= \Big((\mathcal{H}^{g_{int}}_{g_{sc}} \times [f_0, f_1]) \bigcup_{\mathcal{B}_{T_u} \times f_1} (\mathrm{e}^4_1 \cup \ldots \cup \mathrm{e}^4_2) \Big) \bigcup_{\Sigma_{g_{int}} \times [f_0, f_1]} \\
&\quad \bigcup_{\Sigma_{g_{int}} \times [f_0, f_1]} \Big((\mathcal{H}^{g_{tg}}_{g_{int}} \times [f_0, f_1]) \bigcup_{\mathcal{B}_{T_l} \times f_1} (\mathrm{e}^4_1 \cup \ldots \cup \mathrm{e}^4_2) \Big) \\
&= \Big((\mathcal{H}^{g_{int}}_{g_{sc}} \bigcup_{\Sigma_{g_{int}}} \mathcal{H}^{g_{tg}}_{g_{int}}) \times [f_0, f_1] \Big) \bigcup_{(\mathcal{B}_{T_u} \sqcup \mathcal{B}_{T_l}) \times f_1} (\mathrm{e}^4_1 \cup \ldots \cup \mathrm{e}^4_2). \qquad (3.4.1)
\end{aligned}
$$

In other words, $W_{\mathcal{B}_{T_l}} \circ_v W_{\mathcal{B}_{T_u}}$ is given by attaching handles along the surgery diagram $\mathcal{B}_{T_u} \sqcup \mathcal{B}_{T_l}$ to a thickening of the composition of the manifolds $\mathcal{H}^{g_{int}}_{g_{sc}}$ and $\mathcal{H}^{g_{tg}}_{g_{int}}$ over $\Sigma_{g_{int}}$.

This combination of the two link diagrams is divided into six pieces according to the three piece construction in (3.3.4). Specifically, they are given as

$$
\begin{aligned}
&\mathcal{B}_{T_l} \sqcup \mathcal{B}_{T_u} = \\
&\mathcal{B}^+_{g_{tg}, a/b} \bigcup_{I^{i/o}, J^{s/t}} T_l \bigcup_{I^{i/o}, J^{s/t}} \mathcal{B}^-_{g_{int}, a/b} \bigcup \mathcal{B}^+_{g_{int}, a/b} \bigcup_{I^{i/o}, J^{s/t}} T_u \bigcup_{I^{i/o}, J^{s/t}} \mathcal{B}^-_{g_{sc}, a/b} \\
&\subset (\mathcal{H}^+_{g_{tg}} - \mathcal{B}^3) \bigcup S^2_l \times [-1,1] \bigcup_{S^2} (\mathcal{H}^-_{g_{int}} - \mathcal{B}^3) \bigcup_{\Sigma_{int}} (\mathcal{H}^+_{g_{int}} - \mathcal{B}^3) \bigcup \\
&\qquad \bigcup S^2_u \times [-1,1] \bigcup_{S^2} (\mathcal{H}^-_{g_{sc}} - \mathcal{B}^3) \cong \mathcal{H}^{g_{int}}_{g_{sc}} \bigcup_{\Sigma_{int}} \mathcal{H}^{g_{tg}}_{g_{int}}.
\end{aligned}
$$

In this surgery diagram let us analyze further the intermediate two pieces. Here the two opposite standard link diagrams for the same 1-arrow are combined in the gluing of the two corresponding opposite handlebodies, which is by the definition $\mathcal{H}^-_g = S^3 - \mathcal{H}^+_g$ given as S^3 with two balls removed:

$$
\mathcal{B}^-_{g_{int}, a/b} \bigcup \mathcal{B}^+_{g_{int}, a/b} \subset (\mathcal{H}^-_{g_{int}} - \mathcal{B}^3) \bigcup_{\Sigma_{int}} (\mathcal{H}^+_{g_{int}} - \mathcal{B}^3) \cong S^3 - (\mathcal{B}^3 \sqcup \mathcal{B}^3) \cong S^2 \times [r, R].
$$

Putting together Figures 3.10 and 3.11 along the common surface Σ_g this diagram is as depicted on the left in Figure 3.13.

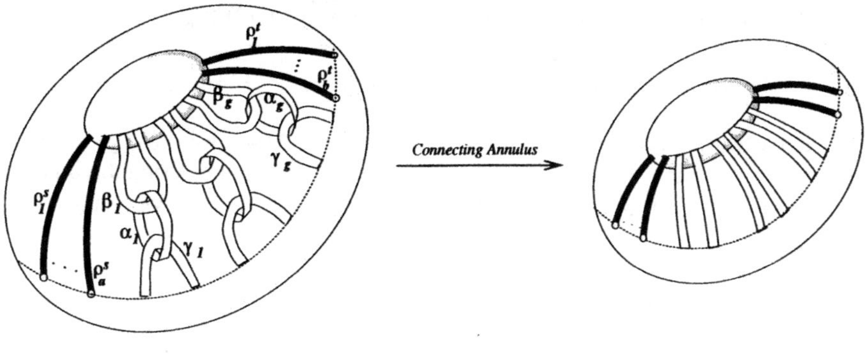

Fig. 3.13. Union of $\mathcal{B}^-_{g,a/b}$ and $\mathcal{B}^+_{g,a/b}$ in $S^3 - (D^3 \sqcup D^3) = S^2 \times (r, R)$

The internal surgery ribbons α_j, β_j, and γ_j link with each other as they link to the handles of the handlebodies, and an external ribbon $\varepsilon_j^{s/t}$ is connected to the corresponding straight piece $\delta_j^{s/t}$ to give an external strip $\rho_j^{s/t} = \varepsilon_j^{s/t} \cup_I \delta_j^{s/t}$.

The strands β_j and γ_j clearly belong to different components. Hence, we can apply the "Connecting Annulus" Move, stated at the end of Section 2.5.1, with the coupon substituted by an annulus according to the TS1-Move from Section 2.3.3. As a result, we obtain the surgery diagram as on the right of Figure 3.13 with parallel strands $\cong I \times [r, R]$ running through $S^2 \times [r, R]$, where I is an interval on S^2.

The intermediate surgery diagram on $S^2 \times [r, R]$ can, thus, be collapsed to S^2 (letting $r \to R$) and, therefore, be omitted from the total surgery diagram. Now the pieces $S_l^2 \times [-1, 1]$ and $S_u^2 \times [-1, 1]$ are glued directly together along the spheres $S_u^2 \times \{+1\} = S_l^2 \times \{-1\}$ such that also the tangles T_u and T_l come together and are joined along their common intervals. In the following identities we use that the composite of the thickened spheres is, clearly, again a thickened sphere, and that the union of the tangles is precisely the vertical stacking composite as defined in Section 2.6.1.

$$\begin{aligned}
\left(\mathcal{B}_{T_l} \sqcup \mathcal{B}_{T_u}\right)_{\beta-\text{Moves}} &= \mathcal{B}^+_{g_{tg},a/b} \bigcup_{I^{i/o},J^{s/t}} T_l \bigcup_{I^{i/o},J^{s/t}} T_u \bigcup_{I^{i/o},J^{s/t}} \mathcal{B}^-_{g_{sc},a/b} \\
&= \mathcal{B}^+_{g_{tg},a/b} \bigcup_{I^{i/o},J^{s/t}} T_l \circ_v T_u \bigcup_{I^{i/o},J^{s/t}} \mathcal{B}^-_{g_{sc},a/b} \\
&\subset (\mathcal{H}^+_{g_{tg}} - \mathring{B}^3) \bigcup S_l^2 \times [-1,1] \bigcup_{S^2} S_u^2 \times [-1,1] \bigcup_{S^2} (\mathcal{H}^-_{g_{sc}} - \mathring{B}^3).
\end{aligned}$$

In summary, we have that the inclusion $(\mathcal{B}_{T_l} \sqcup \mathcal{B}_{T_u})_{\beta-\text{Moves}} \subset \mathcal{H}^{g_{int}}_{g_{sc}} \cup_\Sigma \mathcal{H}^{g_{tg}}_{g_{int}}$ is (canonically, up to an isotopy) homeomorphic to the standard inclusion $\mathcal{B}_{T_l \circ_v T_u} \subset \mathcal{H}^{g_{tg}}_{g_{sc}}$.

By (3.4.1) we know that $W_{\mathcal{B}_{T_l}} \circ_v W_{\mathcal{B}_{T_u}}$ is obtained by attaching handles along the union $\mathcal{B}_{T_l} \sqcup \mathcal{B}_{T_u}$ on the thickened handlebody $\mathcal{H}^{g_{int}}_{g_{sc}}$. Moreover, the Connecting Annulus Move involves g applications of the β-Move, which, by Lemma 3.1.6, correspond to g index-2 surgeries $\mathcal{G}_1, \ldots, \mathcal{G}_g$, on $W_{B_{T_l}} \circ_v W_{B_{T_u}}$

$$(W_{B_{T_l}} \circ_v W_{B_{T_u}})_{\mathcal{G}_1,\ldots,\mathcal{G}_g} = W_{(B_{T_l} \sqcup B_{T_u})_{\beta-\text{Moves}}} = W_{B_{T_l \circ_v T_u}}. \tag{3.4.2}$$

In particular, identity (3.4.2) implies that $W_{B_{T_l}} \circ_v W_{B_{T_u}}$ and $W_{B_{T_l \circ_v T_u}}$ have the same boundary and signature. As this is the only information entering the image of the assignment in Proposition 3.3.2, we see that also the respective images under $\mathfrak{Surg}_2$ are the same.

Although surjectivity and functoriality already imply that vertical identities are mapped to each other, let us verify this fact in more detail.

Lemma 3.4.2. *$\mathfrak{Surg}_2$ maps the class of the identity tangle with straight and parallel vertical strands to the unique vertical identity in $\widetilde{Cob}$.*

Proof. The proof is essentially the same as in [Ker99]. The surgery diagram for the identity tangle contains $2g$ closed internal surgery ribbons on $\mathcal{H}_g^+ \# \mathcal{H}_g^-$, where g of them are the closed ribbons obtained by joining a β_j-ribbon with a γ_j ribbon, and the other g are the α_j-ribbons. Furthermore, we have $a+b$ external strands that connect the holes on opposite surface components by straight segments. In Figure 3.14 it is shown how we can slide handles of the opposite handlebody over the ribbons $\beta_j \cup \gamma_j$ so that they are aligned with handles of the target handlebody.

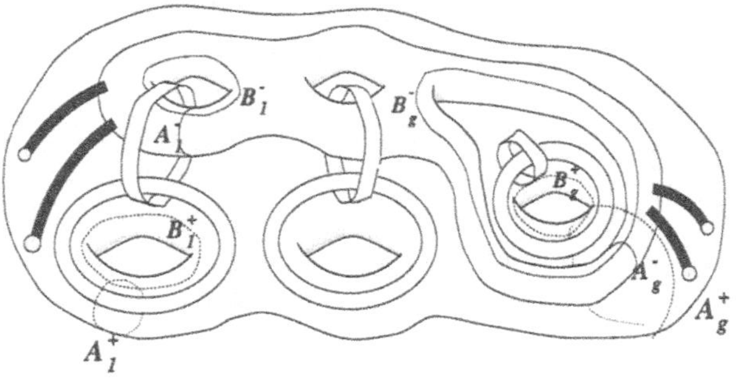

Fig. 3.14. Identity tangle and cobordism.

As a result we obtain a surgery diagram on $\Sigma_g \times [0,1]$. The external strands are identified there with the strips $J_j^{s/t} \times [0,1]$. Also, after the slides, the α_j-ribbons no longer surround the handles but only the ribbons $\beta_j \cup \gamma_j$ so that we can apply β-Moves to remove all internal surgery ribbons. The 3-dimensional cobordism that is represented by the identity tangle is, thus, $\Sigma_{g,a/b} \times [0,1]$, with canonical boundary identifications and, hence, the identity in $\widetilde{Cob}$. The signature extension given by this surgery diagram is also the correct one, since it is easily seen to be an idempotent.

The next lemma is the last missing piece in the proof of Theorem 3.0.6.

Lemma 3.4.3. *The map $\mathfrak{Surg}_2$ is horizontally functorial.*

Proof. As in the vertical case we work out the composition $W_2 \circ_h W_1$ for two 4-manifolds, as defined in Section 1.6.3.H, assuming that they are defined by admissible tangles T_1 and T_2. More precisely, we assign to each tangle T_k the surgery diagram $\mathcal{B}_{T_k} = (\mathcal{A}_k, \mathbf{b}_k)$ as in (3.3.4) and Proposition 3.2.3, and construct the W_k by attaching 4-dimensional handles to the thickened handlebody along $\mathcal{A}_k$ as follows:

$$W_k = \left(\mathcal{H}_{g_k,sc}^{g_k,tg} \times [f_0, f_1]\right) \bigcup_{\mathcal{A}_k \times f_1} \underbrace{\mathrm{e}_2^4 \sqcup \ldots \sqcup \mathrm{e}_1^4}_{N_k} \qquad \text{with } k = 1, 2.$$

Here N_k denotes the total number of attached handles, which is the number of closed ribbons plus the number of ball pairs in $\mathcal{A}_k$.

The first step in the construction of the horizontal composition is the attaching of $2b$ 1-handles e_1^4 to the union $W_2 \sqcup W_2$, which yields the 4-manifold $W_2 \Diamond_h W_1$ as described in (1.6.15). Since all of the attached 1-handles are of the form $\mathrm{e}_1^4 = \mathrm{e}_1^3 \times [f_0, f_1]$, we can use the same argument as in the previous proof for the vertical composition, and exchange the handle attachment with the multiplication by the interval $[f_0, f_1]$

$$W_2 \Diamond_h W_1 = \left(\mathcal{H}_\Diamond \times [f_0, f_1]\right) \bigcup_{(\mathcal{A}_1 \sqcup \mathcal{A}_2) \times f_1} \underbrace{\mathrm{e}_1^4 \sqcup \ldots \sqcup \mathrm{e}_2^4}_{N_1 + N_2}.$$

Here $\mathcal{H}_\Diamond = \mathcal{H}_{g_1,sc}^{g_1,tg} \sqcup \mathrm{e}_1^3 \sqcup \ldots \sqcup \mathrm{e}_1^3 \sqcup \mathcal{H}_{g_2,sc}^{g_2,tg}$ is the union of the two handlebodies with $2b$ 3-dimensional 1-handles attached, as given in (1.6.19). Recall that b of the 1-handles are attached between the $\mathcal{H}_{g_1,tg}^+$ and $\mathcal{H}_{g_2,tg}^+$ parts and the other b handles between $\mathcal{H}_{g_1,sc}^-$ and $\mathcal{H}_{g_2,sc}^-$.

As described in (1.6.19) we move the attachments of the opposite handle into the first 1-handle e^{first} that is attached to the target surfaces. In Figure 3.15 the resulting arrangement of the pieces entering the surgery diagram is schematically depicted.

For example, the annular region bounded by the outer two ellipses on the left indicates $\mathcal{H}_{g_1,tg}^+ - D^3 = \mathcal{H}_1^+ - D^3$ and contains $\mathcal{B}_1^+ = \mathcal{B}_{g_1,tg}^+$ as surgery diagram. We also depict the 1-handles e_1^3 attached to the $\mathcal{H}_{g_{tg}}^+ - D^3$-pieces by the bottom arcs and the ones attached to the $\mathcal{H}_{g_{sc}}^- - D^3$-pieces. Among the former, the first handles e^{first} is depicted as the box area in between the outer ellipses.

The diagram, thus, shows that pushing the attaching areas of the $\mathcal{H}_{g_{sc}}^- - D^3$ as well as the surrounding collar containing the tangles T_1 and T_2, as indicated by the arrows, yields the natural combination of the standard diagrams as in Figure 3.11, with $g_{1,sc} + g_{2,sc}$ internal ribbons α_j and γ_j and $a + 2b + c$ external ribbons. Furthermore, the tangles T_1 and T_2 continue the standard diagrams in juxtaposition inside e^{first}.

In Figure 3.15 we also indicated the b closed external strands $\mathcal{Y}_j$ as defined in Part $\circ$-3) of Section 1.6.3. They are pieced together from ribbons running parallelly

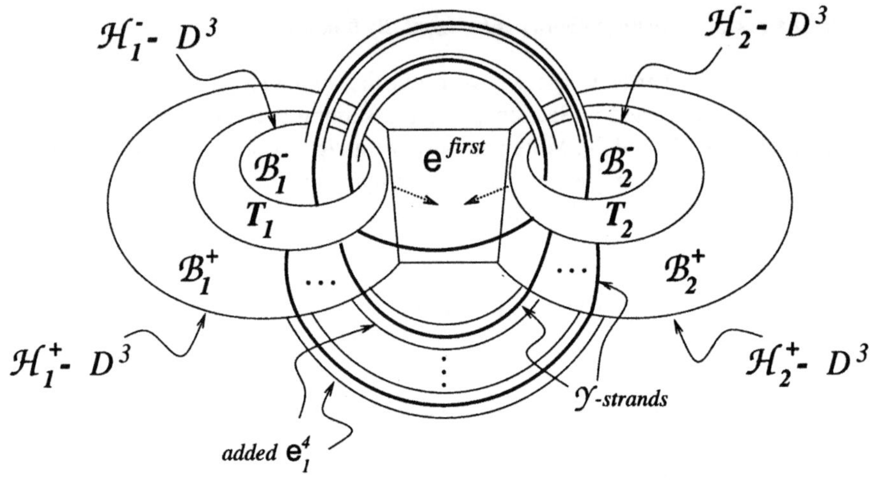

Fig. 3.15. $W_2 \Diamond_h W_1$ as handlebody over $\mathcal{H}_\Diamond$

through the added handles e_1^3, the intermediate straight external ribbons δ_j and ε_j in the standard presentations $\mathcal{B}^{\pm}_{1/2}$, and the b intermediate external ribbons in each T_k.

As the next step, let us add the b 4-dimensional 2-handles e_2^4 to the handlebody boundary piece of $W_2 \Diamond_h W_1$ as given in Part ∘-2) of Section 1.6.3. They are attached along curves $\mathcal{L}_1 \times f_0, \ldots, \mathcal{L}_b \times f_0$ in $\mathcal{H}_\Diamond \times [f_0, f_1] \subset W_2 \Diamond_h W_1$, where the $\mathcal{L}_j \subset \mathcal{H}_\Diamond$ are as depicted in Figure 1.22. Using the same notation as in (1.6.26) for $P = (\mathcal{H}_\Diamond \times [f_0, f_1]) \cup_{\mathcal{L}_1 \times f_0 \ldots \mathcal{L}_b \times f_0} (e_2^4 \cup \ldots \cup e_2^4)$ we can, hence, express the addition of the b 2-handles to the $\Diamond_h$-product as follows:

$$W_2 \Diamond_h W_1 \cup \underbrace{e_2^4 \cup \ldots \cup e_2^4}_{b} = P \bigcup_{(\mathcal{A}_1 \sqcup \mathcal{A}_2) \times f_1} \underbrace{e_1^4 \cup \ldots \cup e_2^4}_{N_1 + N_2} .$$

Recall now that $-P$ can be reexpressed by attaching 2-handles to $\mathcal{H}_o \times [f_0, f_1]$ along curves $\mathcal{L}_j^{opp} \times f_1$ at the opposite side of the thickening as in (1.6.27), where $\mathcal{H}_o = \mathcal{H}^{g_{1,tg}+g_{2,tg}+b-1}_{g_{1,sc}+g_{2,sc}+b-1}$ is the standard handlebody with the $b-1$ intermediate handles in nested instead of paired positions as depicted on the left of Figure 1.3. Putting the attaching curves in general position we can, thus, combine them to one diagram for 1- and 2-handle attachments along the $\mathcal{H}_o \times f_1$ side of the thickened handlebody as follows:

$$W_2 \Diamond_h W_1 \cup \underbrace{e_2^4 \cup \ldots \cup e_2^4}_{b} = (\mathcal{H}_o \times [f_0, f_1]) \bigcup_{(\mathcal{A}_1 \sqcup \mathcal{A}_2 \sqcup \mathcal{L}_1^{opp} \sqcup \ldots \sqcup \mathcal{L}_b^{opp}) \times f_1} \underbrace{e_1^4 \cup \ldots \cup e_2^4}_{N_1 + N_2 + b} . \tag{3.4.3}$$

In the combined surgery diagram $\mathcal{A}_1 \sqcup \mathcal{A}_2 \sqcup \mathcal{L}_1^{opp} \sqcup \ldots \sqcup \mathcal{L}_b^{opp}$ on $\mathcal{H}_o$, many but not all surgery strands are already in the desired standard position. At the handlebody, $\mathcal{H}^-_{g_{1,sc}+g_{2,sc}+b-1}$ in e^{first}, as depicted in the lower part of Figure 1.22, the diagrams

$\mathcal{A}_1$ and $\mathcal{A}_2$ provide the α_j and γ_j type ribbons for the $g_{1,sc}$ leftmost and $g_{2,sc}$ rightmost handles, as defined in Figure 3.11. Also the ribbons $\mathcal{L}_2^{opp}, \ldots, \mathcal{L}_b^{opp}$ are in positions of α type ribbons for the $b-1$ intermediate handles in nested positions. Note that $\mathcal{L}_2^{opp}$ surrounds the first added handle, which is used to connect the two handlebodies, but by itself does not contribute to the handles of $\mathcal{H}^+_{g_{1,sc}+g_{2,sc}+b-1}$ as indicated in Figure 1.3.

Thus, the only missing ingredient for a standard ribbon picture near $\mathcal{H}^-_{g_{1,sc}+g_{2,sc}+b-1}$ are the γ-type ribbons for the $b-1$ intermediate handles. Connected to the diagram so far is the juxtaposition of the tangles T_2 and T_1 from $\mathcal{A}_2$ and $\mathcal{A}_1$, and in the $\mathcal{H}^+_{g_{1,tg}+g_{2,tg}+b-1}$-part we have again the β-type ribbons for the $g_{1,tg}$ leftmost handles as well as for the $g_{2,tg}$ rightmost handles from $\mathcal{A}_2$ and $\mathcal{A}_1$. The corresponding ribbons for the $b-1$ intermediate handles are still missing.

The final step in the construction of the horizontal composition is the attachment of another b 2-handles e_2^4 to the "M_o-parts" as described in (1.6.23) of Section 1.6.3.H2. They are attached along the curves $\mathcal{Y}_j$, which are shrinkings of the tori $\mathcal{T}_j$ as defined in Part $\Diamond$-*3)* of Section 1.6.3.H1. They are attached to the 4-manifold in (3.4.3) to the $\mathcal{H}_o \times f_1$-side after the other 2-handles have been attached. However, in general position the $\mathcal{Y}_j$-curves will be disjoint from all of the other attaching curves and ball pairs so that the order of attachment does not matter anymore. As a result, we can describe the horizontal composite as one combined handle diagram on $\mathcal{H}_o$ as follows:

$$W_2 \circ_h W_1 = W_2 \Diamond_h W_1 \cup \underbrace{e_2^4 \cup \ldots \cup e_2^4}_{2b} \tag{3.4.4}$$

$$= \left(\mathcal{H}_o \times [f_0, f_1]\right) \bigcup_{(\mathcal{A}_1 \sqcup \mathcal{A}_2 \sqcup \mathcal{L}_1^{opp} \sqcup \ldots \sqcup \mathcal{L}_b^{opp} \sqcup \mathcal{Y}_1 \sqcup \ldots \sqcup \mathcal{Y}_b) \times f_1} \underbrace{e_1^4 \cup \ldots \cup e_2^4}_{N_1+N_2+2b} .$$

In Figure 3.15 the path of the $\mathcal{Y}$-strands is schematically included. In the $\mathcal{H}^+_{g_{1,tg}+g_{2,tg}+b-1}$-part of the diagram, the strand $\mathcal{Y}_{\beta^{-1}(j)}$ runs parallelly through the j-th added intermediate handle, with $j = 2, \ldots, b$. These $b-1$ strands, thus, have the positions of the missing β-type ribbons for the intermediate, nested handles. Note, however, that the strand $\mathcal{Y}_{\beta^{-1}(1)}$ runs through e^{first} and, hence, contributes to the juxtaposition of the tangles T_2 and T_1 rather than to the standard diagram.

The strand $\mathcal{Y}_{\beta^{-1}(1)}$ is readily identified with the ribbon C_{tg} in Figure 2.3. The other $b-1$ $\mathcal{Y}$-strands continue now as the other intermediate external strands in either T_2 and T_1. The fact that we attach a 2-handle e_2^4 to each of the $\mathcal{Y}$-strands, thus, implies that these external strands have to be interpreted as internal surgery strands. We have, thus, identified all ingredients in the lower target half of the diagram in Figure 2.3 as well as the target standard diagram.

As indicated in Figure 2.3 and defined in (1.6.20) the $\mathcal{Y}$-strands in the $\mathcal{H}^-$-part also run through the handles e_1^3 added to the opposite handlebodies. The path of each $\mathcal{Y}$ in this region in $\mathcal{H}_\Diamond$ is, therefore, as depicted in Picture (A) of Figure 3.16.

Here the box near the left handlebody indicates other parts of the link diagram. Similarly, there are further components present at the right handlebody, which we shall, however, not depict explicitly in order to keep the illustration simple.

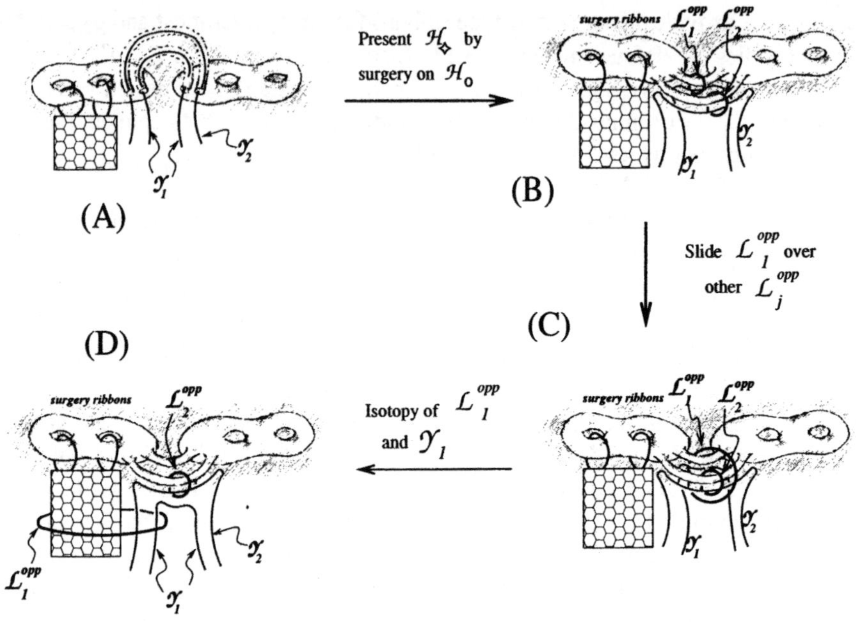

Fig. 3.16. Identifying $\mathcal{L}^{opp}$ and $\mathcal{Y}$ strands as α and γ strands.

In the desired presentation in (3.4.4) the $\mathcal{Y}$-strands are considered on $\mathcal{H}_\circ$. In order to determine their positions there we think of them as running very close to the boundaries ∂e_1^3 of the added handles, and insert them along these surfaces in the presentation of $\mathcal{H}_\diamond$ by surgery along the $\mathcal{L}^{opp}$ in $\mathcal{H}_\circ$ as in Figure 1.23.

As depicted in Picture (B) of Figure 3.16 each $\mathcal{Y}_j$ then runs along the cut-out handles in $\mathcal{H}_\circ$ linking with a corresponding ribbon $\mathcal{L}_j^{opp}$. By slightly deforming the $\mathcal{Y}_j$-strand into loops and identifying the $\mathcal{L}_j^{opp}$ with α-type ribbons it is not hard to see that they attain precisely the desired positions for the missing β-type strands for the nested intermediate handles for $j = 2, \ldots, b$.

We have now identified all ingredients for the standard link diagram near $\mathcal{H}^-_{g_{1,\infty}+g_{2,\infty}+b-1}$, but we also still have additional components that are not part of a standard diagram. Specifically, these are the strands $\mathcal{L}_1^{opp}$ and $\mathcal{Y}_1$. We slide and isotope these from the handlebody region to the tangle region as follows.

As indicated in the step from Picture (B) to Picture (C) we first slide $\mathcal{L}_1^{opp}$ over each of the $\mathcal{L}_j^{opp}$, for $j = 2, \ldots, b$ one-by-one. Since the framing of the ribbons is normal to the handle surfaces, the effect of the sliding is that $\mathcal{L}_1^{opp}$ surrounds each handle after the slides and, eventually, the entire collection of handles.

In the step from Picture (C) to (D) it is shown how we can pull the entire ribbon $\mathcal{L}_1^{opp}$ over the left handlebody and subsequently pull down over the attached tangle diagram. Moreover, since the $\mathcal{Y}_1$-ribbon is now no longer "chained" to the first handle, we can pull it down below all of the other handles and, thus, separate it also from the link diagram near $\mathcal{H}^-_{g_{1,\infty}+g_{2,\infty}+b-1}$.

The diagram in the vicinity of the handlebody is now a pure standard diagram as desired. Furthermore, the loop $\mathcal{L}_1^{opp}$ that is slid over the tangle T_1 is immediately identified with the annulus A in Figure 2.3, and the piece of the strand $\mathcal{Y}_1$ that is pushed into the tangle diagram, clearly, corresponds to the arc $C_{\mathfrak{x}}$. In summary, the presentation from (3.4.4) can be rewritten as

$$W_2 \circ_h W_1 = \left(\mathcal{H}_o \times [f_0, f_1]\right) \bigcup_{(\mathcal{A}_{T_2 \Delta_h T_1}) \times f_1} \underbrace{\mathrm{e}_1^4 \cup \ldots \cup \mathrm{e}_2^4}_{N_1+N_2+2b} .$$

Finally, the transition to a standard handlebody with all paired instead of nested handles is obtained by applying the homeomorphism from (1.1.5). From the diagram in Figure 1.3 it is clear that the suggested standard presentation for nested handles is mapped to the standard presentation for paired handles as in Figures 3.10 and 3.11. The homeomorphism for the handle rearrangement is also easily identified with the braids $\bar{U}_b$ from (2.6.6) that are adjoined to the product $T_2 \Delta_h T_1$ to yield $T_2 \circ_h T_1$.

Hence, we have $W_2 \circ_h W_1 = W_{T_2 \circ_h T_1}$, and also the $a + c$ remaining external strands of $T_2 \circ_h T_1$ provide the braid markings needed in the horizontal composition of 3-cobordisms. The 2-arrow represented by $W_2 \circ_h W_1$ together with the external braids is, thus, the same as the one associated to $T_2 \circ_h T_1$. This completes the proof of horizontal functoriality.

4. Monoidal categories and monoidal 2-categories

In this chapter we provide the definitions of various algebraic structures that will be the building blocks for our construction of TQFT functors. In particular, we will review braided abelian tensor categories (BTC), the properties of Hopf algebras in such BTC's, as well as the construction of a symmetric monoidal 2-category of abelian categories. Our discussion will also include a number of new lemmas that will significantly simplify the proof of topological invariance in Chapter 6.

More specifically, in the discussion of braided tensor categories we give a detailed discussion of rigidity in BTC's, the various ribbon and balancing elements, and their relations. As a result, we obtain that any BTC is equivalent to one, which is strictly rigid, i.e., we have $X = X^{\vee\vee}$ and the canonical balancing is just the identity.

In our review of Hopf algebras in BTC's the results of the theory of their integrals, given as morphisms $\int : \mathrm{Int} \to F$ in the same BTC, are summarized. In particular, we discuss their existence, uniqueness, the invertibility of the object of integrals Int, and duality properties. We also include several criteria for the non-degeneracy of Hopf pairings.

In the last section of this chapter we recall the basic definitions and properties of Deligne's tensor product $\boxtimes$ for abelian categories. We first consider only the 2-category of categories of modules over finite dimensional algebras inside a strict version of the category of vector spaces, for which we ensure that the 2-braiding induces a strict action of the symmetric group S_N on the multifold tensor products $\mathcal{C}_1 \boxtimes \mathcal{C}_2 \boxtimes \ldots \boxtimes \mathcal{C}_N$ of categories of modules.

The TQFT targeted into this model semistrict 2-category is equivalent to a double pseudofunctor with values in the the weak symmetric monoidal 2-category of ordinary abelian categories.

Of great use for computations are graphical calculi that we also introduce in this chapter. One calculus applies to the natural morphisms of a BTC (thin tangles) and the other to the operations of a Hopf algebra in a BTC (thick braided graphs).

4.1 Ribbon monoidal categories

Lemma 1.3.1 states that some topological category which we have to represent is a braided monoidal category (defined below). Furthermore, it is rigid and possesses a balancing as shown in Lemma 1.4.6. Therefore, the target data should include a balanced braided monoidal category $\mathcal{C}$ as an input. Since we are considering a TQFT-like functor, it is natural to assume that $\mathcal{C}$ is additive and $\Bbbk$-linear. Moreover, we assume that $\mathcal{C}$ is abelian, which means existence and good behavior of kernels and cokernels of morphisms, hence, images, quotients etc. Such an assumption is justified when TQFT's are viewed as double functors, since the only suitable for us example of a monoidal 2-category is that of $\Bbbk$-linear abelian categories. Also it is shown in Appendix (Theorem C.2.1) that we have to work only with *bounded* categories, that is, categories equivalent to categories of modules over a finite-dimensional associative $\Bbbk$-algebra.

In this section we recall definitions of various kinds of categories used in the sequel. In a nutshell, a *ribbon* (also *tortile* [Shu94] or *balanced*) category is a braided monoidal category $\mathcal{C}$ [JS91] with a tensor product, $\otimes$, an associativity isomorphism $\mathbf{a}_{XYZ} : X \otimes (Y \otimes Z) \to (X \otimes Y) \otimes Z, \mathbf{a} : \otimes \circ (\mathrm{Id} \times \otimes) \to \otimes \circ (\mathrm{Id} \times \otimes)$, a braiding (commutativity) isomorphism $\mathbf{c}_{X,Y} : X \otimes Y \to Y \otimes X, \mathbf{c} : \otimes \to \otimes \circ P$, and a unit object 1. We also require $\mathcal{C}$ to be rigid, meaning, that for any object $X \in \mathcal{C}$ there are dual objects ${}^\vee X$ and $X^\vee$ with evaluations $\mathrm{ev}' : {}^\vee X \otimes X \to 1$, $\mathrm{ev} : X \otimes X^\vee \to 1$ and coevaluations $\mathrm{coev}' : 1 \to X \otimes {}^\vee X$, $\mathrm{coev} : 1 \to X^\vee \otimes X$. Finally, it possesses a balancing (defined below). The morphisms are required to satisfy standard equations: pentagon equation for associativity, 2 hexagon equations for braiding and associativity, 4 rigidity equations for evaluations and coevaluations etc.

4.1.1 Rigid monoidal categories

We recall here the basic definitions of monoidal categories, monoidal functors and dual objects. For a more detailed exposition on the subject see [Mac88].

Definition 4.1.1. *A* monoidal category $(\mathcal{C}, \otimes, \mathbf{a}, 1, \mathbf{l}, \mathbf{r})$ *is a category* $\mathcal{C}$*, a functor* $\otimes : \mathcal{C} \times \mathcal{C} \to \mathcal{C}$ *(called the* tensor product*), a functorial isomorphism* $\mathbf{a} : X \otimes (Y \otimes Z) \to (X \otimes Y) \otimes Z$ *the associativity constraint, a unit object* 1 *and two functorial isomorphisms* $\mathbf{l} : 1 \otimes X \to X$, $\mathbf{r} : X \otimes 1 \to X$*, such that*

$$\begin{array}{ccccc} X \otimes (Y \otimes (Z \otimes W)) & \xrightarrow{\mathbf{a}} & (X \otimes Y) \otimes (Z \otimes W) & \xrightarrow{\mathbf{a}} & ((X \otimes Y) \otimes Z) \otimes W \\ \downarrow{\scriptstyle X \otimes \mathbf{a}} & & & & \uparrow{\scriptstyle \mathbf{a} \otimes W} \\ X \otimes ((Y \otimes Z) \otimes W) & & \xrightarrow{\qquad \mathbf{a} \qquad} & & (X \otimes (Y \otimes Z)) \otimes W \end{array}$$

commutes (the pentagon equation) and

$$\mathbf{a}_{X,1,Y} = \left(X \otimes (1 \otimes Y) \xrightarrow{X \otimes \mathbf{l}_Y} X \otimes Y \xrightarrow{\mathbf{r}_X^{-1} \otimes Y} (X \otimes 1) \otimes Y\right).$$

Notice that in a monoidal category the equations

$$\mathbf{a}_{1,X,Y} = \left(1 \otimes (X \otimes Y) \xrightarrow{\mathbf{l}} X \otimes Y \xrightarrow{\mathbf{l}_X^{-1} \otimes Y} (1 \otimes X) \otimes Y\right),$$

$$\mathbf{a}_{X,Y,1} = \left(X \otimes (Y \otimes 1) \xrightarrow{X \otimes \mathbf{r}_Y} X \otimes Y \xrightarrow{\mathbf{r}^{-1}} (X \otimes Y) \otimes 1\right)$$

hold as well.

Definition 4.1.2. *A monoidal functor* $(F, \phi, \mathcal{F}) : (\mathcal{C}, \otimes) \to (\mathcal{D}, \otimes)$ *is a functor* $F : \mathcal{C} \to \mathcal{D}$, *a functorial isomorphism* $\phi = \phi_{X,Y} : F(X) \otimes F(Y) \to F(X \otimes Y) \in \mathcal{D}$ *and an isomorphism* $\mathcal{F} : 1 \to F1 \in \mathcal{D}$, *such that*

$$\begin{array}{ccccc} FX \otimes (FY \otimes FZ) & \xrightarrow{1 \otimes \phi} & FX \otimes F(Y \otimes Z) & \xrightarrow{\phi} & F(X \otimes (Y \otimes Z)) \\ \downarrow{\scriptstyle \mathbf{a}} & & & & \downarrow{\scriptstyle F\mathbf{a}} \\ (FX \otimes FY) \otimes FZ & \xrightarrow{\phi \otimes 1} & F(X \otimes Y) \otimes Z & \xrightarrow{\phi} & F((X \otimes Y) \otimes Z) \end{array} \tag{4.1.1}$$

$$\begin{array}{ccc} F1 \otimes FX & \xrightarrow{\phi} & F(1 \otimes X) \\ {\scriptstyle \mathcal{F} \otimes 1}\uparrow & & \downarrow{\scriptstyle F\mathbf{l}} \\ 1 \otimes FX & \xrightarrow{\mathbf{l}} & FX \end{array} \quad , \quad \begin{array}{ccc} FX \otimes F1 & \xrightarrow{\phi} & F(X \otimes 1) \\ {\scriptstyle 1 \otimes \mathcal{F}}\uparrow & & \downarrow{\scriptstyle F\mathbf{r}} \\ FX \otimes 1 & \xrightarrow{\mathbf{r}} & FX \end{array}$$

commute. A morphism of monoidal functors $\lambda : (F, \phi, \mathbf{f}) \to (G, \psi, \mathbf{g})$ *is a functorial morphism* $\lambda : F \to G$ *such that*

$$\begin{array}{ccc} FX \otimes FY & \xrightarrow{\phi} & F(X \otimes Y) \\ {\scriptstyle \lambda \otimes \lambda}\downarrow & & \downarrow{\scriptstyle \lambda} \\ GX \otimes GY & \xrightarrow{\psi} & G(X \otimes Y) \end{array}$$

$$g = (1 \xrightarrow{\mathbf{f}} F1 \xrightarrow{\lambda} G1).$$

The $\mathbf{f}$ datum of a monoidal functor $(F, f, \mathbf{f})$ is uniquely determined by the (F, f) data, so we often denote a monoidal functor as (F, f) or even F.

Definition 4.1.3. *A rigid category* $\mathcal{C}$ *is a monoidal category, in which to every object* $X \in \mathcal{C}$ *dual objects* $X^\vee$ *and* ${}^\vee X \in \mathcal{C}$ *are assigned together with morphisms of evaluation and coevaluation*

$$\mathrm{ev}_X : X \otimes X^\vee \to 1 = {}_X\bigcup{}_{X^\vee} \ , \ \mathrm{ev}'_X \ : {}^\vee X \otimes X \to 1 = {}_{{}^\vee X}\bigcup{}_X \ , \tag{4.1.2}$$

$$\mathrm{coev}_X : 1 \to X^\vee \otimes X = {}^{X^\vee}\bigcap{}^X \ , \ \mathrm{coev}'_X : 1 \to X \otimes {}^\vee X = {}^X\bigcap{}^{{}^\vee X} \ . \tag{4.1.3}$$

The evaluations and coevaluations are chosen such that the compositions

$$X \xrightarrow{\mathbf{r}^{-1}} X \otimes 1 \xrightarrow{1 \otimes \mathrm{coev}} X \otimes (X^\vee \otimes X) \xrightarrow{\mathbf{a}} (X \otimes X^\vee) \otimes X \xrightarrow{\mathrm{ev} \otimes 1} 1 \otimes X \xrightarrow{\mathbf{l}} X,$$

$$X \xrightarrow{\mathbf{l}^{-1}} 1 \otimes X \xrightarrow{\mathrm{coev}' \otimes 1} (X \otimes {}^\vee X) \otimes X \xrightarrow{\mathbf{a}^{-1}} X \otimes ({}^\vee X \otimes X) \xrightarrow{1 \otimes \mathrm{ev}'} X \otimes 1 \xrightarrow{\mathbf{r}} X,$$

$$X^\vee \xrightarrow{\mathbf{l}^{-1}} 1 \otimes X^\vee \xrightarrow{\mathrm{coev}\otimes 1} (X^\vee \otimes X) \otimes X^\vee \to$$
$$\xrightarrow{\mathbf{a}^{-1}} X^\vee \otimes (X \otimes X^\vee) \xrightarrow{1\otimes \mathrm{ev}} X^\vee \otimes 1 \xrightarrow{\mathbf{r}} X^\vee,$$

$$^\vee X \xrightarrow{\mathbf{r}^{-1}} {}^\vee X \otimes 1 \xrightarrow{1\otimes \mathrm{coev}'} {}^\vee X \otimes (X \otimes {}^\vee X) \to$$
$$\xrightarrow{\mathbf{a}} ({}^\vee X \otimes X) \otimes {}^\vee X \xrightarrow{\mathrm{ev}'\otimes 1} 1 \otimes {}^\vee X \xrightarrow{\mathbf{l}} {}^\vee X$$

are all identity morphisms.

By the *opposite tensor product* we mean $\otimes_{\mathrm{op}} = \otimes \circ P : \mathcal{C} \times \mathcal{C} \to \mathcal{C}$, where $P : \mathcal{C} \times \mathcal{C} \to \mathcal{C} \times \mathcal{C}$, $(X, Y) \mapsto (Y, X)$ is the permutation functor.

In a rigid monoidal category $\mathcal{C}$ there is a pairing

$$(X \otimes Y) \otimes (Y^\vee \otimes X^\vee) \xrightarrow{\sim} (X \otimes (Y \otimes Y^\vee)) \otimes X^\vee \to$$
$$\xrightarrow{X\otimes \mathrm{ev}\otimes X^\vee} (X \otimes 1) \otimes X^\vee \xrightarrow{\mathbf{r}\otimes X^\vee} X \otimes X^\vee \xrightarrow{\mathrm{ev}} 1,$$

which induces an isomorphism $j_{+X,Y} : Y^\vee \otimes X^\vee \to (X \otimes Y)^\vee$, such that the above pairing coincides with

$$(X \otimes Y) \otimes (Y^\vee \otimes X^\vee) \xrightarrow{1\otimes j_+} (X \otimes Y) \otimes (X \otimes Y)^\vee \xrightarrow{\mathrm{ev}} 1.$$

The equation

$$\mathrm{coev}_{X\otimes Y} = (1 \xrightarrow{\mathrm{coev}_Y} Y^\vee \otimes Y \simeq Y^\vee \otimes 1 \otimes Y \xrightarrow{1\otimes \mathrm{coev}_X \otimes 1}$$
$$Y^\vee \otimes X^\vee \otimes X \otimes Y \xrightarrow{j_+\otimes 1} (X \otimes Y)^\vee \otimes (X \otimes Y))$$

also holds. Similarly, there is an isomorphism $j_{-X,Y} : {}^\vee Y \otimes {}^\vee X \to {}^\vee(X \otimes Y)$.

There are unique isomorphisms $d : 1 \to 1^\vee$, $d_- : 1 \to {}^\vee 1$, such that

$$\mathbf{r}_1 = (1 \otimes 1 \xrightarrow{1\otimes d} 1 \otimes 1^\vee \xrightarrow{\mathrm{ev}} 1),$$
$$\mathbf{r}_1 = (1 \otimes 1 \xrightarrow{d_-\otimes 1} {}^\vee 1 \otimes 1 \xrightarrow{\mathrm{ev}} 1).$$

Morphisms constructed from braidings and (co)evaluations are often described by tangles. The conventions of [Lyu95a] are listed in Fig. 4.1. The suggested assignment of morphisms in $\mathcal{C}$ to elementary pictures extends to a unique functor Φ from the category of $\mathcal{C}$-colored tangles to the category $\mathcal{C}$ itself [FY92]. With the above interpretation these tangles need not be oriented. Later we shall use the same notation for framed tangles, and the framing will be within the plane.

The maps $\mathrm{Ob}\,\mathcal{C} \to \mathrm{Ob}\,\mathcal{C}$, $X \mapsto X^\vee$ and $X \mapsto {}^\vee X$ extend to contravariant self-equivalences $\mathcal{C} \to \mathcal{C}$, $f \mapsto f^t$ and $f \mapsto {}^t f$. For given f the morphisms f^t and ${}^t f$ can be defined by the following pictures using the assignment from Fig. 4.1.

Fig. 4.1. We denote

a morphism $f : X \to Y$ by

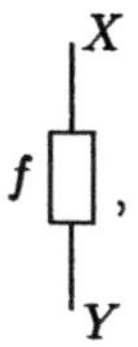

the braiding $c_{X,Y} : X \otimes Y \to Y \otimes X$ by

the inverse braiding $c^{-1} : X \otimes Y \to Y \otimes X$ by

the evaluation $ev_X : X \otimes X^\vee \to 1$ by

the coevaluation $coev_X : 1 \to X^\vee \otimes X$ by

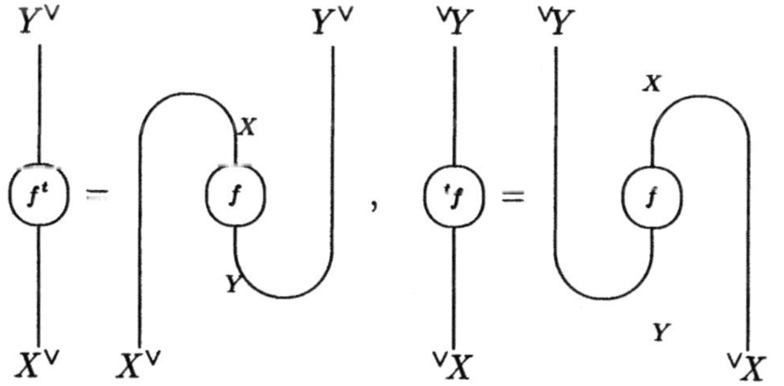

(4.1.4)

It follows easily from the axioms of rigidity that

$$(\text{-}^\vee, j_+, d) : (\mathcal{C}^{op}, \otimes, 1) \to (\mathcal{C}, \otimes_{op}, 1), \quad X \mapsto X^\vee, \quad f \mapsto f^t,$$
$$({}^\vee\text{-}, j_-, d_-) : (\mathcal{C}^{op}, \otimes, 1) \to (\mathcal{C}, \otimes_{op}, 1), \quad X \mapsto {}^\vee X, \quad f \mapsto {}^t f$$

are monoidal equivalences of categories. The "square" of the first one

$$(\text{-}^{\vee\vee}, j_{+2}, d_2) : (\mathcal{C},\otimes,1) \to (\mathcal{C},\otimes,1),\ X\mapsto X^{\vee\vee},\ f\mapsto f^{tt}, \tag{4.1.5}$$
$$j_{+2X,Y} = \big(X^{\vee\vee}\otimes Y^{\vee\vee} \xrightarrow{j_+} (Y^\vee\otimes X^\vee)^\vee \xrightarrow{j_+^{-1t}} (X\otimes Y)^{\vee\vee}\big),$$
$$d_2 = \big(1 \xrightarrow{d} 1^\vee \xrightarrow{d^{-1t}} 1^{\vee\vee}\big),$$

is a monoidal self-equivalence of $\mathcal{C}$.

To simplify notations we assume that the functors $\text{-}^\vee$ and $^\vee\text{-}$ are inverses of each other. We can always achieve this by replacing the category $\mathcal{C}$ by an equivalent one, such that the canonical isomorphisms

$$X \to {}^\vee(X^\vee), \qquad X \to ({}^\vee X)^\vee$$

become identity. Finally, we denote the iterated duals by $X^{(n\vee)} = X^{\vee\cdots\vee}$ (n times) and $X^{(-n\vee)} = {}^{\vee\cdots\vee}X$ (n times) for $n \geqslant 0$.

4.1.2 Braided categories

Here we review the definitions of the braiding isomorphism and further derived isomorphisms. Several basic relations between them are listed. Two important classes of examples of braided categories are given by the categories of modules over quasitriangular Hopf algebras and the categories of tangles. Both classes and their relations are discussed in detail in the book of Kassel, Rosso and Turaev [KRT97].

Definition 4.1.4. *A* braided category $(\mathcal{C},\mathbf{c})$ *is a monoidal category $\mathcal{C}$ equipped with a functorial isomorphism $\mathbf{c} = \mathbf{c}_{X,Y} : X\otimes Y \to Y\otimes X$ – the braiding, or the commutativity constraint – such that the two hexagons commute*

$$\begin{array}{ccccc}
X\otimes(Y\otimes Z) & \xrightarrow{1\otimes\mathbf{c}^{\pm1}} & X\otimes(Z\otimes Y) & \xrightarrow{\mathbf{a}} & (X\otimes Z)\otimes Y \\
\Big\downarrow{\scriptstyle \mathbf{a}} & & & & \Big\downarrow{\scriptstyle \mathbf{c}^{\pm1}\otimes 1} \\
(X\otimes Y)\otimes Z & \xrightarrow{\mathbf{c}^{\pm1}} & Z\otimes(X\otimes Y) & \xrightarrow{\mathbf{a}} & (Z\otimes X)\otimes Y
\end{array}$$

(one for $\mathbf{c}$ *and one for* $\mathbf{c}^{-1}$*).*

The graphical notation for the braiding and its inverse is

$\mathbf{c} = (\mathbf{c}_{X,Y} : X\otimes Y \to Y\otimes X) =$ 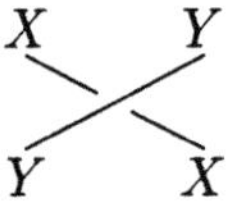

, $\mathbf{c}^{-1} =$ 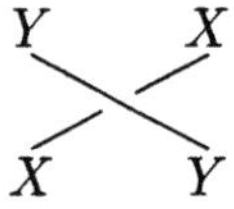

.

In a braided category the following equations hold [Sch92]

$$\mathbf{c}_{X,1} = \big(X\otimes 1 \xrightarrow{\mathbf{r}} X \xrightarrow{\mathbf{l}^{-1}} 1\otimes X\big),$$
$$\mathbf{c}_{1,X} = \big(1\otimes X \xrightarrow{\mathbf{l}} X \xrightarrow{\mathbf{r}^{-1}} X\otimes 1\big).$$

In a rigid braided category we can define [Lyu95a] functorial isomorphisms using again the conventions from Figure 4.1:

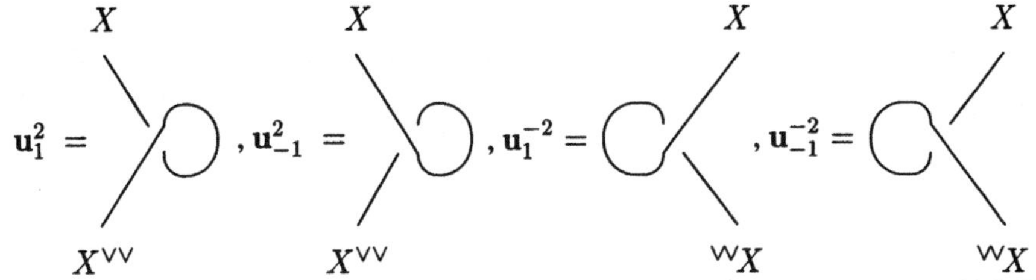

(4.1.6)

The meaning of these isomorphisms in the case of quasitriangular Hopf algebras is explained in Sect. 7.4.4.

There are isomorphisms of monoidal functors (see (4.1.5))

$$\begin{aligned} \mathbf{u}_1^2 : (\mathrm{Id}, \mathbf{c}^{-2}, \mathbb{I}_1) &\longrightarrow (\text{-}^{\vee\vee}, j_2, d_2), \\ \mathbf{u}_{-1}^2 : (\mathrm{Id}, \mathbf{c}^{2}, \mathbb{I}_1) &\longrightarrow (\text{-}^{\vee\vee}, j_2, d_2). \end{aligned}$$

In particular, this implies the commutativity of the diagram

$$\begin{array}{ccc} X \otimes Y & \xrightarrow{\mathbf{c}^{-2}} & X \otimes Y \\ {\scriptstyle \mathbf{u}_1^2 \otimes \mathbf{u}_1^2}\big\downarrow & & \big\downarrow{\scriptstyle \mathbf{u}_1^2} \\ X^{\vee\vee} \otimes Y^{\vee\vee} & \xrightarrow{j_2} & (X \otimes Y)^{\vee\vee} \end{array} \tag{4.1.7}$$

The square of the monoidal functor $(\text{-}^{\vee\vee}, j_2, d_2)$ is

$$(\text{-}^{\vee\vee\vee\vee}, j_4, d_4) : (\mathcal{C}, \otimes, 1) \longrightarrow (\mathcal{C}, \otimes, 1), \qquad X \mapsto X^{\vee\vee\vee\vee}, \quad f \mapsto f^{tttt},$$

where $j_{4X,Y} = \left(X^{\vee\vee\vee\vee} \otimes Y^{\vee\vee\vee\vee} \xrightarrow{j_2} (X^{\vee\vee} \otimes Y^{\vee\vee})^{\vee\vee} \xrightarrow{j_2^{tt}} (X \otimes Y)^{\vee\vee\vee\vee}\right)$, $d_4 = \left(1 \xrightarrow{d_2} 1^{\vee\vee} \xrightarrow{d_2^{tt}} 1^{\vee\vee\vee\vee}\right)$. The natural isomorphism $\mathbf{u}_0^4 = \mathbf{u}_{-1}^2 \circ \mathbf{u}_1^2$ is, in fact, an isomorphism of monoidal functors $\mathbf{u}_0^4 : (\mathrm{Id}, \mathbb{I}, \mathbb{I}) \to (\text{-}^{\vee\vee\vee\vee}, j_4, d_4)$.

4.1.3 Ribbon categories

Now we define balancing and recall some properties of balanced (ribbon) categories.

Definition 4.1.5. *Let $\mathcal{C}$ be a rigid braided category.* A balancing $\beta_X : X \to X^{\vee\vee}$ *is an isomorphism of monoidal functors $\beta : (\mathrm{Id}, \mathbb{I}, \mathbb{I}) \to (\text{-}^{\vee\vee}, j_2, d_2)$, such that $\beta^2 = \mathbf{u}_0^4$ and $\beta_X^t = \beta_{X^\vee}^{-1} : X^{\vee\vee\vee} \to X^\vee$. The category $\mathcal{C}$ equipped with a balancing is called* balanced.

We sometimes also use the notation $\mathbf{u}_0^2 = \beta$. In any balanced category there exists a canonical ribbon twist $\mathbf{v}$. A ribbon twist [JS91, RT90, Shu94] $\mathbf{v} = \mathbf{v}_X : X \to X$, $\mathbf{v} : \mathrm{Id} \to \mathrm{Id}$ is a self-adjoint ($\mathbf{v}_{X^\vee} = \mathbf{v}_X^t$) automorphism of the identity functor such that $\mathbf{c}^2 = (\mathbf{v}_X^{-1} \otimes \mathbf{v}_Y^{-1}) \circ \mathbf{v}_{X \otimes Y}$. It can be determined from the equations

$$\mathbf{u}_0^2 = \mathbf{u}_1^2 \circ \mathbf{v}^{-1} = \mathbf{u}_{-1}^2 \circ \mathbf{v} : X \to X^{\vee\vee},$$
$$\beta^{-1} = \mathbf{u}_0^{-2} = \mathbf{u}_1^{-2} \circ \mathbf{v}^{-1} = \mathbf{u}_{-1}^{-2} \circ \mathbf{v} : X \to {}^{\vee\vee}X.$$

In particular, its square is given by the canonical isomorphism $\mathbf{v}^2 = \mathbf{u}_1^{-2} \circ \mathbf{u}_1^2$. Vice versa, in any rigid braided category with a ribbon twist (called *ribbon category*) there exists a canonical balancing $\mathbf{u}_0^2$ given by the above formulae. Thus, ribbon categories and balanced categories are synonyms.

In the case of $X = 1$ we have that $\beta_1 = d_2 : 1 \to 1^{\vee\vee}$ and $\mathbf{v}_1 = \mathbb{1}_1$.

The following results are used to simplify notations through this book.

Definition 4.1.6. *A ribbon category $\mathcal{D}$ is called strictly rigid if*

(a) $1^\vee = 1$, $d_1 = \mathbb{1}_1 : 1 \to 1^\vee =\!=\!= 1$;

(b) for any object X we have ${}^\vee X = X^\vee$, $X^{\vee\vee} = X$, and $\beta_X = \mathbb{1}_X : X \to X^{\vee\vee} =\!=\!= X$.

(c) for any object X we have $\mathrm{ev}_X = \mathrm{ev}'_{X^\vee} : X \otimes X^\vee \to 1$, and $\mathrm{coev}_X = \mathrm{coev}'_{X^\vee} : 1 \to X^\vee \otimes X$.

Theorem 4.1.7. *For any ribbon category $\mathcal{C}$ there exists a ribbon strictly rigid category $\mathcal{D}$ equivalent to $\mathcal{C}$.*

Proof. First of all, we can replace $\mathcal{C}$ by an equivalent category satisfying (a), so we assume that (a) holds for $\mathcal{C}$.

Set $\mathrm{Ob}\,\mathcal{D} = \mathrm{Ob}\,\mathcal{C} \times \{0,1\}$. Morphisms of $\mathcal{D}$ are defined as

$$\mathcal{D}((X,0),(Y,0)) = \mathcal{C}(X,Y), \qquad \mathcal{D}((X,0),(Y,1)) = \mathcal{C}(X,Y^\vee),$$
$$\mathcal{D}((X,1),(Y,0)) = \mathcal{C}(X^\vee,Y), \qquad \mathcal{D}((X,1),(Y,1)) = \mathcal{C}(X^\vee,Y^\vee).$$

Unit morphisms and the composition in $\mathcal{D}$ are inherited from $\mathcal{C}$. The inclusion functor $\mathcal{I} : \mathcal{C} \hookrightarrow \mathcal{D}$, $X \mapsto (X,0)$ and the projection functor $\mathcal{G} : \mathcal{D} \to \mathcal{C}$, $(X,0) \mapsto X$, $(X,1) \mapsto X^\vee$ are quasi-inverse to each other, $\lambda : \mathcal{I} \circ \mathcal{G} \to \mathrm{Id}_\mathcal{D}$ given by $\lambda_{(X,0)} = \mathbb{1}_X$, $\lambda_{(X,1)} = \mathbb{1}_{X^\vee}$. Therefore, $\mathcal{D}$ is equivalent to $\mathcal{C}$. Notice that $\mathcal{D}(L,M) = \mathcal{C}(\mathcal{G}L,\mathcal{G}M)$.

The tensor product in $\mathcal{D}$ is defined as

$$\otimes' : \mathcal{D} \times \mathcal{D} \xrightarrow{\mathcal{G}\times\mathcal{G}} \mathcal{C} \times \mathcal{C} \xrightarrow{\otimes} \mathcal{C} \xrightarrow{\mathcal{I}} \mathcal{D}.$$

The associativity is chosen as $\mathbf{a}_{L,M,N} = \mathbf{a}_{\mathcal{G}L,\mathcal{G}M,\mathcal{G}N}$ for $L,M,N \in \mathrm{Ob}\,\mathcal{D}$. The unit object is $(1,0)$, and the corresponding isomorphisms are $\mathbf{r}_M = \lambda^{-1} \circ \mathbf{r}_{\mathcal{G}M}$, $\mathbf{l}_M = \lambda^{-1} \circ \mathbf{l}_{\mathcal{G}M}$. The functors

$$(\mathcal{I},\mathbb{1},\mathbb{1}) : (\mathcal{C},\otimes,1) \longrightarrow (\mathcal{D},\otimes',(1,0)),$$
$$(\mathcal{G},\mathbb{1},\mathbb{1}) : (\mathcal{D},\otimes',(1,0)) \longrightarrow (\mathcal{C},\otimes,1)$$

are monoidal equivalences between $\mathcal{C}$ and $\mathcal{D}$. The braiding in $\mathcal{D}$ is chosen as $\mathbf{c}_{M,N} = \mathbf{c}_{\mathcal{G}M,\mathcal{G}N}$ for $M,N \in \mathrm{Ob}\,\mathcal{D}$. The functors $\mathcal{I}$, $\mathcal{G}$ preserve braided structures.

We choose the following rigid structure of $\mathcal{D}$:

$$(X,0)^\vee = (X,1) = {}^\vee(X,0), \qquad (X,1)^\vee = (X,0) = {}^\vee(X,1),$$

$$\mathrm{ev} : (X,0)\otimes(X,1) = (X\otimes X^\vee,0) \xrightarrow{\mathrm{ev}} (1,0),$$

$$\mathrm{coev} : (1,0) \xrightarrow{\mathrm{coev}} (X^\vee\otimes X,0) = (X,1)\otimes(X,0),$$

$$\mathrm{ev} : (X,1)\otimes(X,0) = (X^\vee\otimes X,0) \xrightarrow{X^\vee\otimes\beta} (X^\vee\otimes X^{\vee\vee},0) \xrightarrow{\mathrm{ev}} (1,0), \tag{4.1.8}$$

$$\mathrm{coev} : (1,0) \xrightarrow{\mathrm{coev}} (X^{\vee\vee}\otimes X^\vee,0) \xrightarrow{\beta^{-1}\otimes X^\vee} (X\otimes X^\vee,0) = (X,0)\otimes(X,1). \tag{4.1.9}$$

Calculating $\mathbf{u}^2_{\pm1} : M \to M$, $M \in \mathrm{Ob}\,\mathcal{D}$ with the above duality morphisms we obtain $\mathbf{u}^2_1 = \mathbf{v}$, $\mathbf{u}^2_{-1} = \mathbf{v}^{-1}$, where $\mathbf{v}\big|_{(X,0)} = \mathbf{v}_X$, $\mathbf{v}\big|_{(X,1)} = \mathbf{v}_{X^\vee}$. Indeed,

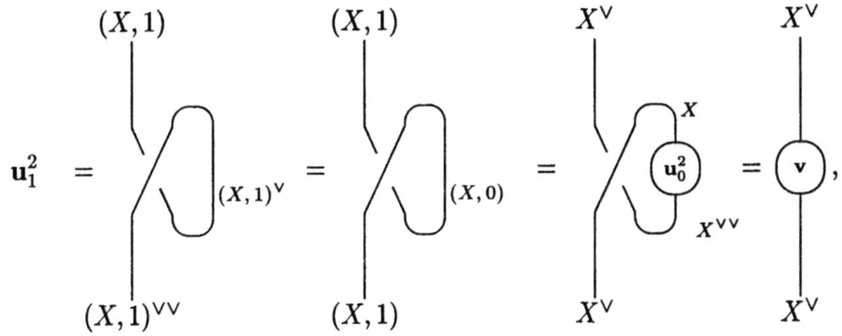

and, similarly, in other cases.

By the general theory from Section 4.1

$$\mathbf{u}^2_1 = \mathbf{v} : (\mathrm{Id}_\mathcal{D}, \mathbf{c}^{-2}, \mathbb{1}_1) \longrightarrow (\mathrm{Id}_\mathcal{D}, j_2, d_2)$$

is an isomorphism of monoidal functors. Hence, diagram (4.1.7) and the properties of ribbon twist yield $j_2 = \mathbb{1}$ and $d_2 = \mathbb{1}$ in $\mathcal{D}$. Also $\mathbf{u}^4_0 = \mathbf{u}^2_{-1} \circ \mathbf{u}^2_1 = \mathbf{v}^{-1} \circ \mathbf{v} = \mathbb{1}$.

Therefore, the general definition of balancing reduces in $\mathcal{D}$ to an isomorphism $\beta : (\mathrm{Id}, \mathbb{1}, \mathbb{1}) \to (\mathrm{Id}, \mathbb{1}, \mathbb{1})$ such that $\beta^2 = 1$ and $\beta^t_M = \beta^{-1}_{M^\vee} : M^\vee \to M^\vee$, $M \in \mathrm{Ob}\,\mathcal{D}$. The choice $\beta = 1$ satisfies these requirements. Hence, $\mathcal{D}$ is a ribbon category and $\mathbf{v}$ is its ribbon twist. The functors $\mathcal{I} : \mathcal{C} \hookrightarrow \mathcal{D}$ and $\mathcal{G} : \mathcal{D} \to \mathcal{C}$ are compatible with the ribbon structure.

Finally, we can replace $\mathcal{D}$ with another category where (a) is satisfied as well as (b) and (c).

Remark 4.1.8. In the category $\mathcal{C} = H$-mod, where H is a ribbon Hopf algebra, the equation $X^\vee = {}^\vee X$ is not necessarily satisfied. Nevertheless, $X^\vee$ is canonically isomorphic to ${}^\vee X$. The same holds in any ribbon category. We identify these objects via $\beta = \mathbf{u}^2_0 : {}^\vee X \to X^\vee$. This allows us to use the right dual objects in place of the left ones. In that rôle the right duals are equipped with the left evaluation and coevaluation from (4.1.8), (4.1.9), called flipped evaluation and coevaluation:

$$\widetilde{\mathrm{ev}} : X^\vee \otimes X \xrightarrow{X^\vee\otimes\beta} X^\vee \otimes X^{\vee\vee} \xrightarrow{\mathrm{ev}} 1,$$

$$\widetilde{\text{coev}} : 1 \xrightarrow{\text{coev}} X^{\vee\vee} \otimes X^{\vee} \xrightarrow{\beta^{-1} \otimes X^{\vee}} X \otimes X^{\vee}.$$

Often they will be denoted simply ev and coev in the theoretical part and should be replaced by $\widetilde{\text{ev}}$ and $\widetilde{\text{coev}}$ in applications. In the Hopf algebra context β is given by the action of a group-like element introduced by Drinfeld [Dri90]. This is discussed in Sect. 7.4.4.

4.2 Hopf algebras in braided categories

Recall that the category of cobordisms of one-holed surfaces is a balanced category by Lemma 1.3.1. Moreover, it has a distinguished object – a 1-hole torus – which has a natural Hopf algebra structure. This implies the importance of braided Hopf algebras in our study. We recall first the notion of a Hopf algebra in a braided monoidal category, and introduce the notion of integrals for such Hopf algebras. Later we shall see that the integrals are related to surgery.

We formulate standard properties of integrals. The relationship between the integrals and the antipode (the Radford formulas) are recalled in the braided setting. Then we prove several lemmas, which give a practical recipe how to find the objects of integrals.

4.2.1 Algebra in a monoidal category

Let $\mathcal{C}$ be a braided monoidal category. Recall that a Hopf algebra $H \in \mathcal{C}$ [Maj93] is an object $H \in \text{Ob}\,\mathcal{C}$ together with an associative multiplication $m : H \otimes H \to H$ and an associative comultiplication $\Delta : H \to H \otimes H$, obeying the bialgebra axiom

$$\begin{aligned}&\big(H \otimes H \xrightarrow{m} H \xrightarrow{\Delta} H \otimes H\big)\\ &= \big(H \otimes H \xrightarrow{\Delta \otimes \Delta} H \otimes H \otimes H \otimes H \xrightarrow{H \otimes c \otimes H} H \otimes H \otimes H \otimes H \xrightarrow{m \otimes m} H \otimes H\big).\end{aligned} \tag{4.2.1}$$

Moreover, H has a unit, $\eta : 1 \to H$, a counit, $\varepsilon : H \to 1$, an antipode, $\gamma : H \to H$ and the inverse antipode $\gamma^{-1} : H \to H$. The defining relations for these are the same as in the classical case. Notice in particular, that the unit is also a morphism. Associativity of multiplication, as well as coassociativity of comultiplication, is formulated with the use of associativity isomorphism (in the non-strict case).

We use the graphical language of [BKLT00] to represent the operations of a Hopf algebra as listed above. The elementary pictures are summed up in Fig. 4.2. To distinguish such tangles-operations from the previously defined one we draw them with fat lines.

For example, the bialgebra axiom (4.2.1) can be drawn as an equation

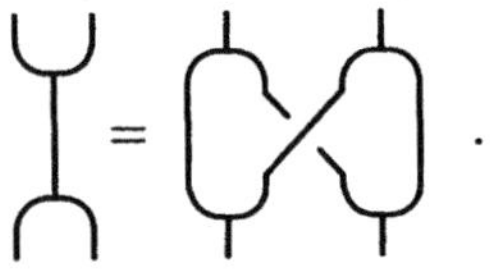

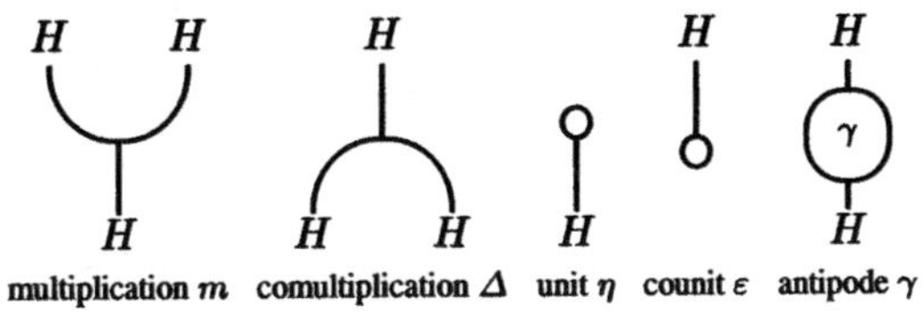

Fig. 4.2. Graphical notation of Hopf algebra operations

Using the graphical language one can prove the following statement.

Proposition 4.2.1. *The antipode γ is an anti-automorphism of a Hopf algebra H in a braided monoidal category, that is, an anti-automorphism of the algebra H:*

$$\big(H \otimes H \xrightarrow{c} H \otimes H \xrightarrow{\gamma\otimes\gamma} H \otimes H \xrightarrow{m} H\big) = \big(H \otimes H \xrightarrow{m} H \xrightarrow{\gamma} H\big)$$

and an anti-automorphism of the coalgebra H:

$$\big(H \xrightarrow{\Delta} H \otimes H \xrightarrow{c} H \otimes H \xrightarrow{\gamma\otimes\gamma} H \otimes H\big) = \big(H \xrightarrow{\gamma} H \xrightarrow{\Delta} H \otimes H\big).$$

Proposition 4.2.2. *Let A, B be Hopf algebras in $\mathcal{C}$ with antipodes γ_A, γ_B. Let $\phi : A \to B$ be a bialgebra homomorphism (a morphism preserving multiplication, comultiplication, unit and counit). Then it commutes with the antipode:*

$$\big(A \xrightarrow{\gamma_A} A \xrightarrow{\phi} B\big) = \big(A \xrightarrow{\phi} B \xrightarrow{\gamma_B} B\big).$$

Proof. The set of morphisms $M = \mathcal{C}(A, B)$ is an associative monoid with respect to the convolution product. For $f, g : A \to B$ their convolution product is $f * g = \big(A \xrightarrow{\Delta} A \otimes A \xrightarrow{f\otimes g} B \otimes B \xrightarrow{m} B\big)$. The unit of this monoid is $\big(A \xrightarrow{\varepsilon} 1 \xrightarrow{\eta} B\big)$. The element $\phi : A \to B$ of M has an inverse $\phi \circ \gamma_A$. However, $\gamma_B \circ \phi$ is also an inverse of ϕ. We conclude that $\phi \circ \gamma_A = \gamma_B \circ \phi$.

4.2.2 Dual Hopf algebras

With each Hopf algebra $(H, m, \eta, \Delta, \varepsilon, \gamma)$ in a rigid braided category $\mathcal{C}$ are associated its dual Hopf algebras $(H^\vee, \Delta^t, \varepsilon^t, m^t, \eta^t, \gamma^t)$ and $({}^\vee H, {}^t\Delta, {}^t\varepsilon, {}^tm, {}^t\eta, {}^t\gamma)$ in $\mathcal{C}$. In a ribbon category the two dual algebras are isomorphic. Specifically, the balancing $\mathbf{u}_0^2 : {}^\vee H \to H^\vee$ is an isomorphism of Hopf algebras.

The relationship between an operation and the dual operation is easy to formulate via Hopf pairings. A pairing $\omega : H \otimes A \to 1$ between two Hopf algebras H and A is called a *Hopf pairing* if it satisfies the following equations:

$$\begin{aligned}\big(H \otimes H \otimes A \xrightarrow{1\otimes 1\otimes\Delta} H \otimes H \otimes A \otimes A \xrightarrow{1\otimes\omega\otimes 1} H \otimes A \xrightarrow{\omega} 1\big)&\\ = \big(H \otimes H \otimes A \xrightarrow{m\otimes 1} H \otimes A \xrightarrow{\omega} 1\big),&\end{aligned}$$

$$\big(H \otimes A \otimes A \xrightarrow{\Delta\otimes 1\otimes 1} H \otimes H \otimes A \otimes A \xrightarrow{1\otimes\omega\otimes 1} H \otimes A \xrightarrow{\omega} 1\big)$$
$$= \big(H \otimes A \otimes A \xrightarrow{1\otimes m} H \otimes A \xrightarrow{\omega} 1\big),$$

$$\big(A = 1 \otimes A \xrightarrow{\eta\otimes 1} H \otimes A \xrightarrow{\omega} 1\big) = \big(A \xrightarrow{\varepsilon} 1\big),$$

$$\big(H = H \otimes 1 \xrightarrow{1\otimes\eta} H \otimes A \xrightarrow{\omega} 1\big) = \big(H \xrightarrow{\varepsilon} 1\big).$$

For instance, $\mathrm{ev} : H \otimes H^\vee \to 1$ and $\mathrm{ev} : {}^\vee H \otimes H \to 1$ are Hopf pairings.

Remark 4.2.3. For any Hopf pairing $\omega : H \otimes A \to 1$ of Hopf algebras H and A we have

$$\big(H \otimes A \xrightarrow{\gamma_H\otimes 1} H \otimes A \xrightarrow{\omega} 1\big) = \big(H \otimes A \xrightarrow{1\otimes\gamma_A} H \otimes A \xrightarrow{\omega} 1\big).$$

Proof. A Hopf pairing $\omega : H \otimes A \to 1$ induces to a bialgebra homomorphism ${}^!\omega : A \to H^\vee$. This homomorphism commutes with the antipodes:

$$\gamma_{H^\vee} \circ {}^!\omega = \gamma_H^t \circ {}^!\omega = {}^!\omega \circ \gamma_A.$$

Hence, the required property of ω.

We call a Hopf pairing $\omega : H \otimes A \to 1$ *side-invertible*, if ${}^!\omega : A \to H^\vee$ is an isomorphism. A *self-dual* Hopf algebra is a Hopf algebra isomorphic to its dual. It is equipped with a side-invertible Hopf pairing $\omega : H \otimes H \to 1$.

There exists a topological illustration of the bialgebra axiom (4.2.1) in a braided category. Kauffman, Saito and Sullivan [KSS97] defined a category of templates, which are a special kind of 2-dimensional stratified pseudomanifolds embedded into a 3-manifold. They are used as models of certain 3-dimensional dynamical systems. Kauffman, Saito and Sullivan [KSS97] show that the category of templates is a free ribbon category, generated by a self-dual object B, equipped with a bialgebra structure – an associative multiplication (possibly without unit) and a coassociative comultiplication (possibly without counit), satisfying the bialgebra axiom (4.2.1), and a bialgebra anti-automorphism γ. Furthermore, it is required that the isomorphism of objects $\alpha : B \xrightarrow{\sim} B^\vee$ induces bialgebra isomorphism $B \xrightarrow{\alpha} B^\vee \xrightarrow{\alpha^{-1\,t}} B^{\vee\vee}$, and, finally, γ^2 is the ribbon twist.

Clearly, a self-dual Hopf algebra H in a ribbon category $\mathcal{C}$, for which the square of the antipode is the ribbon twist, is an example of a bialgebra with the above listed properties. Hence, it gives a functor from the category of templates to $\mathcal{C}$. Such Hopf algebras will be constructed in Sect. 5.2.

4.2.3 Integrals for Hopf algebras

Integrals for finite dimensional Hopf $\Bbbk$-algebras were introduced by Larson and Sweedler in [LS69]. The infinite dimensional case was treated by Sweedler in

[Swe69b]. Integrals for Hopf algebras in braided abelian categories were studied in [Lyu95b]. In this section we define integrals, formulate their standard properties, and recall the relationship between the integrals and the antipode (the Radford formulas) in the braided setting following [BKLT00]. Then we prove several lemmas, which give a practical recipe how to find the object of integrals.

Karoubi studied in [Kar71] categories with the following property. An idempotent, $e = e^2 : X \to X$, in a category, $\mathcal{D}$, is said to be *split* if there exists an object, X_e, and morphisms, $i_e : X_e \to X$ and $p_e : X \to X_e$, such that $e = i_e \circ p_e$ and $\mathbb{1}_{X_e} = p_e \circ i_e$. If every idempotent in $\mathcal{D}$ is split then we say that $\mathcal{D}$ is a *category with split idempotents*.

For a given category $\mathcal{C}$ there exists an embedding, $\mathcal{C} \xrightarrow{i} \widehat{\mathcal{C}}$, such that idempotents in the category $\widehat{\mathcal{C}}$ are split. Moreover, $\widehat{\mathcal{C}}$ can be chosen to be universal in the sense that for any category $\mathcal{D}$ with split idempotents every functor $F : \mathcal{C} \to \mathcal{D}$ factors in the form $F = (\mathcal{C} \xrightarrow{i} \widehat{\mathcal{C}} \xrightarrow{G} \mathcal{D})$ where the functor G is unique up to an isomorphism of functors. The category $\widehat{\mathcal{C}}$ is called the *Karoubi enveloping category of $\mathcal{C}$*. According to Karoubi [Kar71] it may be realized as the category with objects $X_e = (X, e)$, where X is an object in $\mathcal{C}$ and $e : X \to X$ is an idempotent in $\mathcal{C}$. The morphisms in $\widehat{\mathcal{C}}$ are defined by $\widehat{\mathcal{C}}(X_e, Y_f) = \{t \in \mathcal{C}(X, Y) \mid fte = t\}$. The functor i defined by $i(X) = X_{\mathrm{id}_X}$ and $i(f) = f$ is a full embedding, that is, we have $\mathcal{C}(X, Y) = \widehat{\mathcal{C}}(i(X), i(Y))$.

If $\mathcal{C}$ is a (braided) monoidal category, then the category $\widehat{\mathcal{C}}$ can be equipped with a (braided) monoidal structure:

$$1 = (1, \mathbb{1}_1), \qquad X_e \otimes Y_f = (X \otimes Y)_{e \otimes f}, \qquad \mathbf{c}_{X_e, Y_f} = (f \otimes e) \circ \mathbf{c}_{X,Y}.$$

In this case i is a (braided) monoidal functor. Furthermore, if the category $\mathcal{C}$ is rigid, then so is $\widehat{\mathcal{C}}$, and the dual objects of (X, e) are $(X^\vee, e^t)$ and $({}^\vee X, {}^t e)$.

We shall define integrals for Hopf algebras as the output of the following proposition.

Proposition 4.2.4 ([BKLT00]). *Assume that H is a Hopf algebra with an invertible antipode S in a braided monoidal category $\mathcal{C}$ with split idempotents. Then there exist an invertible object* Int H *of $\mathcal{C}$ and the following morphisms*

$$\begin{array}{ll} \int^H, {}^H\!\!\int : \mathrm{Int}\, H \to H & \textit{called left (resp. right) integral-element in } H, \\ \int_H, {}_H\!\!\int : H \to \mathrm{Int}\, H & \textit{called left (resp. right) integral-functional on } H, \end{array}$$

such that

$$\begin{array}{ccc} H \otimes \mathrm{Int}\, H & \xrightarrow{H \otimes \int^H} & H \otimes H \\ {\scriptstyle \varepsilon \otimes \mathrm{Int}\, H} \downarrow & & \downarrow {\scriptstyle m} \\ \mathrm{Int}\, H & \xrightarrow{\int^H} & H \end{array} \qquad \begin{array}{ccc} \mathrm{Int}\, H \otimes H & \xrightarrow{{}^H\!\int \otimes H} & H \otimes H \\ {\scriptstyle \mathrm{Int}\, H \otimes \varepsilon} \downarrow & & \downarrow {\scriptstyle m} \\ \mathrm{Int}\, H & \xrightarrow{{}^H\!\int} & H \end{array} \qquad (4.2.2)$$

$$\begin{array}{ccc} H & \xrightarrow{\int_H} & \operatorname{Int} H \\ \downarrow{\scriptstyle\Delta} & & \downarrow{\scriptstyle\eta\otimes\operatorname{Int} H} \\ H\otimes H & \xrightarrow{H\otimes\int_H} & H\otimes\operatorname{Int} H \end{array} \qquad \begin{array}{ccc} H & \xrightarrow{{}_H\int} & \operatorname{Int} H \\ \downarrow{\scriptstyle\Delta} & & \downarrow{\scriptstyle\operatorname{Int} H\otimes\eta} \\ H\otimes H & \xrightarrow{{}_H\int\otimes H} & \operatorname{Int} H\otimes H \end{array} \tag{4.2.3}$$

are commutative.

Furthermore, any morphism $f : H \to X$ of C such that

$$\big(H \xrightarrow{\Delta} H\otimes H \xrightarrow{\mathbf{1}\otimes f} H\otimes X\big) = \big(H \xrightarrow{f} X \simeq 1\otimes X \xrightarrow{\eta\otimes\mathbf{1}} H\otimes X\big),$$

resp.

$$\big(H \xrightarrow{\Delta} H\otimes H \xrightarrow{f\otimes\mathbf{1}} X\otimes H\big) = \big(H \xrightarrow{f} X \simeq X\otimes 1 \xrightarrow{\mathbf{1}\otimes\eta} X\otimes H\big),$$

admits a unique factorization of the form $H \xrightarrow{\int_H} \operatorname{Int} H \xrightarrow{g} X$ (resp. $H \xrightarrow{{}_H\int} \operatorname{Int} H \xrightarrow{g} X$). Any morphism $f : X \to H$ of C such that

$$\big(H\otimes X \xrightarrow{\mathbf{1}\otimes f} H\otimes H \xrightarrow{m} H\big) = \big(H\otimes X \xrightarrow{\varepsilon\otimes\mathbf{1}} 1\otimes X \simeq X \xrightarrow{f} H\big),$$

resp.

$$\big(X\otimes H \xrightarrow{f\otimes\mathbf{1}} H\otimes H \xrightarrow{m} H\big) = \big(X\otimes H \xrightarrow{\mathbf{1}\otimes\varepsilon} X\otimes 1 \simeq X \xrightarrow{f} H\big),$$

admits a unique factorization of the form $X \xrightarrow{h} \operatorname{Int} H \xrightarrow{\int^H} H$ (resp. $X \xrightarrow{h} \operatorname{Int} H \xrightarrow{{}^H\!\int} H$).

This proposition-definition fixes the *object of integrals* $\operatorname{Int} H$ uniquely up to a unique isomorphism. The integrals $\int_H$, ${}_H\int$, $\int^H$, ${}^H\!\int$ are unique up to an automorphism of $\operatorname{Int} H$.

In the case of an abelian rigid monoidal category C we can define $\operatorname{Int} H$ and the integrals via exact sequences

$$0 \longrightarrow \operatorname{Int} H \xrightarrow{\int^H} H \xrightarrow{\psi_l - \eta\otimes H} H^\vee\otimes H,$$

$$0 \longrightarrow \operatorname{Int} H \xrightarrow{{}^H\!\int} H \xrightarrow{\psi_r - H\otimes\eta} H\otimes{}^\vee H,$$

$${}^\vee H\otimes H \xrightarrow{\alpha_l - \varepsilon\otimes H} H \xrightarrow{\int_H} \operatorname{Int} H \longrightarrow 0,$$

$$H\otimes H^\vee \xrightarrow{\alpha_r - H\otimes\varepsilon} H \xrightarrow{{}_H\int} \operatorname{Int} H \longrightarrow 0,$$

where the (co)actions of dual Hopf algebras on H are

$$\psi_l = \left(H = 1 \otimes H \xrightarrow{\text{coev} \otimes H} H^\vee \otimes H \otimes H \xrightarrow{H^\vee \otimes m} H^\vee \otimes H\right),$$
$$\psi_r = \left(H = H \otimes 1 \xrightarrow{H \otimes \text{coev}} H \otimes H \otimes {}^\vee H \xrightarrow{m \otimes {}^\vee H} H \otimes {}^\vee H\right),$$
$$\alpha_l = \left({}^\vee H \otimes H \xrightarrow{{}^\vee H \otimes \Delta} {}^\vee H \otimes H \otimes H \xrightarrow{\text{ev} \otimes H} 1 \otimes H = H\right),$$
$$\alpha_r = \left(H \otimes H^\vee \xrightarrow{\Delta \otimes H^\vee} H \otimes H \otimes H^\vee \xrightarrow{H \otimes \text{ev}} H \otimes 1 = H\right).$$

Theorem 4.2.5 (see Theorem 3.3 [BKLT00]). *Applying the functor* $-^\vee$ *to left or right integrals on (in)* H *we get right or left integrals in (on)* $H^\vee$. *The four composite morphisms*

$$\textstyle\int_H \circ \int^H, \; \int_H \circ {}^H\!\!\int, \; {}_H\!\!\int \circ \int^H, \; {}_H\!\!\int \circ {}^H\!\!\int \; : \text{Int}\, H \longrightarrow \text{Int}\, H$$

are all invertible. Hence, the four natural pairings $\text{Int}\, H \otimes \text{Int}\, H^\vee \xrightarrow{\int \otimes \int} H \otimes H^\vee \to 1$ *and the four natural copairings* $1 \to H^\vee \otimes H \xrightarrow{\int \otimes \int} \text{Int}\, H^\vee \otimes \text{Int}\, H$ *are isomorphisms.*

In particular, $\text{Int}(H^\vee) \simeq (\text{Int}\, H)^\vee$.

The relationship between integrals and the antipode is clarified by the identities in Fig. 4.3. These formulae from [BKLT00] generalize those of Radford [Rad76].

The reader can prove these identities via the following straightforward lemma.

Lemma 4.2.6. *The maps*

$$b, p : \text{Hom}(H \otimes M, H \otimes N) \to \text{Hom}(H \otimes M, H \otimes N),$$

given below are inverse to each other.

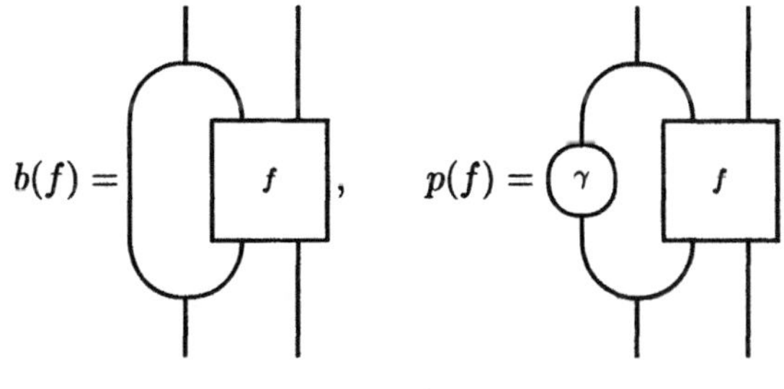

In fact, only one formula from Figure 4.3 is proven in [BKLT00] (using the above lemma), and the remaining three are obtained from it by replacing the Hopf algebra H with the one, having opposite multiplication H^{op} or opposite comultiplication H_{op}. Notice that the identities (a), (c) at Figure 4.3 are transformed into (b), (d) under the action of space π-rotation. It preserves the plane of drawing and fixes a vertical axis, takes a picture into its mirror image, where the left and right are exchanged, but the sign of over/under crossing is *not* changed.

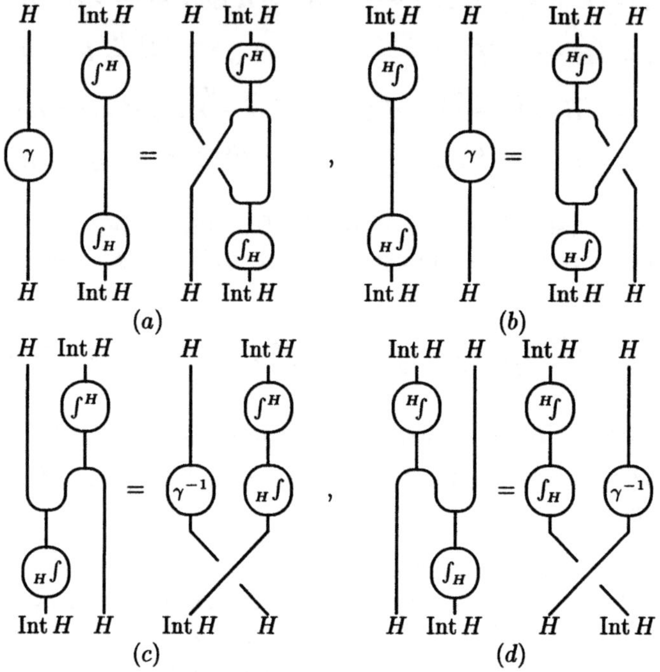

Fig. 4.3. The antipode expressed via integrals

Proposition 4.2.7 (Proposition 4.10 [BKLT00]). *The composition of the antipode with an integral is proportional to another integral:*

$$\gamma \circ {}^H\!\!\int = \mathrm{c}_{\mathrm{Int}\,H,\mathrm{Int}\,H}(\textstyle\int_H \circ {}^H\!\!\int)(\int_H \circ \int^H)^{-1} \cdot \int^H : \mathrm{Int}\,H \to H, \tag{4.2.4}$$

$$\gamma \circ \textstyle\int^H = \mathrm{c}_{\mathrm{Int}\,H,\mathrm{Int}\,H}({}_H\!\int \circ \int^H)({}_H\!\int \circ {}^H\!\!\int)^{-1} \cdot {}^H\!\!\int : \mathrm{Int}\,H \to H, \tag{4.2.5}$$

$${}_H\!\textstyle\int \circ \gamma = \mathrm{c}_{\mathrm{Int}\,H,\mathrm{Int}\,H}({}_H\!\int \circ \int^H)(\int_H \circ \int^H)^{-1} \circ \int_H : H \to \mathrm{Int}\,H, \tag{4.2.6}$$

$$\textstyle\int_H \circ \gamma = \mathrm{c}_{\mathrm{Int}\,H,\mathrm{Int}\,H}(\int_H \circ {}^H\!\!\int)({}_H\!\int \circ {}^H\!\!\int)^{-1} \circ {}_H\!\int : H \to \mathrm{Int}\,H, \tag{4.2.7}$$

where the braiding $\mathrm{c}_{\mathrm{Int}\,H,\mathrm{Int}\,H} : \mathrm{Int}\,H \otimes \mathrm{Int}\,H \to \mathrm{Int}\,H \otimes \mathrm{Int}\,H$ *is viewed as an element of* $\mathrm{Aut}_{\mathcal{C}}\,1$. *The proportionality constant is invertible.*

Proof. The composition of the both sides of the identity at Figure 4.3(a) with ${}^H\!\!\int \otimes 1\!\!1_{\mathrm{Int}\,H}$ gives (4.2.4). The other identities are obtained similarly.

Definition 4.2.8. *Let K be an invertible object and let $A, B \in \mathrm{Ob}\,\mathcal{C}$. A pairing $p : A \otimes B \to K$ is called* side-invertible *if the induced morphism $p^! : A \to K \otimes {}^\vee B$ is invertible, or, equivalently, ${}^!p : B \to A^\vee \otimes K$ is invertible. A copairing $q : K \to A \otimes B$ is called side-invertible, if the induced morphism $q_{!} : K \otimes B^\vee \to A$, is invertible, or, equivalently, ${}_{!}q : {}^\vee A \otimes K \to B$ is invertible.*

Remark 4.2.9. In order to prove that a morphism of an abelian category is invertible, we often use the following elementary observation. If $f : A \to B$ is a monomorphism or epimorphism and $\operatorname{length}(A) = \operatorname{length}(B)$, then f is an isomorphism.

Let $p : A \otimes B \to K$ be a pairing, where K is invertible. Assume that there are copairings $q_1 : K_1 \to D \otimes A$, $q_2 : K_2 \to B \otimes C$ for invertible K_1, K_2 such that

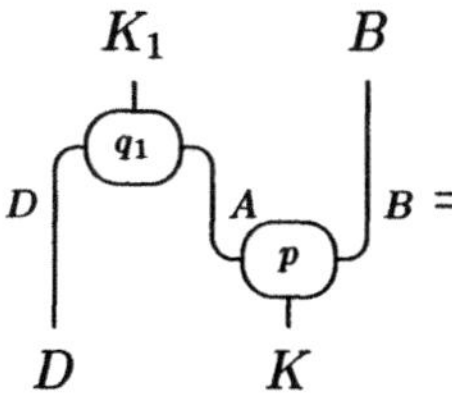

$$= \left(K_1 \otimes B \xrightarrow{q_1 \otimes 1} D \otimes A \otimes B \xrightarrow{1 \otimes p} D \otimes K\right) \tag{4.2.8}$$

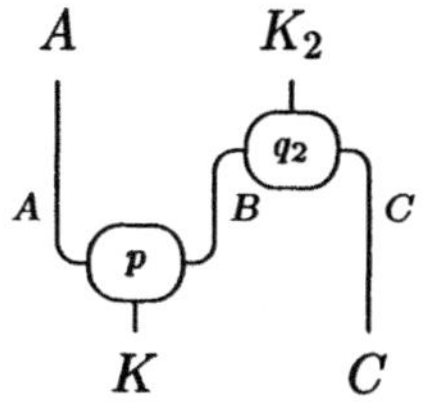

$$= \left(A \otimes K_2 \xrightarrow{1 \otimes q_2} A \otimes B \otimes C \xrightarrow{p \otimes 1} K \otimes C\right) \tag{4.2.9}$$

are isomorphisms. Then p is side-invertible.

Indeed, the transposed morphism to (4.2.8) is invertible, hence, the partially transposed morphism

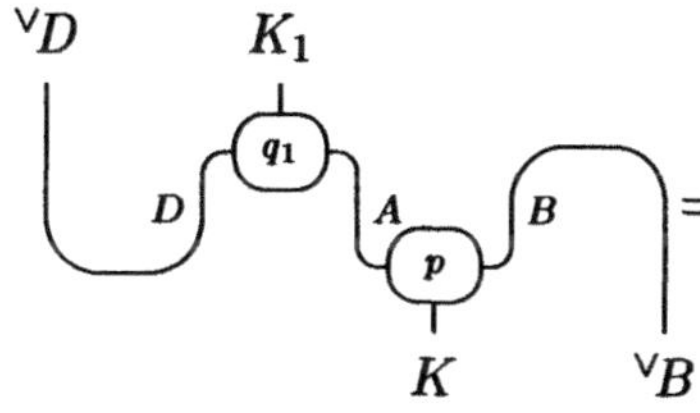

$$= ({}^\vee D \otimes K_1 \xrightarrow{1 \otimes q_1} {}^\vee D \otimes D \otimes A \xrightarrow{\operatorname{ev} \otimes 1} A \xrightarrow{1 \otimes \operatorname{coev}} A \otimes B \otimes {}^\vee B \xrightarrow{p \otimes 1} K \otimes {}^\vee B)$$

is invertible. Applying - $\otimes\, {}^\vee K_2$ to (4.2.9) we get an invertible morphism as well:

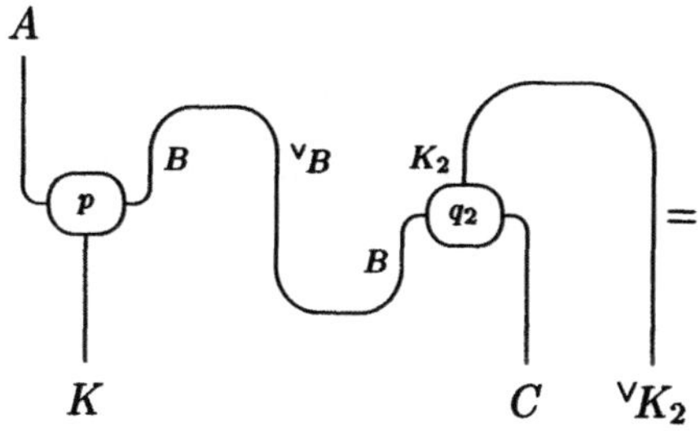

$$= (A \xrightarrow{1 \otimes \operatorname{coev}} A \otimes B \otimes {}^\vee B \xrightarrow{p \otimes 1} K \otimes {}^\vee B \xrightarrow{1 \otimes 1 \otimes \operatorname{coev}} K \otimes {}^\vee B \otimes K_2 \otimes {}^\vee K_2$$
$$\xrightarrow{1 \otimes 1 \otimes q_2 \otimes 1} K \otimes {}^\vee B \otimes B \otimes C \otimes {}^\vee K_2 \xrightarrow{1 \otimes \operatorname{ev} \otimes 1 \otimes 1} K \otimes C \otimes {}^\vee K_2).$$

Thus,

$$p^{!} = \left(A \xrightarrow{1\otimes \mathrm{coev}} A\otimes B\otimes {}^{\vee}B \xrightarrow{p\otimes 1} K\otimes {}^{\vee}B\right)$$

has left and right inverses and is, therefore, invertible.

Dually, let $q : K \to A\otimes B$ be a copairing, where K is invertible. Assume that there are pairings $p_1 : D\otimes A \to K_1, p_2 : B\otimes C \to K_2$, such that K_1 and K_2 are invertible and

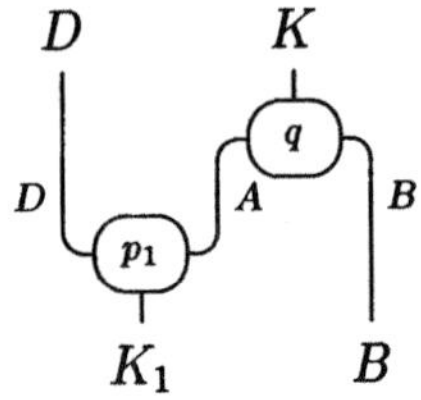

$$= \left(D\otimes K \xrightarrow{1\otimes q} D\otimes A\otimes B \xrightarrow{p_1\otimes 1} K_1\otimes B\right),$$

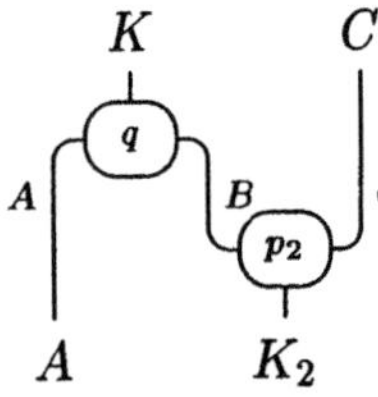

$$= \left(K\otimes C \xrightarrow{q\otimes 1} A\otimes B\otimes C \xrightarrow{1\otimes p_2} A\otimes K_2\right)$$

are isomorphisms. Then q is side-invertible.

Example 4.2.10. Identities at Figure 4.3 show that

$$\beta^H = \left(\mathrm{Int}\, H \xrightarrow{\int^H} H \xrightarrow{\Delta} H\otimes H\right),$$
$$^H\beta = \left(\mathrm{Int}\, H \xrightarrow{^H\!\int} H \xrightarrow{\Delta} H\otimes H\right),$$
$$\beta_H = \left(H\otimes H \xrightarrow{m} H \xrightarrow{\int_H} \mathrm{Int}\, H\right),$$
$$_H\beta = \left(H\otimes H \xrightarrow{m} H \xrightarrow{_H\!\int} \mathrm{Int}\, H\right)$$

are side-invertible.

Lemma 4.2.11. *Let K be an invertible object, and let a morphism $t : H \to K$ be such that the pairing $\phi : H\otimes H \xrightarrow{m} H \xrightarrow{t} K$ is side-invertible. Then* $\mathrm{Int}\, H \simeq K$.

Proof. Since the copairing $\kappa : \mathrm{Int}\, H \xrightarrow{\int^H} H \xrightarrow{\Delta} H\otimes H$ is side-invertible, the induced morphism ${}_!\kappa : {}^{\vee}H\otimes \mathrm{Int}\, H \to H$ is invertible. Hence, the composite

$$\begin{aligned} M &= \left(H\otimes \mathrm{Int}\, H \xrightarrow{H\otimes\kappa} H\otimes H\otimes H \xrightarrow{\phi\otimes H} K\otimes H\right) \\ &= \left(H\otimes \mathrm{Int}\, H \xrightarrow{\phi^{!}\otimes \mathrm{Int}\, H} K\otimes {}^{\vee}H\otimes \mathrm{Int}\, H \xrightarrow{K\otimes {}_!\kappa} K\otimes H\right) \end{aligned}$$

is invertible. In particular,

$$L = \big(H \otimes \operatorname{Int} H \xrightarrow{M} K \otimes H \xrightarrow{K\otimes\varepsilon} K\big)$$

is a split epimorphism with the splitting $\xi = \big(K \xrightarrow{K\otimes\eta} K \otimes H \xrightarrow{M^{-1}} H \otimes \operatorname{Int} H\big)$. We have

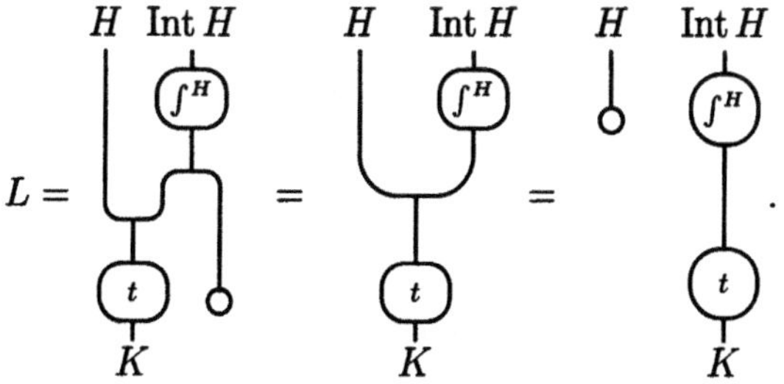

Hence,

$$\begin{aligned}\mathbb{1}_K &= \big(K \xrightarrow{\xi} H \otimes \operatorname{Int} H \xrightarrow{L} K\big)\\ &= \big(K \xrightarrow{\xi} H \otimes \operatorname{Int} H \xrightarrow{\varepsilon\otimes 1} \operatorname{Int} H \xrightarrow{t\circ\int^H} K\big).\end{aligned}$$

Thus, $s = t \circ \int^H : \operatorname{Int} H \to K$ is a split epimorphism with a splitting $r = (\varepsilon \otimes 1)\xi : K \to \operatorname{Int} H$,

$$\big(K \xrightarrow{r} \operatorname{Int} H \xrightarrow{s} K\big) = \mathbb{1}_K.$$

Since K and $\operatorname{Int} H$ are invertible, we deduce that r and s are isomorphisms. Indeed, by tensoring with $(\operatorname{Int} H)^\vee$ we reduce the claim to the case $\big(N \xrightarrow{a} 1 \xrightarrow{b} N\big) = \mathbb{1}_N$, where N is invertible. The object N splits the idempotent $p = \big(1 \xrightarrow{b} N \xrightarrow{a} 1\big)$. Considering the transposed morphisms, we find that $\big(N^\vee \xrightarrow{b^t} 1^\vee \xrightarrow{a^t} N^\vee\big) = \mathbb{1}_{N^\vee}$, and $N^\vee$ splits the idempotent p^t, which is identified with p under the isomorphism $1^\vee \simeq 1$. In particular, $N^\vee \simeq N$. The equation $\mathrm{coev}_N = \big(1 \xrightarrow{\mathrm{coev}} 1^\vee \otimes 1 \xrightarrow{a^t\otimes b} N^\vee \otimes N\big)$ implies that

$$\begin{aligned}&\big(1 \xrightarrow{\mathrm{coev}_N} N^\vee \otimes N \xrightarrow{c} N \otimes N^\vee \xrightarrow{\mathrm{ev}_N} 1\big)\\ &\quad= \big(1 \xrightarrow{\mathrm{coev}} 1^\vee \otimes 1 \xrightarrow{a^t\otimes b} N^\vee \otimes N \xrightarrow{c} N \otimes N^\vee \xrightarrow{\mathrm{ev}_N} 1\big)\\ &\quad= \big(1 \xrightarrow{\mathrm{coev}} 1^\vee \otimes 1 \xrightarrow{c} 1 \otimes 1^\vee \xrightarrow{p\otimes 1} 1 \otimes 1^\vee \xrightarrow{\mathrm{ev}} 1\big) = p.\end{aligned}$$

Since N is invertible, the left hand side is an automorphism, hence, $p = p^2 : 1 \to 1$ is an automorphism. Finally, $\mathbb{1} = p$.

Lemma 4.2.12. *For any Hopf algebra we have*

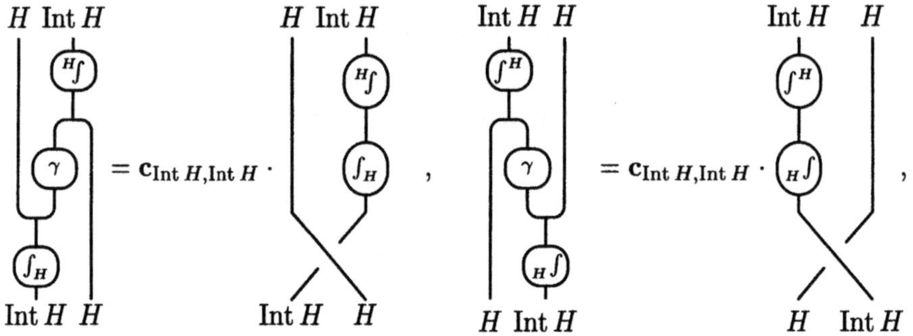

where the braiding $\mathbf{c}_{\mathrm{Int}\,H,\mathrm{Int}\,H}$ *is viewed as an invertible constant.*

Proof. Due to equation (d) from Fig. 4.3 the first equation is equivalent to the following one.

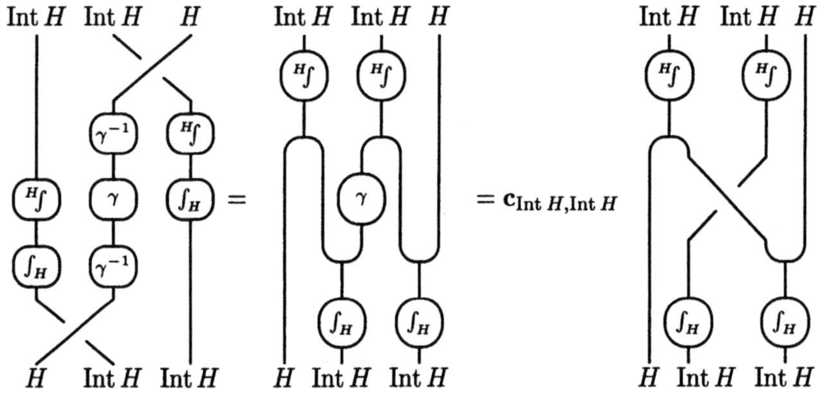

This equation is a corollary of the following:

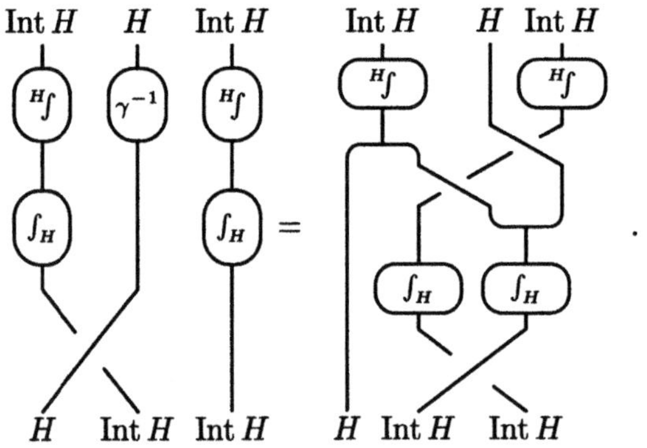

To prove it we use again equation (d) from Fig. 4.3.

The second claim of the lemma (and its proof) is obtained by the space rotation from the first one. So it follows from equation (c) in Fig. 4.3.

In the particular case of $K = 1$, a pairing $p : A \otimes B \to 1$ (or a copairing $q : 1 \to B \otimes A$) is side-invertible if and only if it can be chosen as an evaluation (resp. coevaluation) morphism. Then it can be complemented by a coevaluation q (resp. evaluation p), which is called its side-inverse.

Corollary 4.2.13. *If* $\operatorname{Int} H = 1$, *then* $\beta_H = \big(H \otimes H \xrightarrow{m} H \xrightarrow{\int_H} 1\big)$ *has a side-inverse*

$$(\textstyle\int_H \circ {}^H\!\int)^{-1}\big(1 \xrightarrow{{}^H\beta} H \otimes H \xrightarrow{\gamma\otimes 1} H \otimes H\big),$$

${}^H\beta = \big(1 \xrightarrow{{}^H\!\int} H \xrightarrow{\Delta} H \otimes H\big)$ *has a side-inverse*

$$(\textstyle\int_H \circ {}^H\!\int)^{-1}\big(H \otimes H \xrightarrow{1\otimes\gamma} H \otimes H \xrightarrow{\beta_H} 1\big),$$

${}_H\beta = \big(H \otimes H \xrightarrow{m} H \xrightarrow{{}_H\!\int} 1\big)$ *has a side-inverse*

$$({}_H\textstyle\int \circ \int^H)^{-1}\big(1 \xrightarrow{\beta^H} H \otimes H \xrightarrow{1\otimes\gamma} H \otimes H\big),$$

$\beta^H = \big(1 \xrightarrow{\int^H} H \xrightarrow{\Delta} H \otimes H\big)$ *has a side-inverse*

$$({}_H\textstyle\int \circ \int^H)^{-1}\big(H \otimes H \xrightarrow{\gamma\otimes 1} H \otimes H \xrightarrow{{}_H\beta} 1\big).$$

4.2.4 Self-dual Hopf algebras

Let us give a criterion of side-invertibility (non-degeneracy) of a Hopf pairing.

Proposition 4.2.14. *Let H be a Hopf algebra and let $\omega : H \otimes H \to 1$ be a Hopf pairing.*

(a) Assume that $(\operatorname{Int} H)^\vee \simeq \operatorname{Int} H$. *Suppose that there exist morphisms* $\mu' : {}^\vee(\operatorname{Int} H) \to H$, $\mu'' : (\operatorname{Int} H)^\vee \to H$ *of* $\mathcal{C}$, *such that*

$$\int' = \big(H \simeq 1 \otimes H \xrightarrow{\operatorname{coev}\otimes 1} (\operatorname{Int} H) \otimes {}^\vee(\operatorname{Int} H) \otimes H \xrightarrow{1\otimes\mu'\otimes 1} (\operatorname{Int} H) \otimes H \otimes H \xrightarrow{1\otimes\omega} (\operatorname{Int} H) \otimes 1 \simeq \operatorname{Int} H\big)$$

and

$$\int'' = \big(H \simeq H \otimes 1 \xrightarrow{1\otimes\operatorname{coev}} H \otimes (\operatorname{Int} H)^\vee \otimes \operatorname{Int} H \xrightarrow{1\otimes\mu''\otimes 1} H \otimes H \otimes \operatorname{Int} H \xrightarrow{\omega\otimes 1} 1 \otimes \operatorname{Int} H \simeq \operatorname{Int} H\big)$$

can be chosen for a left or right integral-functional on H (independently). Then ω is side-invertible and the morphisms of C

$\mathsf{S}' =$ 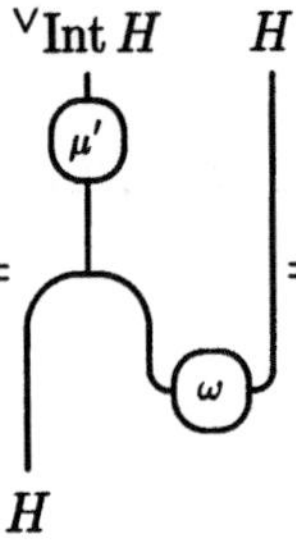

$$= \left({}^{\vee}\mathrm{Int}\,H \otimes H \xrightarrow{\mu'\otimes 1} H \otimes H \xrightarrow{\Delta\otimes 1} H \otimes H \otimes H \xrightarrow{1\otimes\omega} H \otimes 1 \simeq H\right), \quad (4.2.10)$$

$\mathsf{S}'' =$ 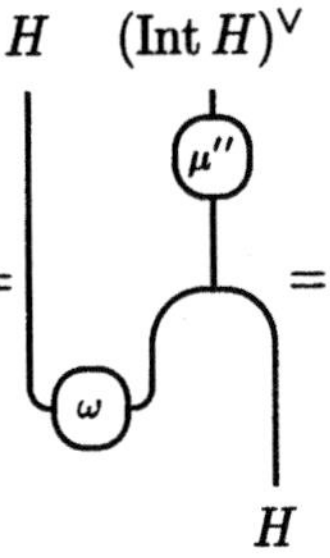

$$= \left(H \otimes (\mathrm{Int}\,H)^{\vee} \xrightarrow{1\otimes\mu''} H \otimes H \xrightarrow{1\otimes\Delta} H \otimes H \otimes H \xrightarrow{\omega\otimes 1} 1 \otimes H \simeq H\right) \quad (4.2.11)$$

are invertible.

(b) If ω is side-invertible, then $(\mathrm{Int}\,H)^{\vee} \simeq \mathrm{Int}\,H$, the composition

$$H \simeq 1 \otimes H \xrightarrow{\mathrm{coev}\otimes 1} (\mathrm{Int}\,H) \otimes (\mathrm{Int}\,H) \otimes H \xrightarrow{1\otimes\int^{H}\otimes 1} (\mathrm{Int}\,H) \otimes H \otimes H \xrightarrow{1\otimes\omega} (\mathrm{Int}\,H) \otimes 1 \simeq \mathrm{Int}\,H$$

or

$$H \simeq H \otimes 1 \xrightarrow{1\otimes\mathrm{coev}} H \otimes (\mathrm{Int}\,H) \otimes (\mathrm{Int}\,H) \xrightarrow{1\otimes\int^{H}\otimes 1} H \otimes H \otimes \mathrm{Int}\,H \xrightarrow{\omega\otimes 1} 1 \otimes \mathrm{Int}\,H \simeq \mathrm{Int}\,H$$

can be chosen for ${}_{H}\!\int : H \to \mathrm{Int}\,H$, and the composition

$$H \simeq 1 \otimes H \xrightarrow{\mathrm{coev}\otimes 1} (\mathrm{Int}\,H) \otimes (\mathrm{Int}\,H) \otimes H \xrightarrow{1\otimes{}^{H}\!\int\otimes 1} (\mathrm{Int}\,H) \otimes H \otimes H \xrightarrow{1\otimes\omega} (\mathrm{Int}\,H) \otimes 1 \simeq \mathrm{Int}\,H$$

or

$$H \simeq H \otimes 1 \xrightarrow{1 \otimes \mathrm{coev}} H \otimes (\operatorname{Int} H) \otimes (\operatorname{Int} H)$$
$$\xrightarrow{1 \otimes {}^{H}\!\!\int \otimes 1} H \otimes H \otimes \operatorname{Int} H \xrightarrow{\omega \otimes 1} 1 \otimes \operatorname{Int} H \simeq \operatorname{Int} H$$

can be chosen for $\int_H : H \to \operatorname{Int} H$.

Proof. (a) Since $\int'$ is an integral-functional, it follows by Example 4.2.10 that

$$\beta' = (H \otimes H \xrightarrow{m} H \xrightarrow{\int'} \operatorname{Int} H) = \big(H \otimes H \xrightarrow{m} H \simeq 1 \otimes H \xrightarrow{\mathrm{coev} \otimes 1}$$
$$(\operatorname{Int} H) \otimes {}^{\vee}(\operatorname{Int} H) \otimes H \xrightarrow{1 \otimes \mu' \otimes 1} (\operatorname{Int} H) \otimes H \otimes H \xrightarrow{1 \otimes \omega} (\operatorname{Int} H) \otimes 1 \simeq \operatorname{Int} H\big)$$

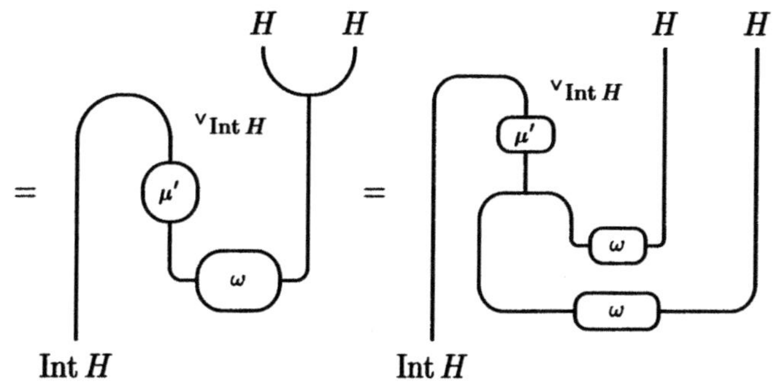

(4.2.12)

is side-invertible. Therefore,

$$\big({}^{\vee}(\operatorname{Int} H) \otimes H\big) \otimes H \xrightarrow{\mathrm{S}' \otimes 1} H \otimes H \xrightarrow{\omega} 1 \qquad (4.2.13)$$

is side-invertible, where S′ is defined by (4.2.10). Similarly,

$$\beta'' = (H \otimes H \xrightarrow{m} H \xrightarrow{\int''} \operatorname{Int} H) =$$

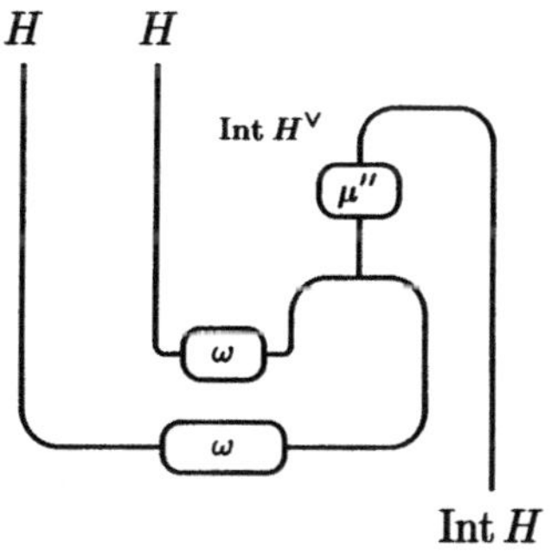

is side-invertible. Therefore,

$$H \otimes \big(H \otimes (\operatorname{Int} H)^{\vee}\big) \xrightarrow{1 \otimes \mathrm{S}''} H \otimes H \xrightarrow{\omega} 1 \qquad (4.2.14)$$

is also side-invertible, where S″ is defined by (4.2.11). Side-invertibility of (4.2.13) and (4.2.14) implies side-invertibility of ω.

Both (4.2.13) and ω can be used as ev'_H. Hence, the canonical morphism S′ relating these pairings is an isomorphism. Similarly, S″ is an isomorphism.

(b) Follows from Theorem 4.2.5.

Assume that ω is side-invertible. Let us denote the morphisms, introduced in Proposition 4.2.14, by

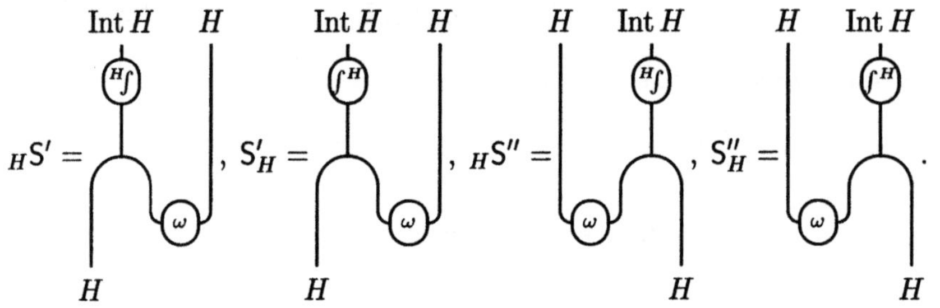

These morphisms are related as follows:

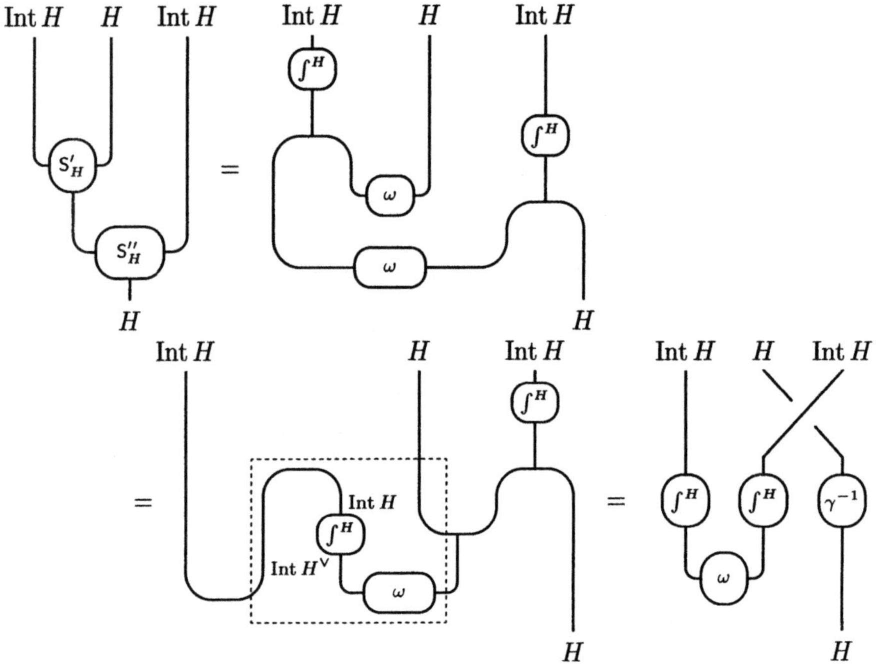

since the dashed part of the third diagram can be chosen for $_H\!\int : H \to (\operatorname{Int} H)^\vee \simeq \operatorname{Int} H$ by Proposition 4.2.14(b) and by Fig. 4.3(c). Similarly,

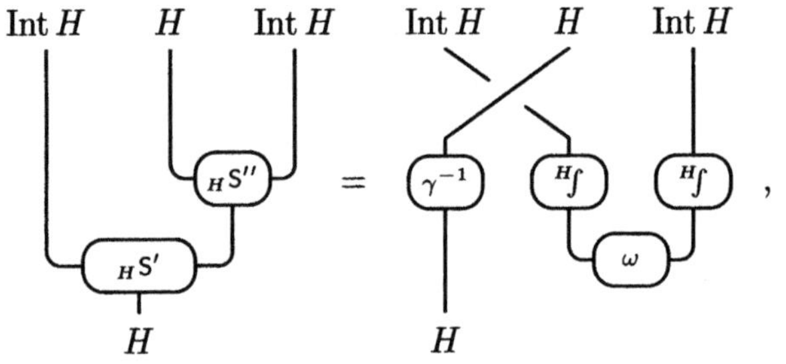

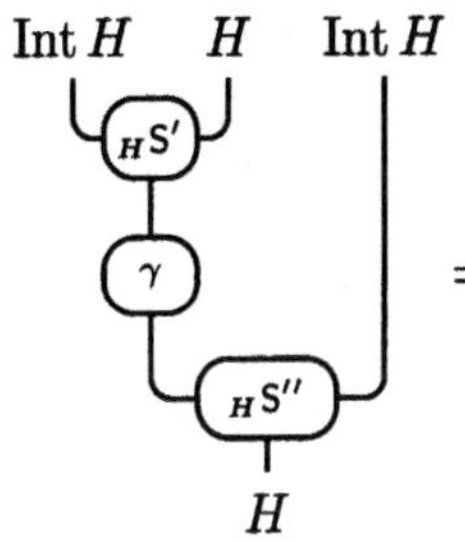

$=\quad c_{\mathrm{Int}\,H,\mathrm{Int}\,H}$

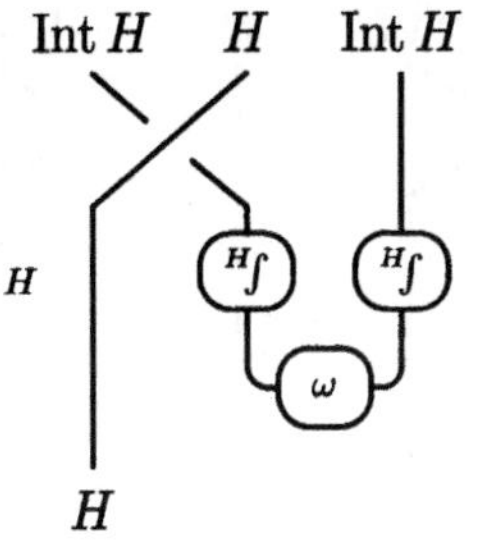

,

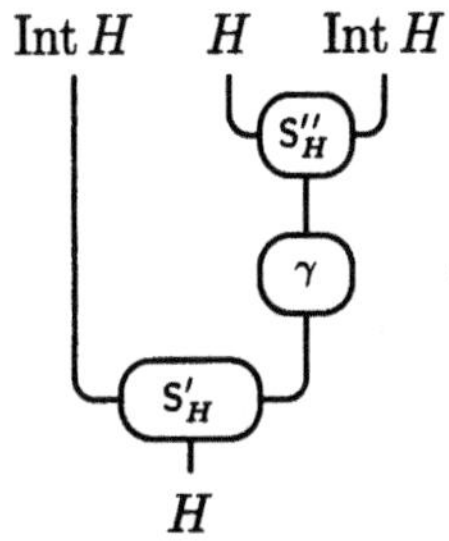

$=\quad c_{\mathrm{Int}\,H,\mathrm{Int}\,H}$

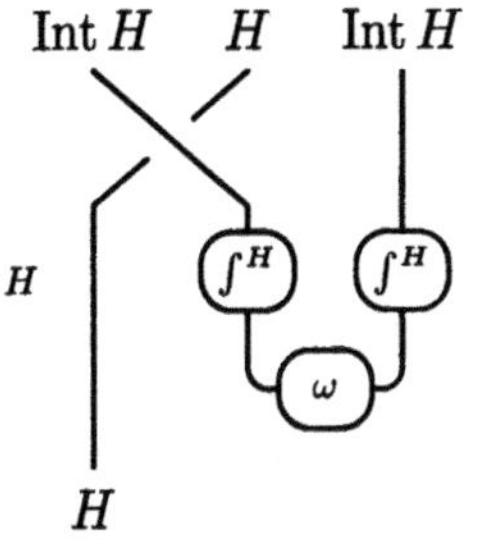

.

In the last two equations Remark 4.2.3 and Lemma 4.2.12 are used.

Lemma 4.2.15. *Suppose that ω is symmetric in the following sense:*

$$\big(H\otimes H \xrightarrow{\mathrm{c}^{-1}} H\otimes H \xrightarrow{\omega} 1\big) = \big(H\otimes H \xrightarrow{\gamma\otimes\gamma} H\otimes H \xrightarrow{\omega} 1\big).$$

Then we have the following relations

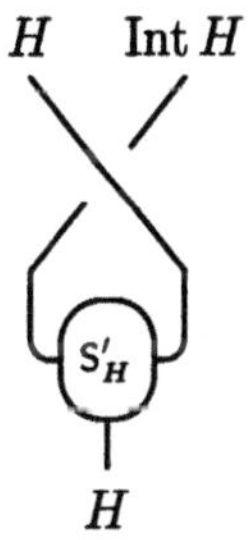

$$= \quad c_{\mathrm{Int}\,H,\mathrm{Int}\,H}\,\frac{{}_H\!\int \circ \int^H}{{}_H\!\int \circ {}^H\!\!\int}$$

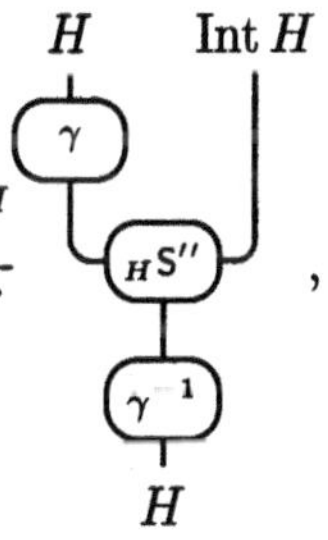

,

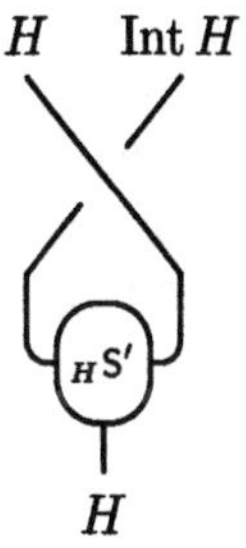

$$= \quad c_{\mathrm{Int}\,H,\mathrm{Int}\,H}\,\frac{\int_H \circ {}^H\!\!\int}{\int_H \circ \int^H}$$

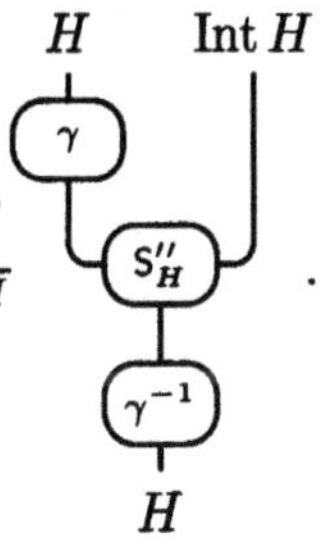

.

Proof. The first equation:

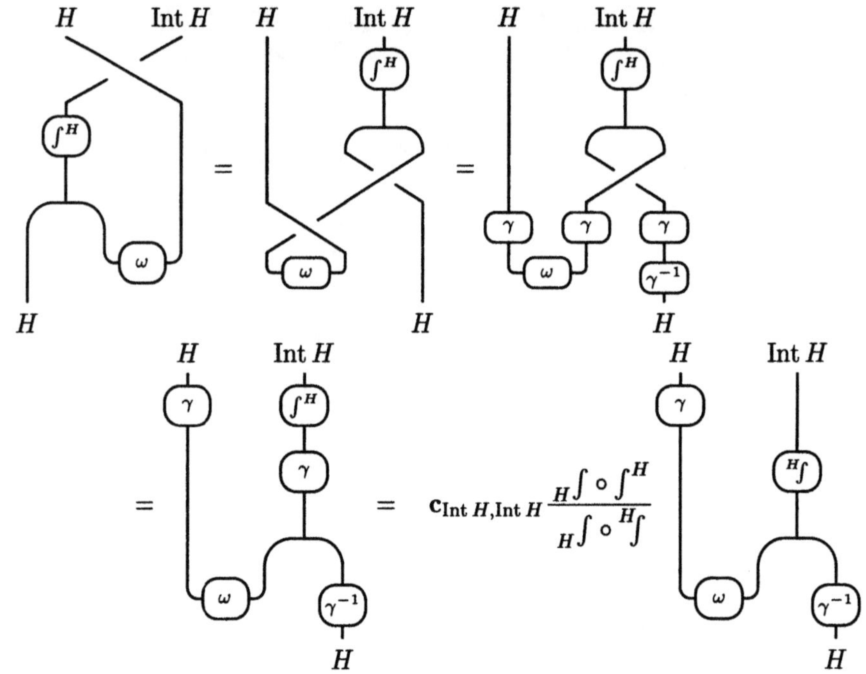

holds due to (4.2.5). The second equation is similar.

Remark 4.2.16. When the integrals are two-sided and/or Int $H \simeq 1$, the above mentioned relations significantly simplify. This holds for certain coends – our main examples of braided Hopf algebras. Operators S' and S'' are analogs of the Fourier transform on a finite abelian group.

4.3 Abelian categories form a monoidal 2-category

In the introduction we proposed to construct TQFT's as double functors with values in the double category of quintets $\mathcal{Q}\mathbf{V}$-$\mathbf{Cat}$-$\mathbf{mod}^l$ in the 2-category $\mathbf{V}$-$\mathbf{Cat}$-$\mathbf{mod}^l$. The latter 2-category is a semistrict version of the weak symmetric monoidal 2-category of bounded abelian categories [Lyu99]. We begin with the description of its monoidal structure given by Deligne's tensor product of $\Bbbk$-linear categories. We define semistrict monoidal 2-categories in general and construct the concrete example, which is obtained from a bounded abelian ribbon category with extra structure, that will enter our construction. Then we add the next ingredient, namely, symmetry.

In our construction of TQFT's we actually need to consider weak symmetric monoidal 2-categories. Nevertheless, the strict version is sufficient to check that the machinery operates correctly. The equivalence of the strict and weak versions of the symmetric monoidal 2-categories of bounded abelian categories allows us to perform the model strict constructions. Technically this is simpler, because a lot of equivalences and isomorphisms become identities in the strict case. That is why we prefer the artificial strict setup to the natural weak one.

4.3.1 Deligne's tensor product of abelian categories

Roughly, abelian categories are additive ones with good behavior of kernels and cokernels, see precise definition for instance in [Fre64]. Besides, we shall not need the most general case. The categories equivalent to categories of modules over finite-dimensional associative $\Bbbk$-algebras are called *bounded*. They are automatically $\Bbbk$-linear and abelian. Such bounded categories with a balanced monoidal structure will be used as input data for the construction of a TQFT. To introduce the exterior tensor product $\boxtimes$ of bounded categories, it suffices to do it for categories of modules. For technical reasons we shall deal with a semistrict version of the symmetric tensor category of $\Bbbk$-vector spaces.

Let us name the main properties of the objects we are dealing with. The 2-category $\mathbf{AbCat}$ (resp. $\mathbf{AbCat}^r$) is formed by
0-morphisms, or objects: essentially small $\Bbbk$-linear abelian categories $\mathcal{A}$ with length,
1-morphisms: $\Bbbk$-linear left (resp. right) exact functors $F : \mathcal{A} \to \mathcal{B}$,
2-morphisms: natural transformations (morphisms of functors).

(See Bénabou [Bén67] for definition of a bicategory.)

Monoidal 2-categories are defined and studied by Kapranov and Voevodsky [KV94]. The more general notion of a monoidal bicategory has been proposed by Gordon, Power and Street [GPS95]. The relationship between these two theories is clarified by Baez and Neuchl [BN96]. The monoidal structure of $\mathbf{AbCat}$ and $\mathbf{AbCat}^r$ is given by Deligne's tensor product of categories $\boxtimes : (\mathcal{A}, \mathcal{B}) \mapsto \mathcal{A} \boxtimes \mathcal{B}$. The tensor product $F \boxtimes G$ of 1-morphisms $F : \mathcal{A} \to \mathcal{C}$ and $G : \mathcal{B} \to \mathcal{D}$ satisfies the condition that the diagram

$$\begin{array}{ccc} \mathcal{A} \times \mathcal{B} & \xrightarrow{F \times G} & \mathcal{C} \times \mathcal{D} \\ \boxtimes \downarrow & & \downarrow \boxtimes \\ \mathcal{A} \boxtimes \mathcal{B} & \xrightarrow{F \boxtimes G} & \mathcal{C} \boxtimes \mathcal{D} \end{array}$$

is commutative up to an isomorphism. However, this requirement determine $F \boxtimes G$ only up to isomorphism and is, hence, not sufficient for building the full monoidal structure.

In order to construct the monoidal 2-structure consider the 2-subcategory of categories of finite dimensional modules over associative finite dimensional algebras and right exact functors. Let F : mod-$A \to$ mod-B be a $\Bbbk$-linear right exact functor and let C be a finite dimensional associative algebra. We have to construct the functor $F \boxtimes C$: mod-$A \otimes C \to$ mod-$B \otimes C$. The composite mod-$A \otimes C \to$ mod-$A \xrightarrow{F}$ mod-B factorises via the underlying functor mod-$B \otimes C \to$ mod-B, due to the algebra homomorphism

$$C \to \operatorname{End}_A M \xrightarrow{F_{M,M}} \operatorname{End}_B FM \tag{4.3.1}$$

for any $A \otimes C$-module M. This determines $(F \boxtimes C)(M) = FM$ as a $B \otimes C$-module. Also $(F \boxtimes C)(f) = Ff$ on morphisms $f \in$ mod-$A \otimes C$.

A natural transformation $\alpha : F \to G : \text{mod-}A \to \text{mod-}B$ determines the natural transformation $\alpha \boxtimes C : F \boxtimes C \to G \boxtimes C : \text{mod-}A \otimes C \to \text{mod-}B \otimes C$, which is given by the linear map $\alpha_M : FM \to GM$ for any $A \otimes C$-module M. Similarly, the functor $C \boxtimes F : \text{mod-}C \otimes A \to \text{mod-}C \otimes B$ and the natural transformation $C \boxtimes \alpha : C \boxtimes F \to C \boxtimes G : \text{mod-}C \otimes A \to \text{mod-}C \otimes B$ are defined.

To give the monoidal structure on this 2-category we also need for given $F : \text{mod-}A \to \text{mod-}B$ and $G : \text{mod-}C \to \text{mod-}D$ an isomorphism of functors

$$\begin{array}{ccc} \text{mod-}A \otimes C & \xrightarrow{A \boxtimes G} & \text{mod-}A \otimes D \\ {\scriptstyle F\boxtimes C}\downarrow & \overset{\boxtimes_{F,G}}{\Longleftarrow} & \downarrow{\scriptstyle F\boxtimes D} \\ \text{mod-}B \otimes C & \xrightarrow{B \boxtimes G} & \text{mod-}B \otimes D \end{array}$$

or $\boxtimes_{F,G} : FGM \to GFM$ for an $A \otimes C$-module M. Clearly, for arbitrary functors such an isomorphism does not exists. However, for right exact functors we have it for the following reason. Consider the exact sequence of A-modules

$$M \otimes A \otimes A \xrightarrow{\alpha \otimes A - M \otimes m} M \otimes A \xrightarrow{\alpha} M \to 0,$$

where $(M, \alpha : M \otimes A \to M)$ is an A-module and its image under F

$$M \otimes A \otimes FA \xrightarrow{\alpha \otimes FA - M \otimes Fm} M \otimes FA \xrightarrow{F\alpha} FM \to 0,$$

which identifies FM with the tensor product $M \otimes_A FA$, where $FA \in A\text{-bimod-}B$. Using such exact sequences one defines $\boxtimes_{F,G}$ on a $A \otimes C$-module M with the actions $\alpha : M \otimes A \to M, \gamma : M \otimes C \to M$ via the diagram

$$\begin{array}{ccccc} M \otimes FA \otimes GC & \xrightarrow{F\alpha \otimes 1} & FM \otimes GC & \xrightarrow{FG\gamma} & (F \boxtimes D)(A \boxtimes G)M \\ {\scriptstyle 1\otimes P}\downarrow & & & & \downarrow{\scriptstyle \boxtimes_{F,G}} \\ M \otimes GC \otimes FA & \xrightarrow{G\gamma \otimes 1} & GM \otimes FA & \xrightarrow{GF\alpha} & (B \boxtimes G)(F \boxtimes C)M \end{array}$$

There are more structure elements [KV94], for instance, the associativity functor $a_{A,B,C} : \text{mod-}A \otimes (B \otimes C) \to \text{mod-}(A \otimes B) \otimes C$, which comes from the isomorphism of algebras $a^{-1} : (A \otimes B) \otimes C \to A \otimes (B \otimes C)$. If one wants to avoid using such data, one can apply the main results of Kapranov and Voevodsky [KV94] and Gordon, Power and Street [GPS95] implying that a monoidal 2-category is equivalent to a semistrict monoidal 2-category.

4.3.2 Semistrict monoidal 2-categories

In order to use the explicit definition of a symmetric monoidal 2-category it should be written down first. The difficulty here is that a weak version of it (a particular case of a weak 6-category) is not so easy to write down. That is why we prefer to work in a model situation and to deal with a strictly symmetric semistrict monoidal

2-category equivalent to the weak one we need. The key point is to replace the symmetric monoidal category of vector spaces (which is not strict) with an equivalent strict symmetric monoidal category. Then we use the monoidal category analogs of finite-dimensional algebras and modules.

The definition of a semistrict monoidal 2-category was given by Gordon, Power and Street in [GPS95] and by Kapranov and Voevodsky in [KV94]. Coincidence of both definitions is elucidated by Baez and Neuchl [BN96].

Definition 4.3.1 ([GPS95, KV94, BN96]). *A* semistrict monoidal 2-category *is a 2-category $\mathfrak{A}$ equipped with*

1. *An object $I \in \mathfrak{A}$ (the unit object);*
2. *For any two objects $A, B \in \mathfrak{A}$ an object $A \boxtimes B$ in $\mathfrak{A}$;*
3. *For any 1-morphism $F : A \to A'$ and any object $B \in \mathfrak{A}$ a 1-morphism $F \boxtimes B : A \boxtimes B \to A' \boxtimes B$;*
4. *For any 1-morphism $G : B \to B'$ and any object $A \in \mathfrak{A}$ a 1-morphism $A \boxtimes G : A \boxtimes B \to A \boxtimes B'$;*
5. *For any object $B \in \mathfrak{A}$ and any 2-morphism $\lambda : F \to G : A \to A'$ a 2-morphism $\lambda \boxtimes B : F \boxtimes B \to G \boxtimes B : A \boxtimes B \to A' \boxtimes B$;*
6. *For any object $A \in \mathfrak{A}$ and any 2-morphism $\lambda : F \to G : B \to B'$ a 2-morphism $A \boxtimes \lambda : A \boxtimes F \to A \boxtimes G : A \boxtimes B \to A \boxtimes B'$;*
7. *For any two 1-morphisms $F : A \to A'$ and $G : B \to B'$ a 2-isomorphism*

$$\begin{array}{ccc} A \boxtimes B & \xrightarrow{A \boxtimes G} & A \boxtimes B' \\ {\scriptstyle F \boxtimes B}\downarrow & \boxtimes_{F,G} \Downarrow & \downarrow{\scriptstyle F \boxtimes B'} \\ A' \boxtimes B & \xrightarrow[A' \boxtimes G]{} & A' \boxtimes B' \end{array}$$

such that the following conditions are satisfied

(i) *For any object $A \in \mathfrak{A}$ we have 2-functors $A \boxtimes - : \mathfrak{A} \to \mathfrak{A}$ and $- \boxtimes A : \mathfrak{A} \to \mathfrak{A}$;*
(ii) *$A \boxtimes I = A = I \boxtimes A$ for any object A,*
$F \boxtimes I = F = I \boxtimes F$ for any 1-morphism $F : A \to B$,
$\alpha \boxtimes I = \alpha = I \boxtimes \alpha$ for any 2-morphism $\alpha : F \to G : A \to B$;
(iii) *The tensor product is associative on objects: $A \boxtimes (B \boxtimes C) = (A \boxtimes B) \boxtimes C$. For any 1-morphism $F : C \to C'$ and any algebras A, B we have $A \boxtimes (B \boxtimes F) = (A \boxtimes B) \boxtimes F$ $A \boxtimes (F \boxtimes B) = (A \boxtimes F) \boxtimes B$, $F \boxtimes (A \boxtimes B) = (F \boxtimes A) \boxtimes B$. If $\alpha : F \to G : C \to C'$ is a 2-morphism then $A \boxtimes (B \boxtimes \alpha) = (A \boxtimes B) \boxtimes \alpha$, $A \boxtimes (\alpha \boxtimes B) = (A \boxtimes \alpha) \boxtimes B$, $(\alpha \boxtimes A) \boxtimes B = \alpha \boxtimes (A \boxtimes B)$;*
(iv) *For any 1-morphisms $F : A \to A'$, $G : B \to B'$ and $H : C \to C'$ in $\mathfrak{A}$ we have $\boxtimes_{A \boxtimes G, H} = A \boxtimes \boxtimes_{G,H}$, $\boxtimes_{F \boxtimes B, H} = \boxtimes_{F, B \boxtimes H}$, $\boxtimes_{F, G \boxtimes C} = \boxtimes_{F,G} \boxtimes C$;*
(v) *For any objects $A, B \in \mathfrak{A}$ we have $1_A \boxtimes B = 1_{A \boxtimes B} = A \boxtimes 1_B$, and for any 1-morphisms $F : A \to A'$, $G : B \to B'$ we have $\boxtimes_{1_A, G} = 1_{A \boxtimes G}$ and $\boxtimes_{F, 1_B} = 1_{F \boxtimes B}$;*

*(vi) For any 1-morphism $F : A \to A'$ and any 2-morphism $\psi : G \to H : B \to B'$
the following two 2-morphisms are equal.*

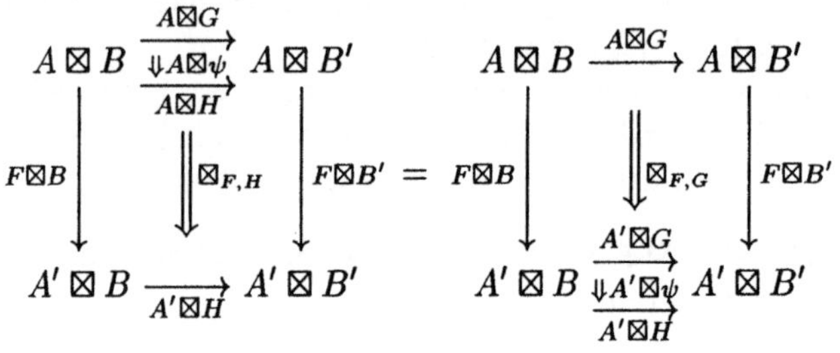

*(vii) For any 1-morphism $G : B \to B'$ and any 2-morphism $\psi : F \to H : A \to A'$
the following two 2-morphisms are equal:*

$$
\begin{array}{ccc}
A \boxtimes B \xrightarrow[F\boxtimes B]{\Uparrow \psi \boxtimes B \; H\boxtimes B} A' \boxtimes B & & A \boxtimes B \xrightarrow{H\boxtimes B} A' \boxtimes B \\
\downarrow A\boxtimes G \quad \Uparrow \boxtimes_{F,G} \quad \downarrow A'\boxtimes G & = & \downarrow A\boxtimes G \quad \Uparrow \boxtimes_{H,G} \quad \downarrow A'\boxtimes G \\
A \boxtimes B' \xrightarrow[F\boxtimes B']{} A' \boxtimes B' & & A \boxtimes B' \xrightarrow[F\boxtimes B']{\Uparrow \psi \boxtimes B' \; H\boxtimes B'} A' \boxtimes B'
\end{array}
$$

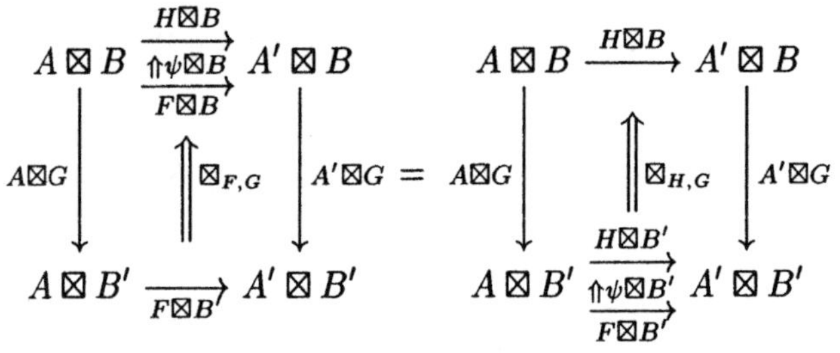

*(viii) For any 1-morphisms $F : A \to A',\ G : B \to C,\ H : C \to D$ the 2-morphism
$\boxtimes_{F,H\circ G}$ equals to the following pasting*

$$
\begin{array}{ccccc}
A \boxtimes B & \xrightarrow{A\boxtimes G} & A \boxtimes C & \xrightarrow{A\boxtimes H} & A \boxtimes D \\
\downarrow F\boxtimes B & \Downarrow \boxtimes_{F,G} & \downarrow F\boxtimes C \quad \Downarrow \boxtimes_{F,H} & & \downarrow F\boxtimes D \\
A' \boxtimes B & \xrightarrow[A'\boxtimes G]{} & A' \boxtimes C & \xrightarrow[A'\boxtimes H]{} & A' \boxtimes D
\end{array}
$$

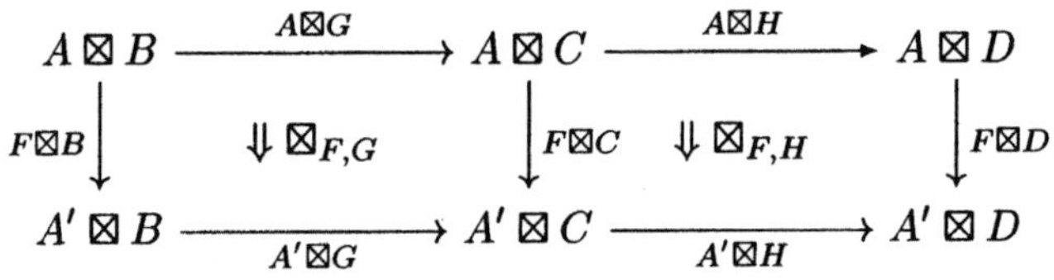

Similarly, the 2-morphism $\boxtimes_{H\circ G,F}$ equals to the pasting of $\boxtimes_{G,F}$ and $\boxtimes_{H,F}$.

4.3.3 A semistrict version of the 2-category of categories of modules

Let us construct in this paragraph a semistrict version of the 2-category of categories of modules. Our example is similar to the examples of monoidal 2-categories
constructed by Day and Street [DS97] via enriched categories. We start with a strict
symmetric monoidal category $\mathbf{V} = (\mathbf{V}, \otimes, 1, P)$, equivalent to $\mathbf{k}$-vect. For instance,
$\mathbf{V}$ is the monoidal category of coordinatized vector spaces $\mathbf{k}\text{-vect}_c$ described by
Kapranov and Voevodsky in [KV94]. Then $\mathbf{V}$ is a closed category and it has internal
homomorphism spaces $\underline{\mathrm{Hom}}(X,Y) = X^\vee \otimes Y$ for any pair of objects $X, Y \in \mathbf{V}$.
  Following Kelly [Kel82] we define a small $\mathbf{V}$-category $\mathcal{C}$ as

1. a set $\mathrm{Ob}\,\mathcal{C}$ of objects, such that;

2. for any pair $X, Y \in \mathrm{Ob}\,\mathcal{C}$ there is an object $\underline{\mathrm{Hom}}(X,Y) \in \mathrm{Ob}\,\mathbf{V}$;
3. for any object $X \in \mathrm{Ob}\,\mathcal{C}$ there is a morphism $1_X : 1 \to \underline{\mathrm{Hom}}(X,X) \in \mathbf{V}$ (the unit);
4. for any triple of objects $X, Y, Z \in \mathcal{C}$ there is a morphism $c_{X,Y,Z} : \underline{\mathrm{Hom}}(X,Y) \otimes \underline{\mathrm{Hom}}(Y,Z) \to \underline{\mathrm{Hom}}(X,Z) \in \mathbf{V}$ (the composition);

such that
the composition c is associative: $c_{X,Y,W} \circ (1 \otimes c_{Y,Z,W}) = c_{X,Z,W} \circ (c_{X,Y,Z} \otimes 1)$;
1_X is a left and right unit for the composition c:

$$(1 \otimes \underline{\mathrm{Hom}}(X,Y) \xrightarrow{1_X \otimes 1} \underline{\mathrm{Hom}}(X,X) \otimes \underline{\mathrm{Hom}}(X,Y) \xrightarrow{c_{X,X,Y}} \underline{\mathrm{Hom}}(X,Y)) = 1$$

(and the left unit property is similar).

A $\mathbf{V}$-functor $F : \mathcal{A} \to \mathcal{B}$ is defined as

1. a map $F : \mathrm{Ob}\,\mathcal{A} \to \mathrm{Ob}\,\mathcal{B}$;
2. for any pair $X, Y \in \mathrm{Ob}\,\mathcal{A}$ a morphism

$$F_{X,Y} : \underline{\mathrm{Hom}}_{\mathcal{A}}(X,Y) \to \underline{\mathrm{Hom}}_{\mathcal{B}}(FX,FY) \in \mathbf{V}$$

coherent with the units, i.e., $1_{FX} = F_{X,X} \circ 1_X : 1 \to \underline{\mathrm{Hom}}_{\mathcal{B}}(FX,FX)$, and with the composition, i.e., $c_{FX,FY,FZ} \circ (F_{X,Y} \otimes F_{Y,Z}) = F_{X,Z} \circ c_{X,Y,Z}$ [Kel82].

A $\mathbf{V}$-natural transformation $\lambda : F \to G : \mathcal{A} \to \mathcal{B}$ is defined as a set of morphisms $\lambda_X : 1 \to \underline{\mathrm{Hom}}(FX,GX) \in \mathbf{V}$ given for all $X \in \mathrm{Ob}\,\mathcal{A}$, such that the analogue of the usual relation holds:

$$\begin{array}{ccc}
\underline{\mathrm{Hom}}_{\mathcal{A}}(X,Y) & \xrightarrow{F_{X,Y} \otimes \lambda_Y} & \underline{\mathrm{Hom}}_{\mathcal{B}}(FX,FY) \otimes \underline{\mathrm{Hom}}_{\mathcal{B}}(FY,GY) \\
{\scriptstyle \lambda_X \otimes G_{X,Y}} \downarrow & & \downarrow {\scriptstyle c_{FX,FY,GY}} \\
\underline{\mathrm{Hom}}_{\mathcal{B}}(FX,GX) \otimes \underline{\mathrm{Hom}}_{\mathcal{B}}(GX,GY) & \xrightarrow{c_{FX,FY,GY}} & \underline{\mathrm{Hom}}_{\mathcal{B}}(FX,GY)
\end{array} \tag{4.3.2}$$

The reader is referred to Kelly [Kel82] for more general definitions, which apply to an arbitrary monoidal category $\mathbf{V}$, which is not necessarily strict. In particular, for $\mathbf{V}' = \Bbbk$-vect we get the usual notions of $\Bbbk$-linear categories, functors and natural transformations. Recall that (essentially) small $\Bbbk$-linear categories form a 2-category $\Bbbk$-Cat. Similarly, $\mathbf{V}$-categories, $\mathbf{V}$-functors and $\mathbf{V}$-natural transformations form a 2-category $\mathbf{V}$-**Cat**.

The objects of $\mathbf{V}$-**Cat** are small $\mathbf{V}$-categories. For any pair, $\mathcal{A}$, $\mathcal{B}$, of $\mathbf{V}$-categories there is a small category, $\mathbf{V}$-**Cat**$(\mathcal{A},\mathcal{B})$, whose objects are $\mathbf{V}$-functors $F : \mathcal{A} \to \mathcal{B}$ and the morphisms $\lambda : F \to G$ are $\mathbf{V}$-natural transformations $\lambda : 1 \to \underline{\mathrm{Hom}}_{\mathcal{B}}(FX,GX)$. The identity morphism $\mathbb{1}_F : F \to F : \mathcal{A} \to \mathcal{B}$ is given by $(\mathbb{1}_F)_X = 1_{FX} : 1 \to \underline{\mathrm{Hom}}_{\mathcal{B}}(FX,FX)$. The composition of $\lambda : F \to G$ and $\mu : G \to H$ is given by

$$1 = 1 \otimes 1 \xrightarrow{\lambda_X \otimes \mu_X} \underline{\mathrm{Hom}}(FX,GX) \otimes \underline{\mathrm{Hom}}(GX,HX) \xrightarrow{c_{FX,GX,HX}} \mathrm{Hom}(FX,HX).$$

The composition in $\mathbf{V}$-**Cat** is given by the functor

$$\mathbf{V}\text{-}\mathbf{Cat}(\mathcal{A},\mathcal{B}) \times \mathbf{V}\text{-}\mathbf{Cat}(\mathcal{B},\mathcal{C}) \longrightarrow \mathbf{V}\text{-}\mathbf{Cat}(\mathcal{A},\mathcal{C})$$
$$(\alpha : F' \to F'', \beta : G' \to G'') \longmapsto \alpha\cdot\beta : F'\cdot G' \to F''\cdot G''$$

$$\begin{aligned}\alpha\cdot\beta &= \big(F'\cdot G' \xrightarrow{\alpha\cdot G'} F''\cdot G' \xrightarrow{F''\cdot\beta} F''\cdot G''\big)\\ &= \big(F'\cdot G' \xrightarrow{F'\cdot\beta} F'\cdot G'' \xrightarrow{\alpha\cdot G''} F''\cdot G''\big),\end{aligned} \tag{4.3.3}$$

where $F\cdot G = GF$ is the composition of $\mathbf{V}$-functors and

$$\alpha\cdot G \overset{\text{def}}{=} \big(1 \xrightarrow{\alpha X} \underline{\mathrm{Hom}}_{\mathcal{B}}(F',F''X) \xrightarrow{G} \underline{\mathrm{Hom}}_{\mathcal{C}}(GF'X,GF''X)\big),$$
$$F\cdot\beta \overset{\text{def}}{=} \big(1 \xrightarrow{\beta_{FX}} \underline{\mathrm{Hom}}_{\mathcal{C}}(G'FX,G''FX)\big).$$

The distributivity (4.3.3) follows by diagram (4.3.2). The unit for this composition is the identity functor. One can verify the associativity axiom easily.

Remark 4.3.2. If $\mathbf{V}$ is monoidally equivalent to $\mathbf{V}'$, then $\mathbf{V}$-**Cat** and $\mathbf{V}'$-Cat are equivalent 2-categories. In particular, for $\mathbf{V}$ equivalent to $\Bbbk$-vect we have the equivalence of 2-categories $\mathbf{V}$-**Cat** $\to$ $\Bbbk$-Cat, $\mathcal{C} \to \underline{\mathcal{C}}$, where $\mathrm{Ob}\,\underline{\mathcal{C}} = \mathrm{Ob}\,\mathcal{C}$, $\underline{\mathcal{C}}(X,Y) = \mathbf{V}(1,\underline{\mathrm{Hom}}_{\mathcal{C}}(X,Y))$. Indeed, the functor $\mathbf{V} \to \Bbbk$-vect, $Z \to \mathbf{V}(1,Z)$ is the canonical equivalence. We shall say that a $\mathbf{V}$-category $\mathcal{C}$ is abelian if $\underline{\mathcal{C}}$ is, that a $\mathbf{V}$-functor $F : \mathcal{A} \to \mathcal{B}$ is exact if $\underline{F} : \underline{\mathcal{A}} \to \underline{\mathcal{B}}$ is, etc. Thus we can transfer the properties of ordinary functors to $\mathbf{V}$-functors. Many results in $\mathbf{V}$-category theory are simple corollaries of ordinary results in this way.

Define the 2-category $\mathbf{V}$-$\mathbf{Cat}^{r}$, whose
objects are abelian $\mathbf{V}$-categories with length,
1-morphisms are right exact $\mathbf{V}$-functors,
2-morphisms are $\mathbf{V}$-natural transformations
and consider its full 2-subcategory $\mathbf{V}$-**Cat-mod** consisting of the categories of modules, defined as follows.

For any morphism in $\mathbf{V}$ choose and fix its kernel. Let A be an associative unital algebra in the category $\mathbf{V}$. Its $\mathbf{V}$-category $\mathbf{V}^{A}$ of right A-modules is defined as the set of unital associative A-modules $(M, \alpha : M\otimes A \to M)$, $M \in \mathrm{Ob}\,\mathbf{V}$, $\alpha \in \mathbf{V}$, equipped with the hom-objects

$$\underline{\mathrm{Hom}}_{A}(M,N) =$$
$$\mathrm{Ker}\big(\underline{\mathrm{Hom}}(\alpha,N) - \underline{\mathrm{Hom}}(M\otimes A,\beta)\circ(-\otimes 1_{A}) : \underline{\mathrm{Hom}}(M,N) \to \underline{\mathrm{Hom}}(M\otimes A,N)\big)$$

for any pair of A-modules (M,α) and (N,β) (recall that kernels are chosen). Here

$$-\otimes 1_{A} = \big(\underline{\mathrm{Hom}}(M,N) = M^{\vee}\otimes N = M^{\vee}\otimes 1\otimes N \xrightarrow{M^{\vee}\,\mathrm{coev}_{A}\otimes N} M^{\vee}\otimes A^{\vee}\otimes A\otimes N$$
$$\xrightarrow{c\otimes c} A^{\vee}\otimes M^{\vee}\otimes N\otimes A \simeq (M\otimes A)^{\vee}\otimes(N\otimes A) = \underline{\mathrm{Hom}}(M\otimes A, N\otimes A)\big).$$

Note that $\mathrm{Hom}_A(M,N) \overset{\mathrm{def}}{=} \mathbf{V}(1, \underline{\mathrm{Hom}}_A(M,N))$ is identified with the set of A-module morphisms, i.e. $f : M \to N \in \mathrm{Mor}\,\mathbf{V}$, such that

$$\big(M \otimes A \xrightarrow{\alpha} M \xrightarrow{f} N\big) = \big(M \otimes A \xrightarrow{f\otimes 1} N \otimes A \xrightarrow{\beta} N\big),$$

where α, β are the actions.

To see that $\mathbf{V}^A$ is a $\mathbf{V}$-category consider the above definition for $\mathbf{V} = \Bbbk$-vect. Clearly, $(\Bbbk\text{-vect})^A$ is the usual category mod-A. We have to construct the composition in $\mathbf{V}^A$, to check that the unit morphisms $1 \to \underline{\mathrm{Hom}}_{\mathbf{V}}(M,M)$ are in the kernel, to verify associativity and unitality of the composition. Applying $\mathbf{V}(1,\text{-})$ to these statements we get valid statements in mod-A. Therefore, they are valid also in $\mathbf{V}^A$.

The following statement is obvious.

Lemma 4.3.3. *Let A be an algebra in $\mathbf{V}$ and let $M \in \mathrm{Ob}\,\mathbf{V}$. A right A-module structure in M is equivalent to a homomorphism of algebras $A \to \underline{\mathrm{End}}\,M \overset{\mathrm{def}}{=} \underline{\mathrm{Hom}}(M,M) = M^\vee \otimes M$.* □

Define a full 2-subcategory $\mathbf{V}$-**Cat-mod** in $\mathbf{V}$-**Cat**^{r} as follows:
The objects are associative unital algebras A in $\mathbf{V}$;
the 1-morphisms $A \to B$ are right exact $\mathbf{V}$-functors $\mathbf{V}^A \to \mathbf{V}^B$;
the 2-morphisms are $\mathbf{V}$-natural transformations.

Remark 4.3.4. Note that a category associated with $\mathbf{V}^A$ is $\underline{\mathbf{V}}^A$, whose morphism sets are $\underline{\mathbf{V}}^A(M,N) = \mathbf{V}(1, \underline{\mathrm{Hom}}_A(M,N)) = \mathrm{Hom}_A(M,N)$ — the sets of A-module morphisms in $\mathbf{V}$. To each $\mathbf{V}$-functor $F : \mathbf{V}^A \to \mathbf{V}^B$ corresponds a unique functor $\underline{F} : \underline{\mathbf{V}}^A \to \underline{\mathbf{V}}^B$ such that

$$\begin{aligned}\underline{F}_{M,N} : \mathrm{Hom}_A(M,N) &= \mathbf{V}(1, \underline{\mathrm{Hom}}_A(M,N)) \\ &\xrightarrow{\mathbf{V}(1,F_{M,N})} \mathbf{V}(1, \underline{\mathrm{Hom}}_B(FM,FN)) = \mathrm{Hom}_B(FM,FN).\end{aligned}$$

A $\mathbf{V}$-natural transformation $\lambda : F \to G : \mathbf{V}^A \to \mathbf{V}^B$ corresponds to a unique natural transformation $\underline{\lambda} : \underline{F} \to \underline{G} : \underline{\mathbf{V}}^A \to \underline{\mathbf{V}}^B$, which is the B-module morphism $\underline{\lambda}_M = \lambda_M \in \mathbf{V}(1, \underline{\mathrm{Hom}}_B(FM,GM)) = \mathrm{Hom}_B(FM,GM)$ given for each A-module $M \in \mathbf{V}^A$.

4.3.4 Construction of the 2-monoidal structure

We shall make $\mathbf{V}$-**Cat-mod** into a semistrict monoidal 2-category in the sense of Kapranov and Voevodsky [KV94], Gordon, Power and Street [GPS95], and made explicit by Baez and Neuchl in [BN96, Lemma 4].

Proposition 4.3.5. $\mathbf{V}$-**Cat-mod** *is a semistrict monoidal 2-category.*

The monoidal structure is built with the following data:

1) The algebra $1 \in \mathbf{V}$ is the unit object of **V-Cat-mod**. Note that $\mathbf{V}^1$ is isomorphic to $\mathbf{V}$.

2) The tensor product of two objects A, B (algebras in **V**) is the algebra $A \otimes B$ with the multiplication

$$A \otimes B \otimes A \otimes B \xrightarrow{(23)} A \otimes A \otimes B \otimes B \xrightarrow{m \otimes m} A \otimes B.$$

Lemma 4.3.6. *Let A, B be algebras in $\mathbf{V}$, and let $M \in \operatorname{Ob} \mathbf{V}$. The structure of a right $A \otimes B$-module in M amounts to any of the following algebra homomorphisms in $\mathbf{V}$*

(i) $A \otimes B \to \underline{\operatorname{End}}\, M$,
(ii) $A \to \underline{\operatorname{End}}_B\, M$,
(iii) $B \to \underline{\operatorname{End}}_A\, M$.

Proof. Straightforward verification. Or one can apply $\mathbf{V}(1,\text{-})$ to the relevant diagrams (the universal proof).

Lemma 4.3.7. *Let $M, N \in \mathbf{V}^{A \otimes B}$. Denote the action of B by $\beta : M \otimes B \to M$, $\gamma : N \otimes B \to N$. Then $\underline{\operatorname{Hom}}_{A \otimes B}(M,N) \to \underline{\operatorname{Hom}}_A(M,N)$ is the equaliser of the pair of morphisms*

$$\underline{\operatorname{Hom}}_A(M,N) \xrightarrow{\operatorname{Hom}(\beta,N)} \underline{\operatorname{Hom}}_A(M \otimes B, N)$$

$$\underline{\operatorname{Hom}}_A(M,N) \xrightarrow{\text{-}\otimes 1_B} \underline{\operatorname{Hom}}_A(M \otimes B, N \otimes B) \xrightarrow{\operatorname{Hom}(M \otimes B,\gamma)} \underline{\operatorname{Hom}}_A(M \otimes B, N)$$

Proof. This holds in $\Bbbk$-vect.

3) Let $F : \mathbf{V}^A \to \mathbf{V}^{A'}$ be a $\mathbf{V}$-functor, and let B be an algebra in $\mathbf{V}$. We want to construct a $\mathbf{V}$-functor $F \boxtimes B : \mathbf{V}^{A \otimes B} \to \mathbf{V}^{A' \otimes B}$. For any module $M \in \mathbf{V}^{A \otimes B}$ there is an algebra morphism

$$\beta' : B \to \underline{\operatorname{End}}_A\, M \xrightarrow{F_{M,M}} \underline{\operatorname{End}}_{A'}\, FM.$$

It makes FM into an $A' \otimes B$-module with the action

$$FM \otimes A' \otimes B \xrightarrow{\alpha'_{FM} \otimes \beta'} FM \otimes \underline{\operatorname{End}}_{A'}\, FM \to FM \otimes \underline{\operatorname{End}}\, FM \to FM.$$

Hence, a map on objects $F \boxtimes B : \operatorname{Ob} \mathbf{V}^{A \otimes B} \to \operatorname{Ob} \mathbf{V}^{A' \otimes B}$ is constructed. We have to extend it to the morphisms.

Proposition 4.3.8. *Let $F : \mathbf{V}^A \to \mathbf{V}^{A'}$ be a $\mathbf{V}$-functor and let $M, N \in \mathbf{V}^{A \otimes B}$. Then $F_{M,N} : \underline{\operatorname{Hom}}_A(M,N) \to \underline{\operatorname{Hom}}_{A'}(FM,FN)$ induces a morphism $(F \boxtimes B)_{M,N} : \underline{\operatorname{Hom}}_{A \otimes B}(M,N) \to \underline{\operatorname{Hom}}_{A' \otimes B}(FM,FN)$ and the collection of such morphisms gives rise to a functor denoted $F \boxtimes B$.*

Proof. Denote the original action of B by $\beta : M \otimes B \to M, \gamma : N \otimes B \to N$ and the induced action by $\beta' : FM \otimes B \to FM, \gamma' : FN \otimes B \to FN$. Equip $M \otimes B$ (resp. $FM \otimes B$) with an A-module (resp. A'-module) structure

$$M \otimes B \otimes A \xrightarrow{(23)} M \otimes A \otimes B \xrightarrow{\alpha \otimes B} M \otimes B,$$

$$FM \otimes B \otimes A' \xrightarrow{(23)} M \otimes A' \otimes B \xrightarrow{\alpha' \otimes B} FM \otimes B,$$

where α, α' are the actions. Then there is an isomorphism of A'-modules $\phi_M : F(M \otimes B) \to FM \otimes B$ due to the fact that $B \simeq 1^n = \oplus_1^n 1$ as an object in $\mathbf{V}$. We use the equivalence $\mathbf{V} \simeq \mathbf{k}$-vect. The homomorphism of A-modules $\beta : M \otimes B \to M$ represented as $\tilde{\beta} : 1 \to \underline{\mathrm{Hom}}_A(M \otimes B, M)$ composed with $F_{M \otimes B, M}$ induces a homomorphism $F\beta : F(M \otimes B) \to FM$ of A'-modules. Using the basis $b_i \in \mathbf{V}(1, B)$ one finds that

$$F\beta = \left(F(M \otimes B) \xrightarrow{\phi} FM \otimes B \xrightarrow{\beta'} FM\right).$$

Compare the following two commutative diagrams

$$\begin{array}{ccc}
\underline{\mathrm{Hom}}_A(M,N) & \xrightarrow{\underline{\mathrm{Hom}}(\beta,N)} & \underline{\mathrm{Hom}}_A(M \otimes B, N) \\
 & & \downarrow{\scriptstyle F_{M \otimes B,N}} \\
\downarrow{\scriptstyle F_{M,N}} & \overset{\underline{\mathrm{Hom}}(F\beta,FN)}{\nearrow} & \underline{\mathrm{Hom}}_{A'}(F(M \otimes B), FN) \\
 & & \downarrow{\scriptstyle \underline{\mathrm{Hom}}(\phi_M^{-1},FN)} \\
\underline{\mathrm{Hom}}_{A'}(FM,FN) & \xrightarrow[\underline{\mathrm{Hom}}(\beta',FN)]{} & \underline{\mathrm{Hom}}_{A'}(FM \otimes B, FN)
\end{array}$$

$$\begin{array}{ccccc}
\underline{\mathrm{Hom}}_A(M,N) & \xrightarrow{-\otimes 1_B} & \underline{\mathrm{Hom}}_A(M\otimes B,N\otimes B) & \xrightarrow{\underline{\mathrm{Hom}}(1,\gamma)} & \underline{\mathrm{Hom}}_A(M\otimes B,N) \\
 & & \downarrow{\scriptstyle F_{M\otimes B,N\otimes B}} & & \downarrow{\scriptstyle F_{M\otimes B,N}} \\
\downarrow{\scriptstyle F_{M,N}} & & \underline{\mathrm{Hom}}_{A'}(F(M\otimes B),F(N\otimes B)) & \xrightarrow{\underline{\mathrm{Hom}}(1,F\gamma)} & \underline{\mathrm{Hom}}_{A'}(F(M\otimes B),FN) \\
 & & \downarrow{\scriptstyle \underline{\mathrm{Hom}}(\phi_M^{-1},\phi_N)} & & \downarrow{\scriptstyle \underline{\mathrm{Hom}}(\phi_M^{-1},FN)} \\
\underline{\mathrm{Hom}}_{A'}(FM,FN) & \xrightarrow{-\otimes 1_B} & \underline{\mathrm{Hom}}_{A'}(FM\otimes B,FN\otimes B) & \xrightarrow{\underline{\mathrm{Hom}}(1,\gamma')} & \underline{\mathrm{Hom}}_{A'}(FM\otimes B,FN)
\end{array}$$

with coinciding left and right columns. Since $r_1 : \underline{\mathrm{Hom}}_{A \otimes B}(M, N) \to \underline{\mathrm{Hom}}_A(M, N)$ is the equaliser of the two upper rows, any path starting with this map and ending in the lower right corner gives the same morphism. Therefore, the composition $F_{M,N} \circ r_1$ factors through $r_1' : \underline{\mathrm{Hom}}_{A' \otimes B}(FM, FN) \to \underline{\mathrm{Hom}}_{A'}(FM, FN)$ — the equaliser of the two lower rows. Functoriality of $F \boxtimes B$ is clear.

4) Let $G : \mathbf{V}^B \to \mathbf{V}^{B'}$ be a $\mathbf{V}$-functor and let A be an algebra in $\mathbf{V}$. For any module $M \in A \otimes B$ there is an algebra morphism

$$\alpha' : A \to \underline{\mathrm{End}}_B\, M \xrightarrow{G_{M,M}} \underline{\mathrm{End}}_{B'}\, GM.$$

It makes GM into an $A \otimes B'$-module. Similarly to 3) we get a functor $A \boxtimes G : \mathbf{V}^{A\otimes B} \to \mathbf{V}^{A\otimes B'}$.

5) For a given $\mathbf{V}$-natural transformation $\lambda : F \to G : \mathbf{V}^A \to \mathbf{V}^{A'}$ we construct a $\mathbf{V}$-natural transformation $\lambda \boxtimes B : F \boxtimes B \to G \boxtimes B : \mathbf{V}^{A\otimes B} \to \mathbf{V}^{A'\otimes B}$ as follows. The natural transformation $\underline{\lambda}$ commutes with the morphisms of A-modules, in particular, with the action of the algebra $\mathbf{V}(1, B)$. This can be written as a commutative diagram

$$\begin{array}{ccc}
B & \longrightarrow & \underline{\mathrm{End}}_A\, M \otimes 1 \\
\downarrow & & \downarrow{\scriptstyle F_{M,M}\otimes\lambda_M} \\
1 \otimes \underline{\mathrm{End}}_A\, M & & \underline{\mathrm{End}}_{A'}\, FM \otimes \underline{\mathrm{Hom}}_{A'}(FM, GM) \\
{\scriptstyle \lambda_M \otimes G_{M,M}}\downarrow & & \downarrow{\scriptstyle c_{FM,FM,GM}} \\
\underline{\mathrm{Hom}}_{A'}(FM, GM) \otimes \underline{\mathrm{End}}_{A'}\, GM & \xrightarrow{c_{FM,GM,GM}} & \underline{\mathrm{Hom}}_{A'}(FM, GM),
\end{array}$$

which transforms to the equality

$$\begin{aligned}
&\big(1 \xrightarrow{\lambda_M} \underline{\mathrm{Hom}}_{A'}(FM, GM) \xrightarrow{\underline{\mathrm{Hom}}(\beta, GM)} \underline{\mathrm{Hom}}_{A'}(FM \otimes B, GM)\big) \\
&= \big(1 \xrightarrow{\lambda_M} \underline{\mathrm{Hom}}_{A'}(FM, GM) \xrightarrow{-\otimes 1_B} \underline{\mathrm{Hom}}_{A'}(FM \otimes B, GM \otimes B) \\
&\qquad \xrightarrow{\underline{\mathrm{Hom}}(FM\otimes B, \gamma)} \underline{\mathrm{Hom}}_{A'}(FM \otimes B, GM)\big),
\end{aligned}$$

where β and γ are actions of B. By Lemma 4.3.7 λ_M factorises via the equaliser as

$$\lambda_M : 1 \xrightarrow{(\lambda\boxtimes B)_M} \underline{\mathrm{Hom}}_{A'\otimes B}(FM, GM) \to \underline{\mathrm{Hom}}_{A'}(FM, GM).$$

This defines a $\mathbf{V}$-natural transformation $\lambda \boxtimes B$.

6) For a given $\mathbf{V}$-natural transformation $\lambda : F \to G : \mathbf{V}^B \to \mathbf{V}^{B'}$ we construct a $\mathbf{V}$-natural transformation $A \boxtimes \lambda : A \boxtimes F \to A \boxtimes G : \mathbf{V}^{A\otimes B} \to \mathbf{V}^{A\otimes B'}$ similar to 5).

7) Let $F : \mathbf{V}^A \to \mathbf{V}^{A'}$, $G : \mathbf{V}^B \to \mathbf{V}^{B'}$ be a right exact $\mathbf{V}$-functors. We look for a $\mathbf{V}$-natural transformation $\boxtimes_{F,G}$

$$\begin{array}{ccc}
\mathbf{V}^{A\otimes B} & \xrightarrow{A\boxtimes G} & \mathbf{V}^{A\otimes B'} \\
{\scriptstyle F\boxtimes B}\downarrow & {\scriptstyle \boxtimes_{F,G}}\Downarrow & \downarrow{\scriptstyle F\boxtimes B'} \\
\mathbf{V}^{A'\otimes B} & \xrightarrow[A'\boxtimes G]{} & \mathbf{V}^{A'\otimes B'}
\end{array}$$

Note that the functors $A \boxtimes G$ and $F \boxtimes B$ exist for arbitrary F and G, not only for right exact ones. However, without the right exactness assumption $(F \boxtimes B')(A \boxtimes G)$ and $(A' \boxtimes G)(F \boxtimes B)$ are not necessarily isomorphic.

Let $(M, \alpha : M \otimes A \to M)$ be an A-module. Then the exact sequence in $\mathbf{V}^A$

$$M \otimes A \otimes A \xrightarrow{\alpha\otimes A - M\otimes m} M \otimes A \xrightarrow{\alpha} M \to 0$$

gives the exact sequence in $\mathbf{V}^{A'}$

$$M \otimes A \otimes S \xrightarrow{\alpha\otimes S - M\otimes\sigma} M \otimes S \xrightarrow{F\alpha} FM \to 0, \tag{4.3.4}$$

where the bimodule $S = {}_A S_{A'} = FA \in {}^A\mathbf{V}^{A'}$ is equipped with the left action $\sigma = Fm_A : A \otimes S \to S$ coming from $A \to \underline{\mathrm{End}}_A(A_A) \xrightarrow{F_{A,A}} \underline{\mathrm{End}}_{A'} S$. Similarly, for a B-module $(N, \beta : N \otimes B \to N)$ we obtain the exact sequence in $\mathbf{V}^{B'}$

$$N \otimes B \otimes Q \xrightarrow{\beta\otimes Q - B\otimes\psi} N \otimes Q \xrightarrow{G\beta} GN \to 0,$$

where the bimodule $Q = {}_B Q_{B'} = GB \in {}^B\mathbf{V}^{B'}$ is equipped with the left action $\psi = Gm_B : B \otimes Q \to Q$.

Now, let $(M, \alpha, \beta) \in \mathbf{V}^{A\otimes B}$. The columns and the lower row of the following commutative diagram in $\mathbf{V}^{A'\otimes B'}$ are exact.

$$\begin{array}{ccccc}
M \otimes A \otimes S \otimes B \otimes Q & \xrightarrow{\alpha\otimes 1 - 1\otimes\sigma\otimes 1} & M \otimes S \otimes B \otimes Q & & \\
{\scriptstyle \overset{(234)}{\beta\otimes A\otimes S\otimes Q} - M\otimes A\otimes S\otimes\psi}\downarrow & & \downarrow{\scriptstyle \overset{(23)}{\beta\otimes S\otimes Q} - M\otimes S\otimes\psi} & & \\
M \otimes A \otimes S \otimes Q & \xrightarrow{\alpha\otimes 1 - 1\otimes\sigma\otimes 1} & M \otimes S \otimes Q & & \\
{\scriptstyle \overset{(234)}{G\beta\otimes A\otimes S}}\downarrow & & \downarrow{\scriptstyle \overset{(23)}{G\beta\otimes S}} & & \\
GM \otimes A \otimes S & \xrightarrow{G\alpha\otimes S - GM\otimes\sigma} & GM \otimes S & \xrightarrow{GF\alpha} & GFM
\end{array}$$

Therefore, there is an exact sequence in $\mathbf{V}^{A'\otimes B'}$

$$M \otimes S \otimes B \otimes Q \oplus M \otimes S \otimes A \otimes Q \xrightarrow{[\beta\otimes 1\circ(23) - 1\otimes\psi]\oplus[\alpha\otimes 1 - 1\otimes\sigma\otimes 1]} M \otimes S \otimes Q$$
$$\xrightarrow{GF\alpha\circ G(\beta\otimes S)\circ(23)} GFM \to 0.$$

A similar exact sequence determines FGM.

Furthermore, (permuted versions of) these sequences are related by a chain map

$$\begin{array}{ccc}
M \otimes B \otimes S \otimes Q \oplus M \otimes A \otimes S \otimes Q & \xrightarrow{[\beta\otimes 1 - 1\otimes\sigma\circ(23)]\oplus[\alpha\otimes 1 - 1\otimes\sigma\otimes 1]} & M \otimes S \otimes Q \\
{\scriptstyle (34)\oplus}\downarrow{\scriptstyle (34)} & & \downarrow{\scriptstyle (23)} \\
M \otimes B \otimes Q \otimes S \oplus M \otimes A \otimes Q \otimes S & \xrightarrow{[\beta\otimes 1 - 1\otimes\sigma\otimes 1]\oplus[\alpha\otimes 1 - 1\otimes\sigma\circ(23)]} & M \otimes Q \otimes S
\end{array}$$

It induces a unique isomorphism $\boxtimes_{F,G}(M)$ between the cokernels FGM and GFM such that

$$\begin{array}{ccccccc} M\otimes S\otimes Q & \xrightarrow{F(\alpha\otimes Q)} & F(M\otimes Q) & \xrightarrow{FG\beta} & FGM & \longrightarrow & 0 \\ \downarrow{\scriptstyle (23)} & & & & \downarrow{\scriptstyle \boxtimes_{F,G}(M)} & & \\ M\otimes Q\otimes S & \xrightarrow{G(\beta\otimes S)} & G(M\otimes S) & \xrightarrow{GF\alpha} & GFM & \longrightarrow & 0 \end{array} \qquad (4.3.5)$$

Clearly, it gives the **V**-natural transformation $\boxtimes_{F,G}$ that we are looking for.

Now we have all structure morphisms, and it remains to verify the axioms of a semistrict monoidal 2-category.

(i) For any algebra $A \in \mathbf{V}$ we have 2-functors $A \boxtimes - : \mathbf{V\text{-}Cat\text{-}mod} \to \mathbf{V\text{-}Cat\text{-}mod}$, $- \boxtimes A : \mathbf{V\text{-}Cat\text{-}mod} \to \mathbf{V\text{-}Cat\text{-}mod}$.

(ii) The unit object 1 viewed as an algebra satisfies the conditions:
$A \otimes 1 = A = 1 \otimes A$ for any algebra A;
$F \boxtimes 1 = F = 1 \boxtimes F$ for any **V**-functor $F : \mathbf{V}^A \to \mathbf{V}^B$;
$\alpha \boxtimes 1 = \alpha = 1 \boxtimes \alpha$ for any **V**-natural transformation $\alpha : F \to G : \mathbf{V}^A \to \mathbf{V}^B$.

(iii) The tensor product is associative on objects: $A \otimes (B \otimes C) = (A \otimes B) \otimes C$. For any **V**-functor $F : \mathbf{V}^C \to \mathbf{V}^{C'}$ and any algebras A, B we have

$$A \boxtimes (B \boxtimes F) = (A \otimes B) \boxtimes F : \mathbf{V}^{A\otimes B\otimes C} \to \mathbf{V}^{A\otimes B\otimes C'},$$

since to give a morphism $A \to \underline{\mathrm{End}}_{B\otimes C} M$ is equivalent to giving a morphism $A \otimes B \to \underline{\mathrm{End}}_C M$. Similarly, $A \boxtimes (F \boxtimes B) = (A \boxtimes F) \boxtimes B$ and $F \boxtimes (A \otimes B) = (F \boxtimes A) \boxtimes B$. If $\alpha : F \to G : \mathbf{V}^C \to \mathbf{V}^{C'}$ is a natural transformation then $A \boxtimes (B \boxtimes \alpha) = (A \otimes B) \boxtimes \alpha$, $A \boxtimes (\alpha \boxtimes B) = (A \boxtimes \alpha) \boxtimes B$, and $(\alpha \boxtimes A) \boxtimes B = \alpha \boxtimes (A \otimes B)$.

(iv) Using the notations from 7) let us assume that $F : \mathbf{V}^A \to \mathbf{V}^{A'}$ and $G : \mathbf{V}^B \to \mathbf{V}^{B'}$ are right exact **V**-functors. Let C be an algebra. We have to prove that $\boxtimes_{F,G\boxtimes C} = \boxtimes_{F,G} \boxtimes C$. The resolution (4.3.4) written for the functor $H = G \boxtimes C : \mathbf{V}^{B\otimes C} \to \mathbf{V}^{B'\otimes C}$ becomes

$$M\otimes B\otimes C\otimes GB\otimes C \xrightarrow{\gamma\otimes 1\circ\beta\otimes 1 - Gm_B\otimes m_C\circ(34)} M\otimes GB\otimes C \xrightarrow{G\beta\circ\gamma\otimes 1\circ(23)} HM \to 0.$$

Denote by $\delta = \gamma\circ\beta\otimes 1 : M\otimes B\otimes C \to M$ the action. From definition (4.3.5) we get the expression for $\boxtimes_{F,H}$ via

$$\begin{array}{ccccccc} M\otimes FA\otimes GB\otimes C & \xrightarrow{F(\alpha\otimes 1)} & F(M\otimes GB\otimes C) & \xrightarrow{FG\delta} & FGM & \longrightarrow & 0 \\ \downarrow{\scriptstyle (432)} & & & & \downarrow{\scriptstyle \boxtimes_{F,H}} & & \\ M\otimes GB\otimes C\otimes FA & \xrightarrow{G(\delta\otimes FA)} & G(M\otimes FA) & \xrightarrow{GF\alpha} & GFM & \longrightarrow & 0 \end{array}$$

Clearly, this diagram projects onto Diagram (4.3.5) using the action $\gamma : M \otimes C \to M$. Therefore, $\boxtimes_{F,H} = \boxtimes_{F,G} \boxtimes C$ as stated. Similarly, $\boxtimes_{A\boxtimes G,H} = A \boxtimes \boxtimes_{G,H}$.

Consider now given right exact **V**-functors $F : \mathbf{V}^A \to \mathbf{V}^{A'}$, $H : \mathbf{V}^C \to \mathbf{V}^{C'}$ and denote $S = FA$, $R = HC$. For any algebra B the **V**-natural transformation $\boxtimes_{F\boxtimes B,H}$ is found from the commutative diagram

$$\begin{array}{ccccccc}
M\otimes S\otimes B\otimes R & \xrightarrow{\beta\otimes 1\circ(23)} & M\otimes S\otimes R & \xrightarrow{F\alpha\otimes 1} & FM\otimes R & \xrightarrow{FH\gamma} & FHM \\
{\scriptstyle (234)}\downarrow & & & \searrow & & & \downarrow{\scriptstyle \boxtimes_{F\boxtimes B,H}} \\
M\otimes R\otimes S\otimes B & \xrightarrow[H\gamma\otimes 1]{} & HM\otimes S\otimes B & \xrightarrow[H\beta\otimes 1\circ(23)]{} & HM\otimes S & \xrightarrow[HF\alpha]{} & HFM
\end{array}$$

and the $\mathbf{V}$-natural transformation $\boxtimes_{F,B\boxtimes H}$ is determined by the commutative diagram

$$\begin{array}{ccccccc}
M\otimes S\otimes B\otimes R & \xrightarrow{F\alpha\otimes 1} & FM\otimes B\otimes R & \xrightarrow{F\beta\otimes 1} & FM\otimes R & \xrightarrow{FH\gamma} & FHM \\
{\scriptstyle (432)}\downarrow & & & \searrow & & & \downarrow{\scriptstyle \boxtimes_{F,B\boxtimes H}} \\
M\otimes B\otimes R\otimes S & \xrightarrow[\beta\otimes 1]{} & M\otimes R\otimes S & \xrightarrow[H\gamma\otimes 1]{} & HM\otimes S & \xrightarrow[HF\alpha]{} & HFM
\end{array}$$

The first rows in these diagrams determine equal composite morphisms. The diagonal arrows also coincide. Therefore, $\boxtimes_{F\boxtimes B,H} = \boxtimes_{F,B\boxtimes H}$.

(v) Let Id^A denote the identity functor $\mathrm{Id} : \mathbf{V}^A \to \mathbf{V}^A$. Clearly, $\mathrm{Id}^A \boxtimes B = \mathrm{Id}^{A\otimes B} = A \boxtimes \mathrm{Id}^B$.

The $\mathbf{V}$-natural transformation $\boxtimes_{\mathrm{Id}^A,G}$ for $G : \mathbf{V}^B \to \mathbf{V}^{B'}$ is found from the commutative diagram

$$\begin{array}{ccccc}
M\otimes A\otimes GB & \xrightarrow{\alpha\otimes 1} & M\otimes GB & \xrightarrow{G\beta} & GM \\
{\scriptstyle (23)}\downarrow & & & & \downarrow{\scriptstyle \boxtimes_{\mathrm{Id}_A,G}} \\
M\otimes GB\otimes A & \xrightarrow{G\beta\otimes 1} & GM\otimes A & \xrightarrow{G\alpha} & GM
\end{array}$$

as the identity morphism $1_{A\boxtimes G}$. Indeed, the above diagram is the image under G of a diagram, which states that actions α and β on M commute. Similarly, $\boxtimes_{F,\mathrm{Id}^B} = 1_{F\boxtimes B}$.

(vi) Let $F : \mathbf{V}^A \to \mathbf{V}^{A'}$ and $\psi : G \to H : \mathbf{V}^B \to \mathbf{V}^{B'}$. We want to show that the following two natural transformations are equal.

$$\begin{array}{ccc}
\mathbf{V}^{A\otimes B} & \overset{A\boxtimes G}{\underset{A\boxtimes H}{\rightrightarrows}}\ {\scriptstyle \Downarrow A\boxtimes\psi} & \mathbf{V}^{A\otimes B'} \\
{\scriptstyle F\boxtimes B}\downarrow & \Downarrow{\scriptstyle \boxtimes_{F,H}} & \downarrow{\scriptstyle F\boxtimes B'} \\
\mathbf{V}^{A'\otimes B} & \xrightarrow[A'\boxtimes H]{} & \mathbf{V}^{A'\otimes B'}
\end{array}
\quad = \quad
\begin{array}{ccc}
\mathbf{V}^{A\otimes B} & \xrightarrow{A\boxtimes G} & \mathbf{V}^{A\otimes B'} \\
{\scriptstyle F\boxtimes B}\downarrow & \Downarrow{\scriptstyle \boxtimes_{F,G}} & \downarrow{\scriptstyle F\boxtimes B'} \\
\mathbf{V}^{A'\otimes B} & \overset{A'\boxtimes G}{\underset{A'\boxtimes H}{\rightrightarrows}}\ {\scriptstyle \Downarrow A'\boxtimes\psi} & \mathbf{V}^{A'\otimes B'}
\end{array}
\qquad (4.3.6)$$

Let $S = FA$, $Q = GB$, $R = HB$ and $M \in \mathbf{V}^{A\otimes B}$. There is a diagram

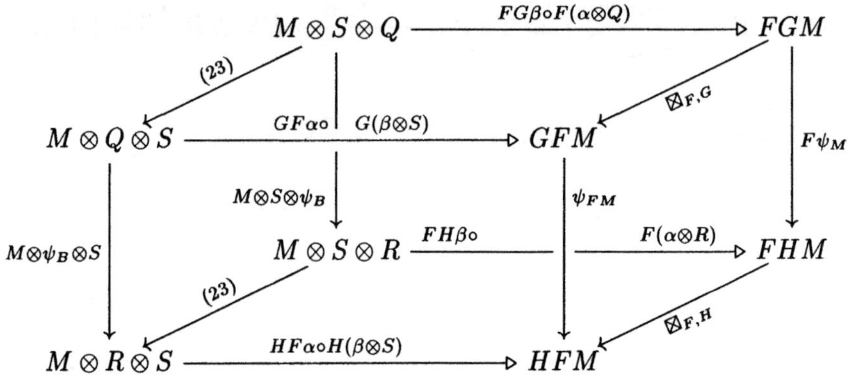

where the upper face is Diagram (4.3.5), and the lower is similar. The left, the front and the back walls commute. Hence, the right wall commutes. Thereby equation (4.3.6) is proven.

(vii) Let $G : \mathbf{V}^B \to \mathbf{V}^{B'}$ and $\psi : F \to H : \mathbf{V}^A \to \mathbf{V}^{A'}$. Similar to (vi) the following two natural transformations are equal:

$$
\begin{array}{ccc}
\mathbf{V}^{A\otimes B} \; \begin{array}{c} \overset{H\boxtimes B}{\longrightarrow} \\ \Uparrow\psi\boxtimes B \\ \underset{F\boxtimes B}{\longrightarrow} \end{array} \; \mathbf{V}^{A'\otimes B} & & \mathbf{V}^{A\otimes B} \xrightarrow{H\boxtimes B} \mathbf{V}^{A'\otimes B} \\
A\boxtimes G \Big\downarrow \quad \Uparrow \boxtimes_{F,G} \quad \Big\downarrow A'\boxtimes G \;=\; A\boxtimes G \Big\downarrow & & \quad \Uparrow \boxtimes_{H,G} \quad \Big\downarrow A'\boxtimes G \\
\mathbf{V}^{A\otimes B'} \xrightarrow{F\boxtimes B'} \mathbf{V}^{A'\otimes B'} & & \mathbf{V}^{A\otimes B'} \; \begin{array}{c} \overset{H\boxtimes B'}{\longrightarrow} \\ \Uparrow\psi\boxtimes B' \\ \underset{F\boxtimes B'}{\longrightarrow} \end{array} \; \mathbf{V}^{A'\otimes B'}
\end{array}
$$

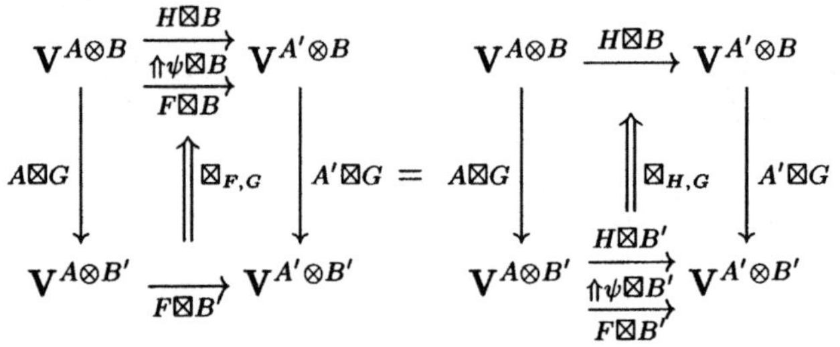

(viii) Let $F : \mathbf{V}^A \to \mathbf{V}^{A'}$, $G : \mathbf{V}^B \to \mathbf{V}^C$ and $H : \mathbf{V}^C \to \mathbf{V}^D$ be right exact $\mathbf{V}$-functors. We want to check that the pasting

$$
\begin{array}{ccccc}
\mathbf{V}^{A\otimes B} & \xrightarrow{A\boxtimes G} & \mathbf{V}^{A\otimes C} & \xrightarrow{A\boxtimes H} & \mathbf{V}^{A\otimes D} \\
F\boxtimes B \downarrow & \Downarrow \boxtimes_{F,G} & \downarrow F\boxtimes C & \Downarrow \boxtimes_{F,H} & \downarrow F\boxtimes D \qquad (4.3.7) \\
\mathbf{V}^{A'\otimes B} & \xrightarrow{A'\boxtimes G} & \mathbf{V}^{A'\otimes C} & \xrightarrow{A'\boxtimes H} & \mathbf{V}^{A'\otimes D}
\end{array}
$$

denoted also as $\mu \circ \lambda$ for $\lambda = \boxtimes_{F,G} * (A' \boxtimes H)$ and $\mu = (A \boxtimes G) * \boxtimes_{F,G}$ equals to $\boxtimes_{F,H\circ G}$.

Let $(M, \alpha, \beta) \in \mathbf{V}^{A\otimes B}$. There are actions $\gamma : GM \otimes C \to GM$ and $\delta : GFM \otimes C \to GFM$. Pasting (4.3.7) fits into the commutative diagram

$$
\begin{array}{ccccccc}
M{\otimes}GB{\otimes}HC{\otimes}FA & \xrightarrow{G\beta\otimes 1} & GM{\otimes}HC{\otimes}FA & \xrightarrow{H\gamma\otimes 1} & HGM{\otimes}FA & \xrightarrow{FHG\alpha} & FHGM \\
(34)\Big\downarrow & & (23)\Big\downarrow & & & & \Big\downarrow\lambda \\
M{\otimes}GB{\otimes}FA{\otimes}HB & \xrightarrow{G\beta\otimes 1} & GM{\otimes}FA{\otimes}HC & \xrightarrow{FG\alpha\otimes 1} & FGM{\otimes}HC & \xrightarrow{HF\gamma} & HFGM \\
(23)\Big\downarrow & & & & \boxtimes_{F,G}(M)\otimes 1 \Big\downarrow & & \Big\downarrow\mu \\
M{\otimes}FA{\otimes}GB{\otimes}HC & \xrightarrow{F\alpha\otimes 1} & FM{\otimes}GB{\otimes}HC & \xrightarrow{GF\beta\otimes 1} & GFM{\otimes}HC & \xrightarrow{H\delta} & HGFM
\end{array}
$$

and it is determined by it. The value of $\boxtimes_{F,HG}$ on M is found from

$$\begin{array}{ccccc}
M \otimes HGB \otimes FA & \xrightarrow{HG\beta\otimes 1} & HGM \otimes FA & \xrightarrow{FHG\alpha} & (F \boxtimes D)(A \boxtimes HG)M \\
{\scriptstyle (23)}\Big\downarrow & & & & \Big\downarrow{\scriptstyle \boxtimes_{F,HG}(M)} \\
M \otimes FA \otimes HGB & \xrightarrow{F\alpha\otimes 1} & FM \otimes HGB & \xrightarrow{HGF\beta} & (A' \boxtimes HG)(F \boxtimes B)M
\end{array}$$

Look at these two diagrams as at the top and the bottom of a parallelepiped diagram, whose walls are made of 6 rectangles; the vertical maps are 3 identity maps in the right half and 3 maps of the form $1 \otimes H\varkappa$ and $1 \otimes H\varkappa \otimes 1$ in the left half, where $\varkappa : GB \otimes C \to GB$ is the action and $H\varkappa : GB \otimes HC \to HGB$. The left, front and back walls are commutative, thereby proving that the right one is as well. That is, $\mu \circ \lambda = \boxtimes_{F,HG}$.

We conclude that **V-Cat-mod** is a semistrict monoidal 2-category.

4.3.5 Braided monoidal 2-structures

We will not give the full definition of a braided semistrict monoidal 2-category, since we deal with a particular case of a *strictly symmetric* one. The meaning is specified in the proof of the following proposition.

Proposition 4.3.9. **V-Cat-mod** *is a braided semistrict monoidal 2-category. Moreover, it is strictly symmetric, that is, braided in the most strict sense (in particular, sylleptic [DS97]) and* $R_{A,B} \cdot R_{B,A} = 1$.

Proof. For any pair of algebras $A, B \in \mathbf{V}$ there is an isomorphism of categories

$$R_{A,B} : \mathbf{V}^{A\otimes B} \to \mathbf{V}^{B\otimes A},$$

$$(M, \alpha : M \otimes A \otimes B \to M) \mapsto (M, M \otimes B \otimes A \xrightarrow{M\otimes P} M \otimes A \otimes B \xrightarrow{\alpha} M).$$

Clearly, $R_{A,B} \cdot R_{B,A} = 1$.

The squares

$$\begin{array}{ccc}
\mathbf{V}^{A\otimes B} & \xrightarrow{F\boxtimes B} & \mathbf{V}^{A'\otimes B} \\
{\scriptstyle R_{A,B}}\Big\downarrow & = & \Big\downarrow{\scriptstyle R_{A',B}} \\
\mathbf{V}^{B\otimes A} & \xrightarrow{B\boxtimes F} & \mathbf{V}^{B\otimes A'}
\end{array}
\quad , \quad
\begin{array}{ccc}
\mathbf{V}^{A\otimes B} & \xrightarrow{A\boxtimes G} & \mathbf{V}^{A\otimes B'} \\
{\scriptstyle R_{A,B}}\Big\downarrow & = & \Big\downarrow{\scriptstyle R_{A,B'}} \\
\mathbf{V}^{B\otimes A} & \xrightarrow{G\boxtimes A} & \mathbf{V}^{B'\otimes A}
\end{array}$$

commute. Also the prisms

$$\begin{array}{ccc}
\mathbf{V}^{A\otimes B} & \overset{F\boxtimes B}{\underset{G\boxtimes B}{\overset{\longrightarrow}{\underset{\longrightarrow}{\Downarrow\lambda\boxtimes B}}}} & \mathbf{V}^{A'\otimes B} \\
{\scriptstyle R_{A,B}}\Big\downarrow & & \Big\downarrow{\scriptstyle R_{A',B}} \\
\mathbf{V}^{B\otimes A} & \overset{B\boxtimes F}{\underset{B\boxtimes G}{\overset{\longrightarrow}{\underset{\longrightarrow}{\Downarrow B\boxtimes\lambda}}}} & \mathbf{V}^{B\otimes A'}
\end{array}
\quad , \quad
\begin{array}{ccc}
\mathbf{V}^{A\otimes B} & \overset{A\boxtimes F}{\underset{A\boxtimes G}{\overset{\longrightarrow}{\underset{\longrightarrow}{\Downarrow A\boxtimes\lambda}}}} & \mathbf{V}^{A\otimes B'} \\
{\scriptstyle R_{A,B}}\Big\downarrow & & \Big\downarrow{\scriptstyle R_{A,B'}} \\
\mathbf{V}^{B\otimes A} & \overset{F\boxtimes A}{\underset{G\boxtimes A}{\overset{\longrightarrow}{\underset{\longrightarrow}{\Downarrow\lambda\boxtimes A}}}} & \mathbf{V}^{B'\otimes A}
\end{array}$$

are commutative. Therefore, $R_{A,B}$ together with identity transformations $R_{F,B} = \mathbb{I}$, $R_{A,G} = \mathbb{I}$ is a pseudonatural transformation.

We have equalities of $\mathbf{V}$-functors:

$$R_{A,B\otimes C} = \left(\mathbf{V}^{A\otimes B\otimes C} \xrightarrow{R_{A,B}\boxtimes C} \mathbf{V}^{B\otimes A\otimes C} \xrightarrow{B\boxtimes R_{A,C}} \mathbf{V}^{B\otimes C\otimes A}\right),$$
$$R_{A\otimes B,C} = \left(\mathbf{V}^{A\otimes B\otimes C} \xrightarrow{A\boxtimes R_{B,C}} \mathbf{V}^{A\otimes C\otimes B} \xrightarrow{R_{A,C}\boxtimes B} \mathbf{V}^{C\otimes A\otimes B}\right).$$

They give identity modifications from the left hand side to the right hand side.

To check that $R_{A,B}$ gives a braiding in the monoidal 2-category **V-Cat-mod** we need to verify the following axioms:

S1) For any $\mathbf{V}$-functors $F : \mathbf{V}^A \to \mathbf{V}^{A'}$, $G : \mathbf{V}^B \to \mathbf{V}^{B'}$ there is a commutative cube

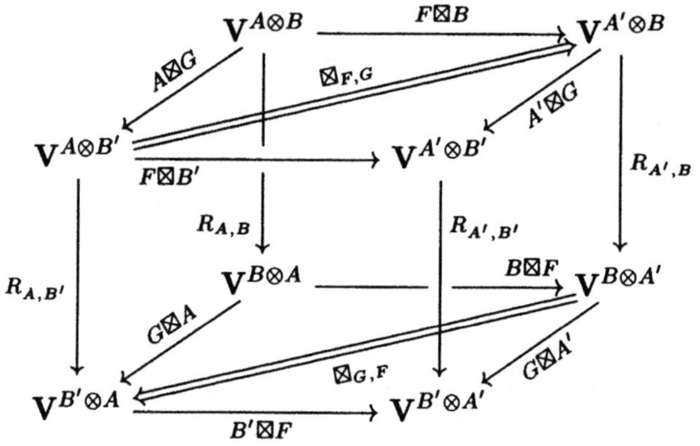

The walls are commutative. The top 2-morphism is determined by Diagram (4.3.5) for any $M \in \mathbf{V}^{A\otimes B}$. We use the notations of 7). The bottom 2-morphism written for

$$M' = R_{A,B}M = \left(M, M\otimes B\otimes A \xrightarrow{(23)} M\otimes A\otimes B \to M\right)$$

is determined by a similar diagram

$$\begin{array}{ccccc} M'\otimes Q\otimes S & \xrightarrow{G\beta\otimes 1} & GM'\otimes S & \xrightarrow{GF\alpha} & GFM' \\ {\scriptstyle (23)}\downarrow & & & & \downarrow{\scriptstyle \boxtimes_{G,F}(M')} \\ M'\otimes S\otimes Q & \xrightarrow{F\alpha\otimes 1} & FM'\otimes Q & \xrightarrow{FG\beta} & FGM' \end{array}$$

Pasting the above diagram and Diagram (4.3.5) and taking into account that $(23)^2 = 1$, we obtain $\boxtimes_{G,F}(M')\circ\boxtimes_{F,G}(M) = 1_{FGM}$. This is the commutativity of the cube.

S2) We have to check that $\boxtimes_{F,R_{B,C}} = 1$. In other words, that the following square commutes:

$$\begin{array}{ccc} \mathbf{V}^{A\otimes B\otimes C} & \xrightarrow{A\boxtimes R_{B,C}} & \mathbf{V}^{A\otimes C\otimes B} \\ {\scriptstyle F\boxtimes B\boxtimes C}\downarrow & = & \downarrow{\scriptstyle F\boxtimes C\boxtimes B} \\ \mathbf{V}^{A'\otimes B\otimes C} & \xrightarrow{A'\boxtimes R_{B,C}} & \mathbf{V}^{A'\otimes C\otimes B} \end{array}$$

Denote $D = B \otimes C$. This algebra acts in $M \in \mathbf{V}^{A\otimes B\otimes C}$ via $\varkappa : M \otimes D \to M$. We have $R_{B,C}D = D$ with the natural action of $C \otimes B$. In our case Diagram (4.3.5) takes the form

$$\begin{array}{ccccc} M \otimes FA \otimes D & \xrightarrow{F\alpha\otimes 1} & FM \otimes D & \xrightarrow{F\varkappa} & FM \\ {\scriptstyle (23)}\downarrow & & & & \Vert \, {\scriptstyle \boxtimes_{F,R_{B,C}}} \\ M \otimes D \otimes FA & \xrightarrow{\varkappa\otimes 1} & M \otimes FA & \xrightarrow{F\alpha} & FM \end{array}$$

The right vertical arrow is the identity map since this diagram is the F-image of a diagram, expressing the fact that M is an $A \otimes D$-module.

S3) Similarly we check that $\boxtimes_{R_{A,B},G} = 1$.

Other numerous axioms of a braided semistrict monoidal 2-category are obvious or follow from the above. Moreover, $R_{B,A} \circ R_{A,B} = 1$, hence **V-Cat-mod** is a symmetric category in a strict sense.

In particular, there is a strict action of the symmetric group S_N in the tensor power $\mathcal{A}^{\boxtimes N} = \mathbf{V}^{A^{\otimes N}}$ of the category $\mathcal{A} = \mathbf{V}^A$. Namely, for any permutation $\sigma \in S_N$ there is a functor $R_\sigma : \mathcal{A}^{\boxtimes N} \to \mathcal{A}^{\boxtimes N}$, which equals $\mathbf{V}^{P_\sigma^{-1}} : \mathbf{V}^{A^{\otimes N}} \to \mathbf{V}^{A^{\otimes N}}$, where $P_\sigma^{-1} : A^{\otimes N} \to A^{\otimes N}$ is the automorphism of the algebra $A^{\otimes N}$ given by the action of the permutation $\sigma^{-1} \in S_N$. Clearly, $R_\sigma \circ R_\tau = R_{\sigma\circ\tau}$ and $R_{(i,i+1)} = A^{\otimes i-1} \boxtimes R_{A,A} \boxtimes A^{\otimes N-i-1}$.

Now we can notice that the 2-category of modules over finite dimensional $\mathbb{k}$-algebras is built from $\mathbb{k}$-vect in the same way as **V-Cat-mod** is built from **V**, taking into account that $\mathbb{k}$-vect is not strict. This makes it a monoidal 2-category with more non-trivial data in comparison with **V-Cat-mod** (some are still trivial, like the pentagon 2-morphism $a_{A,B,C,D}$).

To define the monoidal 2-structure in $\mathbf{AbCat}^r$ (resp. $\mathbf{V}\text{-}\mathbf{Cat}^r$) we recall that any category from Ob $\mathbf{AbCat}^r$ (resp. $\mathbf{V}\text{-}\mathbf{Cat}^r$) is an inductive limit of its subcategories equivalent to mod-A (resp. $\mathbf{V}^A$) [Del91]. For categories mod-A (resp. $\mathbf{V}^A$) the tensor product exists and is again a category of that type [Del91]. In general case Deligne shows the existence of the tensor product of inductive limits from $\mathbf{AbCat}^r$ by proving that the inductive limit of tensor products satisfies the properties of the definition [Del91]. Applying this procedure to $\mathbf{V}\text{-}\mathbf{Cat}^r$ and **V-Cat-mod** we get a weak version of a monoidal 2-category, that is, a monoidal 2-category structure in $\mathbf{V}\text{-}\mathbf{Cat}^r$. It depends on the choice of the above inductive limit. Since the 2-category $\mathbf{AbCat}^r$ is equivalent to $\mathbf{V}\text{-}\mathbf{Cat}^r$ we also get a monoidal 2-category structure in $\mathbf{AbCat}^r$.

Since **V-Cat-mod** is strictly symmetric, the 2-categories $\mathbf{AbCat}^r$ and $\mathbf{V}\text{-}\mathbf{Cat}^r$ are symmetric in a weak sense, see, for instance, [Lyu99].

In order to define the symmetric monoidal 2-structure in **AbCat**, we simply remark that **AbCat** is isomorphic to $\mathbf{AbCat}^r$, and the isomorphism is given by the 2-functor $\mathcal{C} \mapsto \mathcal{C}^{\mathrm{op}}$.

5. Coends and construction of Hopf algebras

In the previous chapter we described properties of Hopf algebras in braided tensor categories. Now we are going to construct an important class of examples of such Hopf algebras. They are built as special inductive limits, namely, coends.

We begin with a discussion of a large class of coends in abelian tensor categories that are determined by an expression with operations $\otimes$, $\boxtimes$, $\_^\vee$, etc. We compute several coends and establish canonical isomorphisms among them.

We review how to construct in any rigid, abelian braided category a natural Hopf algebra using $F = \int^{X \in \mathcal{C}} X \otimes X^\vee$ as an object, if this coend exists in $\mathcal{C}$. This happens precisely when $\mathcal{C}$ is *bounded*, that is, equivalent to the category of finite dimensional modules over a finite dimensional associative $\Bbbk$-algebra. The structure morphisms as well as co-actions of F on $\mathcal{C}$ and a natural Hopf-pairing, are obtained very explicitly from natural transformations associated to special tangles due to the universal property of F as an inductive limit.

We construct a special Hopf pairing, $\omega : F \otimes F \to 1$, for such a Hopf algebra F. When this pairing is non-degenerate the bounded braided category $\mathcal{C}$ is called *modular*. From the modularity we deduce that the object of integrals of F is isomorphic to the unit object. Furthermore, the integrals for F are two-sided. Another fact closely related to one of the basic topological equivalences arises for the natural transformation of the identity functor of $\mathcal{C}$ corresponding to the integral in $\mathrm{Hom}_{\mathcal{C}}(F, 1)$. We use the theory of squared Hopf algebras to show that its image is of the form $1 \oplus \ldots \oplus 1$ for all objects in $\mathcal{C}$.

5.1 The coend

In an extended TQFT formalism a cylinder, viewed as a cobordism from the empty set to two circles, is mapped to an object in $\mathcal{C} \boxtimes \mathcal{C}$. It is shown in [Lyu99, Theorem 1.8.6] that the object representing this cylinder is a special coend, which for our purposes can be written as $\mathbb{F} = \int^{X \in \mathcal{C}} X \boxtimes X^\vee$. Furthermore, this coend will exist as an object of $\mathcal{C} \boxtimes \mathcal{C}$ if and only if $\mathcal{C}$ is bounded. That is why we study coends in this section.

5.1.1 General coends

Let $\mathcal{C}$ be a $\mathbf{k}$-linear abelian category with length. The coend C of a bifunctor $B : \mathcal{P} \times \mathcal{P}^{\mathrm{op}} \to \mathcal{C}$ is defined in [Mac88] as an object of $\mathcal{C}$, which is the inductive limit of the diagram

$$B(X,X) \xleftarrow{B(X,f)} B(X,Y) \xrightarrow{B(f,Y)} B(Y,Y),$$

where $f : X \to Y$ runs over $\operatorname{Mor} \mathcal{P}$. That is, C is equipped with a morphism $i_X : B(X,X) \to C \in \mathcal{C}$ for each $X \in \operatorname{Ob} \mathcal{P}$, the square in the diagram

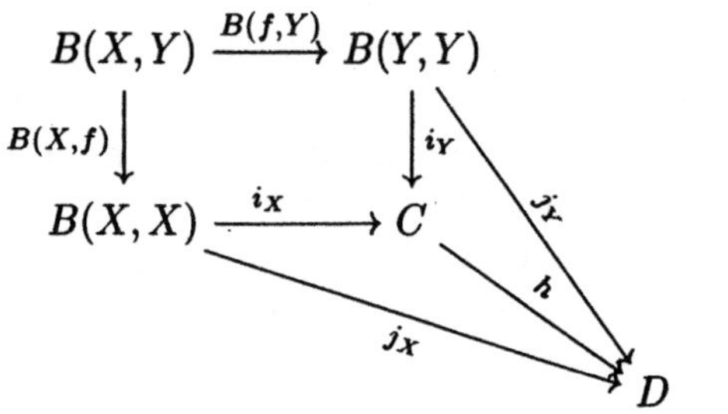

(5.1.1)

is commutative for any $f : X \to Y \in \operatorname{Mor} \mathcal{P}$, and C is universal between such objects. The last condition reads: if the exterior commutes for a system $j_X : B(X,X) \to D$, then there exists a unique $h : C \to D$ making the triangles commutative.

Notation. The coend C is denoted $\int^{X \in \mathcal{P}} B(X,X)$.

If $\mathcal{P}$ is small, we can say that the sequence in the cocompletion $\widehat{\mathcal{C}} = \operatorname{Ind} - \mathcal{C}$

$$\bigoplus_{f:X \to Y \in \operatorname{Mor} \mathcal{P}} B(X,Y) \xrightarrow{B(X,f)-B(f,Y)} \bigoplus_{X \in \operatorname{Ob} \mathcal{P}} B(X,X) \xrightarrow{\oplus i_X} C \to 0$$

is exact. So in this case the coend exists as an object of $\widehat{\mathcal{C}}$. More generally, it exists for essentially small $\mathcal{P}$.

Let us consider the particular case. Let $p : \mathcal{P} \to \mathcal{V}$ be a functor from an essentially small category $\mathcal{P}$, let $\mathcal{C} = \mathcal{V} \boxtimes \mathcal{V}$ and let $B : \mathcal{P} \times \mathcal{P}^{\mathrm{op}} \to \mathcal{C}$, $B(X,Y) = pX \boxtimes (pY)^{\vee}$. The coend is denoted

$$\mathbb{F} = \int^{X \in \mathcal{P}} pX \boxtimes (pX)^{\vee}. \tag{5.1.2}$$

The commutative diagram (5.1.1) can be also expressed as a dinatural transformation $i : B \xrightarrow{\bullet\bullet} C$, where C is the constant bifunctor with the value C. (See Mac Lane [Mac88] for the definition of a dinatural transformation and its properties.) The coend of B is the universal dinatural transformation $i : B \xrightarrow{\bullet\bullet} C$. In particular, any dinatural transformation $j : B \xrightarrow{\bullet\bullet} D$ to an object D factorizes uniquely through the coend as $j : B \xrightarrow{\bullet\bullet} C \xrightarrow{h} D$.

From this definition we immediately find

Proposition 5.1.1 (Mac Lane [Mac88] Proposition IX.7.1). *Let $\gamma : B \to B' : \mathcal{C}^{op} \times \mathcal{C} \to \mathcal{A}$ be a natural transformation, and assume that the coends of B and B' exist. Then there is a unique morphism*

$$\int^X \gamma_{X,X} : \int^{X \in \mathcal{C}} B(X,X) \to \int^{X \in \mathcal{C}} B'(X,X)$$

such that

$$\begin{array}{ccc} B(X,X) & \xrightarrow{\gamma_{X,X}} & B'(X,X) \\ \downarrow{\scriptstyle i_X} & & \downarrow{\scriptstyle i'_X} \\ \int^{X \in \mathcal{C}} B(X,X) & \xrightarrow{\int^X \gamma_{X,X}} & \int^{X \in \mathcal{C}} B'(X,X) \end{array}$$

commutes for any $X \in \operatorname{Ob} \mathcal{C}$.

Multiple coends can be computed consecutively as stated in the following "Fubini theorem". We will have to use them, when dealing with surfaces with many holes.

Proposition 5.1.2 (Mac Lane [Mac88] Proposition IX.8). *Let $B : \mathcal{P}^{op} \times \mathcal{P} \times \mathcal{C}^{op} \times \mathcal{C} \to \mathcal{A}$ be a functor, such that for any pair $P, Q \in \mathcal{P}$ the coend $\int^X B(P,Q,X,X)$ exists. Then it determines a functor $\mathcal{P} \times \mathcal{P}^{op} \to \mathcal{C}$, $(P,Q) \mapsto \int^X B(P,Q,X,X)$. The double coend $\int^{(P,X)\in\mathcal{P}\times\mathcal{C}} B(P,P,X,X)$ and the iterated coend $\int^{P\in\mathcal{P}} \left(\int^{X\in\mathcal{C}} B(P,P,X,X)\right)$ exist simultaneously. They are isomorphic and the isomorphism θ satisfies*

$$\begin{array}{ccc} B(P,P,X,X) & \xrightarrow{i_{P,X}} & \int^{(P,X)\in\mathcal{P}\times\mathcal{C}} B(P,P,X,X) \\ \downarrow{\scriptstyle i_X} & & \downarrow{\scriptstyle \theta} \\ \int^{X\in\mathcal{C}} B(P,P,X,X) & \xrightarrow{i_P} & \int^{P\in\mathcal{P}} \left(\int^{X\in\mathcal{C}} B(P,P,X,X)\right) \end{array}$$

5.1.2 A particular coend

Let $\mathcal{C}$ and $\mathcal{A}$ be additive $\mathbb{k}$-linear categories, such that the $\mathbb{k}$-spaces $\operatorname{Hom}_{\mathcal{C}}(X,Y)$ are finite dimensional. Let $F : \mathcal{C} \to \mathcal{A}$ be a $\mathbb{k}$-linear functor.

We need the notion of a tensor product of an object with a vector space (e.g. Deligne and Milne [DM82]). The easiest way to introduce it is to choose a $\mathbb{k}$-bilinear functor $T : \mathcal{A} \times \mathbb{k}\text{-vect} \to \mathcal{A}$, such that $T(X, \mathbb{k}) = X$, $T(f, \lambda) = \lambda f : X \to Y$ and $f : X \to Y$, where $\mathbb{k}$ is the field viewed as a one-dimensional vector space.

If V is any finite dimensional vector space, it has a basis, that is, a system of maps $e_i : \Bbbk \to V, p_i : V \to \Bbbk, 1 \leqslant i \leqslant n$, such that

$$p_j \circ e_i = \delta_{i,j}, \qquad \sum_i e_i \circ p_i = \mathbb{1}_V. \tag{5.1.3}$$

Hence, there are morphisms $X = T(X, \Bbbk) \xrightarrow{T(1,e_i)} T(X, V)$ and $T(X, V) \xrightarrow{T(1,p_i)} T(X, \Bbbk) = X$, which obey the same relations, thus making $T(X, V)$ into a direct sum $\oplus_1^n X$. It follows immediately that such functors T exist and they are all isomorphic. Now we define the tensor product of an object X with a vector space V as

$$X \otimes V = T(X, V), \qquad f \otimes g = T(f, g).$$

So it is just another notation for the functor T.

Let us apply this to $V = \mathrm{Hom}_{\mathcal{C}}(X, Z)$, with $X, Z \in \mathrm{Ob}\,\mathcal{C}$. Let $e_i : \Bbbk \to \mathrm{Hom}_{\mathcal{C}}(X, Z)$ and $p_i : \mathrm{Hom}_{\mathcal{C}}(X, Z) \to \Bbbk$ be its basis. For any linear map $g : \Bbbk \to \mathrm{Hom}_{\mathcal{C}}(X, Z)$ denote $\bar{g} = g(1) : X \to Z \in \mathcal{C}$, for any morphism $f : X \to Z$ denote $\underline{f} : \Bbbk \to \mathrm{Hom}_{\mathcal{C}}(X, Z)$ the linear map $1 \mapsto f$.

Lemma 5.1.3. *Let*

$$\mathrm{ev}_X = \sum_{j=1}^{n} \left(FX \otimes \mathrm{Hom}(X, Z) \xrightarrow{1 \otimes p_j} FX \otimes \Bbbk = FX \xrightarrow{F\bar{e}_j} FZ \right).$$

Then the following diagram is commutative for any morphism $f : X \to Z$

$$\begin{array}{ccc} FX \otimes \Bbbk & = & FX \\ {\scriptstyle 1 \otimes \underline{f}} \downarrow & & \downarrow {\scriptstyle Ff} \\ FX \otimes \mathrm{Hom}(X, Z) & \xrightarrow{\mathrm{ev}_X} & FZ \end{array}$$

In particular, ev_X *does not depend on the choice of basis.*

Proof. By linearity we can assume that $f = \bar{e}_i : X \to Z$, $\underline{f} = e_i : \Bbbk \to \mathrm{Hom}_{\mathcal{C}}(X, Z)$, and use (5.1.3).

Lemma 5.1.4. *The morphisms* $\mathrm{ev}_X : FX \otimes \mathrm{Hom}(X, Z) \to FZ$, *define a dinatural transformation (see diagram* (5.1.1)*).*

Proof. Let $f : X \to Y$ and $g : Y \to Z$ be morphisms in $\mathcal{C}$. Then in the diagram

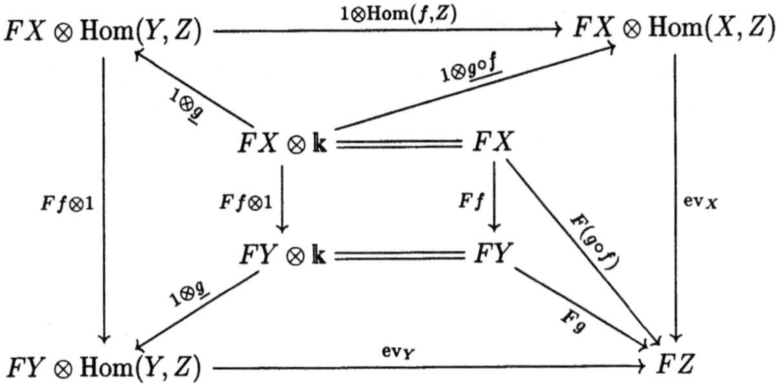

all paths starting in $FX \otimes \mathbf{k}$ and ending in FZ are equal. Hence, the exterior commutes.

Proposition 5.1.5. *The system* (FZ, ev_X) *is the coend* $\int^{X\in\mathcal{C}} FX \otimes \mathrm{Hom}(X, Z)$ *of the bifunctor* $(X, Y) \mapsto FX \otimes \mathrm{Hom}(Y, Z)$.

Proof. Let $i_X : FX \otimes \mathrm{Hom}(X, Z) \xrightarrow{\bullet\bullet} P$ be another dinatural transformation to the constant functor P. Set

$$\kappa = \left(FZ = FZ \otimes \mathbf{k} \xrightarrow{1\otimes \mathbb{1}_Z} FZ \otimes \mathrm{Hom}(Z, Z) \xrightarrow{i_Z} P\right).$$

We have to check that

$$i_X = \left(FX \otimes \mathrm{Hom}(X, Z) \xrightarrow{\mathrm{ev}_X} FZ \xrightarrow{\kappa} P\right). \tag{5.1.4}$$

Set $Y = Z$, $g = \mathbb{1}_Z$ in the diagram of the previous lemma and take any $f : X \to Z \in \mathcal{C}$. In the following diagram

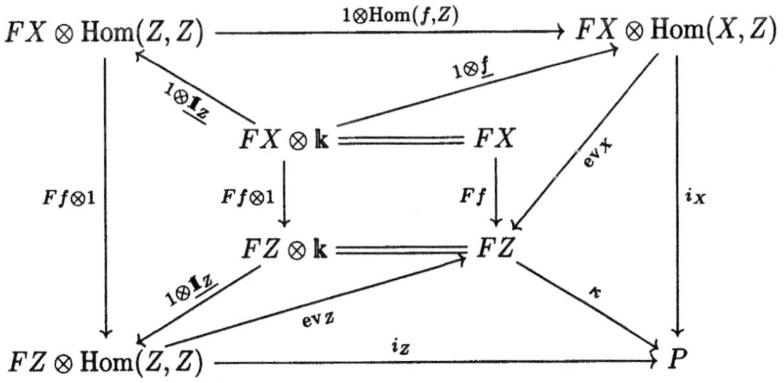

all paths starting from $FX \otimes \mathbf{k}$ and ending in P are equal. In particular,

$$\left(FX \otimes \mathbf{k} \xrightarrow{1\otimes f} FX \otimes \mathrm{Hom}(X, Z) \xrightarrow{i_X} P\right)$$
$$= \left(FX \otimes \mathbf{k} \xrightarrow{1\otimes f} FX \otimes \mathrm{Hom}(X, Z) \xrightarrow{\mathrm{ev}_X} FZ \xrightarrow{\kappa} P\right).$$

This implies (5.1.4).

5.1.3 Coends for bounded categories

Definition 5.1.6. *A set $\mathcal{S} \subset \operatorname{Ob}\mathcal{C}$ p-generates $\mathcal{C}$ if for any $X \in \operatorname{Ob}\mathcal{C}$ there exists an epimorphism $h_X : \oplus_{i\in I} S_i \twoheadrightarrow X$ such that I is finite, $S_i \in \mathcal{S}$ and for any $S \in \mathcal{S}$, $f : S \to X \in \mathcal{C}$ there exists g, such that f factorizes as*

$$f : S \xrightarrow{g} \oplus_{i\in I} S_i \overset{h_X}{\twoheadrightarrow} X.$$

We denote by $\mathcal{S}$ also the full subcategory of $\mathcal{C}$ having $\mathcal{S}$ as its set of objects.

Let $B : \mathcal{C} \times \mathcal{C}^{\mathrm{op}} \to \mathcal{A}$ be a $\Bbbk$-bilinear bifunctor exact in each variable. Denote $B' : \mathcal{S} \times \mathcal{S}^{\mathrm{op}} \to \mathcal{A}$ its restriction to $\mathcal{S}$.

Proposition 5.1.7. *If a full subcategory $\mathcal{S}$ p-generates $\mathcal{C}$, then the coends of B and B' exist in $\mathcal{A}$ simultaneously and the canonical map*

$$\int^{X\in\mathcal{S}} B'(X,X) \to \int^{X\in\mathcal{C}} B(X,X)$$

is an isomorphism.

Proof. Without lack of generality we can assume that $\mathcal{S}$ is additive.

Let $\operatorname{Dinat}(B)$ denote the category of dinatural transformations $i : B \xrightarrow{\bullet\bullet} D$, $D \in \mathcal{A}$. Its morphisms $i \to i'$ are such $f : D \to D' \in \mathcal{A}$ that $i'_X = (B(X,X) \xrightarrow{i_X} D \xrightarrow{f} D')$ for any X in $\mathcal{C}$. We shall prove that the categories of dinatural transformations $\operatorname{Dinat}(B)$ and $\operatorname{Dinat}(B')$ are isomorphic. Namely, the restriction-to-$\mathcal{S}$-functor $r : \operatorname{Dinat}(B) \to \operatorname{Dinat}(B')$ admits an inverse.

The functor r is full and faithful since the image of $i_M : B(M,M) \to D$ is contained in the image of $i_P : B(P,P) \to D$ for $p : P \twoheadrightarrow M$, $P \in \mathcal{S}$. This follows from the diagram

$$\begin{array}{ccc}
B(P,M) & \overset{B(p,1)}{\twoheadrightarrow} & B(M,M) \\
{\scriptstyle B(1,p)}\big\downarrow & & \big\downarrow{\scriptstyle i_M} \\
B(P,P) & \xrightarrow{\quad i_P \quad} & D
\end{array}$$

Now we show that the functor r is bijective on objects. Let $i' : B \xrightarrow{\bullet\bullet} D$ be given. Using a resolution $S \xrightarrow{s} P \xrightarrow{p} M \to 0$ of an arbitrary object $M \in \mathcal{C}$ with $S, P \in \mathcal{S}$ we find a unique arrow $i_M : B(M,M) \to D$, which makes the following diagram commutative

$$\begin{array}{ccccc}
B(S,M) & \xrightarrow{B(s,1)} & B(P,M) & \xrightarrow{B(p,1)} & B(M,M) \quad 0 \\
{\scriptstyle B(1,p)}\big\downarrow & & {\scriptstyle B(1,p)}\big\downarrow & & \\
B(S,P) & \xrightarrow{B(s,1)} & B(P,P) & & \exists!\,\big\downarrow{\scriptstyle i_M} \\
{\scriptstyle B(1,s)}\big\downarrow & & & \searrow{\scriptstyle i'_P} & \\
B(S,S) & & \xrightarrow{\qquad i'_S \qquad} & & D
\end{array}$$

Indeed, the upper row is exact and the left column as well. In particular, for $M = P \in \mathcal{S}$ we use the resolution $0 \to P \to P \to 0$, which implies $i_P = i'_P$.

We have to show that the system $i_M : B(M,M) \to D$ is a dinatural transformation. Note that any morphism $f : M \to N \in \mathcal{C}$ lifts to a morphism of covers $P, Q \in \mathcal{S}$ as in

$$
\begin{array}{ccc}
P & \xrightarrow{\;g\;} & Q \\
{\scriptstyle p}\downarrow & & \downarrow{\scriptstyle q} \\
M & \xrightarrow{\;f\;} & N
\end{array}
$$

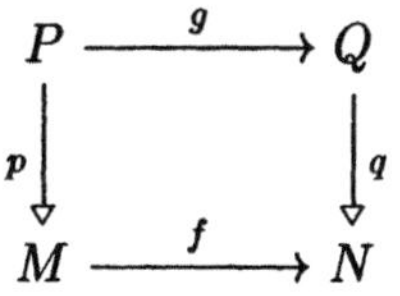

The relation corresponding to f follows from the relation corresponding to g. Indeed, there is a diagram with the commutative exterior

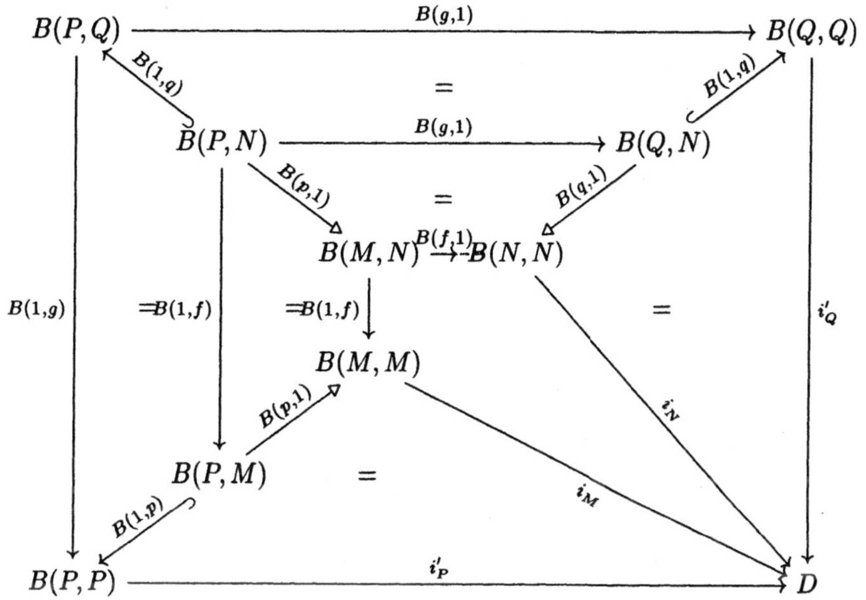

Therefore, $i \in \mathrm{Dinat}(B)$. Moreover, set $f = \mathbf{1} : M \to M$ and choose two different resolutions $P \xrightarrow{\;p\;} M$ and $Q \xrightarrow{\;q\;} M$ of M. The above diagram shows that the two constructed morphisms $i_M : B(M,M) \to D$ coincide. Hence, i depends only on i' and the categories $\mathrm{Dinat}(B)$ and $\mathrm{Dinat}(B')$ are isomorphic. Finally, the discussed coends (if they exist) are initial objects of these categories, and the proposition follows.

Corollary 5.1.8. *If the category $\mathcal{C}$ has a projective generator P, the coend C of a $\Bbbk$-bilinear exact in each variable bifunctor $B : \mathcal{C} \times \mathcal{C}^{\mathrm{op}} \to \mathcal{A}$ exists and is determined by the exact sequence*

$$
\oplus_{j \in J} B(P,P) \xrightarrow{\;B(a_j,1)-B(1,a_j)\;} B(P,P) \to C \to 0,
$$

where $(a_j)_{j \in J}$ is a basis of the vector space $A = \mathrm{End}_{\mathcal{C}} P$.

Corollary 5.1.9. *If $\mathcal{C}$ is semisimple with the list $\{S_j\}_{j\in J}$ of simple objects, then it is bounded if and only if J is finite. In this case $\mathbb{F} = \oplus_{j\in J} S_j \boxtimes S_j^\vee$.*

Note that in the above assumption $\mathcal{C}$ is equivalent to mod-A via the functor $M \mapsto \mathrm{Hom}_{\mathcal{C}}(P, M)$.

Now we apply the general theory to the universal bifunctor $\boxtimes : \mathcal{C} \times \mathcal{C}^{\mathrm{op}} \to \mathcal{C} \boxtimes \mathcal{C}^{\mathrm{op}}$. It follows that the universal coend

$$\int^{X\in\mathcal{C}} X \boxtimes X$$

exists in $\mathcal{C} \boxtimes \mathcal{C}^{\mathrm{op}}$.

Remark 5.1.10. Corollary 5.1.8 holds for bifunctors that are right exact in each variable as well. Indeed, any such bifunctor is isomorphic to one of the form $B : \mathcal{C} \times \mathcal{C}^{\mathrm{op}} \xrightarrow{\boxtimes} \mathcal{C} \boxtimes \mathcal{C}^{\mathrm{op}} \xrightarrow{F} \mathcal{A}$, where F is right exact. Therefore

$$\int^{X\in\mathcal{C}} B(X,X) = F \int^{X\in\mathcal{C}} X \boxtimes X.$$

Let $\mathcal{C}$ be a rigid monoidal abelian $\Bbbk$-linear category with length such that the coend

$$\mathbb{F} = \int^{X\in\mathcal{C}} X \boxtimes X^\vee \simeq \int^{X\in\mathcal{C}} X^\vee \boxtimes X$$

exists in $\mathcal{C} \boxtimes \mathcal{C}$. Then $\mathcal{C}$ is equivalent (as a $\Bbbk$-linear category) to the category A-mod for some finite dimensional unital associative $\Bbbk$-algebra A [Lyu99, Proposition 1.7.5]. We call such $\mathcal{C}$ *bounded*. In a certain sense $\mathbb{F}$ is identified with A^*. In terms of Corollary 5.1.8 we can write an exact sequence

$$\oplus_{j\in J} P \boxtimes P^\vee \xrightarrow{a_j \boxtimes 1 - 1 \boxtimes a_j^t} P \boxtimes P^\vee \to \mathbb{F} \to 0.$$

The exact functor of tensor product $\otimes : \mathcal{C} \times \mathcal{C} \to \mathcal{C}$ decomposes up to an isomorphism as $\otimes \simeq \left(\mathcal{C} \times \mathcal{C} \xrightarrow{\boxtimes} \mathcal{C} \boxtimes \mathcal{C} \xrightarrow{\circledcirc} \mathcal{C}\right)$, where $\circledcirc$ is left exact. Since $\Bbbk$ is perfect, the functor $\circledcirc : \mathcal{C} \boxtimes \mathcal{C} \to \mathcal{C}$ is exact by [Del91, Proposition 5.13(vi)]. If exactness of $\circledcirc$ is known (say, $\mathcal{C}$ is semisimple), we can consider non-perfect $\Bbbk$ as well. Now we prove that $\mathbb{F}$ and $\mathrm{Inv}(- \otimes -)$ obey functor analogs of relations between coev and ev, see Definition 4.1.3. They might be called side-inverse to each other. Instead of equations we get new isomorphisms $\mathbf{z}$ and $\mathbf{n}$, important for what follows.

Lemma 5.1.11 ([Lyu99]). *Denote by* $1 \boxtimes \mathrm{Inv}, \mathrm{Inv} \boxtimes 1 : \mathcal{C} \boxtimes \mathcal{C} \to \mathcal{C}$ *the left exact functors obtained by multiplying with identity the functor* $\mathrm{Inv} : \mathcal{C} \to \Bbbk\text{-vect}$, $X \mapsto \mathrm{Hom}_{\mathcal{C}}(1, X)$, 1 *is the unit object. There are functorial in X isomorphisms*

$$\left(1 \boxtimes (\mathrm{Inv} \circ \otimes)\right)(\mathbb{F} \boxtimes X) = \left(1 \boxtimes (\mathrm{Inv} \circ \otimes)\right) \int^{Y \in \mathcal{C}} Y \boxtimes Y^{\vee} \boxtimes X \xleftarrow{\sim}$$

$$\xleftarrow{\sim} (1 \boxtimes \mathrm{Inv})(\mathbb{F}_{12'} \otimes X_{2''}) = (1 \boxtimes \mathrm{Inv}) \int^{Y \in \mathcal{C}} Y \boxtimes (Y^{\vee} \otimes X) \xrightarrow{\mathbf{z}} X,$$

$$\left((\mathrm{Inv} \circ \otimes) \boxtimes 1\right)(X \boxtimes \mathbb{F}) = \left((\mathrm{Inv} \circ \otimes) \boxtimes 1\right) \int^{Y \in \mathcal{C}} X \boxtimes Y \boxtimes Y^{\vee} \xleftarrow{\sim}$$

$$\xleftarrow{\sim} (\mathrm{Inv} \boxtimes 1)(X_{1'} \otimes \mathbb{F}_{1''2}) = (\mathrm{Inv} \boxtimes 1) \int^{Y \in \mathcal{C}} (X \otimes Y) \boxtimes Y^{\vee} \xrightarrow{\mathbf{n}} X.$$

Proof. Since $1 \boxtimes \otimes$, $\otimes \boxtimes 1$ are exact, these functors commute with the coend over $\mathcal{C}$ (essentially finite inductive limit). This gives the first isomorphisms.

For an injective object X we define $\mathbf{z}_X$ and $\mathbf{n}_X$ as

$$\mathbf{z}_X : (1 \boxtimes \mathrm{Inv}) \int^{Y \in \mathcal{C}} Y \boxtimes (Y^{\vee} \otimes X) = \int^{Y \in \mathcal{C}} Y \otimes \mathrm{Hom}(1, Y^{\vee} \otimes X)$$

$$= \int^{Y \in \mathcal{C}} Y \otimes \mathrm{Hom}(Y, X) \xrightarrow{\mathrm{ev}} X$$

(the morphism $\mathrm{ev} : Y \otimes \mathrm{Hom}(Y, X) \to X$ reduces to $y \otimes f \mapsto f(y)$ for A-modules),

$$\mathbf{n}_X : (\mathrm{Inv} \boxtimes 1) \int^{Y \in \mathcal{C}} (X \otimes Y) \boxtimes Y^{\vee} = (\mathrm{Inv} \boxtimes 1) \int^{Z \in \mathcal{C}} (X \otimes {}^{\vee}Z) \boxtimes Z$$

$$= \int^{Z \in \mathcal{C}} \mathrm{Hom}(1, X \otimes {}^{\vee}Z) \otimes Z = \int^{Z \in \mathcal{C}} \mathrm{Hom}(Z, X) \otimes Z \xrightarrow{\mathrm{ev}} X$$

(the morphism $\mathrm{ev} : \mathrm{Hom}(Z, X) \otimes Z \to X$ reduces to $f \otimes z \mapsto f(z)$ in the case of A-modules). It is shown in Proposition 5.1.5 that $\mathbf{z}_X$ and $\mathbf{n}_X$ are isomorphisms. One deduces that $\mathbf{z}$ and $\mathbf{n}$ are isomorphisms of left exact functors (see the proof of Theorem 1.8.6 [Lyu99]).

Remark 5.1.12. It is explained in [Lyu99, Theorem 1.8.6] that the existence of a side-inverse to $\mathrm{Inv}(- \otimes -)$ is equivalent to the existence of $\mathbb{F}$ in $\mathcal{C} \boxtimes \mathcal{C}$, thus to boundedness. A side-inverse is unique up to an isomorphism.

Remark 5.1.13. $\mathbb{F}$ is one of the main examples of squared Hopf algebras, see [Lyu99].

5.2 Braided function algebra

We assume as everywhere that $\mathbf{k} = \operatorname{End} 1$ is a perfect field. In this section $\mathcal{C}$ will be an abelian category with finite length of objects and finite dimensional $\mathbf{k}$-vector spaces $\operatorname{Hom}_{\mathcal{C}}(A, B)$. One more technical condition: isomorphism classes in $\mathcal{C}$ form a set, that is, $\mathcal{C}$ is *essentially small.*

In such a case there exists a coend $F = \int X \otimes X^\vee \simeq \int X^\vee \otimes X$ as an object of a cocompletion $\hat{\mathcal{C}}$ [Lyu95a] of $\mathcal{C}$ and a coend $\mathbb{F} = \int^{X \in \mathcal{C}} X \boxtimes X^\vee \in \widehat{\mathcal{C} \boxtimes \mathcal{C}}$. The former is obtained from the latter by application of the restriction-to-the-diagonal functor $\otimes : \mathcal{C} \boxtimes \mathcal{C} \to \mathcal{C}$. The general definition of the coend $\mathbb{F}$ specializes here to the universal object satisfying

$$\begin{array}{ccc} A \boxtimes B^\vee & \xrightarrow{f \boxtimes B^\vee} & B \boxtimes B^\vee \\ {\scriptstyle A \boxtimes f^t} \downarrow & & \downarrow {\scriptstyle i_Y} \\ A \boxtimes A^\vee & \xrightarrow{i_A} & \displaystyle\int^{L \in \mathcal{C}} L \boxtimes L^\vee , \end{array} \tag{5.2.1}$$

where $f^t : B^\vee \to A^\vee$ is the transposed to a morphism $f : A \to B$, and similarly for F. In the other words the coend F can be defined via an exact sequence

$$\bigoplus_{f : A \to B \in \mathcal{C}} A \otimes B^\vee \xrightarrow{f \otimes B^\vee - A \otimes f^t} \bigoplus_{L \in \mathcal{C}} L \otimes L^\vee \xrightarrow{\oplus i_L} F \to 0. \tag{5.2.2}$$

$\mathbb{F}$ can be defined similarly.

As $\mathbb{F}$ represents a cylinder, $F = \otimes \mathbb{F}$ corresponds to a 1-holed torus. A 1-holed torus is a Hopf algebra in the cobordism category [BKLT00, Yet97], and so is F in $\hat{\mathcal{C}}$.

We will discuss Hopf operations in details in this section, in particular, the integrals for F. Relations among them will be identified later with elementary topological moves. We define modular categories as the bounded ones, for which the special Hopf pairing $\omega : F \otimes F \to 1$ is non-degenerate. Also we discuss relations between integrals and ω. From the modularity we deduce that the object of integrals is isomorphic to the unit object. We prove that the integrals for F are two-sided.

5.2.1 General properties

Choose a full monoidal subcategory $\mathcal{C}_0 \subset \mathcal{C}$ such that $\mathcal{C}_0$ is equivalent to $\mathcal{C}$ (variant: p-generates $\mathcal{C}$) and $\operatorname{Ob} \mathcal{C}_0$ is a set. As an inductive limit the coend F has a presentation $F = \varinjlim(\Phi : \mathcal{D} \to \mathcal{C})$, where $\operatorname{Ob} \mathcal{D} = \operatorname{Ob} \mathcal{C}_0 \sqcup \operatorname{Mor} \mathcal{C}_0$ and morphisms of $\mathcal{D}$ consists of identity morphisms and two morphisms

$$\operatorname{src}_f : f \to \operatorname{source}(f), \qquad \operatorname{tgt}_f : f \to \operatorname{target}(f)$$

given for each $f \in \operatorname{Mor} \mathcal{C}_0$. For each pair of composable morphisms in $\mathcal{D}$ at least one is an identity, so the composition equals to the other morphism of the pair.

The functor $\Phi : \mathcal{D} \to \mathcal{C}$ looks as follows:

$$\begin{aligned}
\Phi(M) &= M \otimes M^\vee && \text{for } M \in \operatorname{Ob} \mathcal{C}_0 \subset \operatorname{Ob} \mathcal{D} \\
\Phi(f) &= X \otimes Y^\vee && \text{for } (f : X \to Y) \in \operatorname{Mor} \mathcal{C}_0 \subset \operatorname{Ob} \mathcal{D} \\
\Phi(\mathrm{src}_f) &= 1 \otimes f^t : X \otimes Y^\vee \to X \otimes X^\vee \\
\Phi(\mathrm{tgt}_f) &= f \otimes 1 : X \otimes Y^\vee \to Y \otimes Y^\vee
\end{aligned}$$

Since $X \otimes$- (resp. -$\otimes X$) has a right adjoint, this functor commutes with inductive limits. This allows us to present $F \otimes F$ as an inductive limit $\varinjlim(\Phi_2 : \mathcal{D}_2 \to \mathcal{C})$ similar to the presentation of F. Here $\mathcal{D}_2$ is a subcategory of $\mathcal{D} \times \mathcal{D}$

$$\operatorname{Ob} \mathcal{D}_2 = \operatorname{Ob} \mathcal{C}_0 \times \operatorname{Ob} \mathcal{C}_0 \sqcup \operatorname{Ob} \mathcal{C}_0 \times \operatorname{Mor} \mathcal{C}_0 \sqcup \operatorname{Mor} \mathcal{C}_0 \times \operatorname{Ob} \mathcal{C}_0,$$

$\operatorname{Mor} \mathcal{D}_2$ consists of identity morphisms and

$$\begin{aligned}
(M, \mathrm{src}_f) &: (M, f) \to (M, \mathrm{source}(f)) \\
(M, \mathrm{tgt}_f) &: (M, f) \to (M, \mathrm{target}(f)) \\
(\mathrm{src}_f, N) &: (f, N) \to (\mathrm{source}(f), N) \\
(\mathrm{tgt}_f, N) &: (f, N) \to (\mathrm{target}(f), N)
\end{aligned}$$

for $f \in \operatorname{Mor} \mathcal{C}_0$, $M, N \in \operatorname{Ob} \mathcal{C}_0$. The functor Φ_2 is

$$\Phi_2 : \mathcal{D}_2 \hookrightarrow \mathcal{D} \times \mathcal{D} \xrightarrow{\Phi \times \Phi} \mathcal{C} \times \mathcal{C} \xrightarrow{\otimes} \mathcal{C},$$

$$\begin{aligned}
\Phi_2(M, N) &= M \otimes M^\vee \otimes N \otimes N^\vee && \text{for } M, N \in \operatorname{Ob} \mathcal{C}_0 \\
\Phi_2(M, f) &= M \otimes M^\vee \otimes X \otimes Y^\vee && \text{for } (f : X \to Y) \in \operatorname{Mor} \mathcal{C}_0 \\
\Phi_2(f, N) &= X \otimes Y^\vee \otimes N \otimes N^\vee && \text{for } (f : X \to Y) \in \operatorname{Mor} \mathcal{C}_0 \\
\Phi_2(M, \mathrm{src}_f) &= 1 \otimes 1 \otimes 1 \otimes f^t : M \otimes M^\vee \otimes X \otimes Y^\vee \to M \otimes M^\vee \otimes X \otimes X^\vee \\
\Phi_2(M, \mathrm{tgt}_f) &= 1 \otimes 1 \otimes f \otimes 1 : M \otimes M^\vee \otimes X \otimes Y^\vee \to M \otimes M^\vee \otimes Y \otimes Y^\vee \\
\Phi_2(\mathrm{src}_f, N) &= 1 \otimes f^t \otimes 1 \otimes 1 : X \otimes Y^\vee \otimes N \otimes N^\vee \to X \otimes X^\vee \otimes N \otimes N^\vee \\
\Phi_2(\mathrm{tgt}_f, N) &= f \otimes 1 \otimes 1 \otimes 1 : X \otimes Y^\vee \otimes N \otimes N^\vee \to Y \otimes Y^\vee \otimes N \otimes N^\vee
\end{aligned}$$

Note also that the diagram Φ_2 can be embedded into a bigger diagram $\Phi^{(2)}$: $\mathcal{D} \times \mathcal{D} \xrightarrow{\Phi \times \Phi} \mathcal{C} \times \mathcal{C} \xrightarrow{\otimes} \mathcal{C}$ with the same colimit $F \otimes F = \varinjlim(\Phi^{(2)} : \mathcal{D}^2 \to \mathcal{C})$. Similarly, one presents

$$F^{\otimes n} = \varinjlim(\Phi_n : \mathcal{D}_n \to \mathcal{C}) = \varinjlim(\Phi^{(n)} : \mathcal{D}^n \to \mathcal{C}).$$

5.2.2 Braided functions as a Hopf algebra

Here we discuss the structure of the coend F as a Hopf algebra in the category $\hat{\mathcal{C}}$ (see [Lyu95b, LM94, Maj93]).

Comultiplication. The comultiplication in F is uniquely determined by the equation

$$\begin{aligned}&\left(X\otimes X^\vee \xrightarrow{i_X} F \xrightarrow{\Delta} F\otimes F\right)\\ &=\left(X\otimes X^\vee = X\otimes 1\otimes X^\vee \xrightarrow{X\otimes\mathrm{coev}\otimes X^\vee} X\otimes X^\vee\otimes X\otimes X^\vee \xrightarrow{i_X\otimes i_X} F\otimes F\right).\end{aligned}$$

Indeed, the right hand side gives a dinatural transformation with values in $F\otimes F$, and $i_X : X^\vee\otimes X\to F$ is the universal dinatural transformation.

The counit in F is determined by the equation

$$\left(X\otimes X^\vee \xrightarrow{i_X} F \xrightarrow{\varepsilon} 1\right) = \left(X\otimes X^\vee \xrightarrow{\mathrm{ev}} 1\right).$$

Indeed, the right hand side gives a dinatural transformation with values in $\mathbf{1}$. The coalgebra axioms are easily verified.

Multiplication. To construct multiplication for F we use again dinatural transformations in the following form. Assume we found a functor $\underline{m}$ and a natural transformation $\bar{m}$ as in diagram

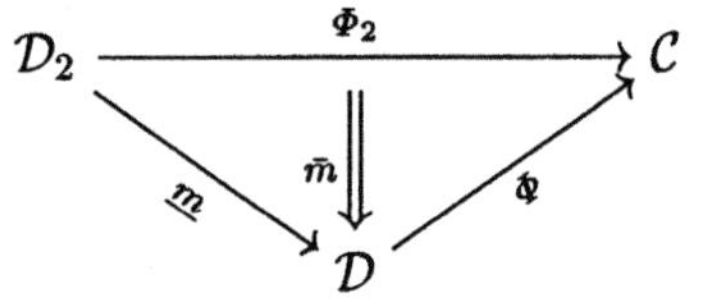

Then the system

$$\left(\Phi_2(U) \xrightarrow{\bar{m}_U} \Phi\underline{m}(U) \xrightarrow{i_{\underline{m}(U)}} \varinjlim\Phi\right)_{U\in\mathrm{Ob}\,\mathcal{D}_2} \tag{5.2.3}$$

forms a cone, because the diagram

$$\begin{array}{ccccc}\Phi_2(U) & \xrightarrow{\bar{m}_U} & \Phi\underline{m}(U) & \xrightarrow{i_{\underline{m}(U)}} & \varinjlim\Phi\\ {\scriptstyle\Phi_2(h)}\downarrow & & {\scriptstyle\Phi\underline{m}(h)}\downarrow & & \|\\ \Phi_2(V) & \xrightarrow{\bar{m}_V} & \Phi\underline{m}(V) & \xrightarrow{i_{\underline{m}(V)}} & \varinjlim\Phi\end{array}$$

is commutative for any morphism $h : U\to V\in\mathcal{D}_2$. By the definition of $\varinjlim\Phi_2$ the system from (5.2.3) induces a morphism $m : \varinjlim\Phi_2\to\varinjlim\Phi$. We use this morphism to construct the multiplication on F.

Concretely, we define $\underline{m} : \mathcal{D}_2\to\mathcal{D}$ on objects as

$$\begin{aligned}\underline{m}(M,N) &= M\otimes N\\ \underline{m}(M,f) &= (M\otimes f : M\otimes X\to M\otimes Y)\\ \underline{m}(f,N) &= (f\otimes N : X\otimes N\to Y\otimes N)\end{aligned}$$

where $M,N\,\mathrm{Ob}\,\mathcal{C}_0$, $(f : X\to Y)\in\mathrm{Mor}\,\mathcal{C}_0$, and on morphisms we set

$$\underline{m}(M, \mathrm{src}_f) = \mathrm{src}_{M\otimes f} : M \otimes f \to M \otimes X$$
$$\underline{m}(M, \mathrm{tgt}_f) = \mathrm{tgt}_{M\otimes f} : M \otimes f \to M \otimes Y$$
$$\underline{m}(\mathrm{src}_f, N) = \mathrm{src}_{f\otimes N} : f \otimes N \to X \otimes N$$
$$\underline{m}(\mathrm{tgt}_f, N) = \mathrm{tgt}_{f\otimes N} : f \otimes N \to Y \otimes N$$

Define $\bar{m} : \Phi_2 \to \Phi \circ \underline{m}$, that is,

$$\bar{m}(M,N) : M \otimes M^\vee \otimes N \otimes N^\vee \to M \otimes N \otimes (M \otimes N)^\vee$$
$$\bar{m}(M,f) : M \otimes M^\vee \otimes X \otimes Y^\vee \to M \otimes X \otimes (M \otimes Y)^\vee$$
$$\bar{m}(f,N) : X \otimes Y^\vee \otimes N \otimes N^\vee \to X \otimes N \otimes (Y \otimes N)^\vee$$

by the same expression

$$\bar{m} = \big(A \otimes B^\vee \otimes C \otimes D^\vee \xrightarrow{(234)^\sim_+} A \otimes C \otimes D^\vee \otimes B^\vee \xrightarrow{1\otimes 1 \otimes j_+} A \otimes C \otimes (B \otimes D)^\vee\big).$$

Then $\bar{m}$ is a morphism of functors and the existence of m follows.

The graphical notation for $\bar{m}$ is given below. Accordingly, the multiplication $m : F \otimes F \to F$ satisfies the diagram below.

$\bar{m} =$ 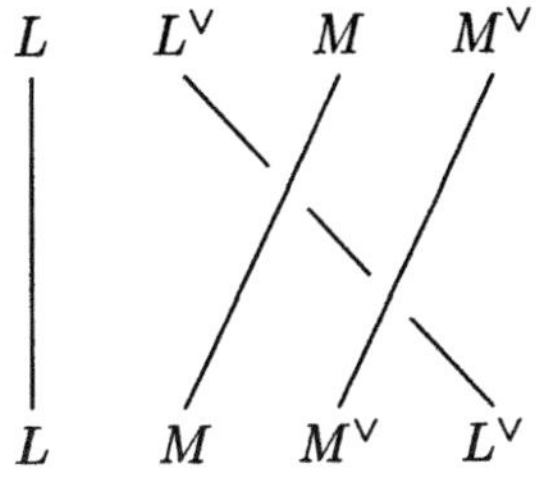

and

$$\begin{array}{ccc} L \otimes L^\vee \otimes (M \otimes M^\vee) & \xrightarrow{i_L \otimes i_M} & F \otimes F \\ {\scriptstyle L\otimes c}\downarrow & & \downarrow{\scriptstyle \exists\, m_F} \\ L \otimes M \otimes (L \otimes M)^\vee & \xrightarrow{i_{L\otimes M}} & F \end{array} \qquad (5.2.4)$$

The unit is given by the morphism

$$\eta : 1 = 1 \otimes 1^\vee \xrightarrow{i_1} F.$$

Associativity follows from the following identity of natural transformations:

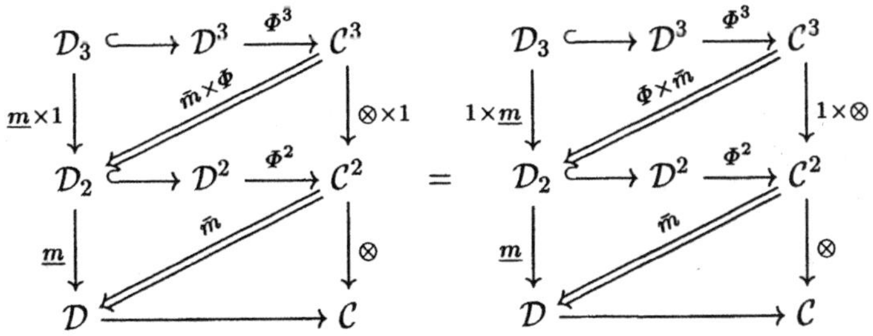

which reduces to an equation in B_6

$$\begin{pmatrix} 123456 \\ 125634 \end{pmatrix}^\sim_+ \circ (432)^\sim_+ = (65432)^\sim_+ \circ (654)^\sim_+ ,$$

where the braid $\sigma^\sim_+ \in B_6$ is the positive lifting of a permutation $\sigma \in S_6$.

Antipode. We define a morphism $\gamma : F \to F$ via the following commutative diagram

$$\begin{array}{ccccc} M \otimes M^{\vee} & \xrightarrow{c} & M^{\vee} \otimes M & \xrightarrow{1 \otimes u_1^2} & M^{\vee} \otimes M^{\vee\vee} \\ {\scriptstyle i_M}\downarrow & & & & \downarrow{\scriptstyle i_{M^{\vee}}} \\ F & & \xrightarrow{\exists \gamma} & & F \end{array}$$

using the fact that the composition of the morphisms in the upper-right path is a dinatural transformation. A straightforward verification shows that γ_F is the antipode of F.

The diagram corresponding to the antipode $\gamma_F : F \to F$ is, hence, given by

$$\gamma_F = $$ 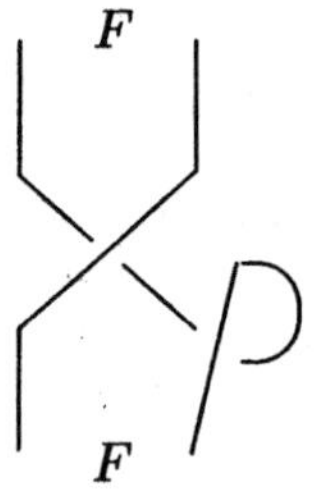

. (5.2.5)

There is a natural coaction of F in objects $X \in \mathcal{C}$

$$\delta_X : X = X \otimes 1 \xrightarrow{X \otimes \mathrm{coev}} X \otimes X^{\vee} \otimes X \xrightarrow{i_X \otimes X} F \otimes X. \quad (5.2.6)$$

Lemma 5.2.1 ([Lyu95b]). *The pairing* $\omega : F \otimes F \to 1$,

$$\omega = $$ 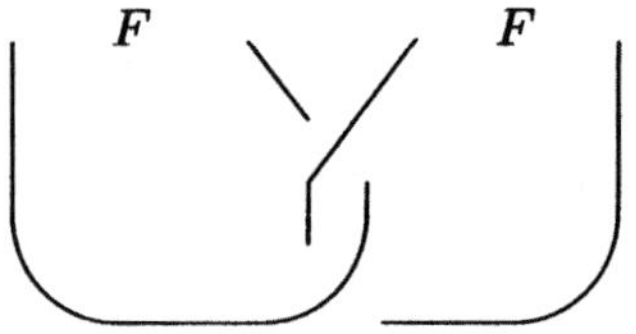

(5.2.7)

is a Hopf pairing. It satisfies

$$\mathrm{Ann}\,\omega \stackrel{def}{=} \mathrm{Ann}^{left}\,\omega = \mathrm{Ann}^{right}\,\omega \in \hat{\mathcal{C}},$$

where the Hopf ideals $\mathrm{Ann}^{left}\,\omega$ *and* $\mathrm{Ann}^{right}\,\omega$ *are the left and right annihilators of* ω*, which in the case of a rigid object* F *coincide with* $\mathrm{Ker}(\omega^{!} : F \to {}^{\vee}F)$ *and* $\mathrm{Ker}({}^{!}\omega : F \to F^{\vee})$ *respectively.*

The antipode γ is symmetric with respect to ω in the following sense

$$\omega \circ (\gamma \otimes 1) = \omega \circ (1 \otimes \gamma) : F \otimes F \to 1 \quad (5.2.8)$$

by Proposition 4.2.2.

Example 5.2.2. Assume that the category $\mathcal{C}$ is semisimple with a finite number of isomorphism classes of simple objects represented by $\{X_j\}_{j \in J}$. Then $F = \oplus_{j \in J} X_j \otimes X_j^{\vee}$.

Example 5.2.3. Assume that the category $\mathcal{C} = H$-mod is the category of finite dimensional modules over a finite dimensional quasitriangular ribbon Hopf algebra H. Then $F = H^*$ as a coalgebra and the H-action in F is the coadjoint action. The multiplication in F differs from the usual one. It is described, e.g., in [Lyu95c]. The coend $\mathbb{F} = \int^X X \boxtimes X^\vee \in \mathcal{C} \boxtimes \mathcal{C} = H \otimes H$-mod is H^* with the two actions of H — by the left translations and by the right translations composed with the antipode.

Lemma 5.2.4. *The pairing* $\omega : F \otimes F \to 1$ *satisfies*

$$\bigl(F \otimes F \xrightarrow{\mathbf{c}^{-1}} F \otimes F \xrightarrow{\omega} 1\bigr) = \bigl(F \otimes F \xrightarrow{\gamma \otimes \gamma} F \otimes F \xrightarrow{\omega} 1\bigr).$$

Proof. Drawing double lines instead of single lines we find that

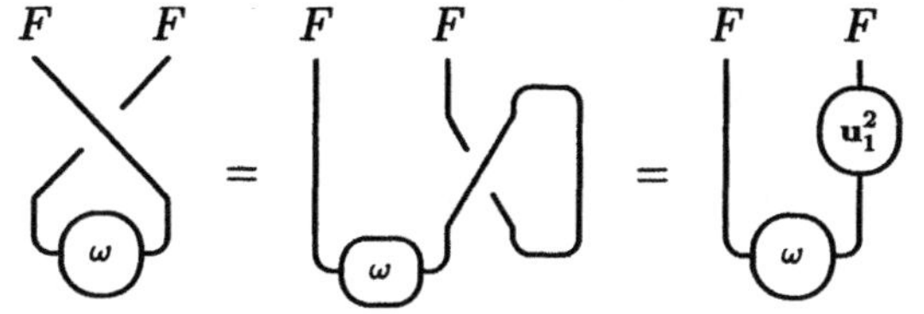

since the transposed morphism to $\mathbf{c}$ is $\mathbf{c}$. This equals

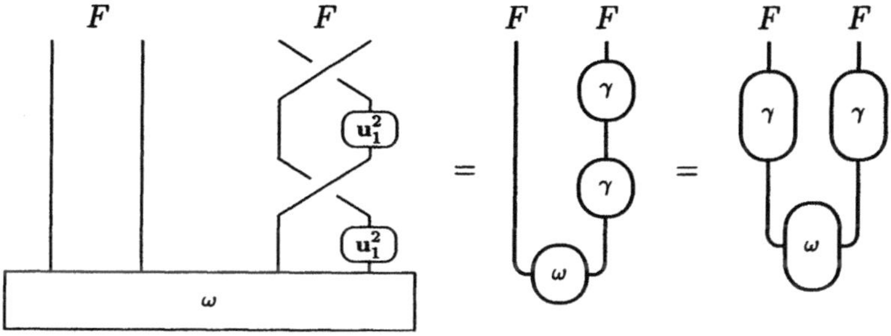

by equation (5.2.8).

Therefore, the hypothesis and the conclusion of Lemma 4.2.15 hold for F.

Endomorphisms of the identity functor. Consider End Id — the commutative $\mathbf{k}$-algebra of endomorphisms of the identity functor $\mathrm{Id} : \mathcal{C} \to \mathcal{C}$. The space of coinvariants $\mathrm{Hom}(F, 1)$ equipped with the convolution product $\phi * \psi = \bigl(F \xrightarrow{\Delta} F \xrightarrow{\phi \otimes \psi} 1 \otimes 1 = 1\bigr)$, $\phi, \psi \in \mathrm{Hom}(F, 1)$ is a $\mathbf{k}$-algebra as well. There are linear maps

$$\tilde{-} : \mathrm{Hom}(F,1) \longrightarrow \mathrm{End}\,\mathrm{Id},$$
$$\psi : F \to 1 \longmapsto \tilde{\psi}_X = \bigl(X \xrightarrow{\delta_X} F \otimes X \xrightarrow{\psi \otimes X} 1 \otimes X = X\bigr),$$

$$\bar{-} : \mathrm{End}\,\mathrm{Id} \longrightarrow \mathrm{Hom}(F,1), \quad \alpha \longmapsto \bar{\alpha} = \bigl(F \xrightarrow{\int \alpha \otimes 1} F \xrightarrow{\varepsilon} 1\bigr)$$

between these spaces.

Proposition 5.2.5 ([Lyu95b]). *The above maps* $\tilde{\ }$ *and* $\bar{\ }$ *are algebra isomorphisms that are inverse to each other.*

Proof. Reduces to standard rigidity relations.

Assume now that $\omega : F \otimes F \to 1$ is non-degenerate. Non-degeneracy of ω implies that there is an isomorphism of vector spaces

$$\begin{aligned} \operatorname{Inv} F = \operatorname{Hom}(1, F) &\longrightarrow \operatorname{Hom}(F, 1) \\ \kappa : 1 \to F &\longmapsto \left(F =\!= 1 \otimes F \xrightarrow{\kappa \otimes F} F \otimes F \xrightarrow{\omega} 1\right). \end{aligned}$$

Due to Lemma 5.2.1 this is an algebra isomorphism.

Corollary 5.2.6 ([Lyu95b]). *The maps* $\tau : \operatorname{Inv} F \to \operatorname{End} \operatorname{Id}$,

$$\tau(\kappa)_X = \left(X =\!= 1 \otimes X \xrightarrow{\kappa \otimes 1} F \otimes X \xrightarrow{\Omega_r} F \otimes X \xrightarrow{\varepsilon \otimes 1} 1 \otimes X =\!= X\right),$$

$\theta : \operatorname{End} \operatorname{Id} \to \operatorname{Inv} F$, $\alpha \mapsto \theta(\alpha)$ *such that*

$$\bar{\alpha} = \left(F =\!= 1 \otimes F \xrightarrow{\theta(\alpha) \otimes 1} F \otimes F \xrightarrow{\omega} 1\right),$$

are algebra isomorphisms, which are inverse to each other.

Here the morphism Ω_r called monodromy is defined via the tangle

$$\Omega_r = \Omega^r_{F,X} : F \otimes X \to F \otimes X =$$

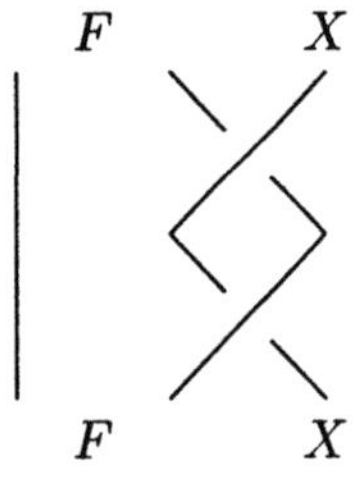

. (5.2.9)

5.2.3 Modular categories

Definition 5.2.7. *A bounded abelian ribbon category* $\mathcal{C}$ *is called* modular, *if*
(PM) the form ω *is non-degenerate:* $\operatorname{Ann} \omega = 0$.

Boundedness implies that F is an object of $\mathcal{C}$ (and not only of a cocompletion $\hat{\mathcal{C}}$). Side-invertibility and non-degeneracy of ω is the same thing.

The case of degenerate form ω was considered in [Lyu95b]. We shall not need it for our purposes.

Lemma 5.2.8. *The Hopf algebra* F *in a modular category* $\mathcal{C}$ *satisfies* $\operatorname{Int} F \simeq 1$.

Proof. Set $t = \left(F \xrightarrow{\int 1 \otimes \mathsf{v}^{-1}} F \xrightarrow{\varepsilon} 1\right)$. Then the pairing $\phi = t \circ m : F \otimes F \to 1$ is given by Figure 5.1. Therefore, it is side-invertible and we can apply Lemma 4.2.11.

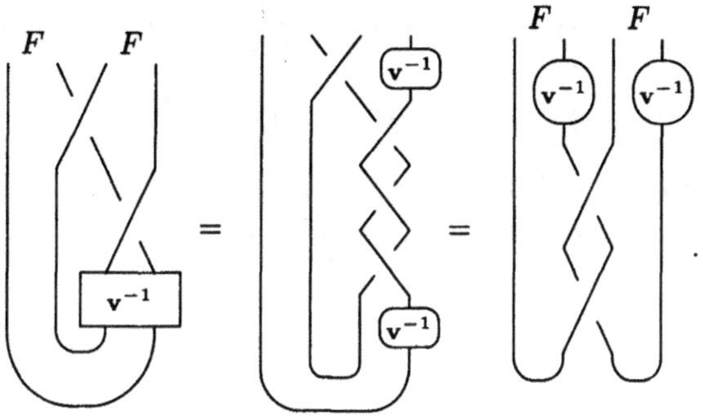

Fig. 5.1. Non-degeneracy of $t \circ m$

Proposition 5.2.9 (criterion of modularity). *Let $\mathcal{C}$ be a bounded abelian ribbon category. Then the following conditions are equivalent:*

(a) the form $\omega : F \otimes F \to 1$ is non-degenerate, thus $\mathcal{C}$ is modular;
(b) there exists a morphism $\mu' : 1 \to F$ of $\mathcal{C}$ such that

$$\int' = \left(F \simeq 1 \otimes F \xrightarrow{\mu' \otimes 1} F \otimes F \xrightarrow{\omega} 1\right) \tag{5.2.10}$$

is a left integral-functional;
(c) there exists a morphism $\mu' : 1 \to F$ of $\mathcal{C}$ such that (5.2.10) *is a right integral-functional;*
(d) there exists a morphism $\mu'' : 1 \to F$ of $\mathcal{C}$ such that

$$\int'' = \left(F \simeq F \otimes 1 \xrightarrow{1 \otimes \mu''} F \otimes F \xrightarrow{\omega} 1\right) \tag{5.2.11}$$

is a left integral-functional;
(e) there exists a morphism $\mu'' : 1 \to F$ of $\mathcal{C}$ such that (5.2.11) *is a right integral-functional.*

Proof. Assume that (a) holds. Then $\operatorname{Int} F \simeq 1$ by Lemma 5.2.8. Taking μ', μ'' to be $\int^F$ or ${}^F\!\int$ we prove (b)–(e) using Proposition 4.2.14(b).

Now let us prove that (b) is equivalent to (e). Assuming that (b) holds, so that $\int' = \int_F$, we set $\mu'' = \left(1 \xrightarrow{\mu'} F \xrightarrow{\gamma} F\right)$. Then

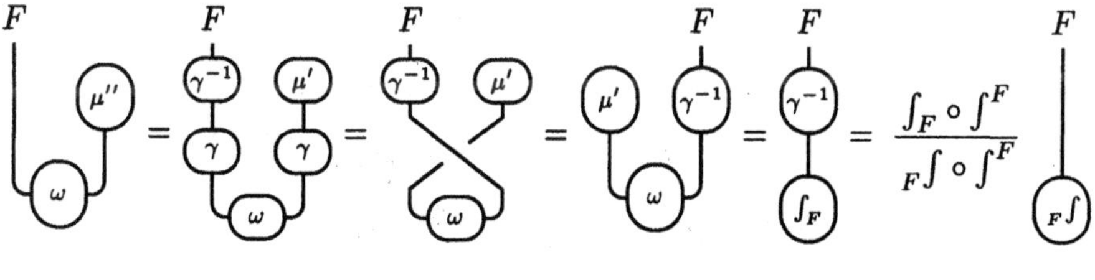

is a right integral-functional by Lemma 5.2.4 and by (4.2.6). Thus, (e) holds. Similarly, (e) implies (b). Conditions (b) and (e) together imply (a) by Proposition 4.2.14(a).

Similarly, (c) is equivalent to (d) and together they imply (a).

Proposition 5.2.10 ([Lyu95b]). *The Hopf algebra F in a modular category $\mathcal{C}$ is unimodular, that is, the integrals-functionals on F are two-sided, $\int_F = {}_F\!\int : F \to 1$.*

Proof. The integral's property (4.2.3) and the non-degeneracy of the form ω gives the following implications:

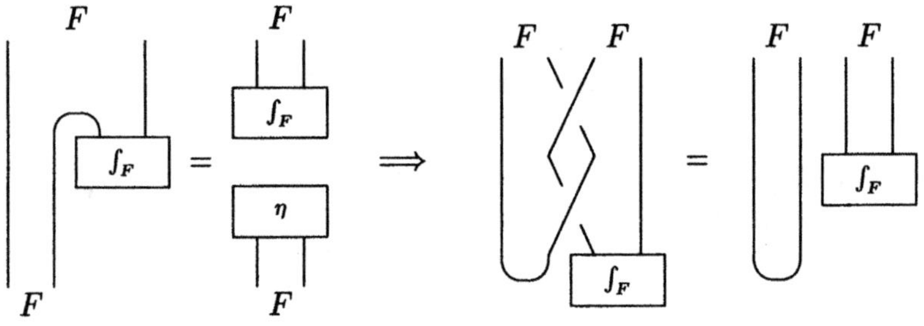

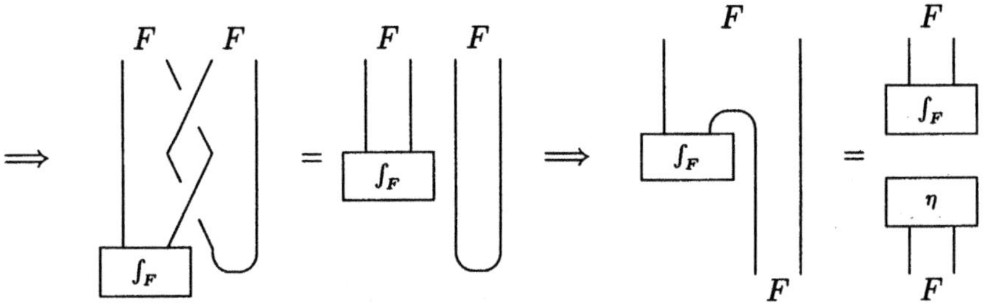

In other notations, we have

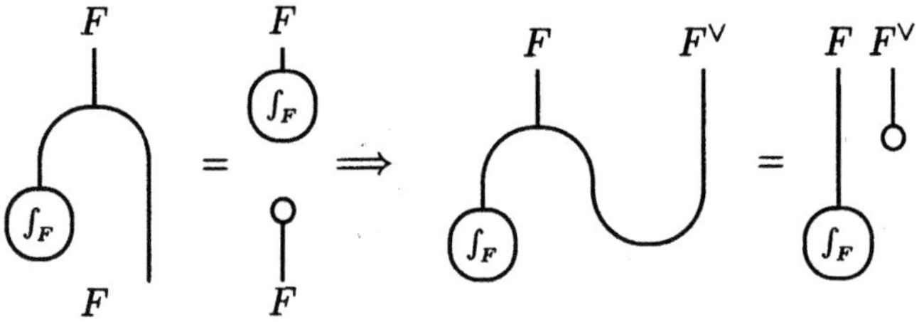

By the definition of the integral ${}_F\!\int : F \to 1$ there exists a constant $\kappa \in \mathbb{k}$, such that $\int_F = \left(F \xrightarrow{{}_F\!\int} 1 \xrightarrow{\kappa} 1\right)$. Since $\kappa \neq 0$, $\int_F$ can be rescaled to give $\kappa = 1$.

Corollary 5.2.11. *The integral-elements in F are two-sided, $\int^F = {}^F\!\!\int : 1 \to F$.*

Proof. The integral-elements in $F^\vee$ are two-sided since we can take

$$^{F^\vee}\!\!\int = \int_F{}^t = {}_F\!\int^t = \int^{F^\vee} : 1 \longrightarrow F^\vee.$$

The Hopf algebra $F^\vee$ is isomorphic to F.

Lemma 5.2.12. *The integrals for F are invariant under the action of the antipode: $\gamma^{\pm 1} \circ \int^F = \int^F$, $\int_F \circ \gamma^{\pm 1} = \int_F$.*

Proof. The left integral-element composed with the antipode $\gamma \circ \int^F : 1 \to F$ is a right integral-element. Hence, it is proportional to $^F\!\!\int$, which by Corollary 5.2.11 coincides with $\int^F$. Therefore, there is a constant $c \in \Bbbk^\times$, such that $\gamma \circ \int^F = c\int^F$.

From the proofs of Lemmas 5.2.8 and 4.2.11 we know that $1 \xrightarrow{\int^F} F \xrightarrow{t} 1$ is an isomorphism (an invertible number), where $t = \left(F \xrightarrow{\int 1\otimes \mathbf{v}^{-1}} F \xrightarrow{\varepsilon} 1\right)$. Therefore,

$$\left(1 \xrightarrow{\int^F} F \xrightarrow{\gamma} F \xrightarrow{t} 1\right) = c\left(1 \xrightarrow{\int^F} F \xrightarrow{t} 1\right) \tag{5.2.12}$$

is also an isomorphism. We have $t \circ \gamma = t : F \to 1$ as the following graphical proof shows.

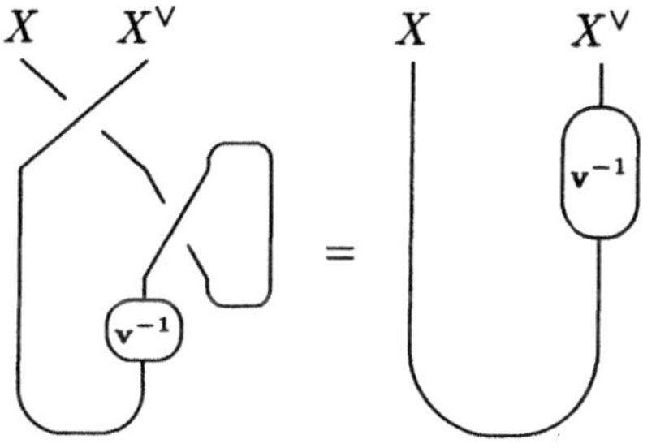

Thus, the left hand side of (5.2.12) is equal to $1 \xrightarrow{\int^F} F \xrightarrow{t} 1$. Comparing this with the right hand side, we find that $c = 1$. Hence, we have proven one of the required equations $\gamma \circ \int^F = \int^F$. The other follow from it.

Notation. Denote $\mu = \int^F = {}^F\!\!\int : 1 \to F$ the integral-element in F.

Notice that $1 =\!=\!= 1 \otimes 1 \xrightarrow{\mu\otimes\mu} F \otimes F \xrightarrow{\omega} 1$ is an isomorphism by Theorem 4.2.5, thus, a number in $\Bbbk^\times$. We assume that this number has a square root in $\Bbbk$, so that we can rescale μ in such a way that

$$\left(1 =\!=\!= 1 \otimes 1 \xrightarrow{\mu\otimes\mu} F \otimes F \xrightarrow{\omega} 1\right) = 1\!\!1_1. \tag{5.2.13}$$

We choose $\mu^t : F^\vee \to 1$ to be the two-sided integral-functional on $F^\vee$. An isomorphism of Hopf algebras

$$\theta : F = 1 \otimes F \xrightarrow{\text{coev}_F \otimes F} F^\vee \otimes F \otimes F \xrightarrow{F^\vee \otimes \omega} F^\vee \otimes 1 = F^\vee$$

allows us to fix a two-sided integral-functional on F via the following equations:

$$\begin{aligned} \lambda = \textstyle\int_F = {}_F\!\int &= \left(F \xrightarrow{\theta} F^\vee \xrightarrow{\mu^t} 1\right) \\ &= \left(F = 1 \otimes F \xrightarrow{\mu \otimes F} F \otimes F \xrightarrow{\omega} 1\right) \\ &= \left(F = 1 \otimes F \xrightarrow{F \otimes \mu} F \otimes F \xrightarrow{\omega} 1\right). \end{aligned}$$

The last equation can be checked graphically.

Notice that (5.2.13) implies

$$\lambda(\mu) = 1.$$

5.2.4 Coupon transformation

Assume that $\mathcal{C}$ is modular. Consider the natural transformation of the identity functor $\tau(\mu) \in \operatorname{End} \operatorname{Id}$. For any $X \in \operatorname{Ob} \mathcal{C}$ decompose the morphism $\tau(\mu)_X : X \xrightarrow{\pi_X} KX \xhookrightarrow{\sigma_X} X$ into an epimorphism and a monomorphism. The map $K : \operatorname{Ob} \mathcal{C} \to \operatorname{Ob} \mathcal{C}$ extends uniquely to a functor $K : \mathcal{C} \to \mathcal{C}$, such that $\pi : \operatorname{Id} \to K$ and $\sigma : K \to \operatorname{Id}$ are natural transformations.

Remark 5.2.13. The natural action of F on an object Y of $\mathcal{C}$

$$act : F \otimes Y \xrightarrow{\Omega_r} F \otimes Y \xrightarrow{\varepsilon \otimes Y} 1 \otimes Y = Y$$

extends $\tau : \operatorname{Inv} F \to \operatorname{End} \operatorname{Id}$ in the sense that

$$\tau(\phi)_Y = \left(Y = 1 \otimes Y \xrightarrow{\phi \otimes Y} F \otimes Y \xrightarrow{act} Y\right).$$

In particular, $\tilde{\lambda} = \tau(\mu)$ can be drawn as

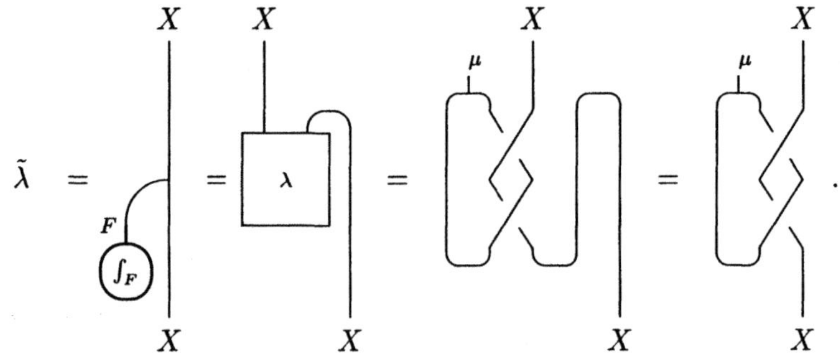

(5.2.14)

Lemma 5.2.14. *For any $X \in \operatorname{Ob} \mathcal{C}$ the object KX is isomorphic to 1^n for some $n \geqslant 0$.*

Proof. Consider the endomorphism $\tilde{\lambda} : \mathrm{Id} \to \mathrm{Id}$ corresponding to the integral $\lambda : F \to 1$. Decompose it into an epimorphism π and a monomorphism σ

$$\tilde{\lambda} : \mathrm{Id} \xrightarrow{\pi} K \overset{\sigma}{\hookrightarrow} \mathrm{Id}.$$

Then

$$\lambda = \left(F \xrightarrow{\int \tilde{\lambda} \otimes 1} F \xrightarrow{\varepsilon} 1\right) =$$ 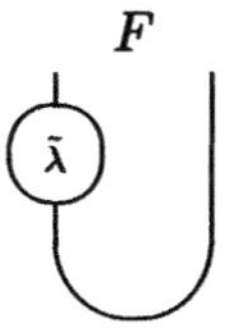

By definition of left-right integral

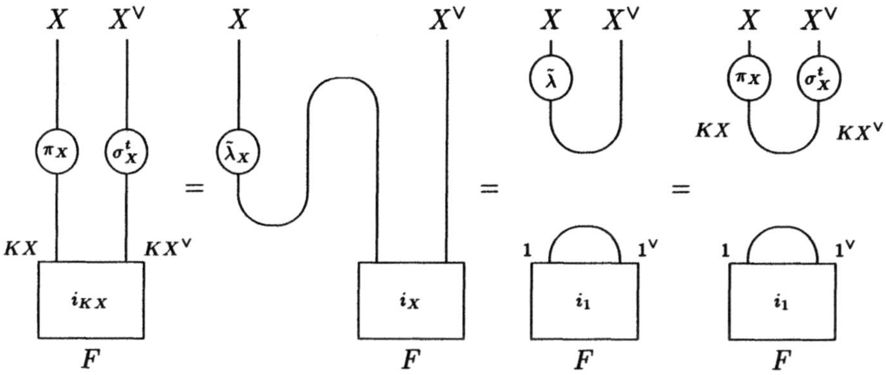

Since π_X and σ^t_X are epi, this equation implies

$$i_{KX} = \left(KX \otimes KX^\vee \xrightarrow{\mathrm{ev}} 1 \xrightarrow{\eta} F\right).$$

Hence, the natural coaction (5.2.6) of F on KX equals

$$\delta_{KX} : KX = 1 \otimes KX \xrightarrow{\eta \otimes KX} F \otimes KX.$$

Theorem 2.7.1 and Corollary 2.7.2 from [Lyu99] state that the category ${}^F\mathcal{C}$ of F-comodules in $\mathcal{C}$ is the tensor product $\mathcal{C} \boxtimes \mathcal{C}$. Namely, the functor $\tilde{\otimes} : \mathcal{C} \times \mathcal{C} \to {}^F\mathcal{C}$, $(X, Y) \mapsto (X \otimes Y, \delta_X \otimes Y : X \otimes Y \to F \otimes X \otimes Y)$, where δ_X is the natural coaction (5.2.6), makes ${}^F\mathcal{C}$ into $\mathcal{C} \boxtimes \mathcal{C}$.

The F-comodule KX belongs to the image $\tilde{\otimes}(\mathcal{C} \times 1)$. On the other hand, by the proof of Theorem 2.7.1 [Lyu99] it follows that it is identified with $1 \boxtimes KX$. Indeed, in notations of loc. cit. KX as an object of $\mathcal{C} \boxtimes \mathcal{C}$ is isomorphic to the following kernel

$$\begin{aligned}
&\mathrm{Ker}(\mathbb{F}_{12'} \otimes \delta_{2''} - \Delta_{12'2''} \otimes KX : \mathbb{F}_{12'} \otimes KX_{2''} \to \mathbb{F}_{12'} \otimes \mathbb{F}_{2''2'''} \otimes KX_{2^4}) \\
&= \mathrm{Ker}(\mathbb{F}_{12'} \otimes \eta_{2''} - \Delta_{12'2''} : \mathbb{F}_{12} \to \mathbb{F}_{12'} \otimes \mathbb{F}_{2''2'''}) \otimes KX \\
&= 1_1 \boxtimes KX_2.
\end{aligned}$$

Therefore, KX belongs to the essential intersection $\mathcal{C} \boxtimes \mathcal{I} \cap \mathcal{I} \boxtimes \mathcal{C}$, where $\mathcal{I} = \langle 1 \rangle$ is equivalent to $\mathbf{k}$-vect. By a result of Deligne [Del91, Proposition 5.14] this essential intersection is equivalent to $\mathcal{I} \boxtimes \mathcal{I}$ and the lemma follows.

6. Construction of TQFT-Double Functors

In Sect. 6.1 of this chapter we give the precise definition of an extended TQFT and state the main result of this book, namely, the existence of a large class of such TQFT's. Then we proceed with the construction of a TQFT from a given modular category. For this we define some functors, representing surfaces with holes, as coends of expressions involving tensor products, the Hopf algebra F and the functor of invariants. In a sequence of lemmas we show that a cobordism between two surfaces determines a natural transformation between the corresponding functors. To achieve this we first associate a natural transformation to an equivalence class of tangles under ambient isotopy. Then we show that it depends on a wider equivalence class, stable under topological moves, which defines the category of cobordisms as a quotient of the category of tangles (Theorem 3.0.6).

In the next three sections we check that the so constructed map from the 2-morphism set of cobordisms to the 2-morphism set of natural transformations is compatible with both compositions. First we check compatibility with vertical compositions for surfaces with colored holes. Then we reformulate these results without coloring. In this setting a surface with a incoming and b outgoing holes corresponds to a functor $\mathcal{C}^{\boxtimes a} \to \mathcal{C}^{\boxtimes b}$. We obtain again the compatibility with the vertical composition. More involved is the compatibility with the horizontal composition, which we prove in Sect. 6.6. The map which takes cobordisms to natural transformations is compatible with horizontal compositions only up to an isomorphism, which is constructed from the braiding. This finishes the construction of the double pseudofunctor (TQFT).

We end up the chapter with remarks which show the necessity of our conditions of modularity.

6.1 Main result

Definition 6.1.1. *A TQFT is a double pseudofunctor* $\mathcal{V} : \widetilde{\mathcal{C}ob} \to \mathcal{Q}\mathbf{AbCat}$ *multiplicative on objects, that is,* $\mathcal{V}(\sqcup_{n+m} S^1)$ *is equivalent to* $\mathcal{V}(\sqcup_n S^1) \boxtimes \mathcal{V}(\sqcup_m S^1)$.

Theorem 6.1.2. *For any modular bounded abelian category* $\mathcal{C}$ *there exists a double pseudofunctor multiplicative on objects (TQFT)*

$$\mathcal{V}_\mathcal{C} = (\mathcal{V}_\mathcal{C}, \alpha, \beta) : \widetilde{\mathcal{C}ob} \to \mathcal{Q}\mathbf{AbCat},$$

where the isomorphism β is obtained from the structure of symmetric monoidal 2-category of **AbCat** *and canonical isomorphisms of coends. The isomorphism α uses also the braiding of $\mathcal{C}$.*

In the model situation, given a modular category $\mathcal{C}$ from **V-Cat-mod**, *one can achieve for the model pseudofunctor*

$$\mathcal{V}_{\mathcal{C}} = (\mathcal{V}_{\mathcal{C}}, \alpha, \beta) : \widetilde{\mathcal{C}ob} \to \mathcal{Q}\mathbf{V}\text{-}\mathbf{Cat}\text{-}\mathbf{mod}$$

that $\beta_{\alpha,\beta} = \mathbb{1}$, so that $\mathcal{V}$ is strict with respect to vertical compositions.

The construction of this pseudofunctor is the subject of this chapter.

6.2 Colorations, Natural Transformations, and Liftings

The first step in building a TQFT functor is the construction of tangle functors. Our strategy is analogous to the way the tangle invariants in [Res88], [Tur88] and [RT90] are defined. The difference is that the objects with which the strands are colored do not need to be irreducible.

To a surface $\Sigma_{g,a/b}$ of genus g with a incoming and b outgoing holes we associate two functors $\tilde{\mathcal{F}}_{g,a/b}$ and $\mathcal{F}_{g,a/b}$, which are equivalent forms of each other. To a diagram of a tangle, representing a cobordism between surfaces we associate a natural transformation between such functors as follows.

First of all, to each surface $\Sigma_{g,a/b}$ there correspond standard graphs

$\tilde{G}_{g,a/b} =$ 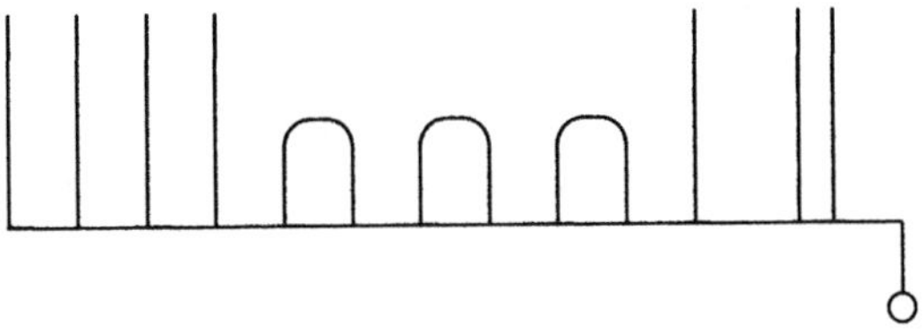

$G_{g,a/b} =$ 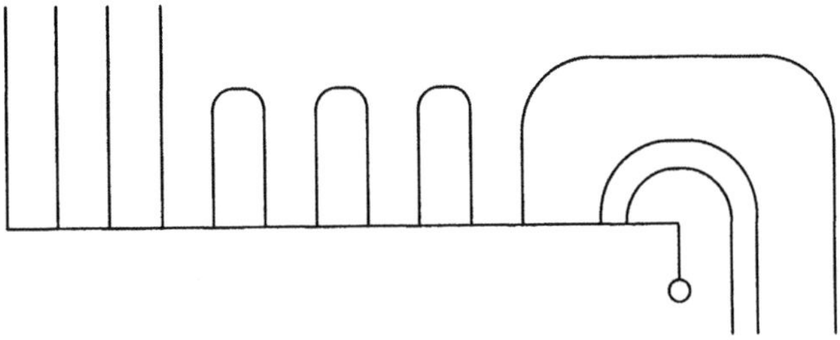

According to Theorem C.2.1 these planar graphs represent functors

$$\tilde{\mathcal{F}}_{g,a/b} : \mathcal{C}^{\boxtimes a} \boxtimes (\mathcal{C}^{\mathrm{op}})^{\boxtimes b} \to \Bbbk\text{-vect},$$

$$\tilde{\mathcal{F}}_{g,a/b}(X_1 \boxtimes \ldots \boxtimes X_a \boxtimes Y_b \boxtimes \ldots \boxtimes Y_1) = \mathrm{Inv}_{\mathcal{C}}\big(X_1 \otimes \ldots \otimes X_a \otimes F^{\otimes g} \otimes Y_b^{\vee} \otimes \ldots \otimes Y_1^{\vee}\big)$$

$$\mathcal{F}_{g,a/b} : \mathcal{C}^{\boxtimes a} \to \mathcal{C}^{\boxtimes b}, \tag{6.2.1}$$

$$\begin{aligned}\mathcal{F}_{g,a/b}&(X_1 \boxtimes \ldots \boxtimes X_a) \\ &= \int^{Y_1,\ldots,Y_b \in \mathcal{C}} \mathrm{Inv}_{\mathcal{C}}\big(X_1 \otimes \ldots \otimes X_a \otimes F^{\otimes g} \otimes Y_b^{\vee} \otimes \ldots \otimes Y_1^{\vee}\big) \otimes Y_1 \boxtimes \cdots \boxtimes Y_b \\ &\simeq \int^{Y_2,\ldots,Y_b \in \mathcal{C}} \big(X_1 \otimes \ldots \otimes X_a \otimes F^{\otimes g} \otimes Y_b^{\vee} \otimes \ldots \otimes Y_2^{\vee}\big) \boxtimes Y_2 \boxtimes \cdots \boxtimes Y_b.\end{aligned}$$

Lemma 6.2.1. *Let $\mathcal{C}$ be a bounded category. If $a, b > 0$ the functor $\mathcal{F}_{g,a/b}$ is exact.*

Proof. Perfectness of $\Bbbk$ implies that the functor $\circledast$ is exact by [Del91, Proposition 5.7]. Hence, $\circledast$ commutes with the coend over $\mathcal{C}$, since it is an essentially finite inductive limit. Therefore, $\mathcal{F}_{g,a/b}$ is isomorphic to the functor

$$\begin{aligned}X_1 \boxtimes \ldots \boxtimes X_a \mapsto X_1 \otimes \ldots \otimes X_a \otimes F^{\otimes g} \otimes \int^{Y_2,\ldots,Y_b \in \mathcal{C}} Y_b^{\vee} \otimes \ldots \otimes Y_2^{\vee} \boxtimes Y_2 \boxtimes \cdots \boxtimes Y_b \\ \stackrel{\text{def}}{=} (\circledast^{a+1} \boxtimes \mathrm{Id})\big(X_1 \boxtimes \ldots \boxtimes X_a \boxtimes F^{\otimes g} \boxtimes \int^{Y_2,\ldots,Y_b \in \mathcal{C}} Y_b^{\vee} \otimes \ldots \otimes Y_2^{\vee} \boxtimes Y_2 \boxtimes \cdots \boxtimes Y_b\big).\end{aligned}$$

The functor $\boxtimes$ is also exact [Del91], hence, the above functor is exact. We conclude that $\mathcal{F}_{g,a/b}$ is exact.

In this section we wish to construct for a given cobordism, $M \in \widetilde{Cob}(a, b)$, between connected surfaces with labeled holes, $M : \Sigma_{g_1,a/b} \to \Sigma_{g_2,a/b}$, a natural transformation between those functors: $\mathcal{V} : \mathcal{F}_{g_1,a/b} \to \mathcal{F}_{g_2,a/b}$. Clearly, it suffices to give a natural transformation $\tilde{\mathcal{V}} : \tilde{\mathcal{F}}_{g_1,a/b} \to \tilde{\mathcal{F}}_{g_2,a/b}$. Specifically, we look for a family of morphisms:

$$\begin{aligned}\tilde{\mathcal{V}}(M|\{X\},\{Y\}) : \mathrm{Inv}_{\mathcal{C}}\big(X_{\alpha(1)} \otimes \ldots \otimes X_{\alpha(a)} \otimes F^{\otimes g_1} \otimes Y_{\beta(b)}^{\vee} \otimes \ldots \otimes Y_{\beta(1)}^{\vee}\big) \longrightarrow \\ \longrightarrow \mathrm{Inv}_{\mathcal{C}}\big(X_1 \otimes \ldots \otimes X_a \otimes F^{\otimes g_2} \otimes Y_b^{\vee} \otimes \ldots \otimes Y_1^{\vee}\big). \quad (6.2.2)\end{aligned}$$

Here $\mathcal{C}$ is an abelian, balanced, braided tensor, strictly rigid category with length and coend $F \in \mathcal{C}$, as defined in Section 4.1. Furthermore, X_j and Y_j are arbitrary objects of $\mathcal{C}$. The permutation of labels is given by $\Pi(M) = \alpha\#\beta$ as defined in (1.2.2). This family obeys a basic naturality property, namely, that for any sequence of morphisms $f_j : X_j \to X_j^{\&}$ and $h_j : Y_j \to Y_j^{\&}$ we have:

$$\begin{aligned}\mathrm{Inv}_{\mathcal{C}}\big((f_{\alpha(1)} \otimes \ldots \otimes f_{\alpha(a)}) \otimes \mathbf{1} \otimes (h_{\beta(b)}^t \otimes \ldots \otimes h_{\beta(1)}^t)\big)\tilde{\mathcal{V}}(M|\{X\},\{Y^{\&}\}) \\ = \tilde{\mathcal{V}}(M|\{X^{\&}\},\{Y\})\,\mathrm{Inv}_{\mathcal{C}}\big((f_1 \otimes \ldots \otimes f_a) \otimes \mathbf{1} \otimes (h_b^t \otimes \ldots \otimes h_1^t)\big). \quad (6.2.3)\end{aligned}$$

The transformations are constructed from a tangle, $T \in \mathcal{Tgl}^{plan;*}_{\mathbb{R}^2}(a,b)$, which represents M. The first step is to associate with a given *coloring* a morphism in $\mathcal{C}$, which is a composite of the natural commutativity, associativity, and rigidity morphisms of $\mathcal{C}$. By a coloring we mean an assignment that attaches an object in $\mathcal{C}$ to every piece of strand between extrema in T in a way that is compatible with (4.1.3). The coloring of a non-closed strand is determined by the color of one of its ends. Observe that the coloring of a split tangle is given exactly by a coloring of the strands emerging at the top-line, since we have only top- and through-ribbons.

To represent a tangle from $\mathcal{Tgl}^{plan;*}_{\mathbb{R}^2}$ we use the conventions of Fig. 4.1. The ribbon twist denoted

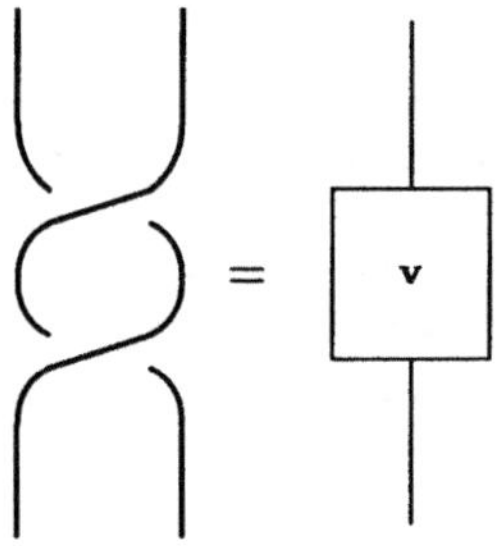

goes to the ribbon twist $\mathbf{v}$.

To a coupon with n vertically penetrating strands we associate the morphism of the respective object, $\lambda(A_1 \otimes \ldots \otimes A_n) \in \mathrm{End}_{\mathcal{C}}(A_1 \otimes \ldots \otimes A_n)$, that belongs to natural transformation λ of the identity on $\mathcal{C}$. Here A_j's is the coloring of the j-th strand. Eventually, we will choose λ to be the transformation obtained from the cointegral of the coend F as defined in Section 5.2.2. We can write

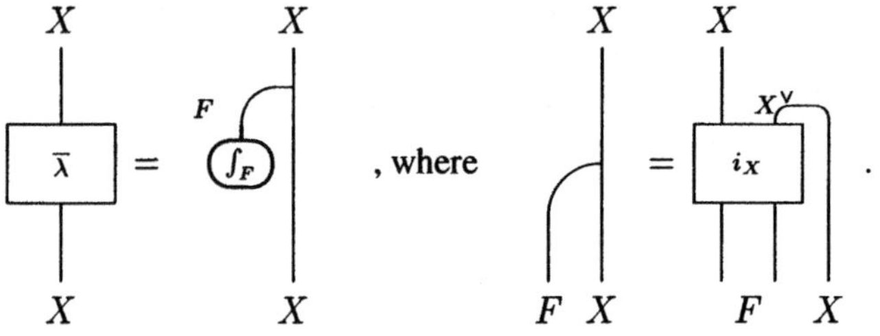

Using the rule for the coaction on a tensor product

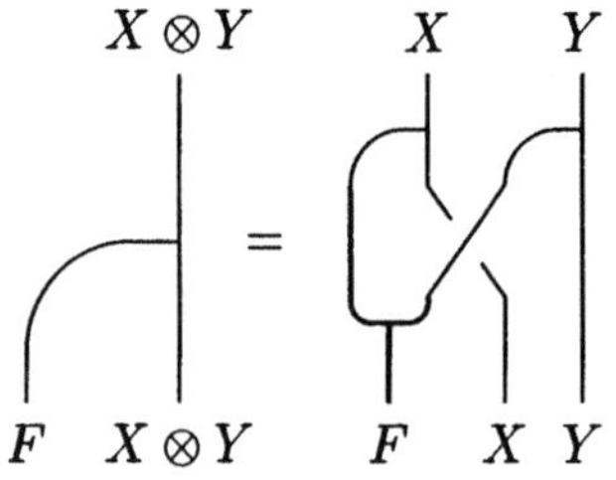

we get the expression for a coupon with two entries

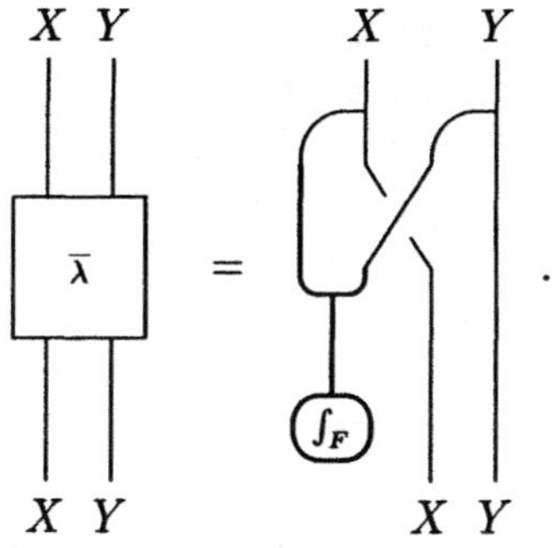

Similarly, we expand expressions for coupons with more entries.

For the construction of the maps in (6.2.2) we shall always color the first a external strands with the objects in the order $X_{\alpha(1)}, \ldots, X_{\alpha(a)}$, and the last b external strands with the objects, $Y_{\beta(b)}, \ldots, Y_{\beta(1)}$. Hence, to the first ordered union of vertical external strands we associate the object

$$\{X_\alpha\} = X_{\alpha(1)} \otimes \ldots \otimes X_{\alpha(a)} \tag{6.2.4}$$

and to the second set of strands the object $\{Y_\beta\}^\vee = Y^\vee_{\beta(b)} \otimes \ldots \otimes Y^\vee_{\beta(1)}$. Also, we color the internal strands of the k-th pair with the same object, P_k, so that the coloring extends uniquely also along top-strands in T^{sp}. The object associated to such a pair is, thus, $P_k^\vee \otimes P_k$, as indicated in (6.2.2). To obtain the object assigned to the collection of all strands, we form the ordered tensor product of all of the objects that appear in (6.2.2).

In general a pair of through strands can originate at any $\{I^i_j, I^o_j\}$ at the top-line and end at any other such pair at the bottom line. In order to keep the notation simple let us assume that the through strands are attached exactly to the first m pairs both at the top- and the bottom-line, i.e., the j-th pair starts at $\{I^i_j, I^o_j\}$ and ends at $\{I^i_{\tau(j)}, I^o_{\tau(j)}\}$, for $j = 1, \ldots, m$, and $\tau \in S_m$. The construction for the general case is literally the same, only that the order, in which the objects of the pairs occur in a tensor product, is permuted. We denote the coloring of the through-strands at the j-th pair by $P_{\tau(j)}$ so that the object for the first $2m$ internal ribbons at the top is

$$[P_\tau] = P^\vee_{\tau(1)} \otimes P_{\tau(1)} \otimes \ldots \otimes P^\vee_{\tau(m)} \otimes P_{\tau(m)}. \tag{6.2.5}$$

Thus, the object at the corresponding $2m$ strands at the bottom is $[P]^\# = L_1 \otimes \ldots \otimes L_m$, where $L_j = P_j^\vee \otimes P_j$ if the through-strand is straight and $L_j = P_j \otimes P_j^\vee$ if it is twisted. We also denote $[P^\#] = (P_1^\#)^\vee \otimes (P_1^\#) \otimes \ldots \otimes (P_m^\#)^\vee \otimes (P_m^\#)$, where $P_j^\# = P_j$ for a straight through-strand and $P_j^\# = P_j^\vee$ for a twisted one.

The balancing isomorphisms, $\beta(P_j) : P_j =\!= P_j^{\vee\vee}$, thus, define a canonical isomorphism,

$$\beta^\#_{[P]} : [P]^\# \longrightarrow [P^\#]. \tag{6.2.6}$$

We denote the coloring of the top-pairs of T by Q_j, with $j = 1, \ldots, g_1 - m$, so that the object associated to the respective set of $2(g_1 - m)$ strands in $[Q]$, defined as in (6.2.5) but without permutations.

For the sake of notational simplicity, we also assume that the splitting ribbons of T occupy the next $g_2 - m$ positions (before the splittings of the closed ones) at the top-line in the same order in which they are attached at the bottom-line. If the coloring at the top-line pairs is given by B_j, with $j = 1, \ldots, g_2 - m$, so that $[B]$ is as in (6.2.5) the object for the respective set of strands, we obtain as a collective object for the bottom pairs $[B]^\#$, i.e., we replace $B_j^\vee \otimes B_j$ by $B_j \otimes B_j^\vee$ if the through-strand resulting from the splitting is a twisted one. Finally, the last n pairs of internal strands at the top-line are the splittings of closed ribbons, with colors $C_j, j = 1, \ldots, n$.

With colorings given in this way we may now associate to a tangle a morphism between the objects at the top- and bottom-line by composing those associated to the singularities in the diagram. Composing the result with the isomorphisms $\beta^\#_{[P]}$ and $\beta^\#_{[B]}$ as defined in (6.2.6), we, thus, find the morphism:

$$\begin{aligned}\Psi\left(T^{sp}|\{X\},[P],[Q],[B],[C]\right) : \{X_\alpha\} \otimes [P_\tau] \otimes [Q] \otimes [B] \otimes [C] \otimes \{Y_\beta\}^\vee \longrightarrow \\ \longrightarrow \{X\} \otimes [P^\#] \otimes [B^\#] \otimes \{Y\}^\vee. \qquad (6.2.7)\end{aligned}$$

The system is *natural* with respect to the colorings. For the colorings $\{X\}$ and $\{Y\}$ this property is expressed precisely by relation (6.2.3). Using the naturality properties of the elementary morphisms we can deduce some relations for the colorings of the internal strands by "pushing" an arbitrary morphism between two colors through the diagram.

For top-strands (with colors Q_j or C_j) we verify what we called in Sect. 5.1.1 dinaturality. For example, if $f : Q_1 \to \tilde{Q}_1$ is any morphism between colors then the following diagram commutes:

$$\begin{array}{ccc} \ldots \otimes \tilde{Q}_1^\vee \otimes Q_1 \otimes \ldots & \xrightarrow{\ldots \otimes f^t \otimes Q_1 \otimes \ldots} & \ldots \otimes Q_1^\vee \otimes Q_1 \otimes \ldots \\ {\scriptstyle \ldots \otimes \tilde{Q}_1^\vee \otimes f \otimes \ldots} \downarrow & & \downarrow {\scriptstyle \Psi(\ldots, Q_1, \ldots)} \\ \ldots \otimes \tilde{Q}_1^\vee \otimes \tilde{Q}_1 \otimes \ldots & \xrightarrow{\Psi(\ldots, \tilde{Q}_1, \ldots)} & \ldots\ldots \end{array} \qquad (6.2.8)$$

Here we indicated the objects that are not changed by dots. For the corresponding relation in the case of through-strands let us consider instead of the general situation, i.e., when we have to deal with maps

$$\Psi(\ldots, P_j, \ldots) : \ldots \otimes P_j^\vee \otimes P_j \otimes \ldots \longrightarrow \ldots \otimes P_j^\vee \otimes P_j \otimes \ldots ,$$

the special case, when we have only one through strand with color P in T and no other strands. If we choose arbitrary morphisms $f : P \to \tilde{P}$ and $g : \tilde{P} \to P$, we find for $\Psi(P) \in \mathrm{End}_{\mathcal{C}}(P^\vee \otimes P)$ the relation

$$\Psi(\tilde{P})(g^t \otimes f) = (g^t \otimes f)\Psi(P), \qquad (6.2.9)$$

for straight pairs of through-strands. In the twisted case we find for $\Psi(P) : P^\vee \otimes P \to P \otimes P^\vee$ the equation

$$\Psi(\tilde{P})(g^t \otimes f) = (f^{tt} \otimes g^t)\Psi(P). \tag{6.2.10}$$

Note, that if a system of morphisms, $\xi_P : P^\vee \otimes P \to Z$, defines a dinatural transformation, then the composite $\xi_P \circ \Psi(P)$ (or $\xi_{P^\vee} \circ \Psi(P)$ for the twisted case) is also a dinatural transformation by (6.2.9) or (6.2.10).

Now, the coend given in (5.2.2) provides us with a universal transformation, $i_P : P^\vee \otimes P \to F$, of this type. Let us denote for $[A] = A_1^\vee \otimes A_1 \otimes \ldots \otimes A_k^\vee \otimes A_k$ the tensor product of the i_{A_j} by $i_{[A]} : [A] \to F$. We may then consider the natural composite of Ψ as in (6.2.7) with $\{X\} \otimes i_{[P\#]} \otimes i_{[B\#]} \otimes \{Y\}^\vee$. This yields the morphism

$$\Psi^*(T^{sp}|\{X\},[P],[Q],[B],[C],\{Y\}) : \{X_\alpha\} \otimes [P_\tau] \otimes [Q] \otimes [B] \otimes [C] \otimes \{Y_\beta\}^\vee \longrightarrow \{X\} \otimes F^{\otimes m} \otimes F^{\otimes(g_2-m)} \otimes \{Y\}^\vee, \tag{6.2.11}$$

which is now dinatural in all internal colorings. As we discussed in Section 5.1.3, dinaturality as in (6.2.8) is equivalent to the existence of a lifting to the coend. For example, if Q is the color of a given pair, we have the following factorization for Ψ^*:

$$\Psi^*(\ldots,Q,\ldots) : \ldots \otimes Q^\vee \otimes Q \otimes \ldots \xrightarrow{\ldots\otimes i_Q \otimes\ldots} \ldots \otimes F \otimes \ldots \xrightarrow{\Psi^*(\ldots,\bullet,\ldots)} \ldots\ \ldots \tag{6.2.12}$$

for a map as in (6.2.8), where $\Psi^*(\ldots,\bullet,\ldots)$ is a morphism that no longer depends on Q. The dots replace as usual colors that are not changed. The lift is, of course, still natural in all of the other colors and can, therefore, again be lifted. Applying this procedure to Ψ^* until all colorings are eliminated we find a morphism,

$$\Psi^{**}(T^{sp}|\{X\},\{Y\}) \equiv \Psi^*(T^{sp}|\{X\},\bullet,\ldots,\bullet,\{Y\}) : \{X_\alpha\} \otimes F^{\otimes m} \otimes F^{\otimes(g_1-m)} \otimes F^{\otimes(g_2-m)} \otimes F^{\otimes n} \otimes \{Y_\beta\}^\vee \to \{X\} \otimes F^{\otimes m} \otimes F^{\otimes(g_2-m)} \otimes \{Y\}^\vee, \tag{6.2.13}$$

which is still natural with respect to the X_j and Y_j.

In summary, we have the following commutative diagram:

$$\begin{array}{ccc}
\{X_\alpha\} \otimes [P_\tau, Q, B, C] \otimes \{Y_\beta\}^\vee & \xrightarrow{\Psi} & \{X\} \otimes [P^\#, B^\#] \otimes \{Y\}^\vee \\
\Big\downarrow {\scriptstyle \{X_\alpha\}\otimes i_{[P_\tau,Q,B,C]}\otimes\{Y_\beta\}^\vee} & & \Big\downarrow {\scriptstyle \{X\}\otimes i_{[P^\#,B^\#]}\otimes\{Y\}^\vee} \\
\{X_\alpha\} \otimes F^{\otimes(g_1+g_2-m+n)} \otimes \{Y_\beta\}^\vee & \xrightarrow{\Psi^{**}} & \{X\} \otimes F^{\otimes g_2} \otimes \{Y\}^\vee
\end{array} \tag{6.2.14}$$

It is clear that for a given system of maps Ψ as in (6.2.7) the existence of such a diagram for all colors is equivalent to the naturality properties of Ψ and that Ψ^{**} is uniquely determined by this diagram for a choice of sufficiently large objects as colors.

Finally, notice that Ψ^* may be included in the diagram (6.2.14) as the arrow going from the upper left to the lower right corner.

6.3 Topological Invariance

In this section we verify that the maps constructed in the previous sections are invariant under the topological moves. Thus, the natural transformations depending on a diagram of a tangle depend, in fact, on a class of equivalence of diagrams. The equivalence relation is generated by the topological moves TI, TD*, and TS*. An equivalence class represents a cobordism by Theorem 3.0.6. So we shall construct the transformations in (6.2.2), which only depend on the represented cobordism.

The following is immediate from the axioms of a rigid, braided tensor category.

Lemma 6.3.1. *The maps Ψ, Ψ^*, and Ψ^{**} are invariant under the isotopy moves TI1 through TI10, for all colorings.*

Proof. For a given coloring the proof for Ψ is standard:

TI1-invariance follows from the equation $\mathbf{u}_{-1}^{-2} = \mathbf{v}^{-1} = \mathbf{u}_{-1}^{2}$ in a strict ribbon category. In the non-strict case the compensating morphisms $\mathbf{u}_0^{\pm 2}$ are introduced, so that the equation $\mathbf{u}_{-1}^{-2} \circ \mathbf{u}_0^{2} = \mathbf{u}_{-1}^{0} = \mathbf{u}_{-1}^{2} \circ \mathbf{u}_0^{-2}$ holds anyway.
TI2-invariance follows from the identity $\mathbf{c}^{-1} \circ \mathbf{c} = 1 = \mathbf{c} \circ \mathbf{c}^{-1}$.
TI3-invariance follows from the hexagon equation in Definition 4.1.4 of a braided category.
TI4-invariance follows from relations between ev and coev in Definition 4.1.3 of a rigid category.
TI5-invariance follows from the properties of the braiding.
TI8- and TI9-invariance follows from the definition of a monoidal category.

Since Ψ^* and Ψ^{**} are constructed directly from Ψ, they also depend on the tangle only up to isotopies.

Lemma 6.3.2. *The following equation holds:*

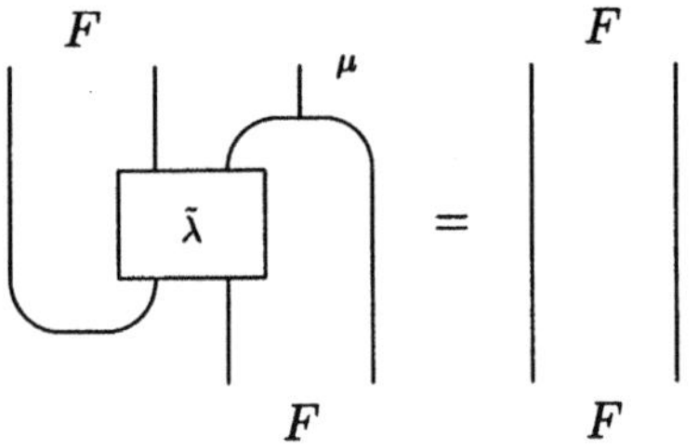

(6.3.1)

Proof. Let us use the definition of the coupon as the natural transformation related with the integral-functional $\lambda : F \to 1$. The left hand side composed with $i_X : X \otimes X^\vee \to F$ equals

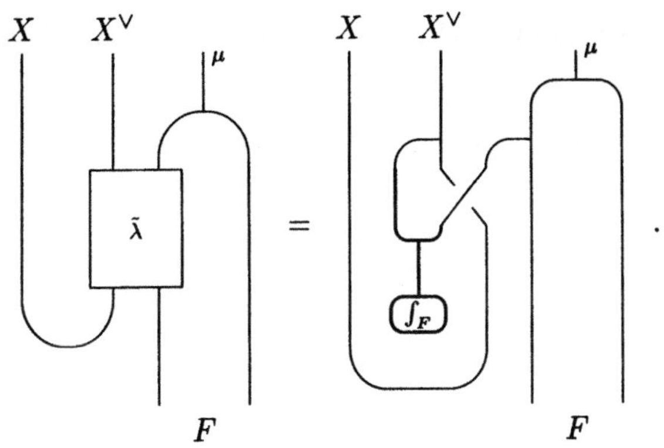

Using the definition of the coaction we reduce this to

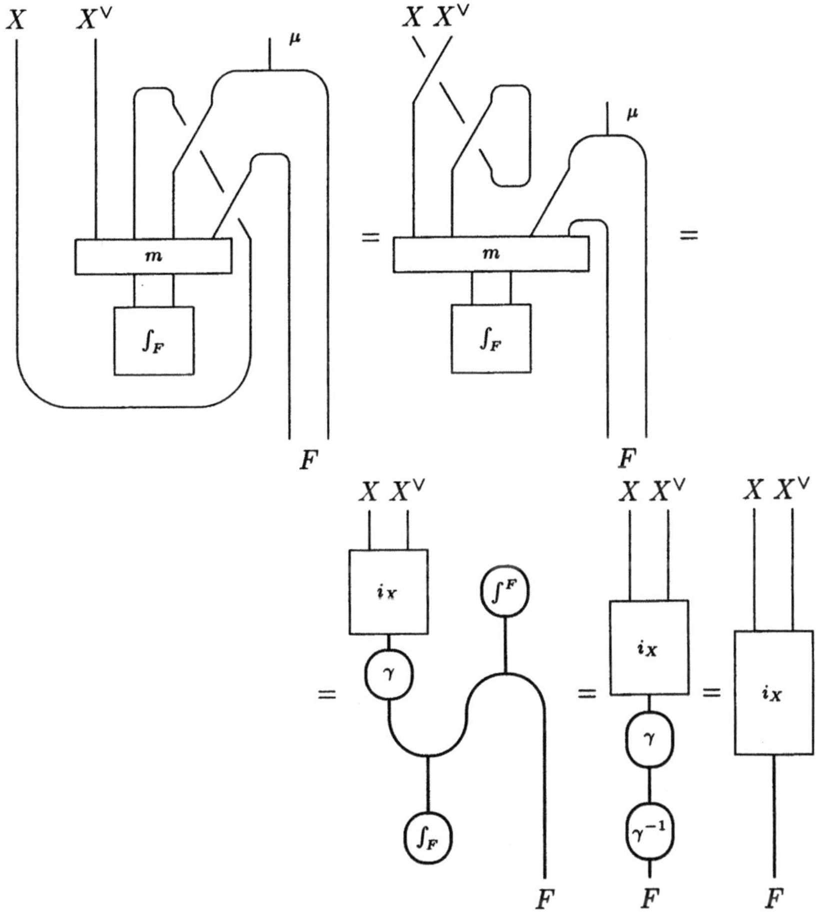

by Fig. 4.3(c) and (5.2.13). The required equation follows.

Lemma 6.3.3. *Let $\mathcal{C}$ be a modular category. Then there are normalizations (unique up to a sign) of the two-sided integrals such that $\hat{\mathcal{V}}(T)$ constructed from them is* *invariant under the moves TI1-TI5, TI8, TI9, TD1, TD2, TD3*-TD5*, TS1*-TS3* and TS4.*

Proof. We have already proven the invariance under TI moves in Lemma 6.3.1. It suffices now to notice the following.

TD1- and TD2-invariance follows from Lemma 5.2.14.
TD3*- and TD4*-invariance follows from the fact that $\operatorname{Int} F = 1$ (Lemma 5.2.8).
TD5*-invariance follows from the fact that $\gamma \circ \int^F = \int^F$ (Lemma 5.2.12).
TS1*-invariance follows from equation (5.2.14).
TS2*-invariance follows from normalization equation (5.2.13).
TS3*-invariance follows from Lemma 6.3.2.

Finally, TS4-invariance is proven as follows. For arbitrary objects $X, Y \in \mathcal{C}$ there is a braiding $\mathbf{c} : X \otimes Y \to Y \otimes X$ and an inverse braiding $\mathbf{c}^{-1} : X \otimes Y \to Y \otimes X$. We have to show that restrictions of these two morphisms to $(\operatorname{Inv} X) \otimes Y$ coincide. That is, for an arbitrary morphism $h : 1 \to X$ we have

$$\big(1 \otimes Y \xrightarrow{h \otimes 1} X \otimes Y \xrightarrow{\mathbf{c}} Y \otimes X\big) = \big(1 \otimes Y \xrightarrow{h \otimes 1} X \otimes Y \xrightarrow{\mathbf{c}^{-1}} Y \otimes X\big).$$

Indeed, by naturality of $\mathbf{c}$ both sides of the equation equal to

$$\big(1 \otimes Y \xrightarrow{\mathbf{c}} Y \otimes 1 \xrightarrow{1 \otimes h} Y \otimes X\big) = \big(1 \otimes Y \xrightarrow{\mathbf{c}^{-1}} Y \otimes 1 \xrightarrow{1 \otimes h} Y \otimes X\big). \square$$

We define the map in (6.2.2) as follows:

$$\tilde{\mathcal{V}}(M|\{X\},\{Y\}) = \operatorname{Inv}_{\mathcal{C}}\big(\hat{\mathcal{V}}(T|\{X\},\{Y\},\overline{\mu})\big). \tag{6.3.2}$$

Summarizing the results of this section, and justifying the notation in (6.3.2), we have:

Proposition 6.3.4. *Suppose the map $\tilde{\mathcal{V}}(M|\{X\},\{Y\})$ is constructed for a modular category with coend F as above, using the true integrals. Then it depends only on the cobordism $M \in \widetilde{\mathcal{C}ob}(a,b)$.*

6.4 Compositions over Colored Surfaces

In this section we verify that $\mathcal{V}$ respects compositions of cobordisms over surfaces, i.e., it is a functor $\tilde{\mathcal{V}} : \widetilde{\mathcal{C}ob} \longrightarrow \Bbbk$-vect for a fixed coloration of the holes. This corresponds to the vertical composition in the double category picture. Following we will eliminate the colorings of the holes by lifting $\mathcal{V}$ to a functor of the form $\widetilde{\mathcal{C}ob}(a,b) \longrightarrow Fun(\mathcal{C}^{\boxtimes a}, \mathcal{C}^{\boxtimes b})$.

In order to show that $\mathcal{V}(M_1)\mathcal{V}(M_2) = \mathcal{V}(M_1 \circ M_2)$ for compatible colorations of the holes, we may use the fact that $\mathcal{V}$ does not depend on the particular split tangle we choose to represent the cobordism. We, therefore, assume that the tangle contains no through strands, which can be arranged by applications of the TS3 Moves. Moreover, we assume that the splitting ribbons of T_1^{sp} are not as in the previous sections to the left but to the right of the last b external strands. If we have two such tangles T_j^{sp} with $j = 1, 2$, representing the cobordisms M_j, then $T_{1\circ 2}^{sp} = T_1^{sp} \circ (T_2^{sp} \sqcup \mathbb{1}_{1sp})$ is a split ribbon whose closure represents $M_2 \circ M_1$. Here $\mathbb{1}_{1sp}$ are parallel strands that extend the pairs of strands, obtained from the splitting ribbons from the top-line of the diagram of T_1^{sp} to the top-line for T_2^{sp}.

For each of the tangles with colorings we can construct the morphisms

$$\Psi_2 : \{X_{\alpha_1 \circ \alpha_2}\} \otimes [Q_2] \otimes [B_2] \otimes [C_2] \otimes \{Y_{\beta_1 \circ \beta_2}\}^\vee \longrightarrow \{X_{\alpha_1}\} \otimes [B_2^{\#}] \otimes \{Y_{\beta_1}\}^\vee$$
$$\text{and} \quad \Psi_1 : \{X_{\alpha_1}\} \otimes [Q_1] \otimes \{Y_{\beta_1}\}^\vee \otimes [B_1] \otimes [C_1] \longrightarrow \{X\} \otimes [B_1^{\#}] \otimes \{Y\}^\vee.$$

Here we have already chosen colorings of the external strands that are compatible. To make the colors of the internal strands also compatible we have to impose the condition $[B_2^{\#}] = [Q_1]$. In this case we can form the composite

$$\Psi_{1\circ 2} = \Psi_1 \circ \big(\Psi_2 \otimes [B_1, C_1]\big) :$$
$$\{X_{\alpha_1 \circ \alpha_2}\} \otimes [Q_2, B_2, C_2] \otimes \{Y_{\beta_1 \circ \beta_2}\}^\vee \otimes [B_1, C_1] \longrightarrow \{X\} \otimes [B_1^{\#}] \otimes \{Y\}^\vee.$$

We now wish to compare this to the morphism $\Psi(T_{1\circ 2}^{sp})$ associated to the composite of the split tangles with colors given by the source objects. Since the assignment of elementary morphisms to singularities is local, the prescriptions for constructing $\Psi(T_{1\circ 2}^{sp})$ and the composite $\Psi_{1\circ 2}$ are the same in the interior of the diagrams for compatible colorations. At the boundary between the diagrams the coloring is consistently continued, and the morphisms of both parts are composed.

Thus, we find:

Lemma 6.4.1. $\Psi(T_{1\circ 2}^{sp}) = \Psi_{1\circ 2}$.

For the tangles T_j^{sp} we obtain with $g_{2,2} = g_{1,1}$ morphisms

$$\Psi_1^{**} : \{X_{\alpha_1}\} \otimes F^{\otimes g_{1,1}} \otimes \{Y_{\beta_1}\}^\vee \otimes F^{\otimes g_{1,2}} \otimes F^{\otimes n_1} \to \{X\} \otimes F^{\otimes g_{1,2}} \otimes \{Y\}^\vee,$$
$$\Psi_2^{**} : \{X_{\alpha_1 \circ \alpha_2}\} \otimes F^{\otimes g_{2,1}} \otimes F^{\otimes g_{2,2}} \otimes F^{\otimes n_2} \otimes \{Y_{\beta_1 \circ \beta_2}\}^\vee$$
$$\to \{X_{\alpha_1}\} \otimes F^{\otimes g_{2,2}} \otimes \{Y_{\beta_1}\}^\vee,$$

which satisfy the relations (6.2.14) with respect to the morphisms Ψ_1 and Ψ_2. Using this and Lemma 6.4.1, we can compute for the morphism obtained from the composite tangle

$$\begin{aligned}
\Psi^*(T^{sp}_{1\circ 2}) &= \Big(\{X\}\otimes i_{[B_1^\#]}\otimes\{Y\}\Big)\Psi(T^{sp}_{1\circ 2}) = \Big(\{X\}\otimes i_{[B_1^\#]}\otimes\{Y\}\Big)\Psi_{1\circ 2}\\
&= \Big(\{X\}\otimes i_{[B_1^\#]}\otimes\{Y\}\Big)\Psi_1(\Psi_2\otimes[B_1,C_1])\\
&= \Psi_1^{**}\Big(\{X_{\alpha_1}\}\otimes i_{[Q_1]}\otimes\{Y_{\beta_1}\}^\vee\otimes i_{[B_1,C_1]}\Big)(\Psi_2\otimes[B_1,B_2])\\
&= \Psi_1^{**}\Big(\{(\{X_{\alpha_1}\}\otimes i_{[Q_1]}\otimes\{Y_{\beta_1}\}^\vee)\Psi_2\}\otimes i_{[B_1,C_1]}\Big)\\
&= \Psi_1^{**}\Big(\{\Psi_2^{**}(\{X_{\alpha_1\circ\alpha_2}\}\otimes i_{[Q_2,B_2,C_2]}\otimes\{Y_{\beta_1\circ\beta_2}\}^\vee)\}\otimes i_{[B_1,C_1]}\Big)\\
= \Psi_1^{**}(\Psi_2^{**}&\otimes F^{\otimes(g_{1,2}+n_1)})(\{X_{\alpha_1\circ\alpha_2}\}\otimes i_{[Q_2,B_2,C_2]}\otimes\{Y_{\beta_1\circ\beta_2}\}^\vee\otimes i_{[B_1,C_1]}).
\end{aligned}$$

Thus, we have found a lifting of $\Psi^*(T^{sp}_{1\circ 2})$ to the product of coends. Since any liftings to coends are unique we conclude:

Lemma 6.4.2. $\qquad \Psi^{**}(T^{sp}_{1\circ 2}) = \Psi^{**}(T^{sp}_1)\circ(\Psi^{**}(T^{sp}_2)\otimes F^{\otimes(g_{1,2}+n_1)}).$

The morphism $\hat{\mathcal{V}}(T^{sp}_{1\circ 2})$ is now given as in (6.7.5) by composing $\Psi^{**}(T^{sp}_{1\circ 2})$ with the following tensor product of integrals:

$$\mu_{all} = \{X_{\alpha_1\circ\alpha_2}\}\otimes F^{\otimes g_{2,1}}\otimes\mu^{\otimes g_{2,2}}\otimes\mu^{\otimes n_2}\otimes\{Y_{\beta_1\circ\beta_2}\}^\vee\otimes\mu^{\otimes g_{1,2}}\otimes\mu^{\otimes n_1}.$$

If we apply this construction to the formula in Lemma 6.4.2, we find immediately

$$\hat{\mathcal{V}}(T^{sp}_{1\circ 2}) = \hat{\mathcal{V}}(T^{sp}_1)\circ\hat{\mathcal{V}}(T^{sp}_2).$$

To this we can apply the $\mathrm{Inv}_{\mathcal{C}}$-functor, in order to derive the same composition law for $\mathcal{V}(T^{sp}_j) = \mathcal{V}(M_j)$.

We can formulate the result of this section more concisely if we introduce the category of colored surfaces $\widetilde{\mathcal{C}ob}_{\mathrm{Col}}$. Its objects are surfaces, $\Sigma^{\{X\}}$, as in $\widetilde{\mathcal{C}ob}$ together with a map that assigns to each hole in Σ an object in $\mathcal{C}$. The set of morphisms $\Sigma_a^{\{X_a\}}\to\Sigma_b^{\{X_b\}}$ consists of all cobordisms $M:\Sigma_a\to\Sigma_b$ in $\widetilde{\mathcal{C}ob}$ such that the colors of the holes that are connected by the cylindrical pieces in ∂M coincide. With these conventions we find the following:

Proposition 6.4.3. *The maps from (6.2.2) define a functor:*

$$\tilde{\mathcal{V}}:\widetilde{\mathcal{C}ob}_{\mathrm{Col}}\longrightarrow \Bbbk\text{-vect}.$$

6.5 Lifting $\mathcal{V}(M)$ to Color-Independent Natural Transformation

In the formulation of Proposition 6.4.3 the definition of the functor still depends on a particular coloring of the holes. This formula also does not intrinsically reflect the property that the resulting maps are natural in the colors.

In this section we eliminate the dependence on the colorings of the holes by reformulating $\mathcal{V}$ as a functor $\widetilde{\mathcal{C}ob}(a,b)\longrightarrow Fun(\mathcal{C}^{\boxtimes a},\mathcal{C}^{\boxtimes b})$. The maps $\tilde{\mathcal{V}}(M)$, which we have constructed above, will, thus, appear as natural transformations between functors that are associated to a $1+1$-cobordism. The compatibility with the vertical compositions holds as well.

The idea of the construction is similar to the lifting a dinatural, color-dependent transformation to a morphism on the coend-object. Only now we lift it to an object in the category of functors and natural transformations.

Specifically, the functor from (6.2.1) associated to the 1+1-cobordism $\Sigma_{g,a/b}$

$$\mathcal{F}_{g,a/b} = \mathcal{V}(\Sigma_{g,a/b}) : \mathcal{C}^{\boxtimes a} \longrightarrow \mathcal{C}^{\boxtimes b},$$

is isomorphic to the following functor.

We can define a natural bifunctor from a b-fold product of a category into the category of functors

$$(\mathcal{C}^{\boxtimes b})^{opp} \times \mathcal{C}^{\boxtimes b} \to Fun(\mathcal{C}^{\boxtimes a}, \mathcal{C}^{\boxtimes b}) : (Y_1 \boxtimes \ldots \boxtimes Y_b) \times (Z_1 \boxtimes \ldots \boxtimes Z_b) \mapsto \mathcal{F}^{\langle Y,Z\rangle}_{g,a/b}, \tag{6.5.1}$$

where the functor is defined by the formula

$$\mathcal{F}^{\langle Y,Z\rangle}_{g,a/b} : X_1 \boxtimes \ldots \boxtimes X_a \mapsto \mathrm{Inv}_{\mathcal{C}}(\{X\} \otimes F^{\otimes g} \otimes \{Y\}^{\vee}) Z_1 \boxtimes \ldots \boxtimes Z_b \tag{6.5.2}$$

with $\{X\}$ defined as in (6.2.4).

We can define a candidate for a coend as follows.

First, set (for a *strict* category):

$$\mathcal{F}_{g,a/b} : \mathcal{C}^{\boxtimes a} \xrightarrow{\otimes^{a-1}} \mathcal{C} \xrightarrow{-\otimes F^{\otimes g}} \mathcal{C} \xrightarrow{\mathcal{F}^0_b} \mathcal{C}^{\boxtimes(b+1)} \xrightarrow{\mathrm{Inv}_{\mathcal{C}} \boxtimes \mathcal{C}^{\boxtimes b}} \mathcal{C}^{\boxtimes b}. \tag{6.5.3}$$

The functor $\mathcal{F}^0_b$ is defined inductively by letting $\mathcal{F}^0_0$ be the identity on $\mathcal{C}$, and

$$\mathcal{F}^0_b : \mathcal{C} \xrightarrow{\mathcal{F}^0_{b-1}} \mathcal{C} \boxtimes \mathcal{C}^{\boxtimes(b-1)} \xrightarrow{\mathcal{C}\boxtimes\mathbb{F}\boxtimes\mathcal{C}^{\boxtimes(b-1)}} \mathcal{C} \boxtimes (\mathcal{C} \boxtimes \mathcal{C}) \boxtimes \mathcal{C}^{\boxtimes(b-1)} \xrightarrow{\otimes\boxtimes\mathcal{C}^{\boxtimes b}} \mathcal{C}^{\boxtimes(b+1)}. \tag{6.5.4}$$

Here $\mathbb{F}$ is the coend from (5.2.1) and we think of it here as a functor $\mathbb{F} : \Bbbk\text{-vect} \to \mathcal{C}^{\boxtimes 2}$. Note that the functor $\mathcal{F}^{\langle Y,Z\rangle}_{g,a/b}$ can be defined in the same way if we replace in (6.5.4) the functor $\mathbb{F}$ with $Y^{\vee} \boxtimes Z : \Bbbk\text{-vect} \to \mathcal{C}^{\boxtimes 2}$. Now the transformations $Y^{\vee} \boxtimes Y \to \mathbb{F}$ can be thought of as natural transformations between constant functors for $Z = Y$. Composing these individual transformations we define a dinatural transformation

$$I_Y : \mathcal{F}^{\langle Y,Y\rangle}_{g,a/b} \xrightarrow{\bullet} \mathcal{F}_{g,a/b}. \tag{6.5.5}$$

For $b \geqslant 1$ it is often more useful to use another isomorphic version of the functor, which we denote by $\mathcal{F}^{\infty}_{g,a/b}$. Similar to (6.5.3) it is defined by the following composite:

$$\mathcal{F}^{\infty}_{g,a/b} : \mathcal{C}^{\boxtimes a} \xrightarrow{\otimes^{a-1}} \mathcal{C} \xrightarrow{-\otimes F^{\otimes g}} \mathcal{C} \xrightarrow{\mathcal{F}^0_{b-1}} \mathcal{C}^{\boxtimes b}. \tag{6.5.6}$$

It represents the graph

$G^{\infty}_{g,a/b} =$ 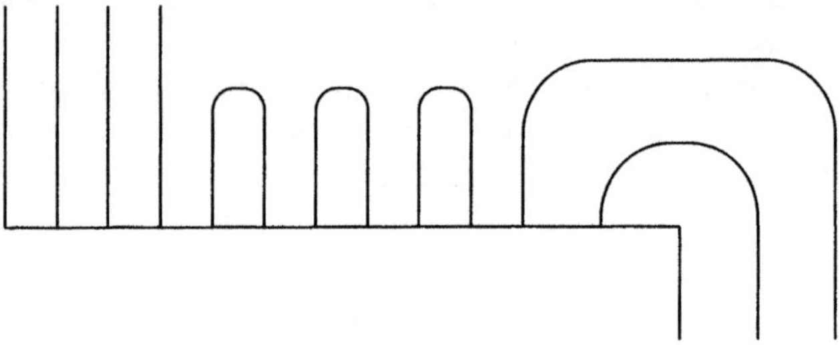 .

If $b = 1$ the functor $\mathcal{F}^{\infty}_{g,a/1}$ looks simpler. Prominent examples are the pair of pants $\Sigma_{0,2/1}$, which yields the functor $\mathcal{F}^{\infty}_{0,2/1} \equiv \otimes$, and also $\mathcal{F}^{\infty}_{g,1/1} \equiv - \otimes F^{\otimes g}$. Another way of writing (6.5.6) is, thus, the following factorization:

$$\mathcal{F}^{\infty}_{g,a/b} \simeq \mathcal{F}^{\infty}_{0,1/b} \circ \mathcal{F}^{\infty}_{g,1/1} \circ \mathcal{F}^{\infty}_{0,a/1}. \tag{6.5.7}$$

We will often extend the notation to $b = 0$ by setting $\mathcal{F}^{\infty}_{0,1/0} = \mathcal{F}_{0,1/0} = \mathrm{Inv}_{\mathcal{C}}$.

As for $\mathcal{F}_{g,a/b}$, we can construct for the latter version of the functor also a system of colored functors as follows:

$$\begin{aligned} \mathcal{F}^{\infty,\langle Y_2,\ldots,Y_b\rangle}_{g,a/b} &: X_1 \boxtimes \ldots \boxtimes X_a \mapsto \\ &\mapsto \left(X_1 \otimes \ldots \otimes X_a \otimes F^{\otimes g_1} \otimes Y_b^{\vee} \otimes \ldots \otimes Y_2^{\vee}\right) \boxtimes Y_2 \boxtimes \ldots \boxtimes Y_b. \end{aligned} \tag{6.5.8}$$

In analogy to (6.5.5) we can define transformations $I^{\infty}_{\hat{Y}} : \mathcal{F}^{\infty,\langle\hat{Y}\rangle}_{g,a/b} \to \mathcal{F}^{\infty}_{g,a/b}$, where we abbreviated $\hat{Y} = Y_2 \boxtimes \ldots \boxtimes Y_b$. Clearly, the evaluation ev_{Y_1} may be used to define the top row that makes the following diagram commutative.

$$\begin{array}{ccc} \mathcal{F}^{\langle Y\rangle}_{g,a/b} & \xrightarrow{\ \mathbf{ev}_{g,a/b}\ } & \mathcal{F}^{\infty,\langle\hat{Y}\rangle}_{g,a/b} \\ {\scriptstyle I_Y}\downarrow & & \downarrow{\scriptstyle I^{\infty}_{\hat{Y}}} \\ \mathcal{F}_{g,a/b} & \xrightarrow{\ \mathbf{n}_{g,a/b}\ } & \mathcal{F}^{\infty}_{g,a/b} \end{array} \tag{6.5.9}$$

For fixed X and $\hat{Y}$, $\mathcal{F}^{\infty,\langle\hat{Y}\rangle}_{g,a/b}(X)$ may be identified with the coend $\int^{Y_1} \mathcal{F}^{\langle Y\rangle}_{g,a/b}(X)$.

In general, if there are functors $\mathcal{F}^{\langle -,-\rangle} : \mathcal{C}^{\mathrm{op}} \times \mathcal{C} \times \mathcal{C}_1 \to \mathcal{C}_2$ and $\mathcal{F} : \mathcal{C}_1 \to \mathcal{C}_2$ and natural transformations $I_Y : \mathcal{F}^{\langle Y,Y\rangle} \to \mathcal{F}$, such that for any object, $X \in \mathcal{C}_1$, we have that $\mathcal{F}(X)$ is the coend $\int^{Y} \mathcal{F}^{\langle Y,Y\rangle}(X)$, then $\mathcal{F}$ is also the coend of $\mathcal{F}^{\langle Y,Y\rangle}$. In order to understand this assertion more concretely let us consider a general functor $\mathcal{G} : \mathcal{C}_1 \to \mathcal{C}_2$ and a system of natural transformations $G_Y : \mathcal{F}^{\langle Y,Y\rangle} \xrightarrow{\ \bullet\ } \mathcal{G}$, which is dinatural in Y, meaning, for which

$$G_V \circ \mathcal{F}^{\langle f,V\rangle} = G_W \circ \mathcal{F}^{\langle W,f\rangle} \quad : \quad \mathcal{F}^{\langle W,V\rangle} \xrightarrow{\ \bullet\ } \mathcal{G}$$

for all $V, W \in \mathcal{C}$ and $f : V \to W$.

The coend $\mathcal{F}$ is characterized as the initial object of the category of such dinatural transformations. This means that there is a unique, natural $h : \mathcal{F} \xrightarrow{\bullet} \mathcal{G}$ such that $G_Y = h \circ I_Y$ for Y, and this h is nothing else but the system of morphisms $h_X : \mathcal{F}(X) \to \mathcal{G}(X)$, $X \in \mathcal{C}_1$ such that $G_{YX} = h_X \circ I_{YX}$ for all $Y \in \mathcal{C}$, which exist by the assumption. The fact that these form a natural transformation is then a consequence of the universality of $\mathcal{F}(X)$ for each X.

In our situation this allows us to view the above coend in Y_1 for fixed X as a coend of functors but still with fixed coloring $\hat{Y}$.

Now the functors $\otimes$, $Y \mapsto Z_X \otimes Y$, and the permutations on $\boxtimes$-products of categories are all exact functors, and, thus, commute with the coend. Applying this to the expressions for $\mathcal{F}^{\infty,\langle\hat{Y}\rangle}(X)$, we see that their coend in $\hat{Y}$ is in fact $\mathcal{F}^{\infty}(X)$, and, hence, also for the functors themselves. Let us summarize this as follows.

Lemma 6.5.1. *For the functors from* (6.5.2), (6.5.3), (6.5.6), *and* (6.5.8), *we have*

$$\mathcal{F}_{g,a/b} = \int^{Y} \mathcal{F}^{\langle Y\rangle}_{g,a/b} \quad \text{and} \quad \mathcal{F}^{\infty}_{g,a/b} = \int^{\hat{Y}} \mathcal{F}^{\infty,\langle\hat{Y}\rangle}_{g,a/b} ,$$

where the corresponding transformations I_Y, and $I^{\infty}_{\hat{Y}}$, are given in (6.5.5) *and* (6.5.9).

Lemma 6.5.1 allows us to lift a color-dependent system of transformation, as it is associated by the functor $\widetilde{Cob_3}(N)^{conn}_{\mathrm{Col}} \to \mathbb{k}\text{-vect}$ in Proposition 6.4.3 to a cobordism, to a color-independent, natural transformation of functors as follows:

We suppose a system of maps $\mathcal{V}(M|\{X\},\{Y\})$ is given as in (6.2.2) and satisfies the relation from (6.2.3). With

$$\mathcal{F}^{\langle Y_\beta, Y_\beta\rangle}_{g_1,a/b} : X_{\alpha(1)}\boxtimes\ldots\boxtimes X_{\alpha(a)} \mapsto \mathrm{Inv}_{\mathcal{C}}\big(\{X_\alpha\}\otimes F^{\otimes g_1}\otimes\{Y_\beta\}^{\vee}\big)Y_{\beta(1)}\boxtimes\ldots\boxtimes Y_{\beta(b)}$$

and

$$\underline{\beta}^{-1} \circ \mathcal{F}^{\langle Y,Y\rangle}_{g_2,a/b} \circ \underline{\alpha} :$$
$$X_{\alpha(1)} \boxtimes \ldots \boxtimes X_{\alpha(a)} \mapsto \mathrm{Inv}_{\mathcal{C}}\big(\{X\} \otimes F^{\otimes g_1} \otimes \{Y\}^{\vee}\big)Y_{\beta(1)} \boxtimes \ldots \boxtimes Y_{\beta(b)}$$

this yields by action on the vector-part a natural transformation

$$\mathcal{V}^{\langle Y\rangle} : \mathcal{F}^{\langle Y_\beta, Y_\beta\rangle}_{g_1,a/b} \xrightarrow{\bullet} \underline{\beta}^{-1} \circ \mathcal{F}^{\langle Y,Y\rangle}_{g_2,a/b} \circ \underline{\alpha}.$$

The composite transformation $G_Y = (\underline{\beta}^{-1} \times I_Y \times \underline{\alpha}) \circ \mathcal{V}^{\langle Y\rangle}$ has the dinaturality properties needed for a lifting to the functor $\mathcal{F}_{g_1,a/b}$ as in Lemma 6.5.1. As a consequence this lift leaves us with a natural transformation

$$\mathcal{V}(M) : \underline{\beta} \circ \mathcal{F}_{g_1,a/b} \xrightarrow{\bullet} \mathcal{F}_{g_2,a/b} \circ \underline{\alpha}.$$

It contains exactly the same information as the system of natural maps (6.2.2).

Moreover, this form of the TQFT-functor reflects exactly the role of a 2-morphism as a transformation in a *square* of horizontal and vertical 1-arrows. Summarily, we have the following.

Lemma 6.5.2. *For a category $\mathcal{C}$ as above we have a map from the squares in $\widetilde{\mathcal{C}ob}(a,b)$ to the squares in* **AbCat**,

$$\begin{array}{ccc} S_a & \xrightarrow{\Sigma_{g_1,a/b}} & S_b \\ {\scriptstyle\alpha}\downarrow & \swArrow\; M & \downarrow{\scriptstyle\beta} \\ S_a & \xrightarrow{\Sigma_{g_2,a/b}} & S_b \end{array} \quad \stackrel{\mathcal{V}}{\longmapsto} \quad \begin{array}{ccc} \mathcal{C}^{\boxtimes a} & \xrightarrow{\mathcal{F}_{g_1,a/b}} & \mathcal{C}^{\boxtimes b} \\ {\scriptstyle\underline{\alpha}}\downarrow & \swArrow\; \mathcal{V}(M) & \downarrow{\scriptstyle\underline{\beta}} \\ \mathcal{C}^{\boxtimes a} & \xrightarrow{\mathcal{F}_{g_2,a/b}} & \mathcal{C}^{\boxtimes b} \end{array}$$

which is strictly functorial under compositions of vertical 1-arrows and the 2-arrows in <u>vertical</u> direction.

In the diagrams S_a is the union of a circles, and the 2-arrow in the left diagram is cobordism, and the one in the right diagram is a natural transformation.

If $\beta(1) = 1$, the natural transformations $\mathcal{V}(M)$ between two functors given in this way can be obtained from morphisms

$$\begin{aligned} \mathcal{V}^{\infty}(T^{\infty}) : \{X_\alpha\} \otimes F^{\otimes g_1} \otimes Y^{\vee}_{\beta(b)} \otimes \ldots \otimes Y^{\vee}_{\beta(2)} \\ \longrightarrow \{X\} \otimes F^{\otimes g_2} \otimes Y^{\vee}_{b} \otimes \ldots \otimes Y^{\vee}_{2}, \end{aligned} \tag{6.5.10}$$

that are constructed from a given tangle in the same way as $\hat{\mathcal{V}}$ from (6.7.5) is. Only here T^{∞} is a tangle in $\mathcal{T}gl^{\infty}(a,b)$, such that $T = \mathcal{I}(T^{\infty}) \in \mathcal{T}gl(a,b)$ represents M, where $\mathcal{I}$ acts on a representing tangle by adjoining another strand to the right. Thus, the morphism $\mathcal{V}(T^{\infty}) \otimes Y_1^{\vee}$ is immediately identified with $\hat{\mathcal{V}}(T)$. From this we easily see that if $\mathcal{V}(M)$ is conjugated by the isomorphism $\mathbf{n} : \mathcal{F} \to \mathcal{F}^{\infty}$ we obtain exactly $\mathcal{V}^{\infty}(M)$. Hence, $\mathcal{V}^{\infty}$ defines in this sense an equivalent functor.

The most basic example for $\mathcal{V}^{\infty}$ is the cobordism, $B : \Sigma_{0,2/1} \to \Sigma_{0,2/1}$ depicted in Figure 1.8, between two pairs of pants. In this case the vertical arrow at the source-holes is the non-trivial generator $\sigma \in S_2$. The associated tangle T^{∞} consists of two external strands with a simple over- or under-crossing. Thus, we have

$$\varepsilon = \mathcal{V}^{\infty}(B) : \textcircled{\scriptsize Y} \xrightarrow{\;\bullet\;} \textcircled{\scriptsize Y} \circ \underline{\sigma}, \tag{6.5.11}$$

i.e., $\mathcal{V}^{\infty} : \widetilde{\mathcal{C}ob_3}(1)^{conn} \longrightarrow \mathcal{F}un(\mathbb{k}\text{-vect}, \mathcal{C}) \cong \mathcal{C}$, strictly preserves the braided morphism, if it is constructed in this way.

Let us describe for selected examples also the construction of natural transformations for the enlarged cobordism 2-category, where a strand of a vertical arrow is allowed to connect a source or target surface to itself. The assignment of local extrema of the strand to functors will be as follows:

$$\begin{aligned} \mathcal{V}(\cup) &: \mathcal{C} \boxtimes \mathcal{C} \longrightarrow \mathbb{k}\text{-vect}, & X \boxtimes Y &\mapsto \mathrm{Inv}_{\mathcal{C}}(X \otimes Y), \\ \mathcal{V}(\cap) &: \mathbb{k}\text{-vect} \longrightarrow \mathcal{C} \boxtimes \mathcal{C}, & \mathbb{k} &\mapsto \mathbb{F} = \textstyle\int X^{\vee} \boxtimes X. \end{aligned} \tag{6.5.12}$$

First, we consider the case where we have cobordant surfaces with $b > 0$ and $b + 2$ target-holes in source and target surface respectively, and the vertical arrow for the source holes is the identity. More generalizations are obtained using sometimes complicated combinations of the functors in (6.5.12), but are, in principle, straightforward.

We further assume that the vertical arrow at the target consists of a simple minimum or maximum following a set of b vertical parallel strands. In the first case we have a cobordism $M^{\cup}$ from $\Sigma_{g_1,a/b+2}$ to $\Sigma_{g_2,a/b}$, in the second case a cobordism of the form $M^{\cap} : \Sigma_{g_1,a/b} \to \Sigma_{g_2,a/b+2}$. We have to explain how we can associate natural transformations

$$\begin{array}{ccc} \mathcal{C}^{\boxtimes a} & \xrightarrow{\mathcal{F}_{g_1,a/b+2}} & \mathcal{C}^{\boxtimes b+2} \\ \Big\| & \swarrow\!\!\!\swarrow \; \mathcal{V}(M^{\cup}) & \Big\downarrow 1\boxtimes\mathcal{V}(\cup) \\ \mathcal{C}^{\boxtimes a} & \xrightarrow{\mathcal{F}_{g_2,a/b}} & \mathcal{C}^{\boxtimes b} \end{array} \quad \text{and} \quad \begin{array}{ccc} \mathcal{C}^{\boxtimes a} & \xrightarrow{\mathcal{F}_{g_1,a/b}} & \mathcal{C}^{\boxtimes b} \\ \Big\| & \swarrow\!\!\!\swarrow \; \mathcal{V}(M^{\cap}) & \Big\downarrow 1\boxtimes\mathcal{V}(\cap) \\ \mathcal{C}^{\boxtimes a} & \xrightarrow{\mathcal{F}_{g_2,a/b+2}} & \mathcal{C}^{\boxtimes b+2} \end{array} \qquad (6.5.13)$$

to these cobordisms.

We begin with a natural isomorphism

$$\left(1^{\boxtimes b} \boxtimes \mathcal{V}(\cup)\right) \circ \mathcal{F}_{g_1,a/b+2} \xrightarrow{\;\cong\bullet\;} \mathcal{F}_{g_1+1,a/b},$$

obtained via Theorem C.2.1. The transformation $\mathcal{V}'(M^{\cup}) : \mathcal{F}_{g_1+1,a/b} \to \mathcal{F}_{g_2,a/b}$ is obtained by interpreting the returning, external ribbon in a tangle $T^{\cup}$ that represents $M^{\cup}$ as an internal top-ribbon. The resulting tangle T (topologically identical to $T^{\cup}$), thus, represents a cobordism $M : \Sigma_{g_1+1,a/b} \to \Sigma_{g_2,a/b}$, and we set $\mathcal{V}'(M^{\cup}) = \mathcal{V}(M)$. Note that the map $M^{\cup} \to M$ is well defined, since the moves in the category of $T^{\cup}$ are also moves in the category of T.

The topological interpretation of M is also quite obvious: the cylinder in $M^{\cup}$, that starts and ends in the source surface, is simply reinterpreted in M as another handle of that surface.

The construction of $\mathcal{V}(M^{\cap})$ is similar.

6.6 Horizontal Compositions

In the previous section we have replaced the vector spaces associated to surfaces by functors, and the linear maps associated to cobordisms by natural transformations. The results from Sect. 6.4 allow us assign to the composition of cobordisms over a surface the composition of the natural transformations. However, we can also form the composite functors and ask how they are related to the functor associated to the composition of surfaces over holes. Furthermore, we have to discuss the composition of natural transformations, associated to two cobordisms glued over the vertical cylindrical boundary components.

For this *horizontal* composition we can not expect to find the same kind of strict functoriality as for the vertical one for the following, simple reason. In order to have a simple presentation of cobordisms we admitted in the class of objects only one surface for every isomorphism class. Thus, the objects on the lowest level can also be identified with integers, and the morphisms, $[g, a/b] : a \to b$, are enumerated by the genus g, with composition $[g_2, b/c] \circ [g_1, a/b] = [g_1 + g_2 + b - 1, a/c]$ for $b \geqslant 1$. Let us denote the combinatorial category defined in this way by $\mathcal{C}_0$.

Clearly, we have a functor $Cob_2 \longrightarrow \mathcal{C}_0$, but it is also obvious that we cannot find a strict functor Σ_*, which goes the other way and inverts the canonical one on $\mathcal{C}_0$. Still we can construct a *pseudofunctor*, if we assign to $[g, a/b]$ a standard surface and choose an isomorphism, $\alpha_*^{(2)}$ as in (1.1.5). In a more formal language, we consider $\mathcal{C}_0$ as a 2-category, whose 2-morphisms are all identities, and the latter equation may then be rewritten for $b \geqslant 1$ as:

$$\alpha_*^{(2)}([g_2, b/c], [g_1, a/b]) : \Sigma_*([g_2, b/c]) \bullet \Sigma_*([g_1, a/b]) \xrightarrow{\cong \bullet} \Sigma_*([g_2, b/c] \circ [g_1, a/b]).$$

Together with an obvious associativity condition this defines a pseudofunctor.

Now the functors we associated to a surface were determined entirely by the genus and number of holes. Hence, in horizontal direction we have an assignment $\mathcal{F}_* : \mathcal{C}_0 \to Fun : [g, a/b] \mapsto \mathcal{F}_{g,a/b}$. On the category of cobordisms we can therefore at most define a pseudofunctor into the category of functors, provided we find a suitable transformation that corresponds to the choice of $\alpha_*^{(2)}$.

Let us begin by constructing a natural isomorphism that makes $\mathcal{F}_*$ into a pseudofunctor, too. Since we want to deal with connected surfaces only, we assume $b \geqslant 1$ so that we can use the functors $\mathcal{F}^\infty_{g,a/b}$ given as coends of the expressions in (6.5.8) and the factorization from (6.5.7). The composition of two functors can be written as

$$\mathcal{F}^\infty_{g_2,b/c} \circ \mathcal{F}^\infty_{g_1,a/b} = \mathcal{F}^\infty_{0,1/c} \circ \mathcal{F}^\infty_{g_2,1/1} \circ \mathcal{F}^\infty_{0,b/1} \circ \mathcal{F}^\infty_{0,1/b} \circ \mathcal{F}^\infty_{g_1,1/1} \circ \mathcal{F}^\infty_{0,a/1}. \tag{6.6.1}$$

Assuming that $\mathcal{C}$ is a strict category, we also have that $\mathcal{F}^\infty_{x,1/1} \circ \mathcal{F}^\infty_{y,1/1} = \mathcal{F}^\infty_{x+y,1/1}$. Hence, a natural transformation of the above composite to $\mathcal{F}^\infty_{g_2+g_1+b-1,a/c}$ is determined by a natural isomorphism of the middle part of the product:

$$\alpha_b^{\mathcal{F}} : \mathcal{F}^\infty_{0,b/1} \circ \mathcal{F}^\infty_{0,1/b} \xrightarrow{\cong \bullet} \mathcal{F}^\infty_{b-1,1/1}. \tag{6.6.2}$$

The composition of functors can be given as the coend of the functor

$$\mathcal{F}^\infty_{0,b/1} \circ \mathcal{F}^{*,\langle Y_b,\ldots,Y_2\rangle}_{0,1/b} : X \longmapsto X \otimes Y_b^\vee \otimes \ldots \otimes Y_2^\vee \otimes Y_2 \otimes \ldots \otimes Y_b. \tag{6.6.3}$$

Now the product of the braids depicted in Figure 1.3 gives rise to the natural braid isomorphism

$$\alpha^{\mathcal{F}}(Y_2, \ldots, Y_b) : Y_b^\vee \otimes \ldots \otimes Y_2^\vee \otimes Y_2 \otimes \ldots \otimes Y_b \to Y_b^\vee \otimes Y_b \otimes \ldots \otimes Y_2^\vee \otimes Y_2. \tag{6.6.4}$$

The composite $(i_{Y_b} \otimes \ldots \otimes i_{Y_2})\alpha(Y_2, \ldots, Y_b)$ then yields a natural transformation from the functor in (6.6.3) to $\mathcal{F}^\infty_{b-1,1/1}$. This can then be lifted to define the isomorphism in (6.6.2).

If we apply this to the decomposition in (6.6.1) we obtain the isomorphism:

$$\alpha^{\mathcal{F}}([g_2, b/c], [g_1, a/b]) = \mathcal{F}^{\infty}_{0,1/c} \circ \mathcal{F}^{\infty}_{g_2,1/1}(\alpha^{\mathcal{F}}_b) :$$

$$\mathcal{F}^{\infty}_{g_2,b/c} \circ \mathcal{F}^{\infty}_{g_1,a/b} \xrightarrow{\cong\bullet} \mathcal{F}^{\infty}_{g_2+g_1+b-1,a/c}. \quad (6.6.5)$$

Due to its form it obeys the basic associativity condition and, thus, makes $\mathcal{F}_*$ into a pseudofunctor.

Note that instead of the braid we used in (6.6.4) to define the isomorphism in (6.6.5) we could have chosen any transformation which permuted the objects properly in order to obtain a pseudo composition law for $\mathcal{F}^{\infty}$. Our choice becomes relevant when we verify compatibility with the $\circ_v$-product for the three-dimensional cobordisms from Section 2.6.1.

Lemma 6.6.1. *Let $\mathcal{C}$ be a modular bounded category. Then the triple $(\mathcal{V}, \alpha^{\mathcal{F}}, \mathbb{1}) : \widetilde{\mathcal{C}ob} \to \mathcal{Q}$***V-Cat-mod** *is a double pseudofunctor in the sense of Definition B.2.1.*

In particular, for any two $\circ_h$-composable cobordisms $M_1 : \Sigma_{g_1^s,a/b} \to \Sigma_{g_1^t,a/b}$ and $M_2 : \Sigma_{g_2^s,b/c} \to \Sigma_{g_2^t,b/c}$ there is a commutative prism, expressed by the following equation

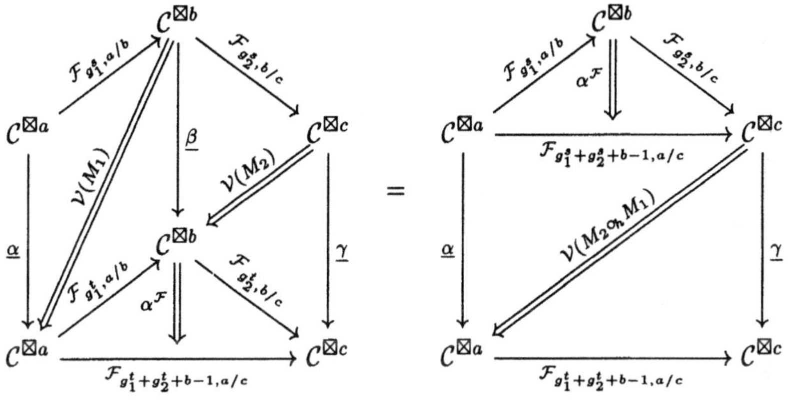

Proof. The considered triple is by construction the composition of two triples:

$$(\mathcal{V}, \alpha^{\mathcal{F}}, \mathbb{1}) : \widetilde{\mathcal{C}ob} \xrightarrow[\sim]{\mathcal{V}'} \mathcal{T}gl \xrightarrow{(\mathcal{V}'', \alpha, \mathbb{1})} \mathcal{Q}\textbf{V-Cat-mod}$$

in the sense of Sect. B.2.1. The first triple is the equivalence of double categories from Chapter 3. We have to prove that the second triple is a double pseudofunctor. The map $\mathcal{V}''$ sends $[g, a/b]$ to

$$\mathcal{F}_{g,a/b}(X_1 \boxtimes \ldots \boxtimes X_a)$$
$$= \int^Y \mathrm{Inv}_{\mathcal{C}}(X_1 \otimes \ldots \otimes X_a \otimes F^{\otimes g} \otimes Y_b^{\vee} \otimes \ldots \otimes Y_1^{\vee}) \otimes Y_1 \boxtimes \ldots \boxtimes Y_b.$$

Since we consider only connected surfaces, we assume $b \geqslant 1$, and to keep notation homogeneous also $a, c \geqslant 1$. The properties (i)–(iv) and (vi) of Definition B.2.1 are obvious. Let us prove (v).

We have shown in Lemma 6.2.1 that the functor $\mathcal{F}$ is exact and, hence, commutes with coends over $\mathcal{C}$ (essentially finite inductive limits). Therefore, repeated application of $\mathcal{F}$ gives repeated coends, which are isomorphic to multiple coends by Mac Lane's "Fubini" theorem for coends [Mac88]. In particular, the composition of two such functors is the double coend. Eliminating Y_1 via Proposition 5.1.5 we reduce it to a single coend. And that is precisely the isomorphism α:

$$
\begin{aligned}
&\mathcal{F}_{g_2^s,b/c}\bigl(\mathcal{F}_{g_1^s,a/b}(X_1 \boxtimes \ldots \boxtimes X_a)\bigr) \\
&\xrightarrow{\sim} \int^{Y,Z} \mathrm{Inv}_{\mathcal{C}}(X_1 \otimes \ldots \otimes X_a \otimes F^{\otimes g_1^s} \otimes Y_b^\vee \otimes \ldots \otimes Y_1^\vee) \otimes \\
&\qquad\qquad \otimes \mathrm{Inv}_{\mathcal{C}}(Y_1 \otimes \ldots \otimes Y_b \otimes F^{\otimes g_2^s} \otimes Z_c^\vee \otimes \ldots \otimes Z_1^\vee) \otimes Z_1 \boxtimes \cdots \boxtimes Z_c \\
&\xrightarrow[\sim]{\beta} \int^{Y_{>1},Z} \mathrm{Inv}_{\mathcal{C}}(X_1 \otimes \ldots \otimes X_a \otimes F^{\otimes g_1^s} \otimes Y_b^\vee \otimes \ldots \otimes Y_2^\vee \otimes \\
&\qquad\qquad \otimes Y_2 \otimes \ldots \otimes Y_b \otimes F^{\otimes g_2^s} \otimes Z_c^\vee \otimes \ldots \otimes Z_1^\vee) \otimes Z_1 \boxtimes \cdots \boxtimes Z_c \\
&\xrightarrow[\sim]{\Upsilon} \int^{Y_{>1},Z} \mathrm{Inv}_{\mathcal{C}}(X_1 \otimes \ldots \otimes X_a \otimes F^{\otimes g_1^s} \otimes Y_2^\vee \otimes Y_2 \otimes \ldots \otimes \ldots \otimes Y_b^\vee \otimes Y_b \otimes \\
&\qquad\qquad \otimes F^{\otimes g_2^s} \otimes Z_c^\vee \otimes \ldots \otimes Z_1^\vee) \otimes Z_1 \boxtimes \cdots \boxtimes Z_c .
\end{aligned}
$$

The statement of the lemma is reduced to commutativity of the diagram in Fig. 6.1, where most $\otimes$ signs are omitted.

The operation $M_1 \triangle_h M_2$ is described in Fig. 2.3. The lower square of the diagram expresses the relationship between $M_1 \triangle_h M_2$ and $M_1 \circ_h M_2$; Υ is the braid used in the definition of the last operation (see equation (2.6.3)). Vertical compositions are denoted $\alpha^{\mathcal{F}}$ in the prism to prove. Thus, we have only to prove commutativity of the upper square.

The proof goes as follows: we present the isomorphisms β and β^{-1} in this diagram via tangles. The isomorphism β is the tangle

$B^+ =$ 

lifted to the coends, see Sect. 6.2. This tangle is a graphical way to write the mapping

$$
\begin{aligned}
&\mathrm{Inv}_{\mathcal{C}}(U \otimes Y_1^\vee) \otimes \mathrm{Inv}_{\mathcal{C}}(Y_1 \otimes W) \to \\
&\qquad \to \mathrm{Inv}_{\mathcal{C}}(U \otimes Y_1^\vee \otimes Y_1 \otimes W) \xrightarrow{\mathrm{Inv}(\mathbf{1} \otimes \mathrm{ev} \otimes \mathbf{1})} \mathrm{Inv}_{\mathcal{C}}(U \otimes W).
\end{aligned}
$$

Another tangle

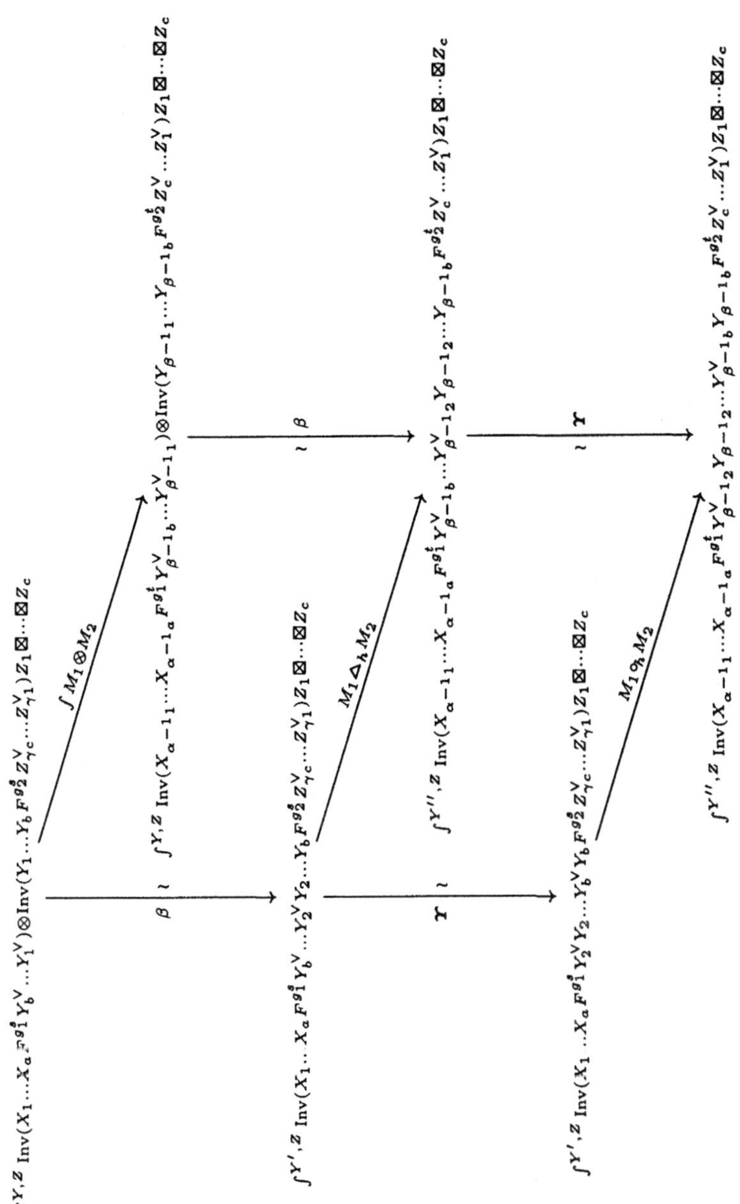

Fig. 6.1. Coherence of $\mathcal{F}$ with horizontal composition

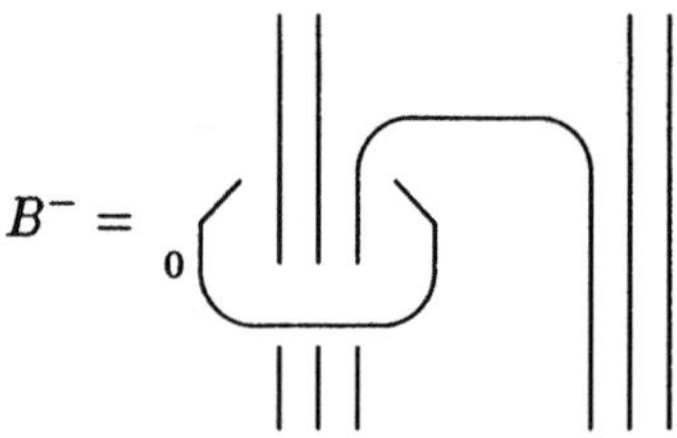

represents a morphism

$$\mathrm{Inv}_{\mathcal{C}}(U \otimes V) \xrightarrow{\mathbf{Inv}(B^-)} \mathrm{Inv}_{\mathcal{C}}(U \otimes F \otimes V).$$

Notice that their product in the order $B^+ \circ B^-$ is the identity map in $\mathcal{T}gl$. Indeed,

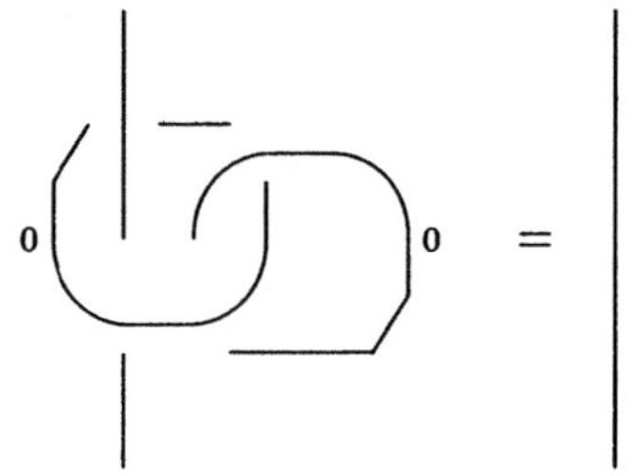

holds in $\mathcal{T}gl$ due to TS1* and TS3* moves of Sect. 2.4.3. It is also an equation between two endomorphisms of the identity functor $\mathrm{Id}_{\mathcal{C}}$. This follows from equation (6.3.1) from Lemma 6.3.2 composed with $\varepsilon : F \to 1$. We deduce that for arbitrary $U, V \in \mathrm{Ob}\,\mathcal{C}$

$$\big(U \otimes V \xrightarrow{B^-} U \otimes F \otimes V \xrightarrow{\mathbb{1} \otimes \varepsilon \otimes \mathbb{1}} U \otimes V\big) = \mathbb{1}.$$

Therefore, B^- is the tangle presentation of β^{-1}. Hence, the product $\beta \circ M_1 \otimes M_2 \circ \beta^{-1}$ is represented by the tangle $B^+ \circ (M_1 \sqcup M_2) \circ B^-$, which is exactly $M_1 \triangle_h M_2$ from Fig. 2.3.

Proof (Theorem 6.1.2). Take the composite in the sense of Proposition B.2.3 of the double pseudofunctor $(\mathcal{V}, \alpha^{\mathcal{F}}, \mathbb{1}) : \widetilde{\mathcal{C}ob} \to \mathcal{Q}\mathbf{V}\text{-}\mathbf{Cat}\text{-}\mathbf{mod}$ from Lemma 6.6.1 with the 2-pseudofunctor $\mathbf{V}\text{-}\mathbf{Cat}\text{-}\mathbf{mod} \to \mathbf{AbCat}$.

6.7 Topological moves imply the properties of integrals and modularity

The assumption that $\mathcal{C} = \mathcal{V}(S^1)$ is bounded abelian is natural, although not necessary. Anyway, it is required by our choice of the target double category for TQFT. On the other hand, the braided monoidal structure of $\mathcal{C}$ is deduced from the topological considerations in Lemma 1.3.1. Also the rigidity (existence of duals) for the monoidal structure of $\mathcal{C}$ is an extra assumption, but existence of the ribbon structure is a corollary (Lemma 1.4.6). The one-holed torus – a Hopf algebra in the category of cobordisms [Ker97, BKLT00, Yet97] – has to be represented by a Hopf algebra in $\mathcal{C}$. Our choice F for this Hopf algebra is deduced from $\mathbb{F}$, assigned to a sphere with two outgoing holes, i.e., a cylinder.

The important result of this section is that invariance under all topological moves of the double functor $\mathcal{V}$ under construction implies that F is modular, that is, $\omega : F \otimes F \to 1$ is non-degenerate. We demonstrate also how the topological moves imply relations between linear maps representing the elementary cobordisms. These relations turn out to be exactly the relations between integrals for the braided Hopf algebra F.

We shall consider separately the moves involving splitting ribbons and the other moves. First, we deal with the second type.

Disregarding the TS4-Move for a moment, the only moves besides the isotopies, which do not involve splitting-ribbons, are, thus, TD1 and TD2. If we consider the equations of morphisms that are obtained from the corresponding equations of tangles for arbitrary colorings, then TD1 and TD2 translate into the relations $\hat{\omega}_{X,Y}\left(\tilde{\lambda}_X \otimes Y\right) = \tilde{\lambda}_X \otimes Y$ and $\mathbf{v}_X\tilde{\lambda}_X = \tilde{\lambda}_X$, respectively, for all X and Y. Here $\hat{\omega}_{X,Y} = \mathbf{c}_{Y,X}\mathbf{c}_{X,Y}$, and $\mathbf{v}_X$ is the ribbon twist.

These conditions can be restated more concisely if we use the isomorphism from Corollary 5.2.5 between the coinvariance of the coend and the set of natural transformations of the identity on $\mathcal{C}$. For the transformation $\tilde{\lambda}_X$ we use in the construction of Ψ we obtain corresponding λ from the following diagrams:

$$\begin{array}{ccc} X^\vee \otimes X & \xrightarrow{X^\vee \otimes \tilde{\lambda}_X} & X^\vee \otimes X \\ {\scriptstyle i_X}\downarrow & & \downarrow{\scriptstyle \mathrm{ev}_X} \\ F & \xrightarrow{\lambda} & 1 \end{array} \tag{6.7.1}$$

Now, using rigidity, the condition for TD1 can be restated by requiring that the morphism

$$f(X,Y) : X^\vee \otimes X \otimes Y^\vee \otimes Y \xrightarrow{X^\vee \otimes \mathrm{coev}_X \otimes X \otimes Y^\vee \otimes Y} X^\vee \otimes X \otimes X^\vee \otimes X \otimes Y^\vee \otimes Y \to$$
$$\xrightarrow{X^\vee \otimes \tilde{\lambda}(X) \otimes X^\vee \otimes \hat{\omega}(X, Y^\vee) \otimes Y} X^\vee \otimes X \otimes X^\vee \otimes X \otimes Y^\vee \otimes Y \xrightarrow{\mathrm{ev}_X \otimes \mathrm{ev}_X \otimes \mathrm{ev}_Y} 1,$$

is equal to $[\mathrm{ev}_X \circ (X^\vee \otimes \tilde{\lambda}_X)] \otimes \mathrm{ev}_Y$ for all X and Y. Since $f(X,Y)$ is dinatural in both arguments, we may, equivalently, consider the condition when lifted to the coend. Using the pairing $\omega : F \otimes F \to 1$, we find

$$\begin{array}{ccc} F \otimes F & \xrightarrow{\Delta \otimes F} & F \otimes F \otimes F \\ {\scriptstyle \lambda \otimes F}\downarrow & & \downarrow{\scriptstyle \lambda \otimes \omega} \\ F & \xrightarrow{\varepsilon = \omega \circ (\eta \otimes F)} & 1 \end{array} \tag{6.7.2}$$

as an equivalent condition for the invariance under TD1 for arbitrary colors. Comparing (6.7.2) with (4.2.3), we see that λ is a right integral-functional on F modulo the kernel of ω. More precisely, the image of the difference of the two maps

$(\lambda \otimes 1) \circ \Delta - \eta \circ \lambda : F \to F$ is contained in $\ker(\omega)$. Let us call an element $\lambda \in \mathrm{Coinv}_{\mathcal{C}}(F)$ with this property an *ω-cointegral.* In the modular case, when ω is non-degenerate and $\lambda \neq 0$, this implies, of course, that λ is a right integral-functional in the precise sense of Theorem 4.2.5.

The condition for TD2 can also be reformulated in this language, if we use the multiplication $*$ on $\mathrm{Coinv}_{\mathcal{C}}(F)$, which is either induced by the coproduct of F or the multiplication of natural transformations. In summary, we have the following statement.

Lemma 6.7.1. *Suppose the transformation $X \mapsto \tilde{\lambda}_X$ used in Section 5.2.4 corresponds to $\lambda \in \mathrm{Coinv}(F)$ as in Proposition 5.2.5. Then the maps Ψ, Ψ^*, and Ψ^{**} are invariant under*

1) *the Coupon-Crossing TD1, that is, for all objects X, Y of $\mathcal{C}$*

$$\left(X \otimes Y \xrightarrow{\tilde{\lambda}_X \otimes 1} X \otimes Y \xrightarrow{c^2} X \otimes Y\right) = \left(X \otimes Y \xrightarrow{\tilde{\lambda}_X \otimes 1} X \otimes Y\right),$$

if and only if λ is an ω-cointegral, that is, Diagram (6.7.2) *commutes;*

2) *the Coupon-Twist TD2, if and only if $\mathbf{v} \circ \tilde{\lambda} = \tilde{\lambda} : X \to X$ for all $X \in \mathrm{Ob}\,\mathcal{C}$.*

Observe that if $\tilde{\lambda}_X : X \to X$ factors through $1 \oplus \ldots \oplus 1$ then the condition for TD1 is automatically fulfilled and the one for TD2 reduces by naturality to

$$\mathbf{v}_1 = 1, \tag{6.7.3}$$

which is satisfied by Section 4.1. Recall also that the coupon-twist only decides about a sign, and its square is already implied by the other moves.

Next we wish to construct maps that do not depend on the splitting of the tangle anymore. As we already explained, the split tangle T^{sp} actually represents the split manifold M^{sp}. Since in the construction of Ψ, Ψ^*, or Ψ^{**} we did nothing to distinguish the split tori from other tori in the boundary of M, we can at most expect that these maps are functions of M^{sp} but are ambiguous with respect to M.

The regluing of a full torus in the surgery operation, which corresponds in the language of tangles to the composition with an arc, may also be viewed as a composition of cobordisms over a connected surface, namely $S^1 \times S^1$. Since we assume the usual TQFT composition law to apply to this situation (see introduction), we have to construct Ψ^{**} in the tensor components of the split ribbons with a morphism associated to the cobordism $\mu : \varnothing \to S^1 \times S^1$ (or more precisely $\mu : D^2 \to S^1 \times S^1 - D^2$), i.e., with the single arc:

$$\mathcal{V}\left(\bigcap\right) = \mu : 1 \longrightarrow F. \tag{6.7.4}$$

Given such μ we get the map for the cobordism with the reglued tori

$$\begin{aligned}\widehat{\mathcal{V}}\big(T\big|\{X\},\{Y\},\mu\big) = \Psi^{**}(T^{sp}|\{X\},\{Y\}) \circ \big(\{X_\alpha\} \otimes F^{\otimes g_1} \otimes \mu^{\otimes(g_1-m+n)} \otimes \{Y_\beta^\vee\}\big) :\\ \{X_\alpha\} \otimes F^{\otimes g_1} \otimes \{Y_\beta\}^\vee \longrightarrow \{X\} \otimes F^{\otimes g_2} \otimes \{Y\}^\vee.\end{aligned} \tag{6.7.5}$$

As suggested in the notation we still have to make a suitable choice for μ, which is consistent with the moves, and we need to prove that $\widehat{\mathcal{V}}$ is actually independent of the splitting of the tangle.

One necessary constraint on μ can be derived immediately when we compose the representing arc with the diagram in TD5* in Section 2.4.3 or in (5.2.5), where it was used to define the braided antipode γ. The resulting tangle is easily deformed into an arc again so that we infer γ-invariance of μ, i.e.,

$$\gamma \circ \mu = \mu : 1 \to F. \tag{6.7.6}$$

Conversely, we have the following invariance property.

Lemma 6.7.2. *For a γ-invariant element $\mu \in \mathrm{Inv}_{\mathcal{C}}(F)$ (such that* (6.7.6) *holds) the morphisms $\widehat{\mathcal{V}}(T|\{X\},\{Y\},\mu)$ are invariant under the moves TD3*, TD4*, and TD5*.*

Proof. The argument for TD5* follows from the above derivation of (6.7.6). Invariance under the special τ-move TD3* follows from the fact that μ is in the invariance of F, and that $\hat{\omega}_{X,Y}$ is natural in both objects and the identity if either X or Y is 1. Specifically, we have:

$$\hat{\omega}_{X,Y}(\mu \otimes Y) = \mu \otimes Y.$$

Since we use the same invariance μ for every splitting, we can deduce consistency with the move TD4* using $\mathbf{c}_{F,F}(\mu \otimes \mu) = \mu \otimes \mu$.

Recall that two splittings of the same tangle can be related by the moves TD3, TD4, and TD5. Lemma 6.7.2, therefore, implies that $\widehat{\mathcal{V}}$ indeed only depends on T.

Following, let us discuss the implications of the internal surgery moves TS1* and TS2* for the special elements $\lambda \in \mathrm{Coinv}_{\mathcal{C}}(F)$ and $\mu \in \mathrm{Inv}_{\mathcal{C}}(F)$. For a color X the (reflected) tangle on the left side of the cancellation TS2* gives rise to the morphism $X \otimes X^\vee \xrightarrow{\tilde{\lambda}_X \otimes 1} X \otimes X^\vee \xrightarrow{\mathrm{ev}_X} 1$, lifted to $\lambda : F \to 1$. Since the ribbon in TS2* is actually a splitting of a closed ribbon, the multiplicative contribution of this isolated subdiagram to $\widehat{\mathcal{V}}$ is $\lambda \circ \mu \in \mathrm{End}_{\mathcal{C}}(1) = \Bbbk$. Invariance under TS2*, thus, imposes that this number is 1.

If Y is the collective object of the vertical strands in the picture for TS1*, and X is the color of the split annulus, then the morphism associated to the left tangle of TS1*-move is given by $Y \otimes X \otimes X^\vee \xrightarrow{\hat{\omega}^{-1}_{Y,X} \otimes 1} Y \otimes X \otimes X^\vee \xrightarrow{Y \otimes \mathrm{ev}_X} Y$. It lifts to a morphism

$$Y \xrightarrow{1 \otimes \mu} Y \otimes F = \int^{X \in \mathcal{C}} Y \otimes X \otimes X^\vee \xrightarrow{\int c^{-2} \otimes 1} \int^{X \in \mathcal{C}} Y \otimes X \otimes X^\vee = Y \otimes F \xrightarrow{1 \otimes \varepsilon} Y$$

and the TS1*-move requires it to be equal to $\tilde{\lambda}_Y$. Since the source of μ is 1, the same morphism can be written as

$$Y \xrightarrow{\mu\otimes 1} F\otimes Y = \int^{X\in\mathcal{C}} X\otimes X^\vee \otimes Y$$

$$\xrightarrow{\int 1\otimes c^2} \int^{X\in\mathcal{C}} X\otimes X^\vee\otimes Y = F\otimes Y \xrightarrow{\varepsilon\otimes 1} Y, \qquad (6.7.7)$$

and the graphical presentation of the equation becomes

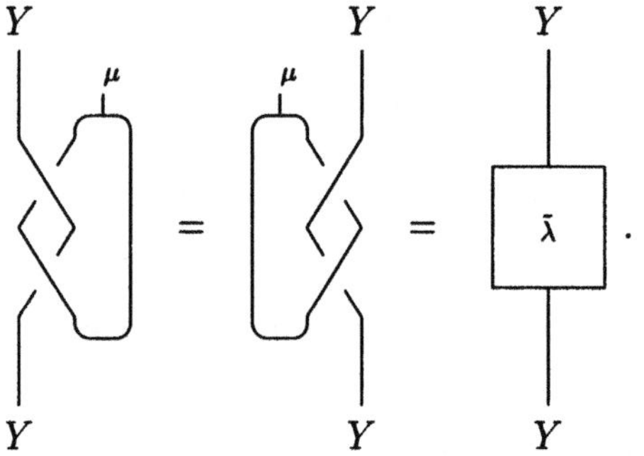

It lifts to F as an equivalent equation

$$\left(F \xrightarrow{\mu\otimes 1} F\otimes F \xrightarrow{\omega} 1\right) = \lambda. \qquad (6.7.8)$$

Let us summarize our findings as follows.

Lemma 6.7.3. *Suppose the morphism* $\widehat{\mathcal{V}}(T|\{X\},\{Y\},\mu)$ *is constructed as above from elements* $\mu \in \mathrm{Inv}_{\mathcal{C}}(F)$ *and* $\lambda \in \mathrm{Coinv}_{\mathcal{C}}(F)$. *Then* $\widehat{\mathcal{V}}$ *is invariant under*

1) the Modification TS1, that is, morphism* (6.7.7) *equals* $\tilde{\lambda}_Y$, *if and only if* (6.7.8) *holds;*

2) the Cancellation TS2, if and only if* $\left(1 \xrightarrow{\mu} F \xrightarrow{\lambda} 1\right) = \mathbb{I}_1$.

Several remarks are in order as to how the conditions in Lemmas 6.7.1, 6.7.2, and 6.7.3 relate to each other:

Rem.1 If λ and μ fulfill the condition for TS1*, then the one for TS2* is simply a condition on the normalization on μ, provided the pairing of μ with itself is non-zero.

Rem.2 If μ is an integral, the latter pairing is never zero by Theorem 4.2.5.

Rem.3 The γ-invariance of μ and (6.7.8) imply that λ is also γ-invariant, since γ is symmetric with respect to ω; see (5.2.8).

Rem.4 γ-invariance of an element $\lambda \in \mathit{Coinv}_{\mathcal{C}}(F)$ implies that it is a left ω-cointegral if and only if it is a right ω-cointegral. Note, that we have made no precise specification in Lemma 6.7.1 in regard to the sidedness of the integral. The two choices distinguish the diagram of TS1* and its reflection.

Rem.5 In terms of the natural transformation associated to λ the condition of γ-invariance is equivalent to

$$\tilde{\lambda}^t_X = \tilde{\lambda}_{X^\vee}.$$

This allows us to slide a coupon through a maximum or a minimum, cf. (4.1.4).

Rem.6 If λ is an ω-cointegral related to μ by (6.7.8) as required by TS1*, then the fact that ω is a Hopf pairing implies that μ is an ω-integral in the obvious, analogous sense.

Thus, all those moves that do not involve the boundary of the diagram and, hence, the boundary of the 3-manifold imply a constraint on λ and μ: namely, they have to be two-sided integrals, dual to each other in the sense of Lemma 6.7.3. This is, thus, enough to construct invariants for closed 3-manifolds. For example, in the construction of the Hennings invariant and in the construction of the Reshetikhin-Turaev invariant for closed 3-folds modularity is not a necessary assumption.

This changes when we require that $\widehat{\mathcal{V}}$ is also invariant under the σ-move TS3*. We shall show next that, given that all previous constraints are fulfilled, this boundary-move is equivalent to modularity of $\mathcal{C}$, and also to the fact that λ and μ are integrals. In order to make the arguments concise let us introduce the following natural (co)pairings

$$\beta : F \otimes F \xrightarrow{m} F \xrightarrow{\lambda} 1 \quad \text{and} \quad \beta^\dagger : 1 \xrightarrow{\mu} F \xrightarrow{\Delta} F \otimes F. \qquad (6.7.9)$$

In the case where λ and μ are left or right integrals, these pairings are side-invertible. This follows from Corollary 4.2.13. For instance, when $\mu = \int^F$, $\lambda = {}_F\!\int$ we have $\beta^\dagger = \beta^F$, $\beta = {}_F\beta$ and Corollary 4.2.13 gives

$$\left(F \xrightarrow{\beta^\dagger \otimes 1} F \otimes F \otimes F \xrightarrow{1 \otimes \gamma \otimes 1} F \otimes F \otimes F \xrightarrow{1 \otimes \beta} F\right) = (\lambda \circ \mu) \cdot \mathbb{1}_F \qquad (6.7.10)$$

with invertible constant $\lambda \circ \mu = {}_F\!\int \circ \int^F$ in the right hand side. Explicitly, this equation reads:

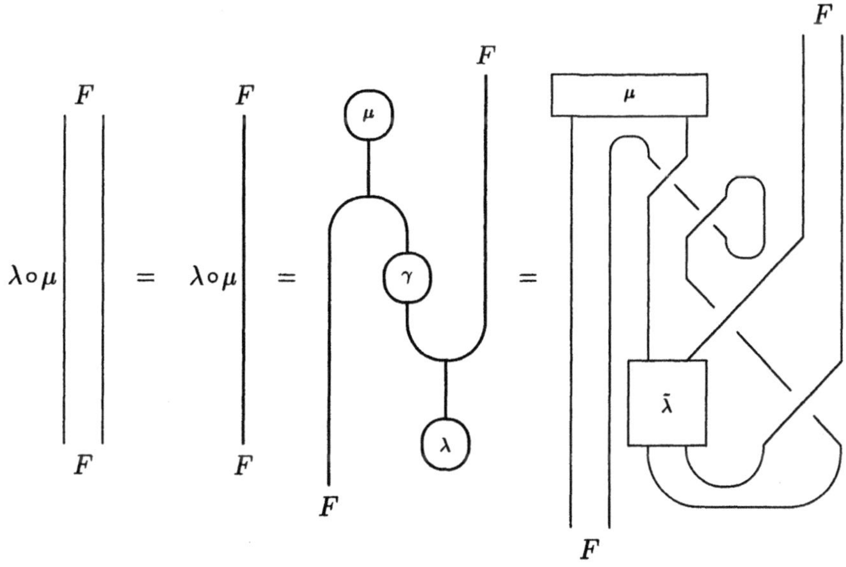

$=$ 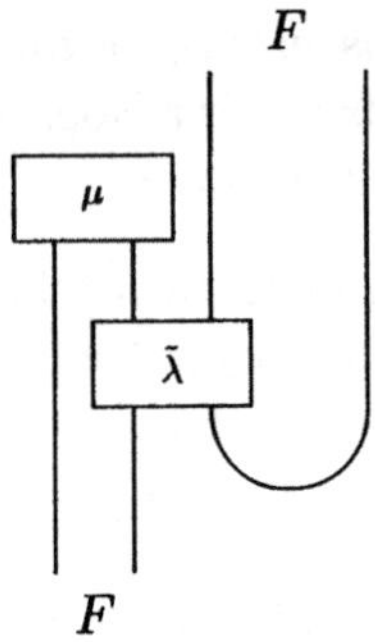 . (6.7.11)

Combined with the equation $\lambda \circ \mu = 1$, this is precisely the condition we have to impose in order to assure invariance under the mirror version of the TS3* move.

Proposition 6.7.4. *Let $\mathcal{C}$ be a bounded ribbon category, and let F be the corresponding braided function Hopf algebra. Assume invariance under the following moves:*

TS3: for some $\mu : 1 \to F$, $\lambda : F \to 1$* (6.7.11) *holds, or, equivalently,* (6.7.10) *holds;*
TS2: $\lambda \circ \mu = 1$;*
TS1: $\lambda = \big(F \xrightarrow{\mu\otimes 1} F \otimes F \xrightarrow{\omega} 1\big)$.*

Then ω is side-invertible.

Proof. Indeed, from (4.2.12) it follows that

$$\beta = \big(F \otimes F \xrightarrow{\mathsf{S}'\otimes 1} F \otimes F \xrightarrow{\omega} 1\big),$$

where

$$\mathsf{S}' = \big(F \simeq 1 \otimes F \xrightarrow{\beta^{\dagger}\otimes 1} F \otimes F \otimes F \xrightarrow{1\otimes\omega} F \otimes 1 \simeq F\big),$$

compare with (4.2.10).

Therefore,

$$\big(F \xrightarrow{\alpha\otimes 1} F \otimes F \otimes F \xrightarrow{1\otimes\omega} F\big) = \mathbb{1},$$

where $\alpha = \big(1 \xrightarrow{\beta^{\dagger}} F \otimes F \xrightarrow{1\otimes\gamma} F \otimes F \xrightarrow{1\otimes\mathsf{S}'} F \otimes F\big)$. Equivalently,

$$\big({}^{\vee}F \xrightarrow{!\alpha} F \xrightarrow{\omega^{!}} {}^{\vee}F\big) = \mathbb{1}_{{}^{\vee}F}.$$

Since the objects F and ${}^{\vee}F$ of $\mathcal{C}$ have the same length, the epimorphism $\omega^{!}$ is invertible.

Remark 6.7.5. Assume that hypotheses of Proposition 6.7.4 hold. If, in addition, λ is an ω-cointegral (that is, the TD1 Move holds, or (6.7.2) commutes), then λ and μ are two-sided integrals for F. Indeed, non-degeneracy of ω implies that λ is a right integral-functional, hence, μ is a left integral-element. The integrals are two-sided by Proposition 5.2.10.

In Section 4.2.3 we asserted that there are unique two-sided integrals, λ and μ, in a modular BTC with a coend F. Thus, if we wish to use exactly the coend F (and not a quotient as in [Lyu95b]) as the object that is associated to the punctured torus, we find from Proposition 6.7.4 that modularity is an essential assumption, and that there is (up to a sign of μ) only one way to construct the functor $\widehat{\mathcal{V}}$.

7. Generalization of a modular functor

In this chapter we are defining enhanced double categories $\widetilde{Cob}^{\cap}$ which are equivalent to Cob but have larger sets of 1-arrows. These 1-arrows, called arc-diagrams, encode the structure of certain coends. The reason to introduce them is to define horizontal compositions that do not require braidings. Hence, it will be also be possible to define a TQFT functor with only canonical isomorphisms. The construction is described in Section 7.3. It requires a modification of the double category of tangles to reflect the modifications in the enhanced cobordism-category.

We describe how the braided Hopf algebra F and its integrals look like in familiar examples. Specifically, we consider the case of a semisimple monoidal category as well as the example of the category of modules of a linear ribbon Hopf algebras in the original sense.

The additional combinatorial structure we introduce here on the surfaces is the minimal one that allows for honest functors. Indeed, the context of conformal field theory suggests much more involved structures, namely spaces of complex curves on which vectors bundles with flat connections are defined. A detailed exposition on this complex-analytic approach to modular functors is given by Kirillov and Bakalov in [BK01].

7.1 Enhanced cobordism categories

The horizontal 1-morphisms of the double category $\overline{Cob}$ were chosen as triples $[g, a/b]$ – classes of homeomorphism of surfaces with marked boundary. Replacing this double category with an equivalent one, we get other classes of horizontal 1-morphisms. One of the possibilities is to take for such class a set of graphs which combinatorially encode certain functors $\mathcal{C}^{\boxtimes a} \to \mathcal{C}^{\boxtimes b}$. Many of these functors are isomorphic, but have distinct definitions. Similarly, the graphs called arc-diagrams with genus g and fixed a/b are all isomorphic in a certain category, but not as plane graphs. Enlarging the set of 1-morphisms we make the third (isomorphism) component of the TQFT double pseudofunctor simpler. Namely, it does not include an explicit braiding isomorphism. The braiding is used only to establish equivalence of the enhanced and the previous double categories. So the enhanced picture looks aesthetically preferable.

7.1.1 Graphs with nested and crossed arcs

For given a and b (the number of incoming and outgoing tentacles) and fixed genus g the corresponding set of 1-arrows is a finite set of combinatorial nature – the set of isomorphism classes of arc-diagrams.

An *arc-diagram* for $[g, a/b]$ consists of base-line with arcs, which are half-circles that start and end in points on the base line. Arcs do not intersect each other, nor do they have a common endpoint. Further, an arc diagram will always have $a-1$ incoming tentacles and $b-1$ outgoing tentacles which are quarter-circles connected to the base-line in one endpoint.

The base-line should be divided into three intervals, such that the incoming (outgoing) tentacles are attached to the left (right) interval without crossings, the arcs are attached to the middle interval. A generic example is given in the following figure.

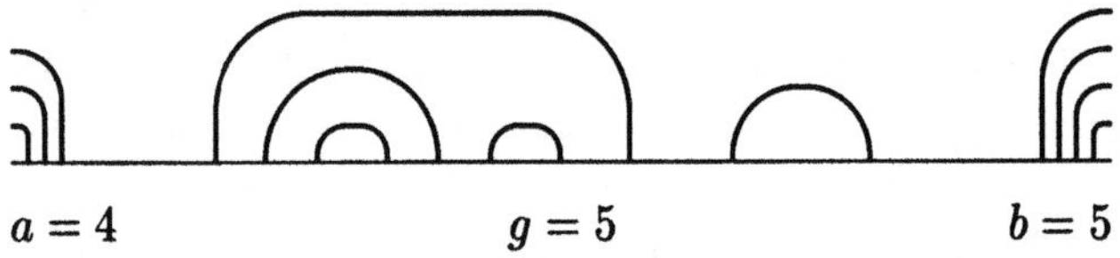

(7.1.1)

Set of these 1-arrows is a category of combinatorial nature. The objects are positive integers a. The composition of two graphs consists of gluing together $b-1$ outgoing tentacles of the first graphs and $b-1$ incoming tentacles of the second graph, so that new arcs are formed. Let $\mathrm{Arc}^{\cap}$ be that category.

7.1.2 Enhanced cobordisms

Instead of choosing one standard surface for each $(g, a/b)$ we choose a pseudofunctor $\mathrm{Arc}^{\cap} \to \mathit{Surf}, G \mapsto \Sigma_G$ in the following way. We take a closed ε-neighborhood $\tilde{G}$ of the graph G in the plane of drawing, so that the 2-dimensional oriented manifold $\tilde{G}$ with boundary is retractable to G. Then we glue together two copies of the thickened graph $\tilde{G} \cup -\tilde{G}$ along the part of the boundary $\partial\tilde{G} - \cup_1^a I_{in}^k - \cup_1^b I_{out}^k$. Any two choices of Σ_G for a given graph are almost canonically homeomorphic in the sense that any two homeomorphisms that fix G combinatorially are isotopic. Clearly, there is an isomorphism

$$\Sigma_G \circ_h \Sigma_H \simeq \Sigma_{G \circ H}$$

canonical up to an isotopy.

Further, we construct a double category $\mathit{Cob}^{\cap}$ out of $\mathrm{Arc}^{\cap}$. Its 2-morphisms are equivalence classes of pairs (M, ψ), described by squares of the form

$$\begin{array}{ccc} a & \xrightarrow{\quad G \quad} & b \\ {\scriptstyle \alpha}\big\downarrow & M & \big\downarrow{\scriptstyle \beta} \\ a & \xrightarrow[\quad H \quad]{} & b, \end{array}$$

where $\alpha \in S_a$ and $\beta \in S_b$ are permutations, G and H are graphs (1-morphisms) from $\mathrm{Arc}^{\cap}$, M is a 3-manifold (a cobordism), and ψ is an isomorphism of the boundary of M with $\Sigma_H \cup -\Sigma_G \cup$ cylinders, joining the boundary holes of Σ_G and Σ_H according to permutations α and β, similarly to (1.2.1).

The double category $\widetilde{\mathit{Cob}}^{\cap}$ is the framing extension of $\mathit{Cob}^{\cap}$. It is defined similarly to Sect. 1.6.2, and has equivalence classes of triples (M, ψ, n), $n \in \mathbb{Z}$, as 2-arrows. The operations repeat literally those of Sections 1.6.3–1.6.6.

7.2 Formulation of TQFT as a double functor in the extended case

In this section we state an extension of Theorem 6.1.2, using the enhanced category $\widetilde{\mathit{Cob}}^{\cap}$ defined in the previous section. This will result in a double TQFT functor, for which the isomorphisms $\delta_{G,H} : \mathcal{F}_G \circ_h \mathcal{F}_H \to \mathcal{F}_{G \circ H}$ are constructed canonically, without the use of braiding. Here we use the coherence proven in Theorem C.2.1.

Theorem 7.2.1. *For any modular bounded abelian category $\mathcal{C}$ there exists a double pseudofunctor multiplicative on objects (TQFT)*

$$\mathcal{V}_{\mathcal{C}} = (\mathcal{V}_{\mathcal{C}}, \delta, \beta) : \widetilde{\mathit{Cob}}^{\cap} \to \mathcal{Q}\mathbf{AbCat}, \qquad (7.2.1)$$

where the isomorphisms δ and β are obtained from the structure of a symmetric monoidal 2-category of **AbCat** *and canonical isomorphisms of coends.*

In the model situation, given a modular category $\mathcal{C}$ from **V-Cat-mod**, *one can make the model double functor strict:*

$$\mathcal{V}_{\mathcal{C}} = (\mathcal{V}_{\mathcal{C}}, \mathbb{1}, \mathbb{1}) : \widetilde{\mathit{Cob}}^{\cap} \to \mathcal{Q}\mathbf{V}\text{-}\mathbf{Cat}\text{-}\mathbf{mod}. \qquad (7.2.2)$$

Remark 7.2.2. Strictness of the functor $\mathcal{V}_{\mathcal{C}}$ is irrelevant for applications. For a TQFT it does not matter whether δ and β are identities or not. Nevertheless, replacing a modular abelian category $\mathcal{C}$ with an equivalent one, and replacing the structure of symmetric monoidal 2-category of **AbCat** with an equivalent one, we can achieve that $\mathcal{V}_{\mathcal{C}}$ is strict in (7.2.1).

Remark 7.2.3. The double pseudofunctor of Theorem 6.1.2 is obtained as a composition of a double pseudofunctor $\widetilde{\mathit{Cob}} \to \widetilde{\mathit{Cob}}^{\cap}$ and the double pseudofunctor $\mathcal{V} : \widetilde{\mathit{Cob}}^{\cap} \to \mathcal{Q}\mathbf{AbCat}$.

7.3 Sketch of the construction of enhanced TQFT

Now we discuss the structure of the double pseudofunctor $\mathcal{V}_{\mathcal{C}}$ (from Theorem 7.2.1), which is precisely the enhanced TQFT. We force $\mathcal{V}_{\mathcal{C}}$ to map an object of $\widetilde{\mathit{Cob}}^{\cap}$, which is a positive integer a, to the category $\mathcal{C}^{\boxtimes a}$. A vertical arrow $\alpha \in S_a$ of $\widetilde{\mathit{Cob}}^{\cap}$

$$\mathcal{V}'_{\mathcal{C}}(G)(X_1,\dots,X_a) =$$
$$\int^{Y_1,\dots,Y_b\in\mathcal{C}} \mathrm{Hom}_{\mathcal{C}}\Big(1, \int^{Z_j\in\mathcal{C}} \mathcal{G}(X_1,\dots,X_a;Z_1,\dots,Z_g;Y_b,\dots,Y_2;G)\otimes Y_1^{\vee}\Big)\otimes$$
$$\otimes\, Y_1\boxtimes Y_2\boxtimes\cdots\boxtimes Y_b.$$

The canonical isomorphism θ between the functors $\mathcal{V}'$ and $\mathcal{V}$ is given by

$$\mathcal{V}'_{\mathcal{C}}(G) \xrightarrow{\sim}$$
$$\int^{Y_1,\dots,Y_b\in\mathcal{C}} \mathrm{Hom}_{\mathcal{C}}\Big(Y_1, \int^{Z_j\in\mathcal{C}} \mathcal{G}(X_1,\dots,X_a;Z_1,\dots,Z_g;Y_b,\dots,Y_2;G)\Big)\otimes Y_1\boxtimes Y_2\boxtimes\cdots\boxtimes Y_b$$
$$\xrightarrow[\text{ev}]{\text{Prop. 5.1.5}} \int^{Y_2,\dots,Y_b\in\mathcal{C}} \left[\int^{Z_j\in\mathcal{C}} \mathcal{G}(X_1,\dots,X_a;Z_1,\dots,Z_g;Y_b,\dots,Y_2;G)\right]\boxtimes Y_2\boxtimes\cdots\boxtimes Y_b$$
$$\xrightarrow{\sim} \int^{Y_2,\dots,Y_b;Z_j\in\mathcal{C}} \mathcal{G}(X_1,\dots,X_a;Z_1,\dots,Z_g;Y_b,\dots,Y_2;G)\boxtimes Y_2\boxtimes\cdots\boxtimes Y_b.$$

The $\mathcal{V}'$ form is useful for defining the TQFT on 2-morphisms.

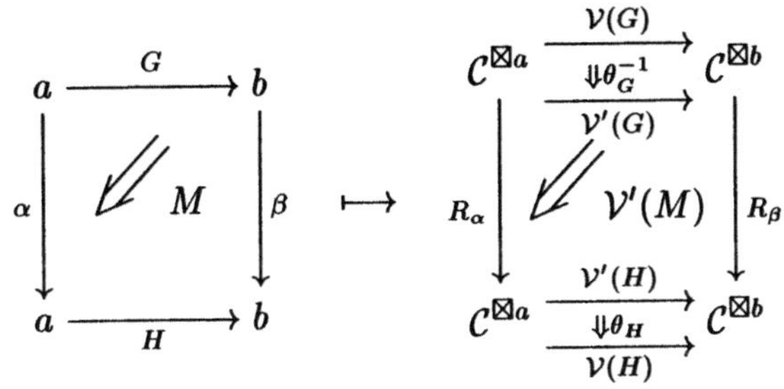

For fixed $X_1,\dots,X_a$ the morphism

$$\mathcal{V}'(M) : \int^{Y} \mathrm{Inv}\Big(\int^{Z} X_1\otimes\cdots\otimes X_a\otimes G(Z,Z^{\vee})\otimes Y^{\vee}_{\beta(b)}\otimes\cdots\otimes Y^{\vee}_{\beta(1)}\Big)\otimes Y_1\boxtimes\cdots\boxtimes Y_b$$
$$\to \int^{Y} \mathrm{Inv}\Big(\int^{W} X_{\alpha^{-1}1}\otimes\cdots\otimes X_{\alpha^{-1}a}\otimes H(W,W^{\vee})\otimes Y_b^{\vee}\otimes\cdots\otimes Y_1^{\vee}\Big)\otimes Y_1\boxtimes\cdots\boxtimes Y_b$$

comes from the natural in X_i, Y_j morphism

$$\mathcal{V}'(M\mid X,Y) : \mathrm{Inv}\Big(\int^{Z} X_1\otimes\cdots\otimes X_a\otimes G(Z,Z^{\vee})\otimes Y^{\vee}_{\beta(b)}\otimes\cdots\otimes Y^{\vee}_{\beta(1)}\Big)$$
$$\to \mathrm{Inv}\Big(\int^{W} X_{\alpha^{-1}1}\otimes\cdots\otimes X_{\alpha^{-1}a}\otimes H(W,W^{\vee})\otimes Y_b^{\vee}\otimes\cdots\otimes Y_1^{\vee}\Big),$$

which is determined by a tangle T, representing the cobordism M. The morphism $\mathcal{V}'(M\mid X,Y)$ itself is defined below as

$$\mathcal{V}'(M \mid X, Y) = \operatorname{Inv}_{\mathcal{C}}(\widehat{\mathcal{V}}(T \mid X, Y, \mu))$$

similarly to (6.3.2).

Let us recall what kind of tangles we are dealing with here. The objects G' of $\mathcal{T}gl^{\mathrm{Arc};*}$ are in bijection with the morphisms of $\mathrm{Arc}^{\cap}$. Precisely, G' is the arc-diagram G with an infinite number of auxiliary arcs added between the middle group of interior arcs and the right group of outgoing tentacles. An auxiliary arc is not placed under any other arc. Interior and auxiliary arcs are distinguished by the rôle (and by thickness in our graphical notations). For example, to G from (7.1.1) corresponds

$G' =$ 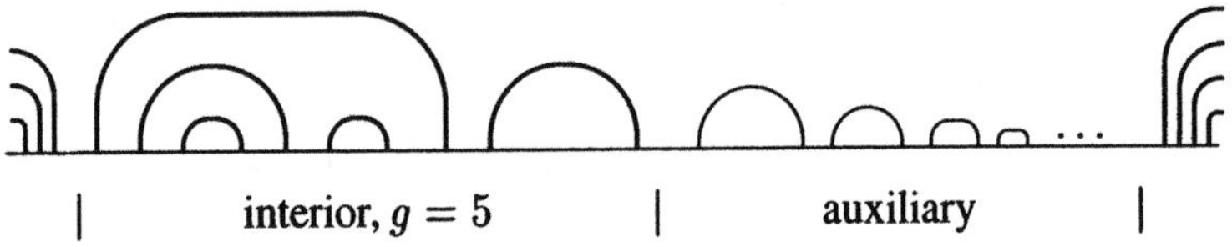

Thus, an object of $\mathcal{T}gl^{\mathrm{Arc};*}$ is consisting of an arc-diagram of the above type and an integer $g \geqslant 0$, the number of interior arcs. A morphism $T : G' \to H'$ is an equivalence class of tangles, consisting of coupons and top and through strands, which are attached to endpoints of arcs, tentacles and the baseline, such that all auxiliary arcs, except a finite number of them, are complemented with a top strand to an untwisted, unknotted and unlinked annulus with a coupon on it.

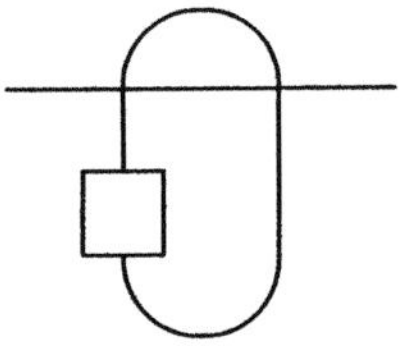

(7.3.1)

We require that (G', T, H') are related as follows. Let us draw the union of G and the plane diagram of the tangle T, isotop the picture in the plane ignoring all coupons, crossing signs and strand twists, and eliminate all closed curves. The result must be the arc-diagram H. The equivalence relation is generated by the moves TI1-TI5, TI8, TI9, TD1, TD2, TD3**–TD5**, TS1**–TS3** and TS4.

The new moves TD3**–TD5**, TS1**–TS3** are versions of the moves TD3*–TD5*, TS1*–TS3*, in which the pair of intervals K_j^- / K_j^+ is replaced with an auxiliary arc.

The map

$$\widehat{\mathcal{V}}(T \mid X, Y, \mu) : \int^Z X_1 \otimes \cdots \otimes X_a \otimes G(Z, Z^\vee) \otimes Y^\vee_{\beta(b)} \otimes \cdots \otimes Y^\vee_{\beta(1)}$$
$$\to \int^W X_{\alpha^{-1}1} \otimes \cdots \otimes X_{\alpha^{-1}a} \otimes H(W, W^\vee) \otimes Y_b^\vee \otimes \cdots \otimes Y_1^\vee$$

is obtained from a dinatural system of maps from the product under the first coend sign to the second coend. This morphism factorizes as

$$X_1 \otimes \cdots \otimes X_a \otimes G(Z, Z^\vee) \otimes Y^\vee_{\beta(b)} \otimes \cdots \otimes Y^\vee_{\beta(1)} \xrightarrow{1_X \otimes 1_G \otimes \mu^{\otimes N} \otimes 1_Y}$$
$$X_1 \otimes \cdots \otimes X_a \otimes G(Z, Z^\vee) \otimes F^{\otimes N} \otimes Y^\vee_{\beta(b)} \otimes \cdots \otimes Y^\vee_{\beta(1)} \xrightarrow{T^*_N}$$
$$\int^W X_{\alpha^{-1}1} \otimes \cdots \otimes X_{\alpha^{-1}a} \otimes H(W, W^\vee) \otimes Y^\vee_b \otimes \cdots \otimes Y^\vee_1 .$$

Here N is a big enough integer, so that Kth auxiliary arc, for each $K > N$, is complemented in T with a top strand to an untwisted, unknotted and unlinked annulus with a coupon on it as in (7.3.1). The tangle T_N is the truncation of T, namely, auxiliary closed ribbons formed by the Kth auxiliary arc and attached top ribbon are removed for $K > N$. The first morphism is obtained via insertion of $1^{\otimes N}$ into the tensor product followed by the morphism $\mu^{\otimes N} : 1^{\otimes N} \to F^{\otimes N}$. The second morphism is realized by the tangle T^*_N as in (6.3.2).

Proposition 7.3.2. *The obtained map $\mathcal{V}'(M \mid X, Y)$ is invariant under the moves TI1-TI5, TI8, TI9, TD1, TD2, TD3**–TD5**, TS1**–TS3** and TS4. Hence, it depends only on the cobordism $M \in \widetilde{\mathcal{C}ob}(a, b)$.*

The proof is similar to the proof of Lemma 6.3.3.

Proof (Theorem 7.2.1). Let us first consider the model functor (7.2.2) for a modular (non-strict monoidal) category $\mathcal{C}$ from **V-Cat-mod**

$$\mathcal{V}_\mathcal{C} = (\mathcal{V}_\mathcal{C}, \delta, \beta) : \widetilde{\mathcal{C}ob}^\cap \to \mathcal{Q}\mathbf{V}\text{-}\mathbf{Cat}\text{-}\mathbf{mod}.$$

Let a graph $G \in \mathrm{Arc}^\cap$ be decomposed into pieces $G_1 \cup G_2 \cup \cdots \cup G_k$ by vertical cuts. Then it is the composition $G_k \circ \cdots \circ G_1$ in the sense of $\mathrm{Arc}^\cap$. Remark C.2.4 implies that

$$\mathcal{V}_\mathcal{C}(G) = \mathcal{V}_\mathcal{C}(G_k) \circ \cdots \circ \mathcal{V}_\mathcal{C}(G_1).$$

Thus, the constructed functor $\mathcal{V}_\mathcal{C}$ is strictly compatible with the composition of 1-morphisms. So we take $\delta = \mathbb{1}$, $\beta = \mathbb{1}$ in the pseudofunctor $(\mathcal{V}_\mathcal{C}, \delta, \beta)$. Obviously, the mapping of 2-morphisms, which is well-defined by Proposition 7.3.2, respects the vertical composition in the strict sense.

Canonicity of the isomorphism $\theta_G : \mathcal{V}'(G) \to \mathcal{V}(G)$ implies the following commutative diagram for composable arc-diagrams G and H

$$\begin{array}{ccc} \mathcal{V}'(G) \circ \mathcal{V}'(H) & \xrightarrow{\sim} & \mathcal{V}'(G \circ H) \\ {\scriptstyle \theta_G \cdot \theta_H} \downarrow & & \downarrow {\scriptstyle \theta_{G \circ H}} \\ \mathcal{V}(G) \circ \mathcal{V}(H) & = & \mathcal{V}(G \circ H) \end{array} .$$

Therefore, the horizontal composition of functors $\mathcal{V}(G)$ is strictly compatible with the composition of arc-diagrams in $\mathrm{Arc}^{\cap}$. Thus, we constructed the double functor (7.2.2).

Now let us consider a modular bounded abelian category $\mathcal{C}$ from **AbCat**. There exists an algebra A from $\mathbf{V}$ such that $\mathcal{C}$ is equivalent to $\underline{\mathbf{V}^A}$, where $\mathcal{C}' = \mathbf{V}^A$ is the category of right A-modules. The equivalence induces a ribbon monoidal structure on $\mathcal{C}'$, which is modular as well. Applying the model construction to $\mathcal{C}'$ we get a strict double functor

$$\mathcal{V}_{\mathcal{C}'} = (\mathcal{V}_{\mathcal{C}'}, \mathbb{I}, \mathbb{I}) : \widetilde{\mathcal{C}ob}^{\cap} \to \mathcal{Q}\mathbf{V}\text{-}\mathbf{Cat}\text{-}\mathbf{mod}.$$

It can be composed with the equivalence of 2-categories

$$\underline{\cdot} : \mathbf{V}\text{-}\mathbf{Cat}\text{-}\mathbf{mod} \to \text{Bounded Abelian } \Bbbk\text{-Categories}, \qquad V^B \mapsto \underline{V^B},$$

described in Remarks 4.3.2 and 4.3.4. It is a strict 2-functor, that is, compatible with the composition of 1-morphisms ($\mathbf{V}$-functors and functors, respectively). We can transport the symmetric monoidal structure from $\mathbf{V}$-$\mathbf{Cat}$-$\mathbf{mod}$ to bounded abelian $\Bbbk$-categories via the functor $\underline{\cdot}$ and its quasi-inverse. This is one of the choices for the monoidal structure. The composition is a strict double functor (7.2.1)

$$\mathcal{V}_{\mathcal{C}'} = (\mathcal{V}_{\mathcal{C}'}, \mathbb{I}, \mathbb{I}) : \widetilde{\mathcal{C}ob}^{\cap} \to \mathcal{Q}\text{Bounded Abelian } \Bbbk\text{-Categories.}$$

This proves Remark 7.2.2. If we use an arbitrary symmetric monoidal structure on **AbCat**, we get a pseudofunctor (7.2.1) for $\mathcal{C}'$ and for $\mathcal{C}$.

7.4 Examples

We illustrate our approach in two familiar examples, where our only contribution is the double category picture. We exhibit explicitly the coends F and $\mathbb{F}$, give formulas for the integral μ of F. The first example treats semisimple categories. For such categories we discuss the Verlinde formula and the $*$-structure of the invariant part of F. The second example concerns the example of an ordinary Hopf algebra, H, with quasi-triangular and ribbon structures. Departing from the categories of modules over such algebras, we produce F, which in this case is nothing else but H^* with modified multiplication. The coend $\mathbb{F}$ is also identified with H. The 3-manifold invariant obtained from this TQFT coincides in this case with the Hennings invariant.

7.4.1 Semisimple Abelian Modular Categories

Reshetikhin and Turaev [RT91] proposed to construct invariants of 3-manifolds via quantum groups. More precisely, they use certain abelian semisimple ribbon categories obtained from quantum groups at roots of unity as trace quotients. One can forget about the origin of these categories and work simply with semisimple modular categories. We shall describe them as input data for our double functor construction.

Let $\mathcal{C}$ be a $\Bbbk$-linear abelian semisimple modular ribbon category. Assume that $\mathcal{C}$ is bounded, that is, the number of isomorphism classes of simple objects is finite. Assume also that 1 is simple and for each simple object L the endomorphisms division algebra $\operatorname{End} L = \Bbbk$. We denote by $\mathcal{S} = \{L_i\}_i$ the list of (representatives of isomorphism classes of) all simple objects.

Under these assumptions many formulas simplify. The coends $F \in \mathcal{C}$ and $\mathbb{F} \in \mathcal{C} \boxtimes \mathcal{C}$ take the form

$$F = \sum_{L \in \mathcal{S}} L \otimes L^\vee, \qquad \mathbb{F} = \sum_{L \in \mathcal{S}} L \boxtimes L^\vee.$$

Any morphism $1 \to F$ is a $\Bbbk$-linear combination of the standard morphisms for $L \in \mathcal{S}$

$\phi_L =$ 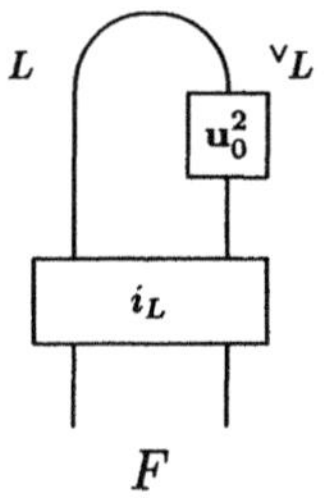

$$: 1 \xrightarrow{\text{coev}} L \otimes {}^\vee L \xrightarrow{1 \otimes \mathbf{u}_0^2} L \otimes L^\vee \xrightarrow{i} F.$$

The morphisms ϕ_L form a basis of the commutative algebra $\operatorname{Inv} F$. The Grothendieck ring of the category $\mathcal{C}$ determines the multiplication law in $\operatorname{Inv} F$ via the algebra isomorphism $\Bbbk \otimes_{\mathbb{Z}} K_0(\mathcal{C}) \to \operatorname{Inv} F$, $[L] \mapsto \phi_L$.

Any morphism $F \to 1$ can be represented as a linear combination of the morphisms

$$\psi_L : F \xrightarrow{\text{pr}_L} L \otimes L^\vee \xrightarrow{\text{ev}_L} 1,$$

where $L \in \mathcal{S}$. The functional $\psi_1 : F \to 1$ satisfies properties (4.2.3) of integrals $\int_F$ and ${}_F\!\int$. Therefore, ψ_1 factors through $\int_F$ as in

$$\psi_1 = \left(F \xrightarrow{\int_F} \operatorname{Int} F \xrightarrow{g} 1\right).$$

The morphism g is a non-zero map between two invertible objects, which are simple by our assumptions. Thus, g is an isomorphism, and ψ_1 can be chosen as a left integral-functional. Similar reasoning proves that ψ_1 can be chosen as a right integral-functional. Thus, ψ_1 is a two-sided integral.

The Verlinde formula. The number

$\dim_q(M) =$ 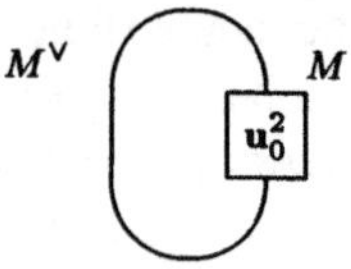

$: 1 \xrightarrow{\text{coev}} M^\vee \otimes M \xrightarrow{1\otimes \mathbf{u}_0^2} M^\vee \otimes M^{\vee\vee} \xrightarrow{\text{ev}} 1$

is called the dimension of an object $M \in \operatorname{Ob} \mathcal{C}$ (Turaev [Tur94]). (The index q reminds that this number coincides with the q-dimension in the case $\mathcal{C} = U_q\mathfrak{g}$-mod.) We have $\dim_q(M^\vee) = \dim_q(M)$.

Definition 7.4.1. *Introduce a biadditive function of two variables* $\mathsf{s} : \operatorname{Ob}\mathcal{C} \times \operatorname{Ob}\mathcal{C} \to \Bbbk$ *on the set of isomorphism classes of* $\mathcal{C}$*:*

$\mathsf{s}_{LM} =$ 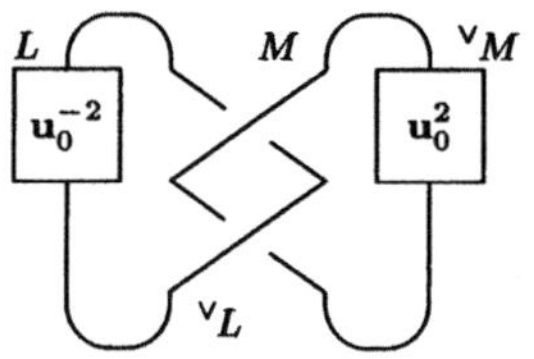

.

In particular, its restriction to $\mathcal{S}$ *is a matrix* $\mathsf{s}|_{\mathcal{S}} : \mathcal{S} \times \mathcal{S} \to \Bbbk$*, denoted again by* $\mathsf{s} = (\mathsf{s}_{LM})_{L,M\in\mathcal{S}}$ *by abuse of notation; here* L *and* M *run over simple objects.*

Notice that $\mathsf{s}_{LM} = \mathsf{s}_{ML}$, so the matrix s is symmetric. Let us consider the $\Bbbk$-algebra $\operatorname{Inv} F = \operatorname{Hom}_{\mathcal{C}}(1, F)$. It has the basis ϕ_L, $L \in \mathcal{S}$, hence, it is n-dimensional, where $n = \operatorname{Card} \mathcal{S}$. The form ω on F induces a bilinear form

$$\omega' : \operatorname{Inv} F \times \operatorname{Inv} F \xrightarrow{\otimes} \operatorname{Hom}(1, F \otimes F) \xrightarrow{\operatorname{Hom}(1,\omega)} 1.$$

The matrix (s_{LM}) is the matrix of the form ω' in the basis (ϕ_L).

Lemma 7.4.2 (The Verlinde formula). *For any simple* $L \in \mathcal{S}$ *and any objects* M *and* N *of* $\mathcal{C}$ *we have*

$$\mathsf{s}_{L1} = \dim_q(L), \qquad \mathsf{s}_{L1}\mathsf{s}_{L,M\otimes N} = \mathsf{s}_{LM}\mathsf{s}_{LN}.$$

Proof. The first formula is straightforward. Since

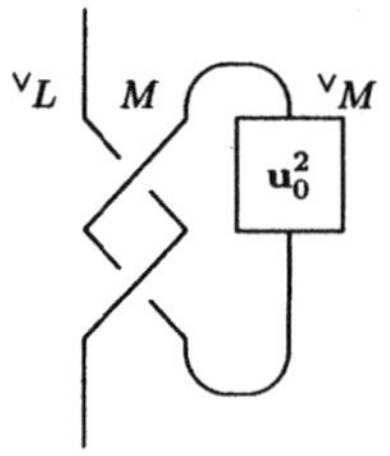

$\in \operatorname{End} {}^\vee L \simeq \Bbbk$

is a number, we can move it from the second factor to the first in the following computation:

$\mathsf{s}_{L1}\mathsf{s}_{L,M\otimes N} =$

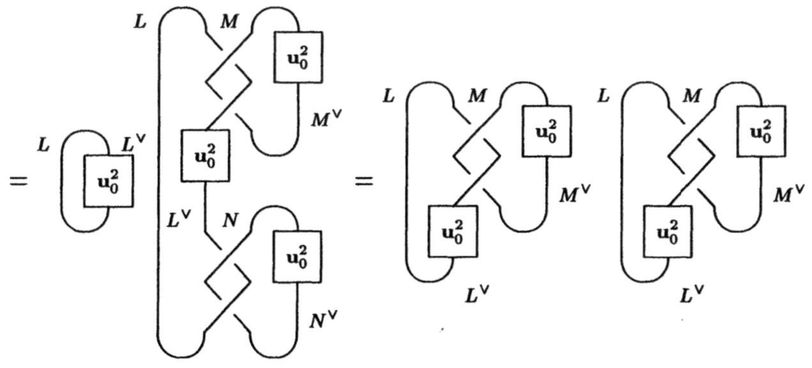

$= \mathsf{s}_{LM}\mathsf{s}_{LN}.$

This proves the second formula.

The criterion of modularity of Proposition 5.2.9 gives in our case the following.

Proposition 7.4.3. *In the assumptions of Sect. 7.4.1 the following conditions are equivalent:*

(i) $\mathcal{C}$ is modular (ω is non-degenerate);
(ii) the matrix $(\mathsf{s}_{LM})_{L,M\in\mathcal{S}}$ is non-degenerate;
(iii) for all $L \in \mathcal{S}$ $\dim_q L \neq 0$ and there exist numbers μ'_M, $M \in \mathcal{S}$, such that for all $L \in \mathcal{S}$ we have $\sum_{M\in\mathcal{S}} \mathsf{s}_{LM}\mu'_M = \delta_{L1}$.

Proof. (i) $\Longrightarrow$ (ii). Semisimplicity implies that the restriction ω' of the non-degenerate form ω to $\operatorname{Inv} F$ is also non-degenerate.

(ii) $\Longrightarrow$ (iii). If the dimension $\dim_q(L) = \mathsf{s}_{L1}$ of a simple object L vanishes, the Verlinde formula implies that $\mathsf{s}_{LM}^2 = 0$ for all $M \in \operatorname{Ob}\mathcal{C}$. This contradicts to the assumption of non-degeneracy of (s_{LM}).

(iii) $\Longrightarrow$ (i). For any simple object $L \in \mathcal{S}$ there exists a number $\kappa_L \in \mathbf{k}$ such that

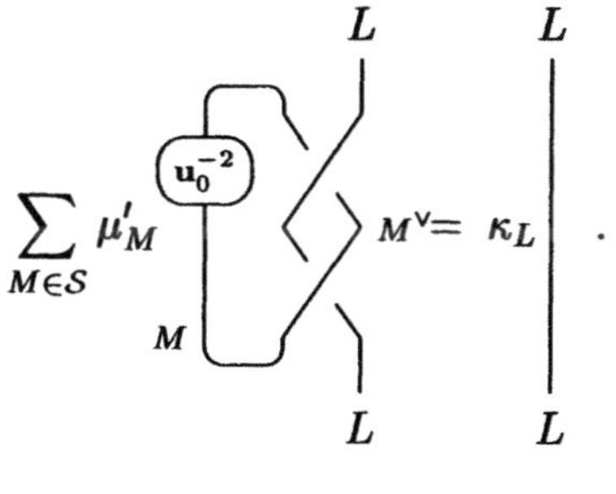

(7.4.1)

Composing (7.4.1)$\otimes 1\!\!1_{L^\vee}$ with ϕ_L and ev_L, we get

$$\sum_{M \in \mathcal{S}} \mathsf{s}_{ML} \mu'_M = \kappa_L \cdot \dim_q L.$$

Since $\mathsf{s}_{ML} = \mathsf{s}_{LM}$, we have

$$\kappa_L = (\dim_q L)^{-1} \delta_{L1} = \delta_{L1}.$$

Composing (7.4.1)$\otimes 1\!\!1_{L^\vee}$ with ev_L, we get the integral

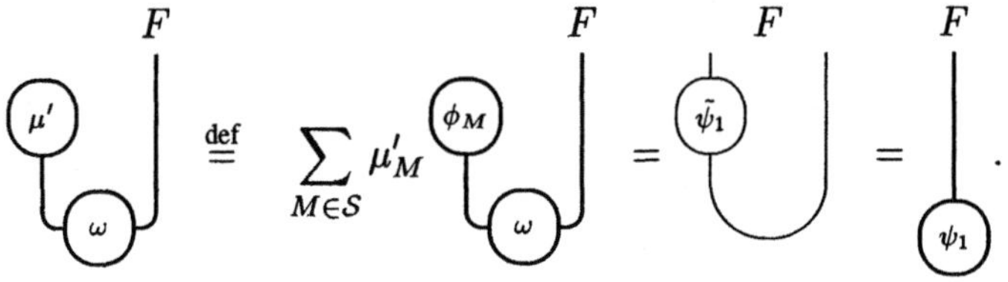

The modularity criterion of Proposition 5.2.9(b) implies that ω is side-invertible and $\mathcal{C}$ is modular.

Remark 7.4.4. Property (ii) of Proposition 7.4.3 is used by Turaev [Tur94] to define modular categories. Thus, his and our definitions agree for semisimple bounded ribbon categories.

Remark 7.4.5. Specializing equation (5.2.14) and Lemma 6.3.3 to the semisimple situation, we see that property (iii) is precisely the algebraic translation of the TS1-Move from Section 2.3.3. Reshetikhin and Turaev find a similar linear equation from the Fenn-Rourke move, which involves besides the matrix (s_{LM}) also ribbon numbers $\mathbf{v}_L \in \mathrm{End}(L) = \Bbbk$ as defined in Section 4.1.3.

Let us determine the coefficients μ_M of the integral-element

$$\mu = \sum_{M \in \mathcal{S}} \mu_M \phi_M : 1 \to F.$$

We use equation (6.3.1) in the form

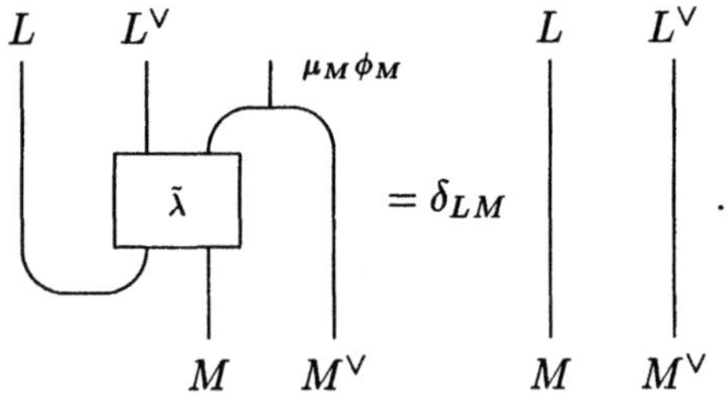

The two-sided integral λ is proportional to ψ_1. Here L and M vary over $\mathcal{S}$. The right hand side is the identity morphism if $L = M$, and vanishes otherwise. Substituting the definition of ϕ_M, we rewrite the equation as follows

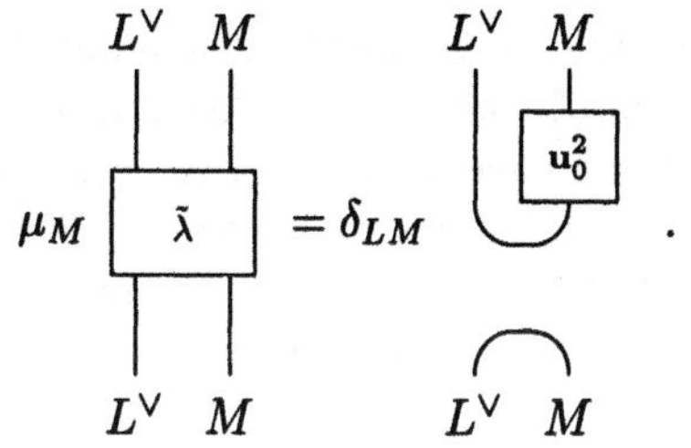

$$\mu_M \boxed{\tilde{\lambda}} = \delta_{LM} \boxed{u_0^2} \qquad (7.4.2)$$

For $L = 1$ we get

$$\mu_M \cdot \tilde{\lambda}_M = \delta_{1M} \cdot \mathbb{1}_M : M \to M. \qquad (7.4.3)$$

Recall that $M \not\cong 1$ implies $\tilde{\lambda}_M = 0$. So (7.4.3) tells essentially that

$$\mu_1 \cdot \tilde{\lambda}_1 = \mathbb{1}_1 : 1 \to 1. \qquad (7.4.4)$$

Now return to (7.4.2) with $L = M$. If we compose that equation with coev : $1 \to M^\vee \otimes M$ we obtain:

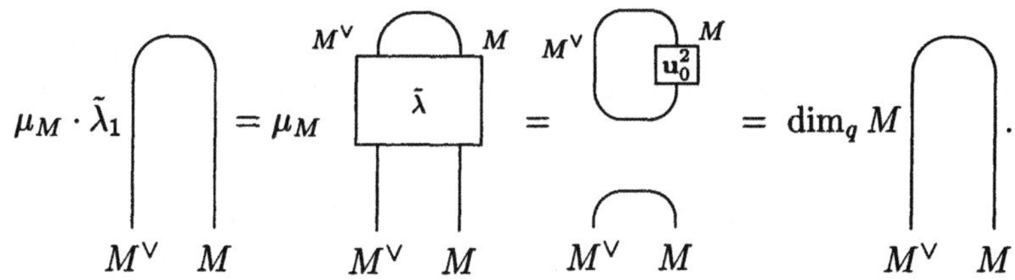

$$\mu_M \cdot \tilde{\lambda}_1 \quad = \mu_M \boxed{\tilde{\lambda}} \quad = \boxed{u_0^2} \quad = \dim_q M \qquad (7.4.5)$$

Multiplying both sides of (7.4.5) with μ_1 we find

$$\mu_M = \mu_1 \cdot \dim_q(M).$$

The normalization is fixed by equation (7.4.4), which we can write as

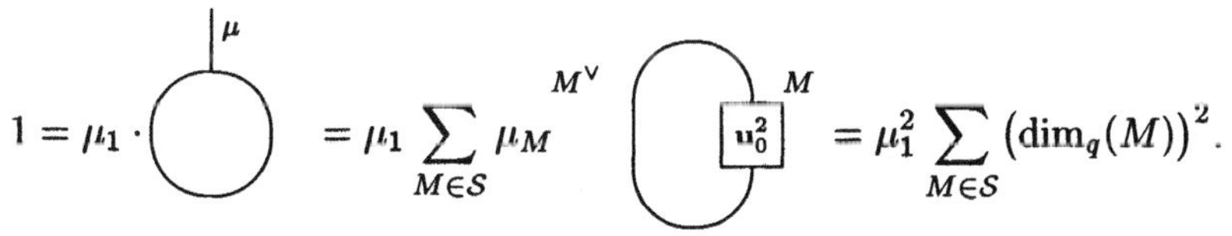

$$1 = \mu_1 \cdot \bigcirc^{\mu} = \mu_1 \sum_{M \in \mathcal{S}} \mu_M \boxed{u_0^2} = \mu_1^2 \sum_{M \in \mathcal{S}} \left(\dim_q(M)\right)^2.$$

Hence,

$$(\mu_1)^2 = \left(\sum_{M \in \mathcal{S}} \left(\dim_q(M)\right)^2\right)^{-1}. \qquad (7.4.6)$$

At least for an algebraically closed field $\Bbbk$ we find μ_1, unique up to a sign.

Remark 7.4.6. We may add one more equivalent condition to the condition list of Proposition 7.4.3. In assumptions of Sect. 7.4.1 $\mathcal{C}$ is modular if and only if
 (iv) for each simple $L \not\cong 1$ we have $\sum_{M \in \mathcal{S}} s_{LM} \dim_q M = 0$ and $\dim_q L \neq 0$.

Remark 7.4.7. As an application, we compute the invariant of a closed 3-manifold, presented by a framed link $\mathcal{L}$ with $n = |\mathcal{L}|$ components. A link is a special case of a tangle with only closed internal components. For a coloration of the link by objects in $\mathcal{C}$ we obtain from (6.2.7) the morphism $\Psi(\mathcal{L}^{sp}, [C]) : [C] \to 1$ as well as the lifted morphism $\Psi^*(\mathcal{L}^{sp}, [C]) : F^{\otimes n} \to 1$. The invariant of the link, defined in [RT90] for the given coloration, is expressed as

$$\begin{aligned} \tau_{C_1 \ldots C_n}(\mathcal{L}) &= (\mathrm{coev}_{C_1} \otimes \ldots \mathrm{coev}_{C_n}) \Psi(\mathcal{L}^{sp}, [C]) \\ &= (\phi_{C_1} \otimes \ldots \phi_{C_n}) \Psi^*(\mathcal{L}^{sp}) \in \mathrm{End}(1) = \mathbb{C}. \end{aligned}$$

Inserting the expression $\mu = \sum_{C \in \mathcal{S}} \mu_C \phi_C$ into (6.7.5), we get the invariant of the (2-framed) 3-manifold:

$$\begin{aligned} \tau(M_{\mathcal{L}}) &= \widehat{\mathcal{V}}(\mathcal{L}^{sp}) \\ &= (\mu \otimes \ldots \otimes \mu) \Psi^*(\mathcal{L}^{sp}) \\ &= \mu_1{}^n \sum_{C_1 \ldots C_n} \dim_q(C_1) \ldots \dim_q(C_n) \tau_{C_1 \ldots C_n}(\mathcal{L}). \end{aligned}$$

This expression coincides with the formula from [RT91] except for an additional factor for the signature correction. A similar calculation reproduces the formulae from [Tur94] for TQFT's.

Complex case. Assume that $\Bbbk = \mathbb{C}$. From the Verlinde formula (Lemma 7.4.2) we conclude that the commutative $\mathbb{C}$-algebra $\mathrm{Inv}\, F$ possesses homomorphisms

$$\chi_L : \mathrm{Inv}\, F \to \mathbb{C}, \quad \phi_M \mapsto (\dim_q(L))^{-1} \mathrm{s}_{LM} = \mathrm{s}_{LM} / \mathrm{s}_{L1}.$$

The matrix s is invertible so that its columns can not be proportional. Hence, all χ_L are different characters. Their number is $n = \mathrm{Card}\, \mathcal{S} = \dim_{\mathbb{C}} F$, hence, there is an isomorphism of $\mathbb{C}$-algebras

$$\chi : \mathrm{Inv}\, F \to \mathbb{C} \times \cdots \times \mathbb{C} = \mathbb{C}^n, \quad \phi \mapsto (\chi_1(\phi), \ldots, \chi_n(\phi)).$$

Now we show that the dimensions $\dim_q(M)$ are real numbers so that also μ_1 is a real number. It is natural to take for μ_1 the positive root of the right hand side of (7.4.6). Positiveness fixes μ_1 uniquely.

One can introduce in $\mathrm{Inv}\, F$ an antilinear involution

$$\text{-}^* : \mathrm{Inv}\, F \to \mathrm{Inv}\, F, \qquad (\phi_L)^* = \phi_{L^\vee}$$

and a scalar (Hermitian) product

$$(\phi_L | \phi_M) = \delta_{LM}, \qquad L, M \in \mathcal{S}.$$

Then $\mathrm{Inv}\, F$ becomes a finite dimensional commutative Hilbert algebra. Indeed,

$$(\phi_L \phi_M | \phi_N) = \dim \mathrm{Hom}(L \otimes M, N) = \dim \mathrm{Hom}(L, M^\vee \otimes N) = (\phi_L | \phi_M^* \phi_N).$$

From the theory of finite dimensional commutative Hilbert algebras we know that idempotents in the algebra Inv F are self-adjoint (only in that case the scalar product can be positive definite). Hence, χ is a *-morphism, that is, $\chi_L(\phi^*) = \overline{\chi_L(\phi)}$. Therefore, $s_{LM^\vee}/s_{L1} = \overline{s_{LM}}/\overline{s_{L1}}$. In particular, for $L = 1$, we have $s_{11} = 1$ and

$$\dim_q(M) = \dim_q(M^\vee) = s_{1M^\vee} = \overline{s_{1M}} = \overline{\dim_q(M)}.$$

This proves the following

Proposition 7.4.8. *For any $M \in \mathcal{C}$ its dimension* $\dim_q(M)$ *is a real number.*

Corollary 7.4.9. *We can choose μ_1 positive.*

7.4.2 Examples of Semisimple Modular Categories

In their original paper [RT91] Reshetikhin and Turaev use as algebraic input data the representation theory of the quantum deformation $U = U_q(sl_2)$ of the Lie algebra $sl(2, \mathbb{C})$, where q is a root of unity. They construct the invariant as a trace over U-equivariant morphisms, and prove the necessary modularity condition concerning the non-degeneracy of the braided pairing.

The general picture is drawn by Turaev in his book [Tur94], where 3-manifold invariants and TQFT's in the sense of Atiyah are constructed from semisimple modular categories. He shows how to obtain the latter as quotients of certain subcategories of representations of a modular Hopf algebra by the ideal of trace-negligible morphisms.

The heuristically defined partition function of the Witten-Chern-Simons theory at level k and with a compact, connected and simply connected gauge group G is identified, e.g., via the corresponding S-matrix data, with the rigorously constructed invariant $\tau_U(M)$ for $U = U_q(\mathfrak{g})$. Here $q = \exp(\pm\frac{2\pi i}{k+h_\mathfrak{g}})$, $\mathfrak{g}$ is the Lie algebra of G, and $h_\mathfrak{g}$ is the dual Coxeter number of $\mathfrak{g}$.

To make the verification of modular properties easier, Turaev and Wenzl introduce in [TW93] the notion of a quasimodular Hopf algebra, show that such an algebra produces invariants of 3-manifolds similar to a modular one, and prove that $U_q\mathfrak{g}$ at a root of unity is quasimodular for a Lie algebra $\mathfrak{g}$ of the series A, B, C, and D. The proof uses the structure of the algebra, generated by the braiding automorphisms of tensor powers of some modules. For the A series this is the Hecke algebra; for the B, C, D series the generalized Hecke algebras appear, which are now called Birman-Wenzl-Murakami algebras, see also [BW89] and [Mur87]. The results of Andersen [And92], combined with the results of Turaev and Wenzl [TW93], prove that $U_q\mathfrak{g}$ is modular for $\mathfrak{g}$ of series A, B, C, and D in the sense of [Tur94].

In our setting $\mathcal{C}$ does not have to be related to any Hopf algebra at all so that, given such examples, our theory does in fact imply new invariants and TQFT's. Constructions that do not use Hopf algebras can be found in the field of operator algebras, such as the invariants by Ocneanu, Evans and Kawahigashi [EK95] or by Xu [Xu], who use the theory of subfactors and algebraic quantum field theory.

In all cases, however, the invariants are identified with ours via a unitary modular category. Finally, there are generalized Hopf algebras, for instance, quantum groupoids, whose categories of modules are modular and might be used as an input for our construction. They naturally appear in the classification of subfactors in some von Neumann algebras, see Nikshych and Vainerman [NV00]. Further examples of semisimple modular categories can be obtained by applying various quotient and orbit constructions to known categories, see [FK93].

7.4.3 Further Related Constructions

A natural refinement of the extension in (0.3.1) is obtained by considering framings rather than 2-framings of the 3-manifolds with corners. This implies via $[M, SO(3)] \to H^1(M, \mathbb{Z}/2)$ a spin structure on M. Hence, M is naturally presented by a bounding 4-manifold, with a compatible spin structure and, therefore, even intersection number. Thus, the extension is restricted to a subgroup of index 16, resulting in the following map between short exact sequences:

$$\begin{array}{ccccccccc} 1 & \longrightarrow & \Omega_4^{spin} & \longrightarrow & {}^{spin}\widetilde{Cob}^{\cap} & \longrightarrow & Cob^{\cap} & \longrightarrow & 1 \\ & & {\scriptstyle 16}\big\downarrow & & \big\downarrow & & {\scriptstyle =}\big\downarrow & & \\ 1 & \longrightarrow & \Omega_4 & \longrightarrow & \widetilde{Cob}^{\cap} & \longrightarrow & Cob^{\cap} & \longrightarrow & 1 \end{array}$$

Here, Ω_4^{spin} denotes the spin cobordism group and ${}^{spin}\widetilde{Cob}^{\cap}$ the category of framed 3-cobordisms. A surgery calculus for manifolds with spin structures can be drawn from [KM91] and was made explicit in [Saw99]. Our constructions of presentations and TQFT functors generalize without much complication to cobordisms with spin structures. The inclusion of spin structures allows us to work with modular categories with relaxed assumptions on the balancing property. The theory developed in [BHMV95] also leads to constructions of Spin TQFT's as given in [Mas97] and [Bel98].

Let us also mention here a number of more combinatorial approaches that give interesting computational, technical as well as geometrical insights into the Witten-Reshetikhin-Turaev invariants. Blanchet, Habegger, Masbaum and Vogel construct the WRT-invariants from the Kauffman bracket entirely in [BHMV95]. In their article [KM91] Kirby and Melvin manage to show that for $U = U_q(sl_2)$ and small k the invariant τ_U can be given as summations over spin structures of the 3-manifold and is congruent to classical invariants such as the Rohlin or Casson invariants. A topological quantum field theory, based on a concrete group $G = SO(3)$, is constructed by Frohman and Kania-Bartoszyńska in [FKB96]. The generalization of the identifications with the Casson invariant presented by Murakami in [Mur94] provides the first link of the Witten-Reshetikhin-Turaev invariants to general finite type invariants as defined by Ohtsuki in [Oht96] and by Le, H. Murakami, J. Murakami and Ohtsuki in [LMMO95]. This is presently under investigation by many people, see for example Garoufalidis and Levine [GL98] and references therein.

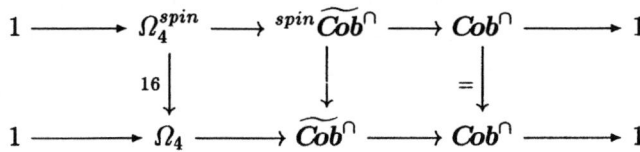

goes to the permutation functor $R_\alpha : \mathcal{C}^{\boxtimes a} \to \mathcal{C}^{\boxtimes a}$. Now we are going to assign functors to graphs from $\mathrm{Arc}^\cap$, which are horizontal arrows of $\widetilde{\mathcal{C}ob}^\cap$.

A graph $G \in \mathrm{Arc}^\cap$ is read from left to right. More familiar interpretation is obtained if we rotate the base line by $-\frac{\pi}{2} + \varepsilon$, where $\varepsilon > 0$ is very small. Thus, the incoming tentacles (left or top one) do not contain maxima. Each arc in the middle interval and each outgoing tentacle (right or bottom one) contains one maximum. For each $(-\frac{\pi}{2} + \varepsilon)$-rotated graph, denoted $\tilde{G}$, we assign to the local maxima and to the endpoints of arcs, belonging to the baseline, the following functors:

 $\longmapsto$ $\mathrm{Coend} : \mathbf{k}\text{-vect} \to \mathcal{C} \boxtimes \mathcal{C}$,

 $\longmapsto$ $\otimes : \mathcal{C} \boxtimes \mathcal{C} \to \mathcal{C}$.

The composite functor is assigned to G. This is nothing else but the functor, corresponding to the planar thick tangle $\tilde{G}$ under the trifunctor $\mathcal{PTT} \to \mathbf{V\text{-}Cat}$ from Theorem C.2.1.

The value of the functor $\mathcal{V}_\mathcal{C}(G) : \mathcal{C}^{\boxtimes a} \to \mathcal{C}^{\boxtimes b}$ on an object $X_1 \boxtimes \cdots \boxtimes X_a$ can be represented in a coend form:

$$\int^{Z_1,\dots,Z_g,Y_2,\dots,Y_b \in \mathcal{C}} \mathcal{G}(X_1,\dots,X_a;Z_1,\dots,Z_g;Y_b,\dots,Y_2;G) \boxtimes Y_2 \boxtimes \cdots \boxtimes Y_b,$$

$$\begin{aligned}\mathcal{G}(X_1,\dots,X_a;Z_1,\dots,Z_g;Y_b,\dots,Y_2;G) \\ = X_1 \otimes \cdots \otimes X_a \otimes Z_{i_1} \otimes \cdots \otimes Z_{i_{2g}}^\vee \otimes Y_b^\vee \otimes \cdots \otimes Y_2^\vee,\end{aligned}$$

where the list of indices $(i_1,\dots,i_{2g})$ is a permutation of the sequence $(1,1,2,2,\dots,g,g)$, such that j appears earlier than k in the list, if $j < k$. To the first occurrence of j corresponds Z_j, to the second corresponds $Z_j^\vee$. The position of the pair (j,j) is determined by the position of endpoints of j-th arc in G. Indeed, there are canonical mappings of the integrand to $\mathcal{V}_\mathcal{C}(G)(X)$, depending on Z_j and Y_k dinaturally. They are built from elementary isomorphisms, listed in Appendix C.1.

Considering a modular category $\mathcal{C}$ from **AbCat** we can assume that it is strict rigid monoidal. In the case of a modular category $\mathcal{C}$ from **V-Cat-mod** it cannot be made strict in general, so we may indicate the parentheses for the tensor product.

Example 7.3.1. To the graph (7.1.1) corresponds the functor, representable as

$$\mathcal{V}_\mathcal{C}(G)(X_1 \boxtimes \cdots \boxtimes X_4) = \int^{Z_1,\dots,Z_5,Y_2,\dots,Y_5 \in \mathcal{C}} X_1 \otimes \cdots \otimes X_4 \otimes$$

$$\otimes Z_1 \otimes Z_2 \otimes Z_3 \otimes Z_3^\vee \otimes Z_2^\vee \otimes Z_4 \otimes Z_4^\vee \otimes Z_1^\vee \otimes Z_5 \otimes Z_5^\vee \otimes Y_5^\vee \otimes \cdots \otimes Y_2^\vee \boxtimes Y_2 \boxtimes \cdots \boxtimes Y_5.$$

There is another isomorphic form of the functor $\mathcal{V}_\mathcal{C}(G)$:

7.4.4 Quasi-Triangular Hopf algebras

Let H be a finite dimensional Hopf $\Bbbk$-algebra. Assume that H has an *R-matrix*, which is an invertible element $R \in H \otimes H$, satisfying the relations of Drinfeld [Dri87]

$$\begin{aligned}(\Delta \otimes 1)R &= R^{13}R^{23},\\ (1 \otimes \Delta)R &= R^{13}R^{12},\\ \Delta^{\mathrm{op}}a &= R\Delta(a)R^{-1}\end{aligned}$$

for any $a \in H$, so (H, R) is *quasi-triangular*. It makes $\mathcal{C} = H$-mod $=$ comod-H^* into a braided category. The braiding is $\mathbf{c}(x \otimes y) = R^{21}.y \otimes x$.

For a finite dimensional H-module V as for any vector space there is a canonical linear map $v_0^2 : V \to V^{\vee\vee}$ such that $\langle v, y\rangle = \langle y, v_0^2(v)\rangle$ for $v \in V$, $y \in V^\vee$. Its square gives $v_0^4 = (V \xrightarrow{v_0^2} V^{\vee\vee} \xrightarrow{v_0^2} V^{(4\vee)})$. On the other hand, in $\mathcal{C}$ as in any rigid braided category there are morphisms $\mathbf{u}_0^{-4} = (V \xrightarrow{\mathbf{u}_1^{-2}} {}^{\vee\vee}V \xrightarrow{\mathbf{u}_{-1}^{-2}} V^{(-4\vee)})$. Composing them, we get linear bijections

$$g_V : V \xrightarrow{\mathbf{u}_0^{-4}} V^{(-4\vee)} \xrightarrow{v_0^4} V.$$

They are decomposable into a product of the two bijections

$$u_1 : V \xrightarrow{v_0^2} V^{\vee\vee} \xrightarrow{\mathbf{u}_{-1}^{-2}} V, \qquad u_4 : V \xrightarrow{v_0^2} V^{\vee\vee} \xrightarrow{\mathbf{u}_1^{-2}} V.$$

Theorem 7.4.10 (Drinfeld [Dri90]). *The maps u_1 and u_4 are given by the action of the following elements:*

$$u_1 = \gamma(R'')R', \qquad u_4 = \gamma^2(R')R'' = \gamma(u_1)^{-1}, \qquad g = u_1u_4.$$

The element g is grouplike ($\varepsilon(g) = 1$, $\Delta g = g \otimes g$), and for any $a \in H$ we have $gag^{-1} = \gamma^4(a)$.

7.4.5 Ribbon Hopf algebras

The notion of a ribbon Hopf algebra was proposed by Reshetikhin and Turaev [RT90]. Let us recall it, starting from the category of modules. Assume that $\mathcal{C} = H$-mod has a ribbon structure. Then there is a morphism $\mathbf{u}_0^{-2} = \mathbf{u}_{-1}^{-2}\mathbf{v} = \mathbf{u}_1^{-2}\mathbf{v}^{-1} : V \to {}^{\vee\vee}V$ for any finite dimensional H-module V. One can prove that the map

$$\varkappa_V : V \xrightarrow{\mathbf{u}_0^{-2}} {}^{\vee\vee}V \xrightarrow{v_0^2} V$$

commutes with all morphisms and satisfies $\varkappa_{X\otimes Y} = \varkappa_X \otimes \varkappa_Y$ and $\varkappa_X^2 = g_X$. If H is finite dimensional, we deduce that $\varkappa_V$ is the action of a grouplike element $\varkappa$ of H. The following definition is a version of the original definition given by Reshetikhin and Turaev [RT90].

Definition 7.4.11. *A ribbon Hopf algebra $(H, R, \varkappa)$ is a quasi-triangular Hopf algebra (H, R) and a grouplike element $\varkappa \in H$ such that*

$$\varkappa^2 = g, \qquad \varkappa a \varkappa^{-1} = \gamma^2(a)$$

for any $a \in H$.

In the category of finite dimensional modules over a ribbon Hopf algebra we have canonical isomorphisms $\mathbf{u}_0^2 : V \to V^{\vee\vee}$, $\mathbf{u}_0^2(v) = v_0^2(\varkappa v) = \varkappa v_0^2(v)$, which we use to identify these modules.

The following shows that, indeed, ribbon Hopf algebras produce ribbon categories.

Theorem 7.4.12 (cf. [KR93]). *If $(H, R, \varkappa)$ is a ribbon Hopf algebra, then the category H-mod is a ribbon braided category with the ribbon twist given by multiplication by the central element*

$$\nu = \gamma^2(R')R''\varkappa^{-1} = R''\gamma^2(R')\varkappa = R''\varkappa R' = R'\varkappa^{-1}R''.$$

The following holds

$$\nu^{-1} = R'\gamma(R'')\varkappa = \gamma(R'')R'\varkappa^{-1},$$

$$\varepsilon(\nu) = 1, \qquad \gamma(\nu) = \nu, \qquad \Delta\nu = (R^{21}R^{12}) \cdot \nu \otimes \nu. \tag{7.4.7}$$

7.4.6 The braided Hopf algebra F and the coend $\mathbb{F}$

Here we describe explicitly the braided Hopf algebra F and the coend $\mathbb{F}$ for our category $\mathcal{C} = H$-mod. Let H be a finite dimensional ribbon Hopf algebra and let H^* be its dual algebra [Swe69a]. The pairing $\langle , \rangle : H^* \otimes H \to \Bbbk$ satisfies

$$\langle fg, x\rangle = \langle f, x_{(1)}\rangle\langle g, x_{(2)}\rangle, \qquad \langle f, xy\rangle = \langle f_{(1)}, x\rangle\langle f_{(2)}, y\rangle$$

for $f, g \in H^*$, $x, y \in H$, where $\Delta f = f_{(1)} \otimes f_{(2)}$ is the coproduct in H^*.

Consider linear maps $i_L : {}^{\vee}L \otimes L \to H^*$, $l^a \otimes l_b \mapsto t^a_{Lb}$, where t^a_{Lb} is the matrix element of the H-module L with a basis (l_b), that is, t^a_{Lb} is a linear function on H given by $\langle u, t^a_{Lb}\rangle = \langle l^a, u.l_b\rangle$ for $u \in H$. The maps i_L become homomorphisms of H-modules if H^* is equipped with the coadjoint H-module structure

$$u \triangleright f = \langle \gamma(f_{(1)})f_{(3)}, u\rangle f_{(2)}$$

for $u \in H$, $f \in H^*$. The vector space H^* with this H-module structure will be denoted F. The maps i_L are homomorphisms of $H \otimes H$-modules if H^* is equipped with the following $H \otimes H$-module structure

$$(u \otimes v).f = \langle \gamma(f_{(1)}), u\rangle\langle f_{(3)}, v\rangle f_{(2)}$$

for $u, v \in H$, $f \in H^*$. The vector space H^* with this $H \otimes H$-module structure will be denoted $\mathbb{F}$.

Theorem 7.4.13 (cf. [Del91, Saa72, Sch92, Yet]). *The family* $(i_L : {}^\vee L \otimes L \to F)_{L \in \mathcal{C}}$ *is a coend of the bifunctor* $\mathcal{C}^{\text{op}} \times \mathcal{C} \to \mathcal{C}$, $(A, B) \mapsto {}^\vee A \otimes B$, *so we can write* $F = \int^L {}^\vee L \otimes L$. *The family* $(i_L : {}^\vee L \otimes L \to \mathbb{F})_{L \in \mathcal{C}}$ *is a coend of the bifunctor* $\mathcal{C}^{\text{op}} \times \mathcal{C} \to H \otimes H$*-mod,* $(A, B) \mapsto {}^\vee A \boxtimes B = {}^\vee A \otimes B$, *so we can write* $F = \int^L {}^\vee L \boxtimes L$.

The braided Hopf structure of F is described as follows. As a coalgebra, F coincides with H^*. The multiplication m_F in F is expressed via the multiplication m in H^*:

$$m_F(f \otimes g) = f_{(2)} \cdot g_{(2)} \langle \gamma(f_{(1)}) f_{(3)} \otimes \gamma(g_{(1)}), R \rangle.$$

The unit of F is the same as the unit of H^*. The antipode γ_F of F is expressed via the antipode γ of H^*:

$$\gamma_F(f) = \gamma(f_{(3)}) \langle \gamma(f_{(1)}) \otimes f_{(4)} \gamma^{-1}(f_{(2)}), R \rangle$$

for $f \in F$. All the structure maps $\Delta, \varepsilon, m_F, \gamma_F$ are homomorphisms of H-modules.

Now let us assume that H is modular, that is, H-mod is modular. The integral $\mu : 1 \to F, 1 \mapsto \mu$ is determined by the following properties.

1. The element μ is H-invariant, that is,

$$\mu_{(2)} \otimes \gamma(\mu_{(1)}) \mu_{(3)} = \mu \otimes 1.$$

2. It is a right integral-element in F, that is,

$$m_F(\mu \otimes g) = \varepsilon(g) \mu.$$

Indeed, these equations imply

$$\mu g = \varepsilon(g) \mu.$$

In other words, μ is a right integral-element in ordinary Hopf algebra H^*, so it is unique up to a multiplicative constant.

Since coalgebra structure and the unit for F and H^* are the same, the two-sided integral-functional λ on F is also a two-sided integral-functional $\lambda : H^* \to \Bbbk$ on the ordinary Hopf algebra H^*. To relate the integrals μ and λ, we use the endomorphism $\tilde{\lambda} : X \to X$ given by (5.2.14) for an arbitrary comodule $X \in$ comod- H^*.

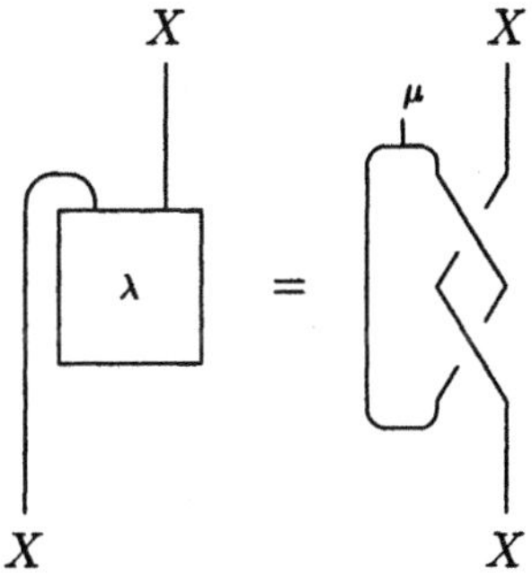

Explicitly it is given by the expression

$$\tilde{\lambda}(x) = x_{(0)} \langle \mu \otimes x_{(1)}, R_{12}^{-1} R_{21}^{-1} \rangle,$$

where $\delta(x) = x_{(0)} \otimes x_{(1)}$ is the coaction. This implies

$$\lambda(f) = \langle \mu \otimes f, R_{12}^{-1} R_{21}^{-1} \rangle.$$

The normalization equation reads

$$\lambda(\mu) = \langle \mu \otimes \mu, R_{12}^{-1} R_{21}^{-1} \rangle = 1.$$

A computation similar to the above gives

$$\omega(f \otimes g) = \langle f \otimes \gamma(g), R_{21} R_{12} \rangle.$$

Therefore, a finite dimensional ribbon Hopf algebra is modular, that is, the pairing ω is non-degenerate, if and only if the tensor $R_{21}R_{12}$ is non-degenerate, in other words, side-invertible.

7.4.7 The Hennings invariant

Let H be a finite dimensional modular Hopf algebra. We construct a 3-manifold invariant, $\tau(M)$, by taking the modular category $\mathcal{C} = H$-mod as data. Calculating $\tau(L, \mu)$ for some link L, we get an expression involving R-matrices (as many as there are crossings in $\bar{D}_L$), elements μ and the powers of κ (as many as there are components in L) under the sign of counit. It turns out that the result of calculation coincides with the Hennings invariant [Hen96] defined in an unoriented setting by Kauffman and Radford [KR95] up to change of conventions. Indeed, the calculation of $\tau(L, \mu)$ can be performed using the graphical rules of Hennings–Kauffman–Radford. They extend rules given by Reshetikhin [Res90] for invariants of tangles and are given as follows:

– change all crossings in $\bar{D}_L$ to the composite of R-matrix and the permutation

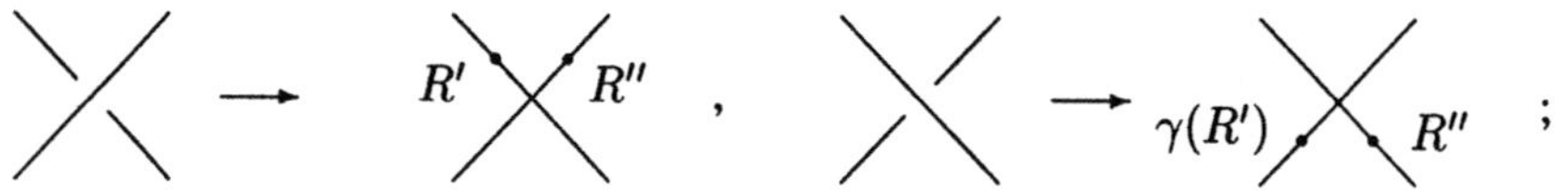

– add to each component such a power of the morphism $\mathbf{u}_0^2 = v_0^2 \cdot \kappa = \kappa \cdot v_0^2$ or its inverse $\mathbf{u}_0^{-2} = v_0^{-2} \cdot \kappa^{-1}$

so that after cancellations of v_0^2's with v_0^{-2}'s the diagram becomes a disjoint union of unknots;

– slide all elements put on the diagram to the left of the chosen points e_i via the rule

Now each component looks like:

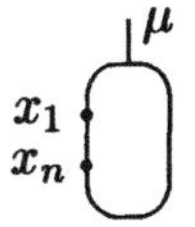

with $x_1, \dots, x_n \in H$ attached to it, and its contribution is

$$\varepsilon(\underline{x_n \dots x_1 \otimes 1}(\mu)) = \varepsilon(\langle x_n \dots x_1, \mu_{(1)} \rangle \mu_{(2)}) = \langle x_n \dots x_1, \mu \rangle .$$

We can view μ as the right integral $\mu : H \to \Bbbk$ on the algebra H. So

$$\tau(L, \mu) = \sum \mu(x_n \dots x_1) \dots \mu(z_m \dots z_1),$$

where the arguments are either powers of κ or come from R-matrices acted upon by some powers of the antipode. This is the knot invariant, used in the definition of the Hennings invariant [Hen96] up to a change of the sign of the braiding.

It has been shown in [Ker97] that the Hennings invariant for a modular Hopf algebra coincides with our invariant for $\mathcal{C} = H$-mod, and, in the case where H is semisimple, also with the Reshetikhin-Turaev invariant. Moreover, [Ker97] contains an extension of the combinatorial procedure of Hennings to the construction of topological quantum field theories.

It is also interesting to note that the TQFT, constructed from intersection homological data of representation varieties of Frohman and Nicas [FN91, FN94], can be constructed as a non-semisimple Hennings TQFT and, thus, is a special example of our theory, see [Ker00].

7.4.8 Quantum Invariants via Cell Decompositions

Finally let us mention a family of quantum invariants and TQFT's that are constructed using cell decompositions of 3-manifolds meaning triangulations or Heegaard splittings. The most prominent one of these is the model of Turaev and Viro [TV92] which uses 6-j-symbols of quantum groups. An earlier attempt of Ponzano and Regge [PR68] using 6-j-symbols of classical groups included ill defined infinite summations. The vector spaces of the Turaev–Viro theory are worked out in [KS93] and generalizations of the type of monoidal categories that can be used are given in [GK96].

In [Kup91] and [Kup96] Kuperberg succeeds in defining quantum invariants directly from a (not necessarily semisimple) Hopf algebra without using its representation theory but starting from a Heegaard diagram of the 3-manifold. In [BW95]

Barrett and Westbury show that in the semisimple case Kuperberg's invariant coincides with the Turaev–Viro invariant.

These invariants are closely related to the one considered here. Particularly it is known that the Reshetikhin–Turaev invariant of $(-M)\#M$ is the same as the Turaev–Viro invariant for the same modular category. In [Rob95] Roberts gives a proof of this fact using the "chain mail" version of the cellular invariant.

There is also strong evidence that a cellular invariant for a monoidal category $\mathcal{C}$ is the same as the surgery invariant for the Drinfeld double $D(\mathcal{C})$ of this category but the same 3-manifold. A problem in this conjecture is to find the correct additional framing structures for the 3-manifolds considered. Also we do not know any convenient representations of cobordisms in terms of cellular decompositions that would lend themselves for explicit constructions of TQFT's as the tangle presentations.

A. From Quantum Field Theory to Axiomatics

In the first part of this appendix we review the basics of the Chern-Simons topological quantum field theory in three dimensions and rational conformal field theory in two dimensions. The notions of Wilson lines in general three dimensional spacetime and charges on a surface are briefly explained. We argue that continuous sewing of surfaces over charged punctures translates to gluings of three dimensional spacetimes along neighborhoods of Wilson lines. In the second part we discuss how the conditions on the physical models can be cast into a more formal axiomatic framework. We start with an outline of a few previous approaches to extended TQFT's, and discuss their relations amongst each other. Finally, we explain how the double category picture summarizes and extends a formulation of extended TQFT's given by Kazhdan and Reshetikhin.

A.1 Witten-Chern-Simons Theory and Conformal Field Theory

In the conventional field theoretic setting a large class of interesting TQFT's in dimension three is obtained from a generally covariant action, namely the integral over the well known Chern-Simons term:

$$S_{CS}(A) = \frac{k}{4\pi} \int_M \mathrm{Tr}(A \wedge dA + \frac{2}{3} A \wedge A \wedge A). \qquad \text{(A.1.1)}$$

Here, M is an oriented 3-manifold, $k \in \mathbb{Z}$ an integer, and A is a connection on the trivial principal G-bundle over M, where G is a compact simple Lie group. The connection A may also be seen as a Lie algebra valued 1-form, and Tr stands for a multiple of the Cartan-Killing form of the Lie algebra of G. Classically, this theory is mod-$\mathbb{Z}$-invariant under general gauge transformations.

In his seminal work [Wit89] Witten describes the Chern-Simons theory as a TQFT in the axiomatic sense of Atiyah, and gives an interpretation of the Jones polynomial [Jon87] for knots and links. The 3-manifold M is interpreted as the "space-time" of the theory. The "physical space" at equal times is given by a 2-dimensional surface. In the case of $M = \Sigma \times [0,1]$, where Σ is a closed surface, we have a distinguished time coordinate along the interval, and the physical space $\Sigma_t = \Sigma$ is the same at all times t. The analysis in [Wit89] of the canonically quantized Hamiltonian theory on this product space identifies the "state space" with the space $\mathcal{H}_\Sigma$ of chiral conformal blocks of the corresponding conformal field theory on Σ.

For a general 3-dimensional space-time M we no longer have a well defined time coordinate but at best a Morse function $T : M \to [0,1]$, which may have singularities at which the physical space changes its topology. We suppose that M has boundary $\partial M = -\Sigma_0 \sqcup \Sigma_1$, with $T^{-1}(0) = \Sigma_0$ and $T^{-1}(1) = \Sigma_1$. Then we can still define heuristically a propagator $\mathcal{H}_M : \mathcal{H}_{\Sigma_0} \to \mathcal{H}_{\Sigma_1}$ by using functional integration

$$\langle \chi_1, \mathcal{H}_M \chi_0 \rangle = \int \mathcal{D}\bar{A} \; \exp(iS_{CS}) \,,$$

where the measure $\mathcal{D}\bar{A}$ is formally written on the set of gauge classes on M with boundary conditions on the Σ_j, characterized by the states χ_j. (Note that $\mathcal{H}_M \equiv Z(M)$ in [Wit89]). Unlike the Hamiltonian picture, the "propagator" $\mathcal{H}_M$ no longer has to be an isomorphism. Since the Chern-Simons action is generally covariant, we expect $\mathcal{H}_M$ to be dependent only on the homeomorphism class of M.

If we consider another 3-manifold, N, with boundary $\partial N = -\Sigma_1 \sqcup \Sigma_2$, we may think of it as a continuation of the space-time with time function $T : N \to [1,2]$. Hence, $M \cup_{\Sigma_1} N$ can be viewed as a total space time over the time interval $[0,2]$. The functional integral suggests that the total propagator is formally given by a summation over the intermediate states in $\mathcal{H}_{\Sigma_1}$, and, thus, by the composition $\mathcal{H}_{M\cup_{\Sigma_1}N} = \mathcal{H}_N \circ \mathcal{H}_M$.

Consequently, a Chern-Simons quantum field theory has to assign vector spaces to surfaces and linear maps to homeomorphism classes of cobordisms so that the assignment is compatible with compositions, given by gluing over an intermediate surface for the cobordisms, and the usual compositions of linear maps. It is, therefore, also a TQFT in the axiomatic sense of Atiyah [Ati88] as defined in the introduction of this book.

The content of any quantum field theory is determined by its "observables" and "charges", and their "expectation values". In [Wit89] the observables of a Chern-Simons theory are given by Wilson lines or loops defined as path ordered integrals of the gauge connection along a path $\mathcal{K} \subset M$:

$$\mathcal{W}_{\mathcal{K},\rho} \;=\; \mathrm{Tr}_\rho\Big(P\text{–}\exp\big(\int_{\mathcal{K}} A^\rho\big)\Big)$$

Here, the gauge field is taken in a particular, simple representation $\rho \in Rep(G) = G$–mod of the gauge group G. The expectation value $\langle \mathcal{W}_{\mathcal{K}_1,\rho_1} \cdot \ldots \cdot \mathcal{W}_{\mathcal{K}_b,\rho_b} \rangle$ for disjoint Wilson loops naturally constitutes a topological invariant of the link $\mathcal{K}_1 \sqcup \ldots \sqcup \mathcal{K}_b$ in M. For $G = SU(2)$ and the fundamental representation ρ this invariant is, in essence, given by the Jones polynomial. Wilson lines may be interpreted as trajectories of charges moving in the space-time M. They puncture a generic time level surface $\Sigma_t = T^{-1}(t)$ in points $p_1, \ldots, p_m \subset \Sigma_t$. To each of these points we assign the representation $\rho_j \in G$–mod of the respective Wilson line, with $j = 1, \ldots, m$.

In the corresponding conformal field theory (CFT) these punctures are identified with "inserted charges". In an axiomatic approach to CFT, proposed by Graeme Segal, a *modular functor* assigns to such configurations generalizations of conformal blocks $\mathcal{H}_{\Sigma^*,[\rho_1,\ldots,\rho_m]}$. We denote by $\Sigma^* = \Sigma - (U(p_1) \sqcup \ldots \sqcup U(p_m))$ the surface with parametrized boundary obtained by removing neighborhoods $U(p_j)$ of the punctures, and ρ_j is the representation associated to the j-th hole. Segal's axioms also include a gluing rule, which implies, as a special case, a composition rule similar to that in the Atiyah axioms:

$$\begin{aligned}&\mathcal{H}_{\Sigma_1^* \cup_{bS^1} \Sigma_2^*,[\rho_1,\dots,\rho_m,\mu_1,\dots,\mu_n]} \\ &= \bigoplus_{\nu_1,\dots,\nu_b \in \mathcal{R}(G,k)} \mathcal{H}_{\Sigma_1^*,[\rho_1,\dots,\rho_m,\nu_1,\dots,\nu_b]} \otimes \mathcal{H}_{\Sigma_2^*,[\mu_1,\dots,\mu_n,\nu_1^\vee,\dots,\nu_b^\vee]}. \end{aligned} \tag{A.1.2}$$

Here, the gluing $\Sigma_1^* \cup_{bS^1} \Sigma_2^*$ of two surfaces over b boundary components may be understood as a composition of 2-dimensional cobordisms between 1-manifolds. The summation, however, is not over vectors in a vector space anymore, but over the set of inequivalent simple objects $\mathcal{R}(G,k)$ in the category G_k-mod of representations that are associated to the WZW conformal field theory with group G at level $k \in \mathbb{Z}$.

Thus the categorical interpretation of a modular functor is rather different from that of a TQFT functor on the algebraic side. Moore and Seiberg extracted a set of building blocks, called "basic data", for the construction of a modular functor [MS89]. It became apparent that the structural data had to be the one of a semisimple (braided) tensor category, and that special 2-cobordisms played the role of functors. For example, the tensor product functor corresponds to a pair of pants.

Many technical details, related to modular functors with minimal basic data and several algebraic relations to the Chern-Simons TQFT, have been discovered and worked out in [MS89] as well as the later article of Walker [Wal]. Further development of these works calls for a unification of axioms of Atiyah for 3-dimensional Chern-Simons TQFT with the axioms of Segal, Moore and Seiberg for the related CFT's. Such a theory has to allow us to define propagators $U_{M,\mathcal{W}}$ between the extended spaces $\mathcal{H}_{\Sigma^*,[\rho_1,\dots,\rho_m]}$ that are associated to cobording manifolds M with Wilson lines $\mathcal{W} \subset M$.

Observe that the operation of gluing surfaces together extends naturally to cobording manifolds with Wilson lines $(M_1, \mathcal{W}_1)$ and $(M_2, \mathcal{W}_2)$ since we can sew surfaces $T^{-1}(t) \cap M_1$ and $T^{-1}(t) \cap M_2$ together along their punctures for every generic time $t \in [0,1]$, where $T : M_1 \sqcup M_2 \to [0,1]$ is a common Morse function. A more direct way to realize this operations is to cut out tubular neighborhoods from respective components $\mathcal{W}_\nu$ of the Wilson lines in M_1 and M_2 and glue them together along the newly created boundary pieces $\mathcal{W}_\nu \times S^1$. The unifying set of axioms should also declare the nature of a corresponding binary operation between the maps $U_{M_1,\mathcal{W}_1}$ and $U_{M_2,\mathcal{W}_2}$, that is, a horizontal composition. In [Kon] Kontsevich explains why a modular functor in the sense of Segal and [MS89] should imply a topological quantum field theory by applying the elements of the basic data set to the succession of level surfaces of a Morse function.

A.2 Developing the Axiomatics for Extended TQFT's

In the following years the field theories described above led to many proposals to generalize the notion of topological quantum field theories to manifolds with corners. In particular, the simplicial lattice gauge theoretic description of Chern-Simons given by Dijkgraaf and Witten in [DW90] was the starting point for a number of more detailed investigations. Specializing this to finite gauge groups provided a toy model that allowed a rather detailed study by Freed and Quinn [FQ93] and exposed a variety of implied algebraic structures. Subsequently, in [Qui95] Quinn proposes the notion of *modular domain categories*, which considers gluings of manifolds with corners over boundary pieces that may have themselves boundaries. The definition of a *modular TQFT* given there implies in the three dimensional case that to each 1-manifold one associates a coalgebra, A_S, to every surface, Σ, an $A_{\partial\Sigma}$-comodule, V_Σ, and the gluing of two surfaces Σ_1 and Σ_2 over boundaries R is given as $V_{\Sigma_1} \otimes_{A_R} V_{\Sigma_2}$. Hence, this formalism presents a framework that allows us to avoid assigning "charges" to the boundaries of surfaces.

Freed [Fre94] further studied extensions of these types of axiomatic TQFT's for manifolds with corners in more general dimensions and codimensions. Here the notion of n-categories of n-vector spaces is introduced. A similar set of axioms, including n-vector spaces, emerged in [Law93] from the study of the structures of simplicial decompositions. A general discussion of n-categorical structures for manifolds with corners in arbitrary dimensions can be found in the work of Baez and Dolan [BD95]. In four dimensions this leads to the formulation of interesting higher structures such as Hopf categories, see [CY93, CF94, CKY97]. Unfortunately, nontrivial examples of such TQFT's for higher dimensional manifolds with corners are very difficult to find.

The program of Moore and Seiberg [MS89] for the construction of TQFT's with corners departing from so called *basic data* is developed into a mathematically more coherent theory by Walker [Wal] and also Gelca [Gel98]. As it turns out the basic data can be viewed as the structure coefficients of a semisimple braided category plus some extra data.

Guided by the physical axiomatic theories above and rigorous constructions of 3-manifold invariants via quantum groups, Kazhdan and Reshetikhin[1] suggest to cast the formalism given in [MS89] into a more abstract system of axioms, in which one assigns to a 1-manifold an abelian category. For an outline of this philosophy see also [Ker94]. Let us list a few more assumptions and consequences of such an assignment in Table A.1. The essential assumptions are marked with a star.

Let us relate the properties listed above to the previous discussion of models and theories, and comment on some basic consequences and redundancies:

In *(i)** we avoid attaching to each puncture a "charge", given by representations of G in the conformal theories, by associating to the boundary of a surface the category, from whose objects we obtain the label set of "charges". In the CFT example we have $\mathcal{C} = \mathcal{C}_{S^1} = G_k$−mod, and the label set $\mathcal{R}(G,k)$ consists of equivalence

[1] Private communication

Table A.1. A topology – category theory dictionary

	Topological Structure		Algebraic Structure
*(i)**	1-manifold S	$\mapsto$	Abelian category $\mathcal{C}_S$
(ii)	Unions of 1-manifolds	$\mapsto$	Deligne products of categories
(iii)	Surface Σ with boundary	$\mapsto$	Object $X_\Sigma \in \mathcal{C}_{\partial\Sigma}$
*(iv)**	Surfaces cobounding two 1-manifolds	$\mapsto$	Left exact functors between abelian categories
*(v)**	Empty 1-manifold	$\mapsto$	Category of vector spaces
(vi)	Closed surface Σ	$\mapsto$	Vector space V_Σ
*(vii)**	Relative, framed 3-cobordism of 2-cobordisms	$\mapsto$	Natural transformation of corresponding functors
(viii)	Relative 3-cobordism of surfaces Σ_1 and Σ_2 with boundary S	$\mapsto$	Morphism in $\mathrm{Hom}_{\mathcal{C}_S}(X_{\Sigma_1}, X_{\Sigma_2})$
(ix)	Ordinary framed 3-cobordism	$\mapsto$	Linear map between vector spaces
*(x)**	Empty surface	$\mapsto$	Ground field $\mathbf{k}$
(xi)	3-manifold M with boundary	$\mapsto$	Vector in $V_{\partial M}$
(xii)	Closed 3-manifold	$\mapsto$	Number in $\mathbf{k}$

classes of simple objects of $\mathcal{C}$. For m punctures the possible charge assignments are clearly given by elements in $\mathcal{R}(G,k) \times \ldots \times \mathcal{R}(G,k)$ (m-times). This motivates property *(ii)* since this product is also the set of inequivalent simple objects of $\mathcal{C}^{\boxtimes m} = \mathcal{C}_{\sqcup^m S^1} = \mathcal{C} \boxtimes \ldots \boxtimes \mathcal{C}$ (m-times), where $\boxtimes$ is Deligne's product of abelian categories as defined in [Del91]. In the case of axiomatic CFT's we obtain the object that is associated according to *(iii)* to a surface Σ^* with boundary by interpreting the conformal blocks as multiplicity spaces as follows:

$$X_{\Sigma^*} = \bigoplus_{\rho_1,\ldots,\rho_m \in \mathcal{R}(G,k)} \mathcal{H}_{\Sigma^*,[\rho_1,\ldots,\rho_m]} \otimes \rho_1 \boxtimes \ldots \boxtimes \rho_m \in (G_k\text{–mod})^{\boxtimes m}. \tag{A.2.1}$$

Note, that an abelian category $\mathcal{C}$ also satisfies the basic condition of a 2-vector space, namely an inner product with values in the category of vector spaces:

$$\langle\!\langle\ \ \rangle\!\rangle_{\mathcal{C}} : \mathcal{C}^{opp} \boxtimes \mathcal{C} \longrightarrow \mathrm{Vect} : \quad X \boxtimes Y \mapsto \mathrm{Hom}(X,Y)$$

Moreover, we easily see that *(v)** and *(iv)** already imply *(iii)*. A surface Σ^* can be understood as a cobordism $\varnothing \to \partial\Sigma^*$ so that we have by *(iv)** a functor $\mathcal{F}_{\Sigma^*}$: $\mathbf{k}$-vect $= \mathcal{C}_\varnothing \to \mathcal{C}_{\partial\Sigma^*}$, which is completely determined by the image of the ground field $X_{\Sigma^*} = \mathcal{F}_{\Sigma^*}(\mathbf{k}) \in \mathcal{C}_{\partial\Sigma^*}$. Conversely, if $\Sigma : S_1 \to S_2$ is a 2-cobordism, and $X_{S_2 \sqcup -S_1} \in \mathcal{C}_{S_2} \boxtimes \mathcal{C}_{S_1}^{opp}$ we can construct a functor $\mathcal{F}_\Sigma : \mathcal{C}_{S_1} \to \mathcal{C}_{S_2}$ via the following contraction formula:

$$\mathcal{F}_\Sigma(Y) = \mathcal{C}_{S_2} \boxtimes \langle\!\langle\ \ \rangle\!\rangle_{\mathcal{C}_{S_1}} (X_{S_2 \sqcup -S_1} \boxtimes Y). \tag{A.2.2}$$

It is, in fact, an easy exercise to show that Segal's gluing axiom for conformal blocks in (A.1.2) is equivalent to functoriality $\mathcal{F}_{\Sigma''} \circ \mathcal{F}_{\Sigma'} \cong \mathcal{F}_{(\Sigma'' \cup_{\sqcup^b S^1} \Sigma')}$ of the map from 2-cobordisms to functors.

In a later version [Seg98] of his lecture notes Graeme Segal states the conditions *(i)* through *(iv)* in the abstract setting as axioms for "category-valued field theories". He also proves that the strictly functorial version of an extended TQFT implies that the modular category has to be semisimple, see also Quinn [Qui95].

The category in [FQ93] is given as $\mathcal{C}_{S^1} = A^*_{S^1}-\text{mod}$, where $A^*_{S^1} = D(G)$ is the Drinfel'd double of the finite group G as in [DPR90]. Since this category is semisimple, the gluing formula of [Qui95] can be derived in the same way as in the case of an axiomatic CFT.

Property *(vi)* is obviously also a consequence of *(iii)* and *(v)**. The dependence of the properties *(vii)** and *(viii)* is completely analogous to that between *(iii)* and *(iv)**. Moreover, *(v)**, *(vi)* and *(viii)* clearly implies *(ix)* so that Atiyah's axioms are, indeed, a special case of this list of properties. (We may assume that the equivalence Vect ⊠ Vect ≅ Vect is given by the tensor product $\otimes_{\mathbf{k}}$).

Recall that Segal's gluing axiom is implied by the functoriality of $\Sigma \mapsto \mathcal{F}_\Sigma$. However, Segal's axioms of a modular functor also involve the morphisms of the category of surfaces. These are given by diffeomorphisms $f : \Sigma_s \xrightarrow{\sim} \Sigma_t$ up to isotopy, and the modular functor has to assign maps also to them. We know, however, by Proposition 1.4.2, that these can be identified with the *invertible* relative 3-cobordisms between Σ_s and Σ_t. More precisely, we found a group isomorphism $\pi_0(\mathcal{D}iff^+(\Sigma)) \cong \mathrm{Aut}_{\mathcal{C}ob(\partial\Sigma)}(\Sigma)$ between the mapping class group of a surface Σ and the group of invertible self cobordisms. By *(viii)* this group is, thus, presented in $\mathrm{Aut}_{\mathcal{C}_{\partial\Sigma}}(X_\Sigma)$. Compatibility of the modular assignment under composition of diffeomorphisms is, thus, implied by compatibility of the TQFT assignment under composition of cobordisms.

Hence, a system of axioms with the listed properties does in fact include both the Atiyah and the Segal axioms.

The fact that *(xi)* and *(xii)* are implied by *(ix)* and *(x)** follows from the standard arguments in the Atiyah theory and is analogous to the way that *(iii)* and *(vi)* are implied by *(iv)** and *(v)**.

In summary, all of the listed relations are consequences of the properties *(i)**, *(iv)** and *(vii)** together with conventions *(v)** and *(x)** as well as assumptions on tensor products. Hence, the Kazhdan-Reshetikhin philosophy can be distilled down to three independent types of assignments that can be organized hierarchically as in the following table.

Categorical level	Topological morphism	Algebraic morphism
object	1-manifold	abelian category
1-arrow	2-cobordism	left exact functor
2-arrow	relative 3-cobordism	natural transformation

The topological morphisms listed above admit two compositions obtained by gluing two 3-cobordisms over bounding 2-cobordisms or over cylinders over the 1-manifolds, see Chapter 1. The operational structure gives rise to a double category of relative cobordisms. For the algebraic data we have the usual the 2-category of abelian categories, left exact functors, and natural transformations.

In the physical context of Section A.1 the assignment of the left exact functors to surfaces is obtained from the conformal blocks via (A.2.2) and (A.2.1). The natural transformations of these functors, associated to cobordisms with Wilson lines, are similarly obtained from the action of the generalized propagators $U_{M,\mathcal{W}}$ on the conformal blocks. As explained in the end of Section A.1, we require them to be compatible with compositions along the surfaces as well as excised neighborhoods of Wilson lines.

On the level of categories, functors and natural transformations the compatibility translates precisely into the functoriality of the above assignment with respect to horizontal and vertical compositions in a 2-category or a double category. The extended TQFT, therefore, implies a double functor as proposed in Theorem 0.4.2. Axiom *(ii)* in the above list also implies compatibility with a tensor product of categories.

A.3 Generalized TQFT's in Gauge Theory

For a Lie group G and a topological space X let $R(X) = \mathrm{Hom}(\pi_1(X), G)/G$ be the moduli space of homomorphisms modulo the adjoint action of G. It is natural for the construction of TQFT type structures so study the intersection theory and homology of the representation varieties $R(X)$.

As a first example [FN91] Frohman and Nicas consider the abelian case $G = U(1)$ and construct a TQFT where the vector space associated to a surface Σ is the cohomology ring $H^*(R(\Sigma), \mathbb{Z}) \cong \bigwedge^* H_1(\Sigma, \mathbb{Z})$. For a cobordism M with $\partial M = -\Sigma_{in} \sqcup \Sigma_{out}$ they consider a Heegaard splitting $M = A \circ B$ where A and B are given by attaching 1-handles and 2-handles respectively to the cylinder over a surface and are glued together over an intermediate surface Σ_{med}. The morphism for M is obtained from the intersection number of $J(A) \times J(\Sigma_{out})$ and $J(\Sigma_{in}) \times J(B)$ in $J(\Sigma_{in}) \times J(\Sigma_{med}) \times J(\Sigma_{out})$. In case M has interior homology, meaning $\beta_1(M) > \frac{1}{2}\beta_1(\partial M)$, these varieties do not intersect transversally so that the associated morphism is zero as well.

Consequently the Frohman–Nicas TQFT is an example of a *non-semisimple* TQFT as defined in [Ker98b] using a modified composition law expressing functoriality in the case that connected cobordisms are glued over unconnected surfaces. The non-semisimple modular category associated to the TQFT in [FN91] is given in [Ker00].

Similar non-semisimple phenomena occur for the TQFT's that Frohman and Nicas construct in [FN94] using intersection homology in order to obtain $SU(n)$ Casson type invariants for knots. They circumvent the deviation from the traditional composition law by restricting to "monotonous" subcategories of cobordisms. Interestingly it turns out that the higher order knot invariants can be computed from the Alexander polynomial obtained from the $U(1)$-TQFT, suggesting that a similar assertion holds for the TQFT's themselves.

In [Don99] Donaldson uses similar TQFT methods to give an interpretation of a Casson invariant of homology circles, which is closely related to their Alexander polynomial. Instead of $R(\Sigma)$ he considers a moduli space $M(\Sigma)$ of $SO(3)$-connections in a *non-trivial* bundle and defines the vector spaces as their homology groups. Also here we have the non-semisimple characteristic that cobordisms with non-trivial interior homology vanish. Moreover this TQFT essentially decomposes into a sum of $U(1)$-theories. Similar TQFT's are considered to relate the Seiberg-Witten invariant to Milnor's torsion.

Inspired by ideas of Donaldson and Segal Fukaya [Fuk99] defines the relative Floer homology groups of a 3-manifold with boundary using the an analogous double category picture as the one we use here, only in one dimension higher. To a surface Σ he associates a category $\mathcal{C}_0(\Sigma) = \mathcal{L}ag(R(\Sigma))$ of Lagrangian submanifolds with $G = SU(2)$ (or $\mathcal{L}ag(M(\Sigma))$ with $G = SO(3)$). The morphisms associated to the 3-cobordisms between two Lagrangian manifolds are given by the Floer homology groups $HF(L_1, L_2)$ of the corresponding 3-manifold. The required category is then the A^∞-category $\mathcal{C}(\Sigma) = \mathcal{F}unc(\mathcal{C}_0(\Sigma), \mathcal{C}h)$, where $\mathcal{C}h$ is the category of chain complexes. The relative Floer homology is subsequently constructed via the solutions of certain anti-selfdual equations on the bounding 4-manifolds with corners.

B. Double Categories and Double Functors

B.1 Double Categories

Double categories have been introduced by Ehresmann [Ehr63a] as a generalization of the notion of a 2-category and, hence, also of strict monoidal categories. They are defined as a class of morphisms equipped with two compositions, which are required to be distributive with respect to each other, such that all structure maps of one category are functors with respect to another. Specifically, the formal definition is as follows:

Definition B.1.1. *A* double category *is a class $\mathfrak{D}$ equipped with two multiplication operations, $(a, b) \mapsto a \circ_h b$ and $(a, b) \mapsto a \circ_v b$, called the horizontal and vertical composition respectively, such that*

1. *$(\mathfrak{D}, \circ_h)$ is a class of morphisms of a category, denoted $\mathfrak{D}_h$. Similarly for $(\mathfrak{D}, \circ_v)$.*
2. *The subclass of* vertical 1-morphisms *defined as $U^h_{\mathfrak{D}} = \{\text{units of } \mathfrak{D}_h\}$ is stable under vertical composition. Similarly, the subclass of* horizontal 1-morphisms *$U^v_{\mathfrak{D}} = \{\text{units of } \mathfrak{D}_v\}$ is stable under horizontal compositions.*
3. ***(Interchange law)** If all products $b \circ_h a$, $d \circ_h c$, $c \circ_v a$, and $d \circ_v b$ are defined, then the following expressions are defined and equal*
$$(d \circ_h c) \circ_v (b \circ_h a) = (d \circ_v b) \circ_h (c \circ_v a).$$
4. *Denote by* source_h *and* target_h *(resp.* source_v *and* target_v*) the maps $\mathfrak{D} \to \mathfrak{D}$, associating with each morphism its source or target in $\mathfrak{D}_h$ (resp. in $\mathfrak{D}_v$). Then*
$$\begin{aligned}
\mathrm{source}_h \circ \mathrm{source}_v &= \mathrm{source}_v \circ \mathrm{source}_h, \\
\mathrm{target}_h \circ \mathrm{source}_v &= \mathrm{source}_v \circ \mathrm{target}_h, \\
\mathrm{source}_h \circ \mathrm{target}_v &= \mathrm{target}_v \circ \mathrm{source}_h, \\
\mathrm{target}_h \circ \mathrm{target}_v &= \mathrm{target}_v \circ \mathrm{target}_h .
\end{aligned}$$

This definition suggests the following diagrammatic presentation for $f \in \mathfrak{D}$, which is called a 2-morphism, or a 2-arrow, or a 2-cell.

$$\begin{array}{ccc}
\mathrm{source}_h(\mathrm{source}_v\, f) & \xrightarrow{\mathrm{source}_v\, f} & \mathrm{target}_h(\mathrm{source}_v\, f) \\
{\scriptstyle \mathrm{source}_h\, f}\big\downarrow & f & \big\downarrow{\scriptstyle \mathrm{target}_h\, f} \\
\mathrm{source}_h(\mathrm{target}_v\, f) & \xrightarrow[\mathrm{target}_v\, f]{} & \mathrm{target}_h(\mathrm{target}_v\, f)
\end{array}$$

The interchange law 3) assumes a situation as indicated in the diagram below. It asserts that we obtain the same element in $\mathfrak{D}$, whether we first perform all horizontal multiplication and then the vertical one or whether we start with the compositions in vertical direction and then the remaining horizontal one.

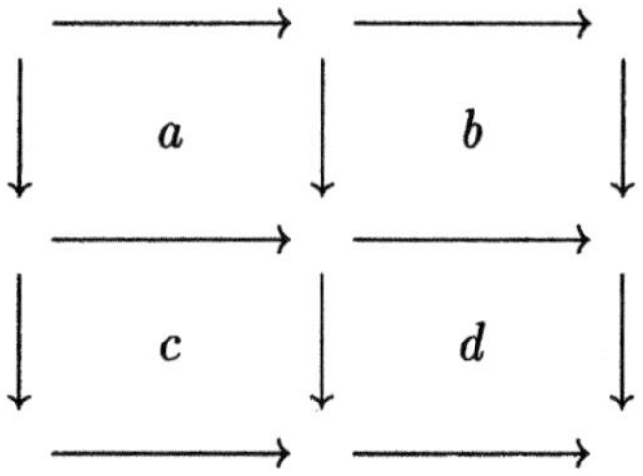

Notation. The diagrammatic notation is easier to translate into formulae using the opposite multiplication laws $b \bullet_h a = a \circ_h b$ and $b \bullet_v a = a \circ_v b$. We shall use these notations as synonyms. For instance, the interchange law reads

$$(a \bullet_h b) \bullet_v (c \bullet_h d) = (a \bullet_v c) \bullet_h (b \bullet_v d).$$

For a vertical 1-morphism, $g \in U_{\mathfrak{D}}^h$, (resp. a horizontal 1-morphism, $j \in U_{\mathfrak{D}}^v$) the diagrammatic presentation reduces to

$$\text{source}_v\, g \xequal{\text{source}_v\, g} \text{source}_v\, g \qquad \text{source}_h\, j \xrightarrow{j} \text{target}_h\, j$$

$$g \downarrow \qquad g \qquad \downarrow g \qquad \qquad \text{source}_h\, j \Vert \qquad j \qquad \Vert \text{target}_h\, j$$

$$\text{target}_v\, g \xequal{\text{target}_v\, g} \text{target}_v\, g \qquad \text{source}_h\, j \xrightarrow[j]{} \text{target}_h\, j$$

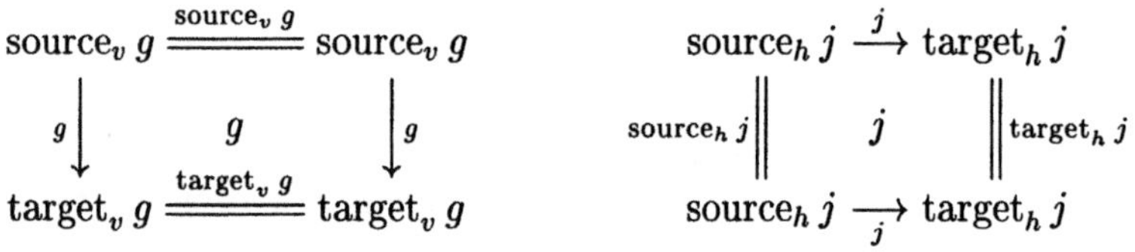

The 2-cell for g (resp. j) collapses to a vertical (resp. horizontal) arrow.

Elements $x \in U_{\mathfrak{D}}^h \cap U_{\mathfrak{D}}^v$ which are both horizontal and vertical units are called *objects*. For objects the 2-cell collapses to a vertex.

Definition B.1.2. *A 2-category is a double category for which all vertical 1-morphisms are units with respect to vertical composition (i.e. objects).*

Let $\mathfrak{A}$ be a 2-category. Ehresmann [Ehr63b] associates with it the double category of quintets $\mathcal{Q}\mathfrak{A}$. It has the class of morphisms

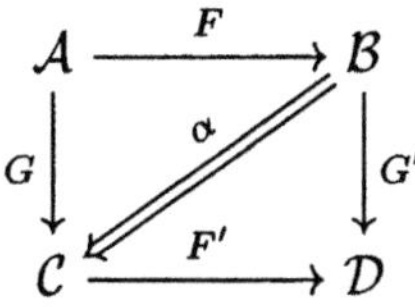

where α is a 2-morphism (natural transformation) from $\mathfrak{A}$. There are two compositions on this class of morphisms, namely, horizontal and vertical pasting.

In particular, we use the double category of quintets $\mathcal{Q}\mathbf{V}\text{-}\mathbf{Cat}\text{-}\mathbf{mod}^l$ in the 2-category $\mathbf{V}\text{-}\mathbf{Cat}\text{-}\mathbf{mod}^l$. We are interested in $G = R_\alpha$ and $G' = R_\beta$ representing the action of the symmetric group defined in Section 4.3.3. The double subcategory with $G = 1$ and $G' = 1$ is the 2-category $\mathbf{V}\text{-}\mathbf{Cat}\text{-}\mathbf{mod}^l$ itself.

B.2 Double pseudofunctors

A (strict) *double functor* between two double categories is a map compatible with horizontal and vertical source and target, and with horizontal and vertical compositions. There is no weak version of a double category. However, double functors admit weak versions. We are going to describe them here, as well as related notions: horizontal and vertical natural transformations etc. Our main objects of studies, namely, extended TQFT's, are a priori double functors in the weak sense.

Definition B.2.1. *Let $\mathfrak{C}$ and $\mathfrak{D}$ be double categories. A* (double) pseudofunctor $\Phi = (\Phi, \alpha, \beta) : \mathfrak{C} \to \mathfrak{D}$ *is a triple consisting of*

(1) a map $\Phi : \mathfrak{C} \to \mathfrak{D}$ commuting with the maps source_h, target_h, source_v *and* target_v *(in particular, Φ maps horizontal or vertical 1-morphisms to 1-morphisms of the same kind and objects to objects);*

(2) for any pair of $\circ_h$-composable horizontal 1-morphisms $\mathcal{A} \xrightarrow[hor]{F} \mathcal{B} \xrightarrow[hor]{G} \mathcal{C}$ a vertically invertible 2-cell $\alpha_{F,G}$ with $\text{source}_h(\alpha_{F,G}) = \Phi(\mathcal{A})$, $\text{target}_h(\alpha_{F,G}) = \Phi(\mathcal{C})$, $\text{source}_v(\alpha_{F,G}) = \Phi F \bullet_h \Phi G$, $\text{target}_v(\alpha_{F,G}) = \Phi(F \bullet_h G)$. *It is visualized as*

$$\begin{array}{ccc} \Phi\mathcal{A} & \xrightarrow{\Phi F \bullet_h \Phi G} & \Phi\mathcal{C} \\ \| & \alpha_{F,G} & \| \\ \Phi\mathcal{A} & \xrightarrow{\Phi(F \bullet_h G)} & \Phi\mathcal{C} \end{array} \quad = \quad \begin{array}{ccccc} \Phi\mathcal{A} & \xrightarrow{\Phi F} & \Phi\mathcal{B} & \xrightarrow{\Phi G} & \Phi\mathcal{C} \\ \| & & \alpha_{F,G} & & \| \\ \Phi\mathcal{A} & & \xrightarrow{\Phi(F \bullet_h G)} & & \Phi\mathcal{C} \end{array} \;;$$

(3) for any pair of $\circ_v$-composable vertical 1-morphisms $\mathcal{A} \xrightarrow[ver]{F} \mathcal{B} \xrightarrow[ver]{G} \mathcal{C}$ a horizontally invertible 2-morphism $\beta_{F,G}$ with $\text{source}_h(\beta_{F,G}) = \Phi F \bullet_v \Phi G$, $\text{target}_h(\beta_{F,G}) = \Phi(F \bullet_h G)$, $\text{source}_v(\beta_{F,G}) = \Phi(\mathcal{A})$, *and* $\text{target}_v(\beta_{F,G}) = \Phi(\mathcal{C})$. *It is visualized as*

$$\begin{array}{ccc} \Phi\mathcal{A} & = & \Phi\mathcal{A} \\ {\scriptstyle \Phi F \bullet_v \Phi G}\downarrow & \beta_{F,G} & \downarrow{\scriptstyle \Phi(F \bullet_v G)} \\ \Phi\mathcal{C} & = & \Phi\mathcal{C} \end{array} \quad = \quad \begin{array}{ccc} \Phi\mathcal{A} & = & \Phi\mathcal{A} \\ {\scriptstyle \Phi F}\downarrow & & \\ \Phi\mathcal{B} & \beta_{F,G} & \downarrow{\scriptstyle \Phi(F \bullet_v G)} \\ {\scriptstyle \Phi G}\downarrow & & \\ \Phi\mathcal{C} & = & \Phi\mathcal{C} \end{array} \;;$$

such that

(i) the 2-cell $\alpha_{F,G}$ collapses to a horizontal morphism $\alpha_{F,G} = \Phi G$ (resp. $\alpha_{F,G} = \Phi F$) whenever F (resp. G) is an object;

(ii) the 2-cell $\beta_{F,G}$ collapses to a vertical morphism $\beta_{F,G} = \Phi G$ (resp. $\beta_{F,G} = \Phi F$) whenever F (resp. G) is an object;

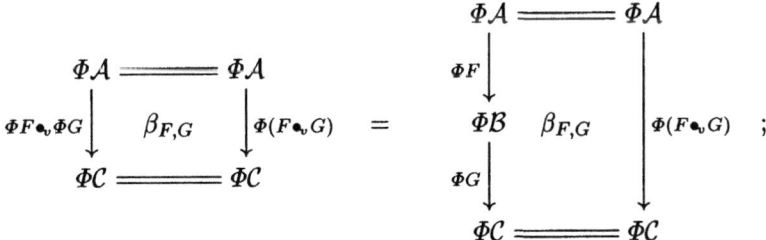

(iii) for any three $\circ_h$*-composable horizontal 1-morphisms* $\mathcal{A} \xrightarrow[hor]{F} \mathcal{B} \xrightarrow[hor]{G} \mathcal{C} \xrightarrow[hor]{H} \mathcal{D}$ *we have the cocycle property*

$$(\alpha_{F,G} \bullet_h \Phi H) \bullet_v \alpha_{F \bullet_h G,H} = (\Phi F \bullet_h \alpha_{G,H}) \bullet_v \alpha_{F,G \bullet_h H},$$

which is visualized as

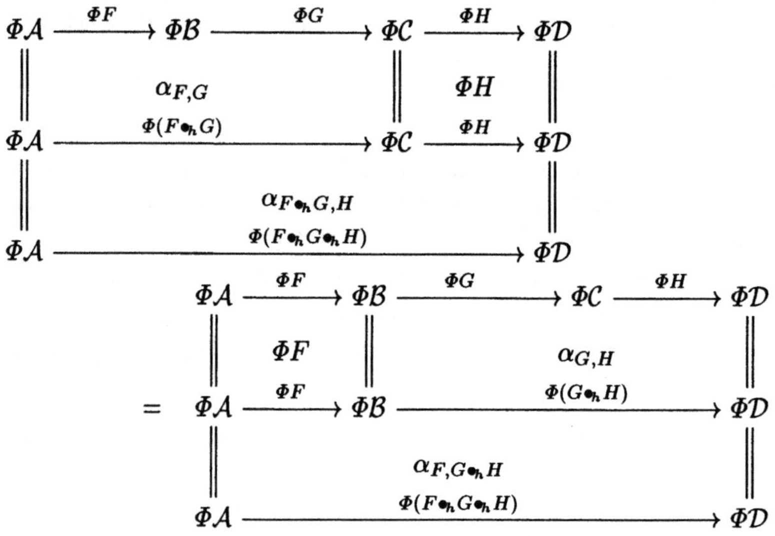

(iv) for any three $\circ_v$*-composable vertical 1-morphisms* $\mathcal{A} \xrightarrow[ver]{F} \mathcal{B} \xrightarrow[ver]{G} \mathcal{C} \xrightarrow[ver]{H} \mathcal{D}$ *we have the cocycle property*

$$(\beta_{F,G} \bullet_v \Phi H) \bullet_h \beta_{F \bullet_v G,H} = (\Phi F \bullet_v \beta_{G,H}) \bullet_h \beta_{F,G \bullet_v H},$$

which is visualized as the mirror reflection of the previous diagram with respect to the diagonal;

(v) for any pair (a, b) *of* $\circ_h$*-composable 2-morphisms*

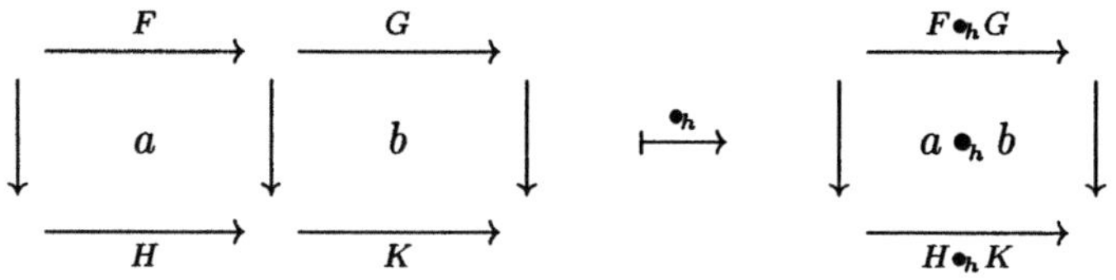

we have the naturality property

$$(\Phi a \bullet_h \Phi b) \bullet_v \alpha_{H,K} = \alpha_{F,G} \bullet_v \Phi(a \bullet_h b),$$

which is visualized as

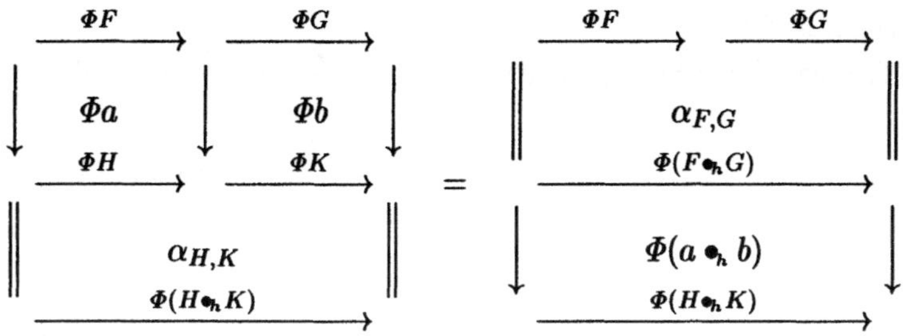

(vi) for any pair (a, b) of $\circ_v$-composable 2-morphisms

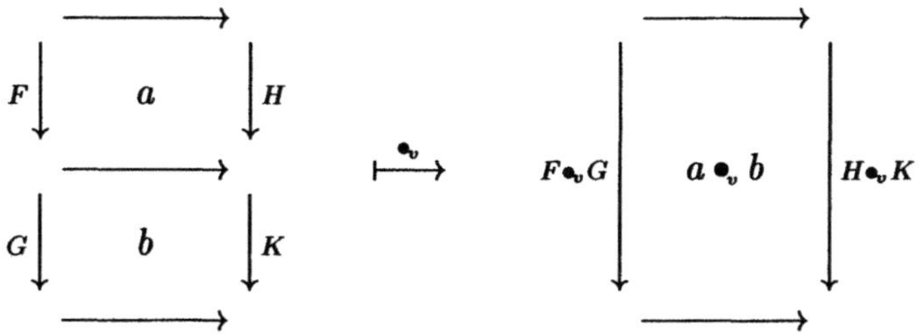

we have the naturality property

$$(\Phi a \bullet_v \Phi b) \bullet_h \beta_{H,K} = \beta_{F,G} \bullet_h \Phi(a \bullet_v b),$$

which is visualized as

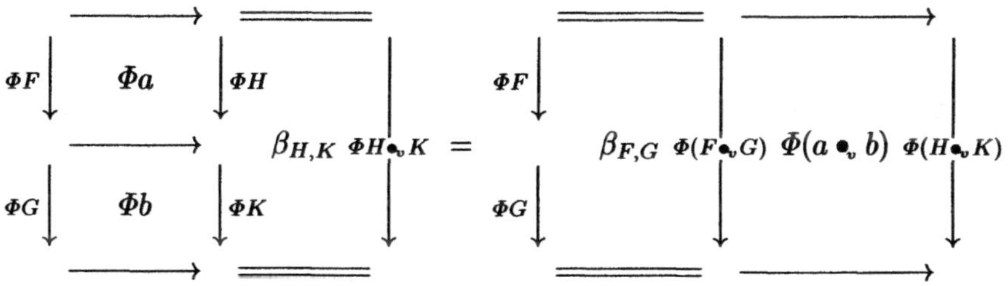

Remark B.2.2. When $\mathfrak{C}$ is a 2-category, and $(\Phi, \alpha, \beta) : \mathfrak{C} \to \mathfrak{D}$ is a double pseudofunctor then by (ii) $\beta = \mathbb{I}$, since all vertical morphisms are objects. Hence, the notion of a double pseudofunctor $\Phi : \mathfrak{C} \to \mathfrak{D}$ coincides with the notion of a 2-pseudofunctor $\Phi : \mathfrak{C} \to \mathfrak{A}$ [Bén67], where the 2-category $\mathfrak{A} \subset \mathfrak{D}$ consists of 2-morphisms whose horizontal source and target are objects.

B.2.1 Compositions of double pseudofunctors

Given a pair of double pseudofunctors $\Phi' : \mathfrak{A} \to \mathfrak{B}$ and $\Phi'' : \mathfrak{B} \to \mathfrak{C}$, their composition

$$(\Phi, \alpha, \beta) = \left(\mathfrak{A} \xrightarrow{(\Phi',\alpha',\beta')} \mathfrak{B} \xrightarrow{(\Phi'',\alpha'',\beta'')} \mathfrak{C}\right)$$

is defined as follows.

The map Φ is the composition of maps $\mathfrak{A} \xrightarrow{\Phi'} \mathfrak{B} \xrightarrow{\Phi''} \mathfrak{C}$. Other components are

$$\alpha_{F,G} = \alpha''_{\Phi' F,\Phi' G} \bullet_v \Phi''(\alpha'_{F,G}) = \begin{array}{ccc} \Phi''\Phi' A & \xrightarrow{\Phi''\Phi' F \bullet_h \Phi''\Phi' G} & \Phi''\Phi' C \\ \Vert & \alpha''_{\Phi' F,\Phi' G} & \Vert \\ \Phi''\Phi' A & \xrightarrow{\Phi''(\Phi' F \bullet_h \Phi' G)} & \Phi''\Phi' C \\ \Vert & \Phi''(\alpha'_{F,G}) & \Vert \\ \Phi''\Phi' A & \xrightarrow{\Phi''\Phi'(F \bullet_h G)} & \Phi''\Phi' C \end{array}$$

$$\beta_{F,G} = \beta''_{\Phi' F,\Phi' G} \bullet_h \Phi''(\beta'_{F,G}) =$$

$$= \begin{array}{ccccc} \Phi''\Phi' A & = & \Phi''\Phi' A & = & \Phi''\Phi' A \\ {\scriptstyle \Phi''\Phi' F \bullet_v \Phi''\Phi' G}\downarrow & \beta''_{\Phi' F,\Phi' G} & \downarrow{\scriptstyle \Phi''(\Phi' F \bullet_v \Phi' G)} & \Phi''(\beta'_{F,G}) & \downarrow{\scriptstyle \Phi''\Phi'(F \bullet_h G)} \\ \Phi''\Phi' C & = & \Phi''\Phi' C & = & \Phi''\Phi' C \end{array}$$

The proof that these 2-cells satisfy to cocycle and naturality conditions is left to the reader.

B.2.2 Functors from double categories to 2-categories

A particular case, when the target double category is the category of quintets, is especially interesting for us. Let us specialize the general definition to this case. Let $\mathfrak{D}$ be a double category, and let $\mathfrak{A}$ be a 2-category. A *(double) pseudofunctor* $\Phi = (\Phi, \alpha, \beta) : \mathfrak{D} \to Q\mathfrak{A}$ is a triple consisting of

- a map $\Phi : \mathfrak{D} \to Q\mathfrak{A}$ commuting with the maps source_h, target_h, source_v and target_v,
- a 2-isomorphism $\alpha_{F,G} : \Phi(F) \circ \Phi(G) \xrightarrow{\sim} \Phi(F \circ_h G)$ is given for any pair of $\circ_h$-composable horizontal 1-morphisms F and G,
- a 2-isomorphism $\beta_{F,G} : \Phi(F) \circ \Phi(G) \xrightarrow{\sim} \Phi(F \circ_v G)$ is given for any pair of $\circ_v$-composable vertical 1-morphisms F and G,

such that

(i) $\alpha_{F,G} = 1$ whenever F or G is an object;
(ii) $\beta_{F,G} = 1$ whenever F or G is an object;
(iii) for any three $\circ_h$-composable horizontal 1-morphisms $\mathcal{A} \xrightarrow{F} \mathcal{B} \xrightarrow{G} \mathcal{C} \xrightarrow{H} \mathcal{D}$ there is a commutative tetrahedron:

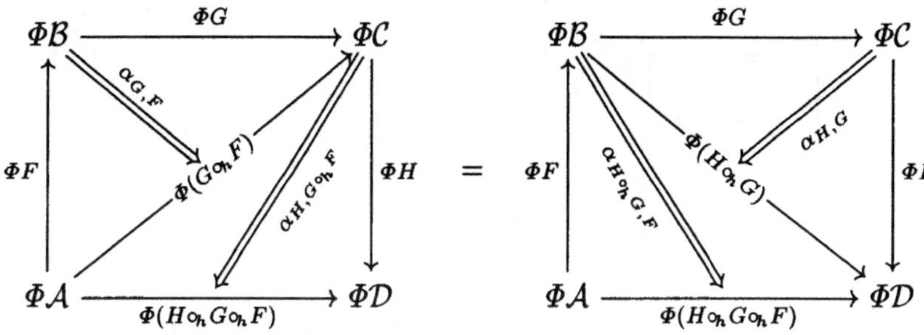

(iv) similar axiom with horizontal, $\circ_h$, α replaced with vertical, $\circ_v$, β;

(v) for any pair (a, b) of $\circ_h$-composable 2-morphisms

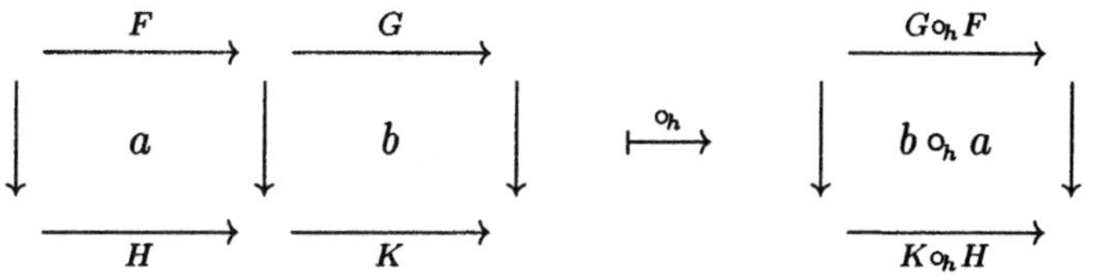

we have a commutative prism:

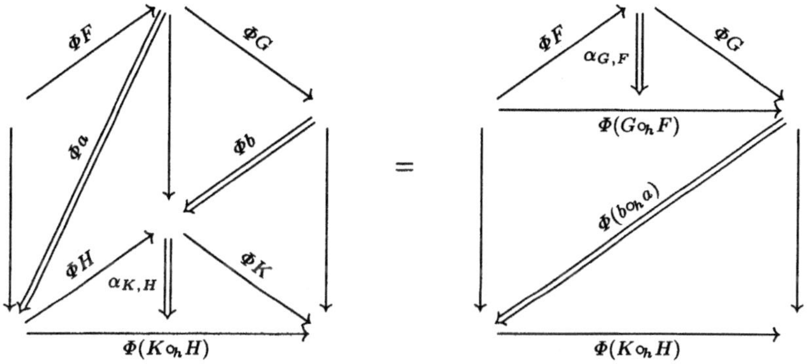

(vi) for any pair (a, b) of $\circ_v$-composable 2-morphisms

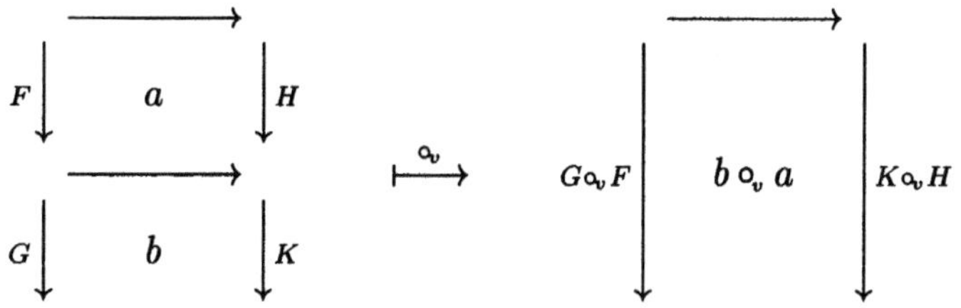

we have a commutative prism:

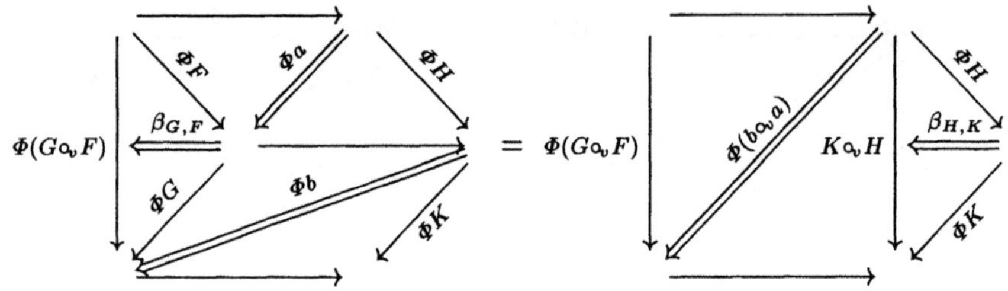

Proposition B.2.3. *Suppose that $\mathfrak{D}$ is a double category, $\mathfrak{A}$ and $\mathfrak{B}$ are 2-categories, $(\Phi, \alpha, \beta) : \mathfrak{D} \to \mathcal{Q}\mathfrak{A}$ is a double pseudofunctor, and $(\Psi, \delta) : \mathfrak{A} \to \mathfrak{B}$ is a 2-pseudofunctor. Then the composite $(\Xi, \alpha', \beta') : \mathfrak{D} \to \mathcal{Q}\mathfrak{B}$ is a double functor given by the following pastings:*

1.

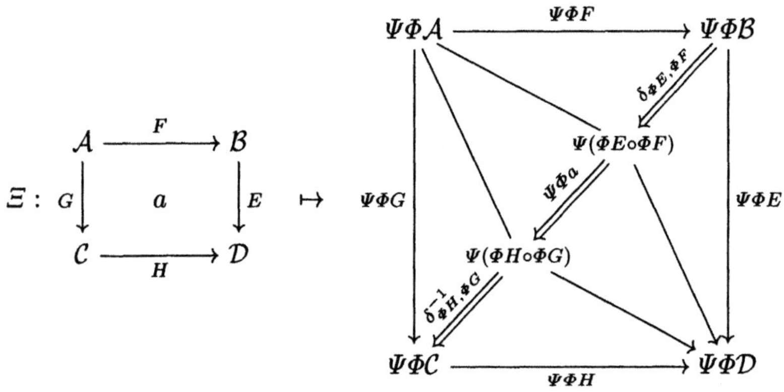

2. *For any pair of $\circ_h$-composable horizontal 1-morphisms $\mathcal{A} \xrightarrow{G} \mathcal{B} \xrightarrow{F} \mathcal{C}$ the 2-isomorphism $\alpha'_{F,G}$ is given by the pasting*

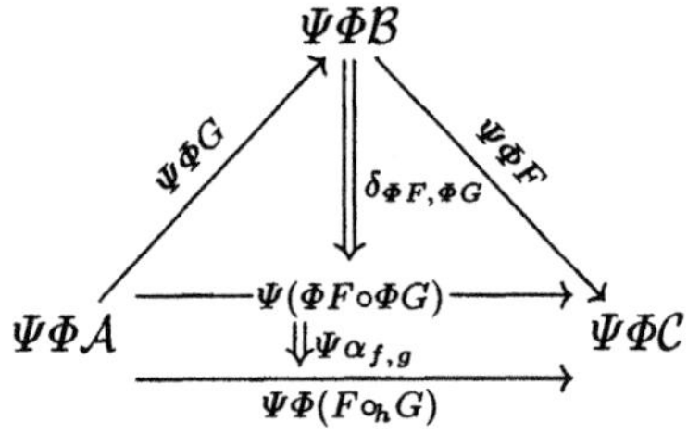

3. *$\beta'_{F,G}$ is similar to the above with horizontal, $\circ_h$, $\Psi\alpha_{F,G}$ replaced with vertical, $\circ_v$, $\Psi\beta_{F,G}$.*

Proof. The proof is analogous to the case of 2-pseudofunctors [Bén67].

Corollary B.2.4. *Let $\mathfrak{A}$ and $\mathfrak{B}$ be 2-categories. Let $(\Psi, \delta) : \mathfrak{A} \to \mathfrak{B}$ be a 2-pseudofunctor, $\delta_{F,G} : \Psi(F) \circ \Psi(G) \xrightarrow{\sim} \Psi(F \circ G)$. Then there is a double pseudofunctor $(\Phi, \delta, \delta) : \mathcal{Q}\mathfrak{A} \to \mathcal{Q}\mathfrak{B}$*

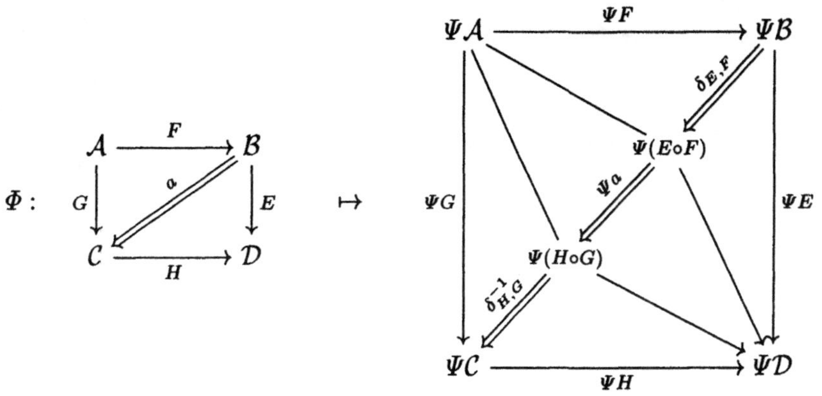

Proof. Apply Proposition B.2.3 to the identity double (pseudo)functor $(\Phi, \alpha, \beta) = (\mathrm{Id}, \mathbb{1}, \mathbb{1}) : \mathcal{Q}\mathfrak{A} \to \mathcal{Q}\mathfrak{A}$.

B.2.3 Double transformations

There are two kinds of natural transformations between double pseudofunctors, namely, horizontal and vertical ones. Let us consider here only the horizontal transformations. The vertical ones are defined similarly.

Definition B.2.5. *Let $(\Phi, \alpha^{\Phi}, \beta^{\Phi})$ and $(\Psi, \alpha^{\Psi}, \beta^{\Psi})$ be double functors $\mathfrak{C} \to \mathfrak{D}$. A* horizontal transformation $\lambda : \Phi \xrightarrow[hor]{} \Psi : \mathfrak{C} \to \mathfrak{D}$ *consists of*

(i) a horizontal 1-morphism $\lambda : \Phi\mathcal{A} \xrightarrow[hor]{} \Psi\mathcal{A}$ for every object $\mathcal{A}$ of $\mathfrak{C}$;
(ii) a 2-cell λ_F for each vertical 1-morphism $F : \mathcal{A} \xrightarrow[ver]{} \mathcal{B}$

$$\begin{array}{ccc} \Phi\mathcal{A} & \xrightarrow{\lambda_{\mathcal{A}}} & \Psi\mathcal{A} \\ {\scriptstyle \Phi F}\downarrow & \lambda_F & \downarrow{\scriptstyle \Psi F} \\ \Phi\mathcal{B} & \xrightarrow{\lambda_{\mathcal{B}}} & \Psi\mathcal{B} \end{array} ,$$

such that

(a) if $F = \mathcal{A} = \mathcal{B}$ is an object, then $\lambda_F = \lambda_{\mathcal{A}}$ is a horizontal 1-morphism;
(b) for every 2-cell a of $\mathfrak{C}$

$$\begin{array}{ccc} \mathcal{A} & \xrightarrow{H} & \mathcal{C} \\ {\scriptstyle F}\downarrow & a & \downarrow{\scriptstyle G} \\ \mathcal{B} & \xrightarrow[K]{} & \mathcal{D} \end{array}$$

naturality property holds

$$
\begin{array}{ccccc}
\Phi\mathcal{A} & \xrightarrow{\Phi H} & \Phi\mathcal{C} & \xrightarrow{\lambda_{\mathcal{C}}} & \Psi\mathcal{C} \\
{\scriptstyle \Phi F}\downarrow & \Phi a & \downarrow{\scriptstyle \Phi G} & \lambda_G & \downarrow{\scriptstyle \Psi G} \\
\Phi\mathcal{B} & \xrightarrow[\Psi K]{} & \Phi\mathcal{D} & \xrightarrow[\lambda_{\mathcal{D}}]{} & \Psi\mathcal{D}
\end{array}
\quad = \quad
\begin{array}{ccccc}
\Phi\mathcal{A} & \xrightarrow{\lambda_{\mathcal{A}}} & \Psi\mathcal{A} & \xrightarrow{\Psi H} & \Psi\mathcal{C} \\
{\scriptstyle \Phi F}\downarrow & \lambda_F & \downarrow{\scriptstyle \Psi F} & \Psi a & \downarrow{\scriptstyle \Psi G} \\
\Phi\mathcal{B} & \xrightarrow[\lambda_{\mathcal{B}}]{} & \Psi\mathcal{B} & \xrightarrow[\Psi K]{} & \Psi\mathcal{D}
\end{array}
$$

(c) for $\mathcal{A} \xrightarrow[ver]{F} \mathcal{B} \xrightarrow[ver]{G} \mathcal{C}$ *we have*

$$
\begin{array}{ccccc}
\Phi\mathcal{A} & \xrightarrow{\lambda_{\mathcal{A}}} & \Psi\mathcal{A} & = & \Psi\mathcal{A} \\
{\scriptstyle \Phi F}\downarrow & \lambda_F & \downarrow{\scriptstyle \Psi F} & & | \\
\Phi\mathcal{B} & \xrightarrow{\lambda_{\mathcal{B}}} & \Psi\mathcal{B} & \beta^{\Psi}_{F,G} & {\scriptstyle \Psi(F\bullet_v G)} \\
{\scriptstyle \Phi G}\downarrow & \lambda_G & \downarrow{\scriptstyle \Psi G} & & \downarrow \\
\Phi\mathcal{C} & \xrightarrow{\lambda_{\mathcal{C}}} & \Psi\mathcal{C} & = & \Psi\mathcal{C}
\end{array}
\quad = \quad
\begin{array}{ccccc}
\Phi\mathcal{A} & = & \Phi\mathcal{A} & \xrightarrow{\lambda_{\mathcal{A}}} & \Psi\mathcal{A} \\
{\scriptstyle \Phi F}\downarrow & & | & & | \\
\Phi\mathcal{B} & \beta^{\Phi}_{F,G} & {\scriptstyle \Phi(F\bullet_v G)} & \lambda_{F\bullet_v G} & {\scriptstyle \Psi(F\bullet_v G)} \\
{\scriptstyle \Phi G}\downarrow & & \downarrow & & \downarrow \\
\Phi\mathcal{C} & = & \Phi\mathcal{C} & \xrightarrow{\lambda_{\mathcal{C}}} & \Psi\mathcal{C}
\end{array}
$$

Definition B.2.6. *A* modification *of horizontal transformations* $m : \lambda \to \mu : \phi \to \psi : \mathcal{A} \to \mathcal{B}$ *is such a collection of 2-morphisms*

$$
\begin{array}{ccc}
\Phi\mathcal{A} & \xrightarrow{\lambda_{\mathcal{A}}} & \Psi\mathcal{A} \\
\| & m_{\mathcal{A}} & \| \\
\Phi\mathcal{A} & \xrightarrow[\mu_{\mathcal{A}}]{} & \Psi\mathcal{A}
\end{array}
$$

that for every vertical 1-morphism $F : \mathcal{A} \xrightarrow[ver]{} \mathcal{B}$ *we have* $m_{\mathcal{A}} \bullet_v \mu_F = \lambda_F \bullet_v m_{\mathcal{B}}$, *or, in detail,*

$$
\begin{array}{ccc}
\Phi\mathcal{A} & \xrightarrow{\lambda_{\mathcal{A}}} & \Psi\mathcal{A} \\
\| & m_{\mathcal{A}} & \| \\
\Phi\mathcal{A} & \xrightarrow{\mu_{\mathcal{A}}} & \Psi\mathcal{A} \\
{\scriptstyle \Phi F}\downarrow & \mu_F & \downarrow{\scriptstyle \Psi F} \\
\Phi\mathcal{B} & \xrightarrow[\mu_{\mathcal{B}}]{} & \Psi\mathcal{B}
\end{array}
\quad = \quad
\begin{array}{ccc}
\Phi\mathcal{A} & \xrightarrow{\lambda_{\mathcal{A}}} & \Psi\mathcal{A} \\
{\scriptstyle \Phi F}\downarrow & \lambda_F & \downarrow{\scriptstyle \Psi F} \\
\Phi_{\mathcal{B}} & \xrightarrow{\lambda_{\mathcal{B}}} & \Psi\mathcal{B} \\
\| & m_{\mathcal{B}} & \| \\
\Phi\mathcal{B} & \xrightarrow[\mu_{\mathcal{B}}]{} & \Psi\mathcal{B}
\end{array}
\quad .
$$

C. Thick tangles

C.1 Monoidal bicategory of thick tangles

It is a rather involved task to describe the relations between isomorphisms between various functors which use coends. To make them accessible we introduce the monoidal bicategory of thick tangles as a kind of a free monoidal bicategory with duals generated by a self-dual object. An analogy would be to introduce the category of unoriented tangles as the free braided category, generated by a self-dual object. Naturally, the bicategory of thick tangles is much simpler than its braided counterpart – the category of 2-tangles proposed by Baez and Langford [BL98a]. They prove in [BL98b] that the category of 2-tangles is a free semistrict *braided* monoidal 2-category with duals on one unframed self-dual object. Very similar geometric 2-categories appear in the theory of knotted surfaces in 4-dim space as developed by Carter, Saito, Fischer, and others, see for example [CRS97], [CS98], and [Fi94].

The category of tangles can be presented by generators and relations. Similarly, the monoidal bicategory of thick tangles is presented by generators, which include the generator-object 1, 1-morphism generators, 2-morphism generators between compositions of a few 1-morphism generators, and, finally, there is a list of relations between 2-morphism generators. This list of relations enables us to present this monoidal category combinatorially. At the next step we modify the set of 1-generators. That allows us to reduce the list of 2-generators and the list of relations. The latter form of the category (equivalent to the previous ones) is especially easy to represent in the 2-category of abelian categories. To give such a representation it suffices to give a bounded abelian monoidal category. To achieve this result we prove a few lemmas about coends.

Definition C.1.1. *The monoidal bicategory of Planar Thick Tangles $\mathcal{PTT}$ has non-negative integers as objects. The 1-morphisms from k to l are smooth oriented compact surfaces X with boundary ∂X equipped with disjoint distinguished intervals $i^s_j : I \hookrightarrow \partial X$, $1 \leqslant j \leqslant k$, $i^t_m : I \hookrightarrow \partial X$, $1 \leqslant m \leqslant l$, which admits a smooth embedding $d : X \hookrightarrow \mathbb{R} \times [0,1]$ such that*

$$d^{-1}(\mathbb{R}\times 0) = I^s_1 \sqcup I^s_2 \sqcup \cdots \sqcup I^s_k, \quad I^s_j = i^s_j(I), \quad d(I^s_j) = [j - \tfrac{1}{3}, j + \tfrac{1}{3}] \times 0,$$

$$d^{-1}(\mathbb{R}\times 1) = I^t_1 \sqcup I^t_2 \sqcup \cdots \sqcup I^t_l, \quad I^t_j = i^t_j(I), \quad d(I^t_j) = [j - \tfrac{1}{3}, j + \tfrac{1}{3}] \times 1.$$

The image $d(X)$ is called a diagram of a thick tangle.

The 2-morphisms $\phi : X \to Y : k \to l$ of the bicategory $\mathcal{PTT}$ are isotopy classes of oriented homeomorphisms $f : X \to Y$, which preserve the distinguished intervals, i.e., $i^{s/t}_{jY} = \big(I \xrightarrow{i^{s/t}_{jX}} X \xrightarrow{f} Y\big)$. (Each homeomorphism $f_t : X \to Y$, $t \in [0,1]$ in the isotopy family also preserves distinguished intervals.) Composition $Y \circ X$ of 1-morphisms $k \xrightarrow{X} l \xrightarrow{Y} m$ is defined by sewing of surfaces at boundary intervals $I^t_j(X)$ and $I^s_j(Y)$. The unit 1-morphism $1_k : k \to k$ is the union $\bigsqcup_{j=1}^k [j - \frac{1}{3}, j + \frac{1}{3}] \times [0,1]$. The isomorphisms $1_l \circ X \xrightarrow{\sim} X$ and $X \circ 1_k \xrightarrow{\sim} X$ are obtained by taking a neighborhood $(U, I^s_j) \simeq ([0,1] \times [0,1[, [0,1] \times 0)$ of the distinguished interval $I^s_j \subset X$ and by taking any isomorphism $[0,1] \times [0,1] \bigcup_{[0,1]\times 1 \sim I^s_j} U \simeq U$. The tensor product is the disjoint union. The unit object is 0. The associativity constraints are obvious.

Example C.1.2. Let Σ be a disk with n holes and with $k + l > 0$ distinguished intervals. Then $\mathrm{Aut}_{\mathcal{PTT}}\, \Sigma \simeq B_n$.

We wish to present the monoidal bicategory $\mathcal{PTT}$ by tensor generators and relations. The generating object is 1.

When we consider 1-morphisms, we may assume that the diagrams $d : X \hookrightarrow \mathbb{R} \times [0,1]$ are in generic position. In particular, the function $pr_2 \circ d : \partial X \to [0,1]$ is a Morse function. Cutting X in pieces corresponding to level lines of the function $pr_2 \circ d : X \to [0,1]$ results in decomposing X into a product of elementary pieces

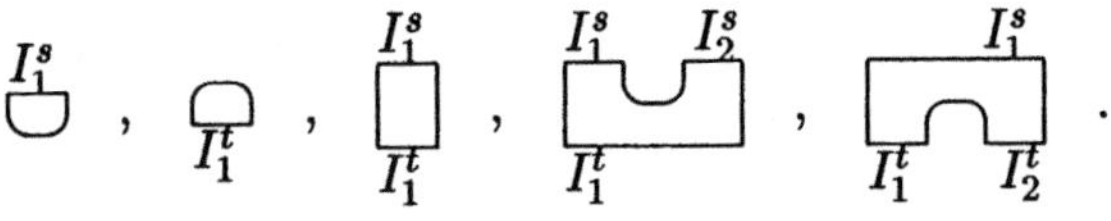

(C.1.1)

When we consider 2-morphisms, we deal with generic families of diagrams $d_t : X \hookrightarrow \mathbb{R} \times [0,1]$, $t \in [0,1]$. So we may assume that all d_t are as above except for finite number of t, where $pr_2 \circ d : \partial X \to [0,1]$ has a codimension 1 singularity. This corresponds to two types of singularities:

– coincidence of critical values

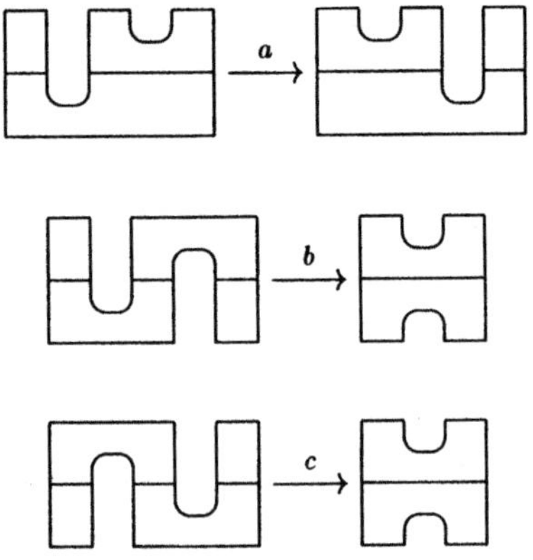

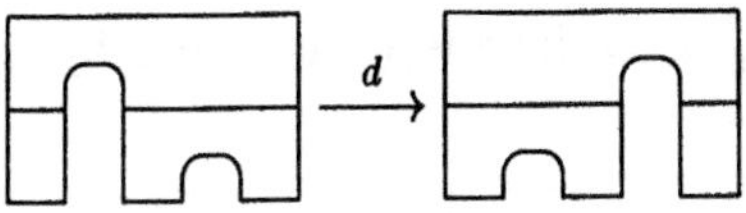

and 4 kinds of morphisms $\boxtimes_{\sqcup,\sqcup}$, $\boxtimes_{\sqcup,\sqcap}$, $\boxtimes_{\sqcap,\sqcup}$, $\boxtimes_{\sqcap,\sqcap}$:

$\boxtimes_{\sqcup,\sqcup}$:

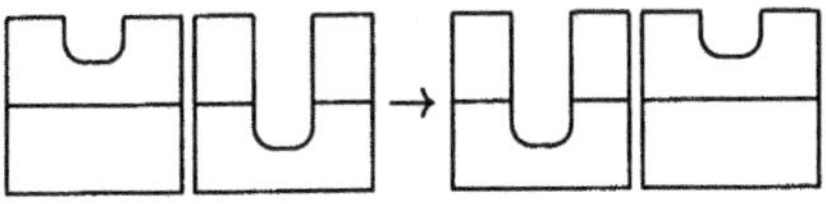

$\boxtimes_{\sqcup,\sqcap}$:

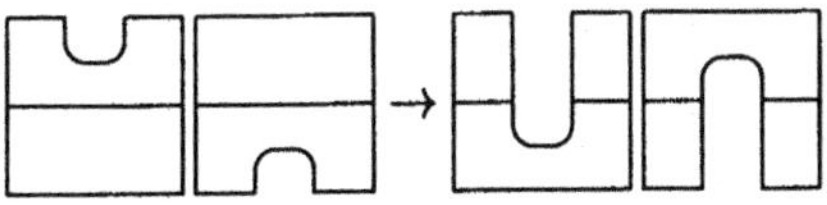

$\boxtimes_{\sqcap,\sqcup}$:

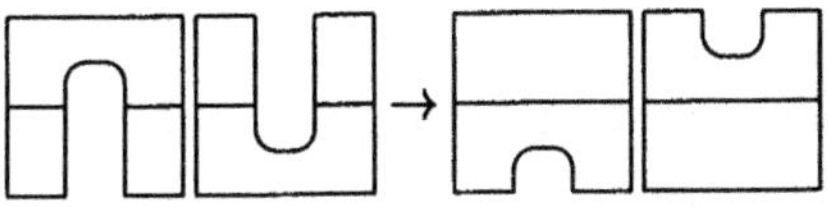

$\boxtimes_{\sqcap,\sqcap}$:

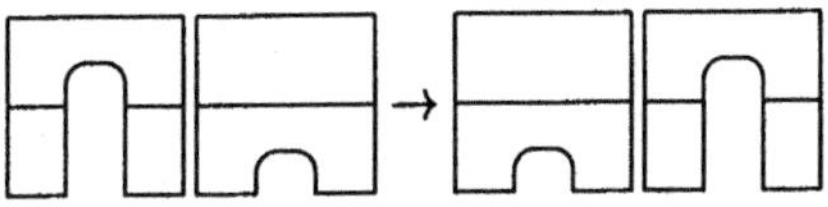

describing moving one extremum of the boundary curve past another extremum;
– birth–death of a pair maximum–minimum

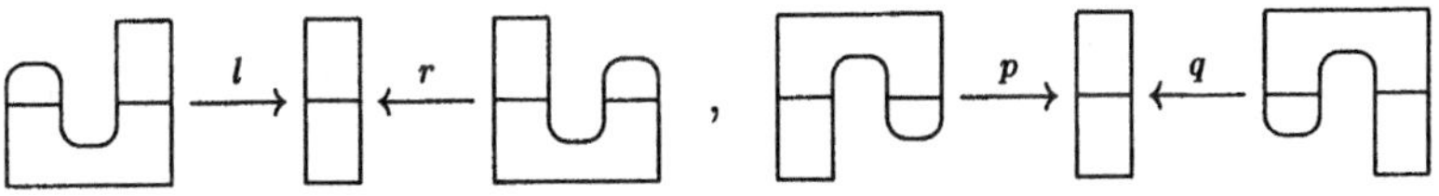

The above pictures are to be understood as follows: the same surface (disk with 2 or 4 distinguished intervals) is mapped to $\mathbb{R} \times [0, 1]$ in many different ways. The picture is the image of this mapping. The levels separating critical values are shown explicitly. This must be kept in mind when dealing with non-simply connected surfaces.

By Cerf theory, all relations between the above 2-morphisms follow from the 20 equations below:

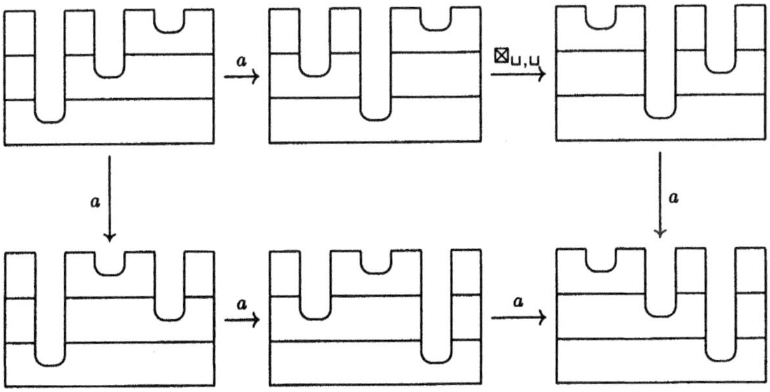

an equation concerning d obtained from the above by reflection with respect to horizontal axis,

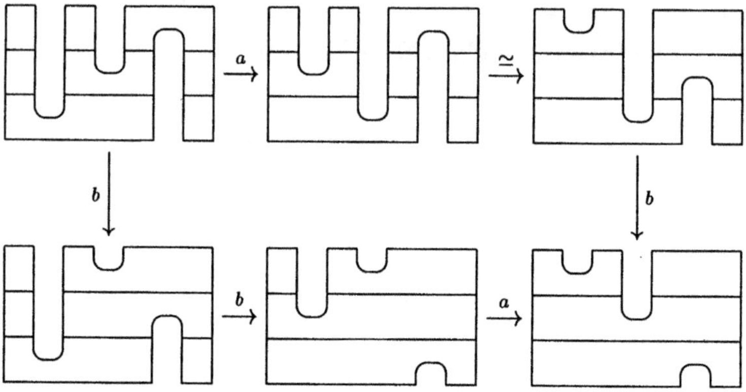

3 more equations obtained from the above by reflections with respect to horizontal and vertical axes,

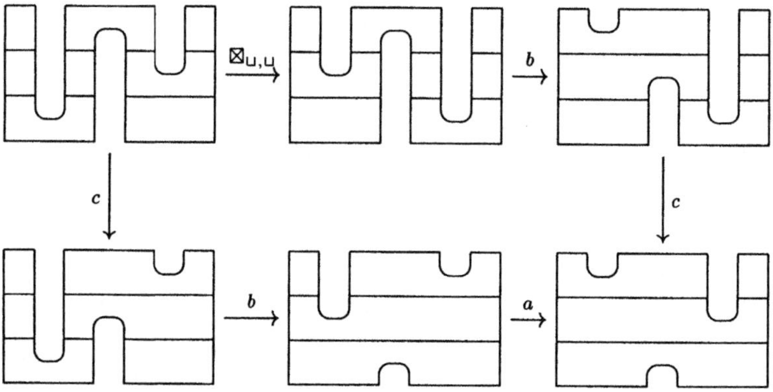

one more equation obtained from the above by reflection with respect to horizontal axis,

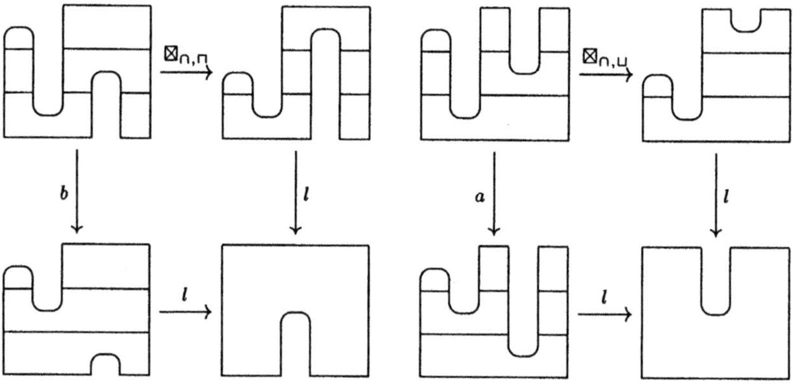

together with 6 more equations obtained by reflections of each from the above with respect to horizontal and vertical axes,

$a =$ 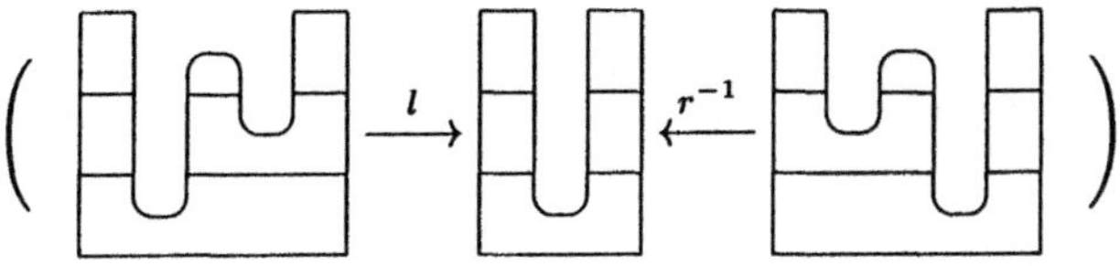

one more equation obtained from the above by reflection with respect to horizontal axis,

$r =$ 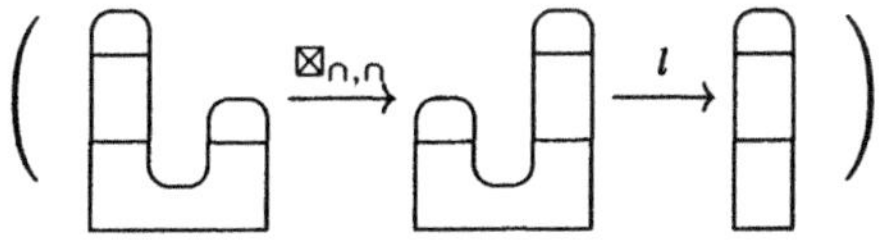

and, finally, one more equation obtained from the above by reflection with respect to the horizontal axis.

We get:

Proposition C.1.3. *The monoidal bicategory $\mathcal{PTT}$ is triequivalent to the monoidal bicategory $\mathcal{MPTT}$ of Marked Planar Thick Tangles, whose objects are non-negative integers, 1-morphisms are planar thick tangles X with marking — isotopy class of collections of lines drawn on X with endpoints in $\partial X \cup_j I_j^{s/t}$, cutting X into the pieces (C.1.1). The 2-morphisms are generated by the marking preserving homeomorphisms by a, b, c, d, l, r, p, q, $\boxtimes_{\sqcup,\sqcup}$, $\boxtimes_{\sqcup,\sqcap}$, $\boxtimes_{\sqcap,\sqcup}$ and $\boxtimes_{\sqcap,\sqcap}$, subject to the above 20 relations and to relations which say that $\boxtimes_{-,-}$ are structural morphisms of the bicategory.*

Definition C.1.4. *The semistrict monoidal 2-category $\mathcal{PTG}$ of Planar Thick Graphs is defined as follows. The objects are non-negative integers, 1-morphisms $k \to l$ are planar graphs oriented by an ordering of vertices having valency 1,2 or 3, with k incoming and l outgoing legs. Allowed orientations around a vertex are*

and each 3-vertex is equipped with a cyclic order on the set of adjacent edges, which will always coincide with the orientation of the plane. Furthermore, a graph defines a 1-morphism only if it can be drawn in the strip $\mathbb{R} \times [0,1]$ so that ends of incoming legs were on $\mathbb{R} \times 0$, ends of outgoing legs were on $\mathbb{R} \times 1$, and all edges had positive projections on $[0,1]$. The composition of 1-morphisms is defined in an obvious way with the following modification: composition of any graph X with an identity morphism equals X (extra vertices are removed).

The 2-morphisms are generated over isomorphisms of oriented graphs by

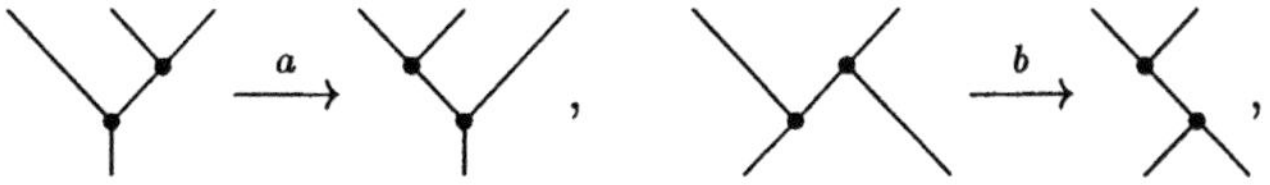

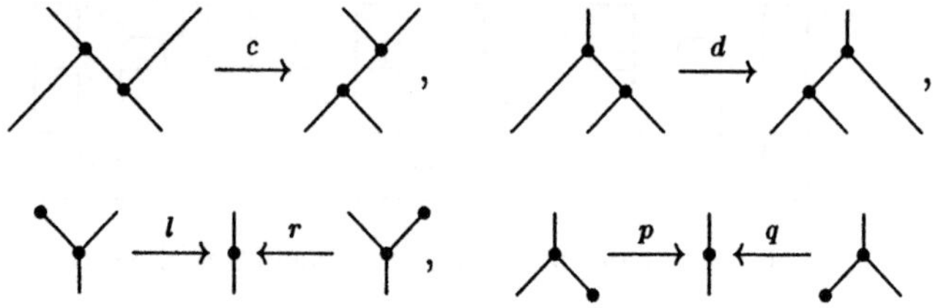

and by elementary transpositions of the order of vertices of the form

$\boxtimes_{\sqcup,\sqcup}$: 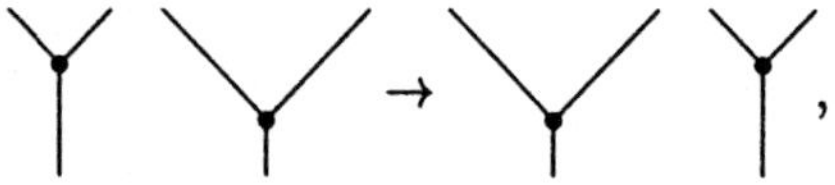 ,

$\boxtimes_{\sqcup,\sqcap}$:  ,

$\boxtimes_{\sqcap,\sqcup}$:  ,

$\boxtimes_{\sqcap,\sqcap}$: 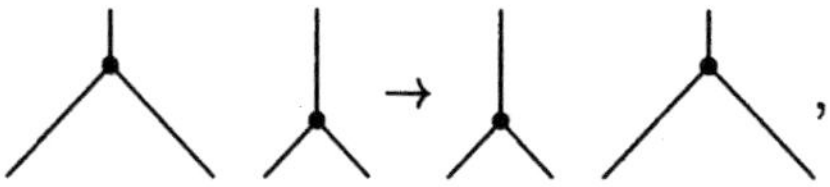 ,

subject to 20 relations:

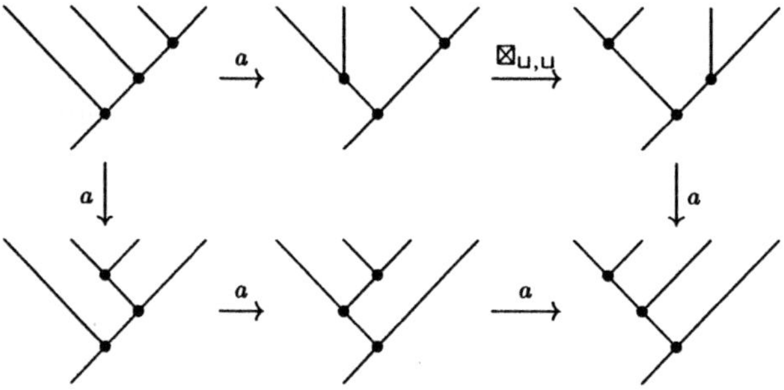

(C.1.2)

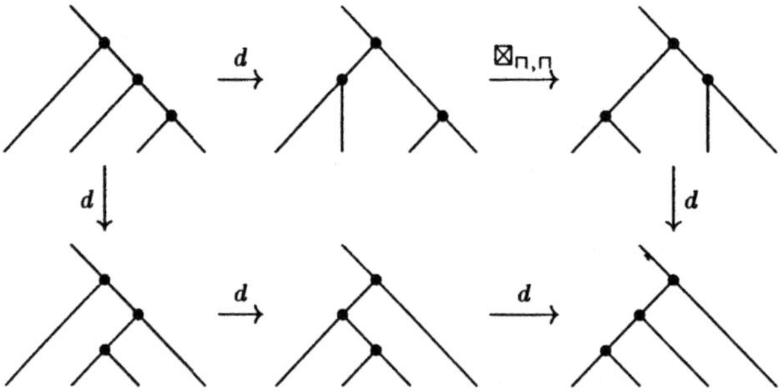

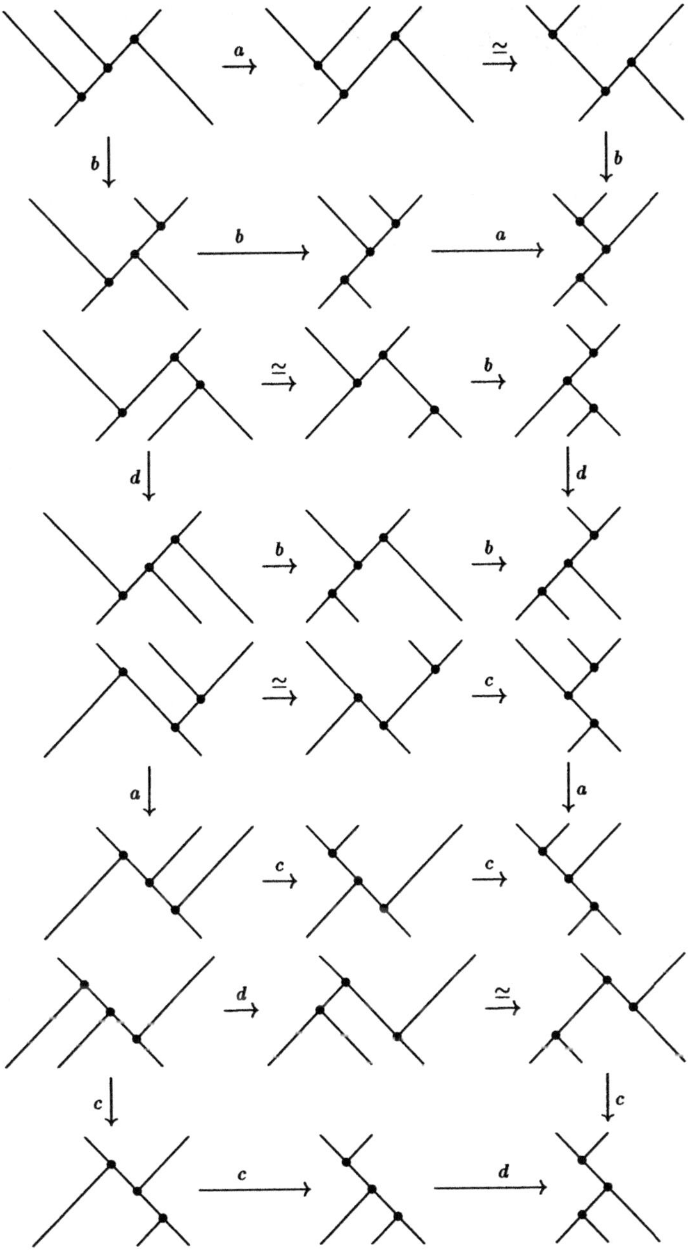

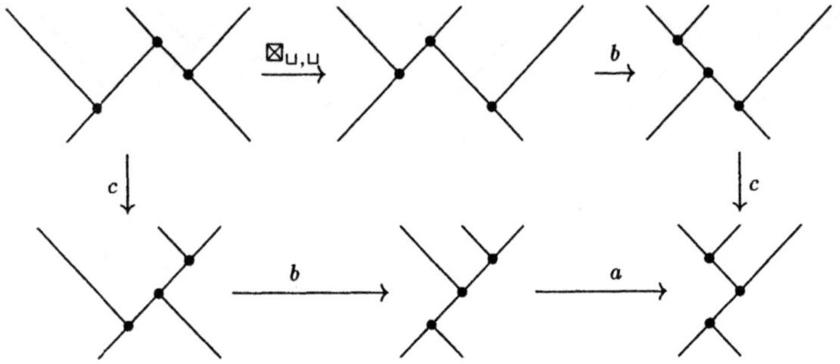

$\boxtimes_{\sqcup,\sqcup}$
b
c
c
b
a

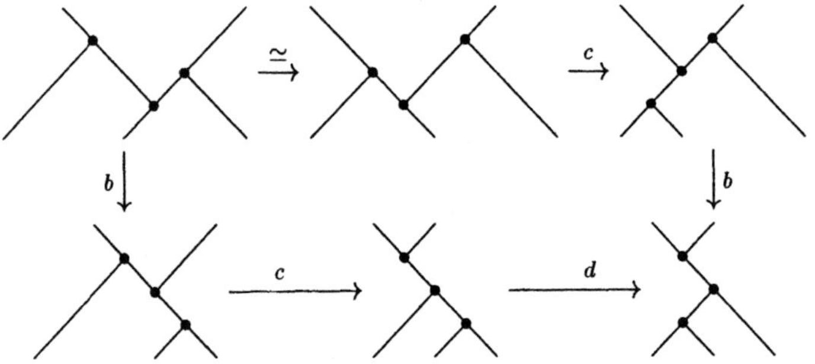

$\simeq$
c
b
b
c
d

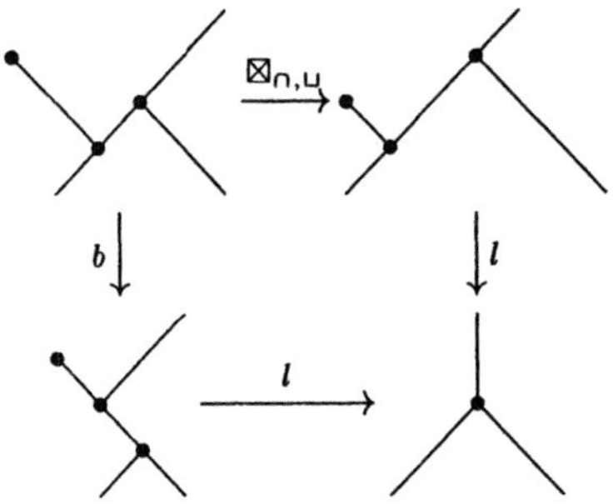

$\boxtimes_{\cap,\sqcup}$
b
l
l

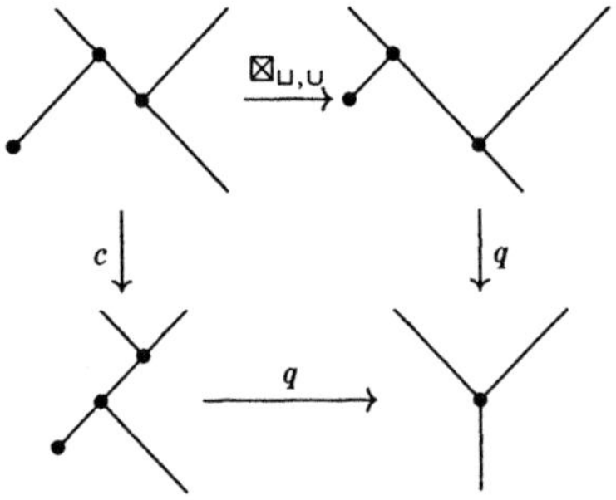

$\boxtimes_{\sqcup,\cup}$
c
q
q

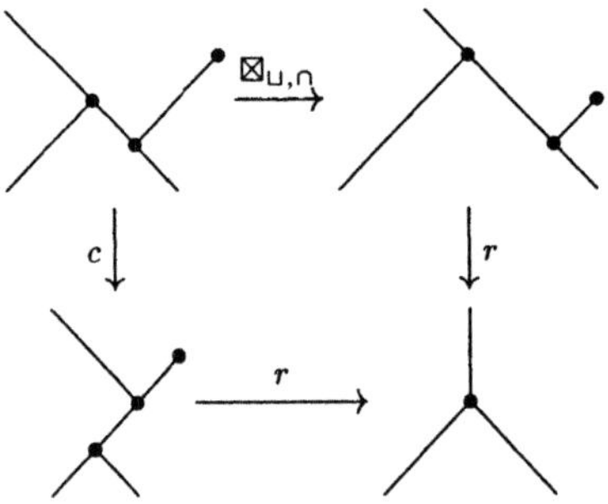

$\boxtimes_{\sqcup,\cap}$
c
r
r

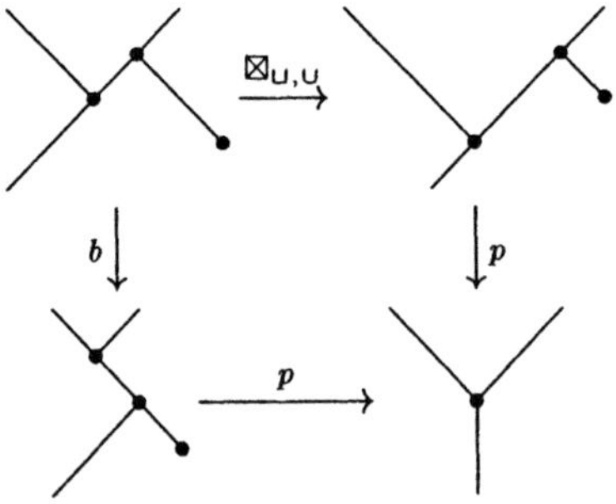

$\boxtimes_{\sqcup,\cup}$
b
p
p

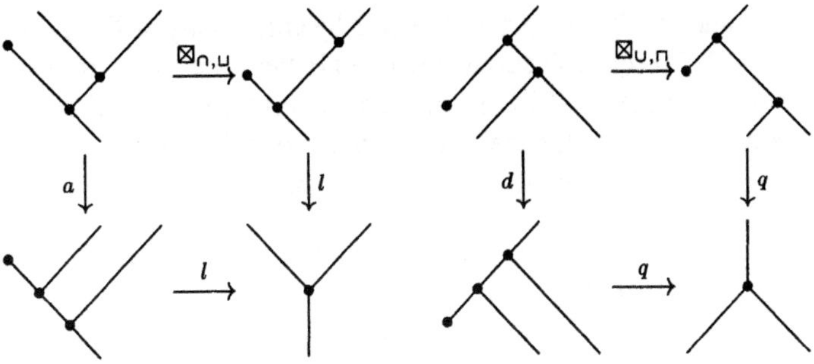

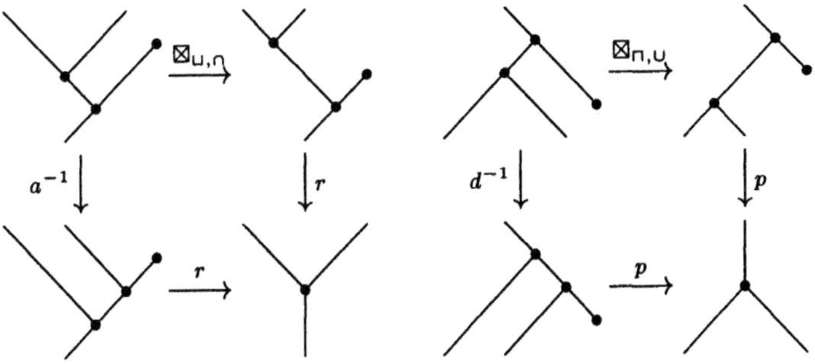

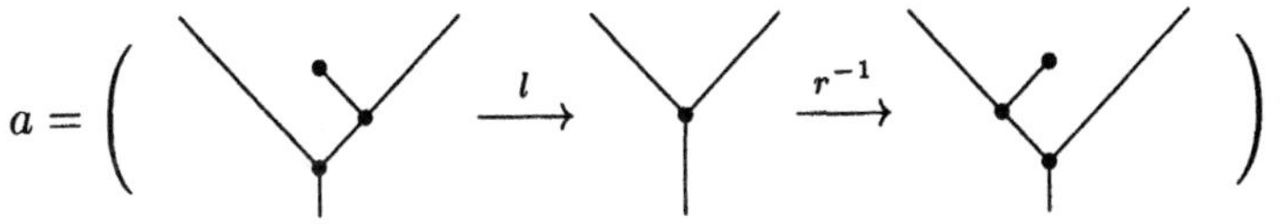

(C.1.3)

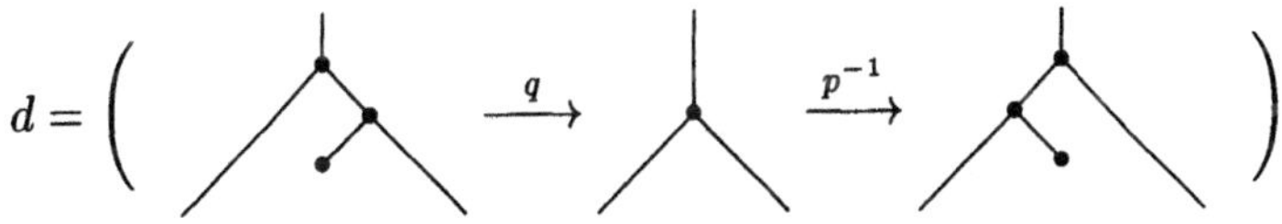

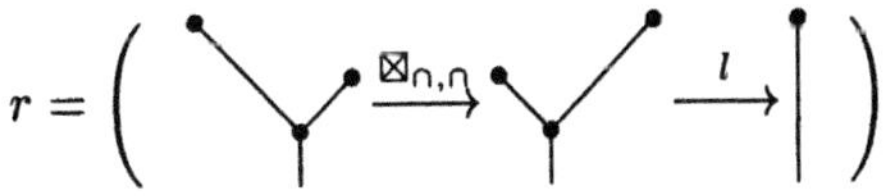

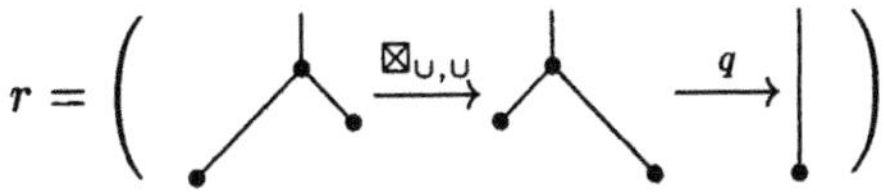

as well as to relations required by setting $\boxtimes_{-,-}$ *to structure morphisms of the monoidal 2-category.*

Proposition C.1.5. *The monoidal bicategories* $\mathcal{MPTT}$ *and* $\mathcal{PTG}$ *are triequivalent.*

Proof. The trifunctor $\mathcal{PTG} \to \mathcal{MPTT}$ exists by universality of $\mathcal{PTG}$. The trifunctor $\mathcal{MPTT} \to \mathcal{PTG}$ sends all surfaces marked in the same way to the same object, determined by the marking. That is, all homeomorphisms from $\mathcal{MPTT}$, which preserve the marking, go to the identity morphism of a thick graph. The generators a, b, c, d, l, r, p, q go to respective a, b, c, d, l, r, p, q.

For a connected graph X_g of genus g with non-zero number of incoming and outgoing legs $\operatorname{Aut} X_g \simeq B_g$.

From the practical point of view this system of generators and relations is not the best. As we can choose

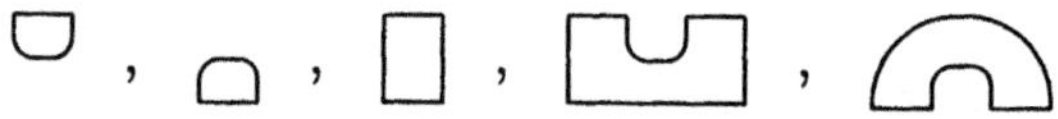

for the set of generators of $\mathcal{MPTT}$, we introduce another category of planar graphs.

Definition C.1.6. *The semistrict monoidal 2-category $\mathcal{UPTG}$ of Useful Planar Thick Graphs has non-negative integers as objects and the same 1-morphisms as $\mathcal{PTG}$ except that the allowed orientations of adjacent edges to a vertex are*

2-morphisms are generated by the isomorphisms

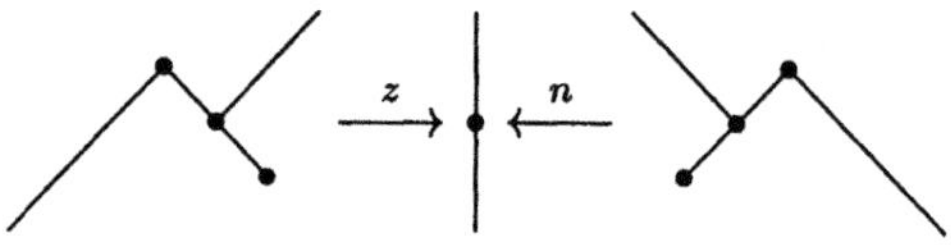

subject to (C.1.2), (C.1.3) *and the following equations*

$\boxtimes_{\wedge,\wedge} =$

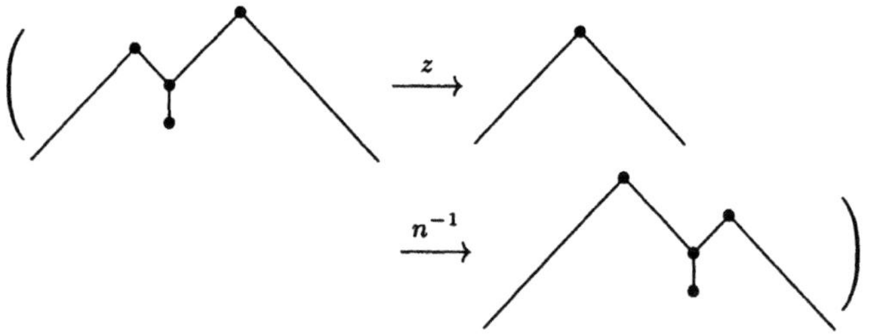

(C.1.4)

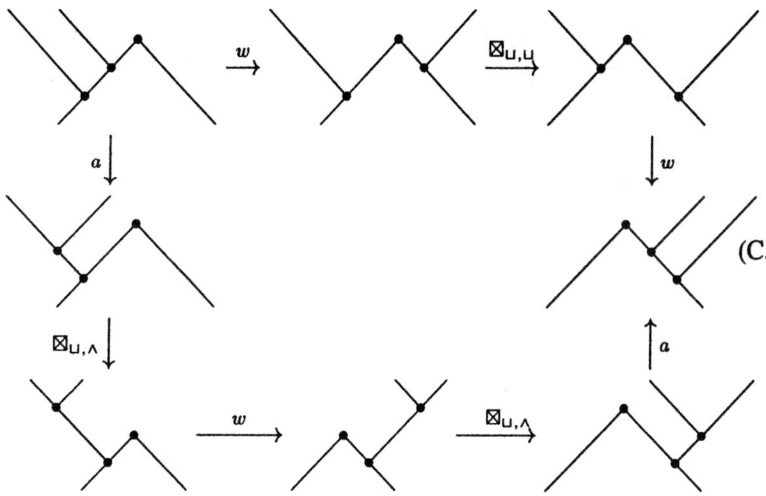

(C.1.5)

where

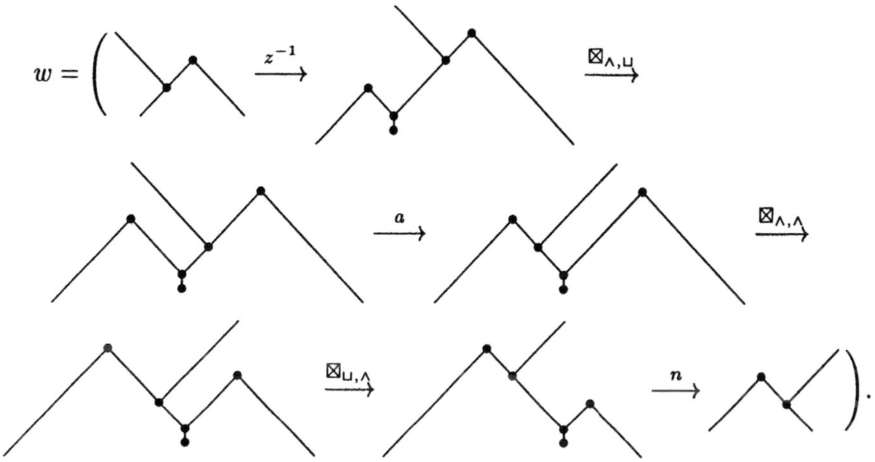

Theorem C.1.7. *The semistrict monoidal 2-categories $\mathcal{PTG}$ and $\mathcal{UPTG}$ are triequivalent. There are monoidal bifunctors (in the strict sense)*

$$\Phi : \mathcal{PTG} \longrightarrow \mathcal{UPTG}$$

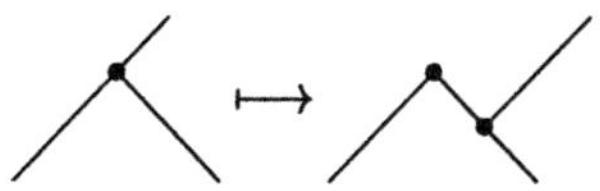

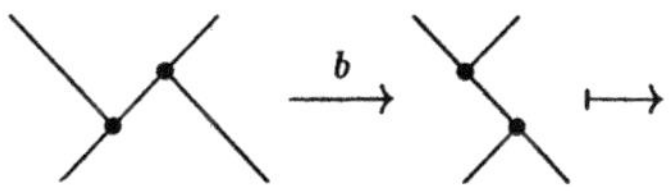

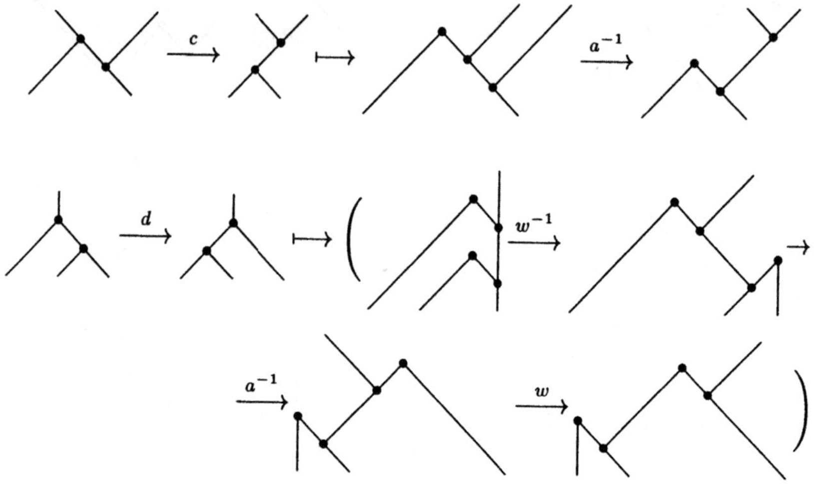

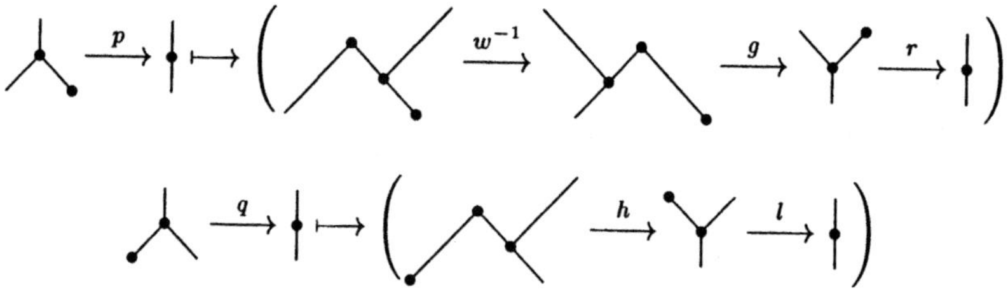

where

$g =$ 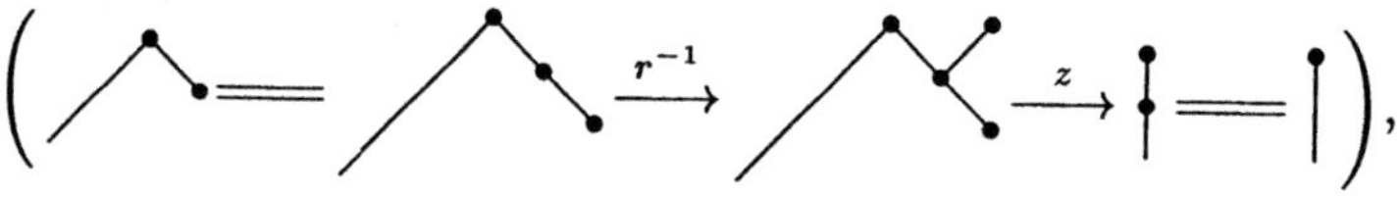

$h =$ 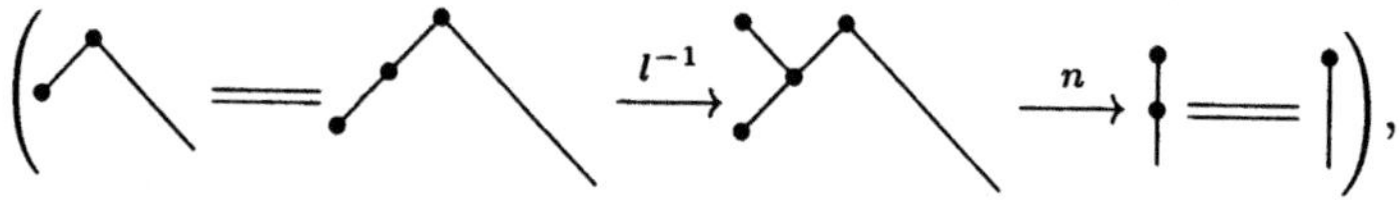

and

$$\Psi : \mathcal{UPTG} \longrightarrow \mathcal{PTG}$$

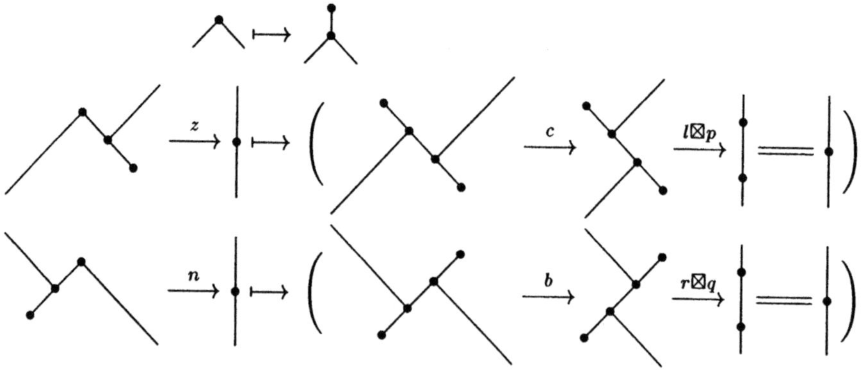

The bifunctors Φ and Ψ induce identity maps on the set of objects $\mathbb{Z}_{\geqslant 0}$. The functors $\Phi : \mathcal{PTG}(k,l) \to \mathcal{UPTG}(k,l)$ and $\Psi : \mathcal{UPTG}(k,l) \to \mathcal{PTG}(k,l)$ are equivalences of categories, and quasi-inverse to each other.

C.2 Representation of thick tangles by abelian categories

Now we associate with a rigid monoidal, bounded abelian category $\mathcal{C} = (\mathcal{C}, \otimes, \mathbf{a}, 1, \mathbf{r}, \mathbf{l})$ a bifunctor from thick graphs to bounded abelian categories.

Theorem C.2.1. *There is a trifunctor*

$$\mathcal{UPTG} \longrightarrow \mathbf{V}\text{-}\mathbf{Cat}$$

$$m \longmapsto \mathcal{C}^{\boxtimes m}$$

 $\longmapsto \mathrm{Id} : \mathcal{C} \to \mathcal{C}$

 $\longmapsto \mathrm{Unit} : \mathbb{k}\text{-vect} \to \mathcal{C}, \quad \mathbb{k} \mapsto 1$

 $\longmapsto \mathrm{Inv} : \mathcal{C} \to \mathbb{k}\text{-vect}, \quad X \mapsto \mathrm{Hom}(1, X)$

 $\longmapsto \otimes : \mathcal{C} \boxtimes \mathcal{C} \to \mathcal{C}$

 $\longmapsto \mathrm{Coend} : \mathbb{k}\text{-vect} \to \mathcal{C} \boxtimes \mathcal{C}, \ \mathbb{k} \mapsto \int^{X \in \mathcal{C}} X \boxtimes X^\vee$

 $\xrightarrow{a}$  $\longmapsto$

$$\begin{array}{ccc} \mathcal{C} \boxtimes \mathcal{C} \boxtimes \mathcal{C} & \xrightarrow{\otimes \boxtimes 1} & \mathcal{C} \boxtimes \mathcal{C} \\ {\scriptstyle 1 \boxtimes \otimes} \downarrow & \overset{\mathbf{a}}{\Longrightarrow} & \downarrow {\scriptstyle \otimes} \\ \mathcal{C} \boxtimes \mathcal{C} & \xrightarrow{\otimes} & \mathcal{C} \end{array}$$

 $\xrightarrow{l}$  $\longmapsto$

$$\begin{array}{ccc} \mathbb{k}\text{-vect} \boxtimes \mathcal{C} & =\!=\!= & \mathcal{C} \\ {\scriptstyle \mathrm{Unit} \boxtimes 1} \downarrow & \overset{\mathbf{l}}{\Longrightarrow} & \downarrow {\scriptstyle \mathrm{Id}} \\ \mathcal{C} \boxtimes \mathcal{C} & \xrightarrow{\otimes} & \mathcal{C} \end{array}$$

 $\xrightarrow{r}$  $\longmapsto$

$$\begin{array}{ccc} \mathcal{C} \boxtimes \mathbb{k}\text{-vect} & =\!=\!= & \mathcal{C} \\ {\scriptstyle 1 \boxtimes \mathrm{Unit}} \downarrow & \overset{\mathbf{r}}{\Longrightarrow} & \downarrow {\scriptstyle \mathrm{Id}} \\ \mathcal{C} \boxtimes \mathcal{C} & \xrightarrow{\otimes} & \mathcal{C} \end{array}$$

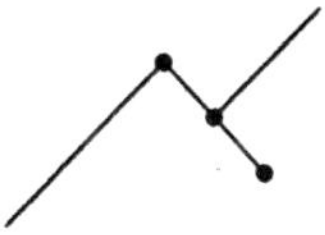 $\xrightarrow{z}$  $\longmapsto (1 \boxtimes \mathrm{Inv}) \int^{Y \in \mathcal{C}} Y \boxtimes (Y^\vee \otimes X) \xrightarrow{\mathbf{z}} X$

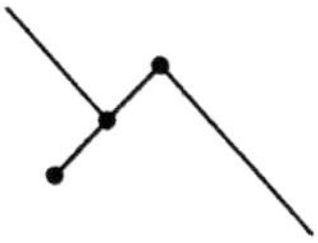 $\xrightarrow{n}$  $\longmapsto (\mathrm{Inv} \boxtimes 1) \int^{Y \in \mathcal{C}} (X \otimes Y) \boxtimes Y^\vee \xrightarrow{\mathbf{n}} X$

The proof is based on the following lemmas.

Lemma C.2.2. *The isomorphism $w : F' \to F'' : \mathcal{C} \to \mathcal{C} \boxtimes \mathcal{C}$,*

$$F'(X) = \int^{Y\in\mathcal{C}} (X \otimes Y) \boxtimes Y^\vee, \qquad F''(X) = \int^{Y\in\mathcal{C}} Y \boxtimes (Y^\vee \otimes X)$$

is given by the following morphism

$$\int^{Y\in\mathcal{C}} (X \otimes Y) \boxtimes Y^\vee \xrightarrow{1\boxtimes(1_{Y^\vee}\otimes \mathrm{coev}_X)} \int^{Y\in\mathcal{C}} (X \otimes Y) \boxtimes (Y^\vee \otimes X^\vee) \otimes X$$
$$\simeq \int^{Y\in\mathcal{C}} (X \otimes Y) \boxtimes (X \otimes Y)^\vee \otimes X \longrightarrow \int^{Z\in\mathcal{C}} Z \boxtimes (Z^\vee \otimes X).$$

In detail, the upper row in

$$\begin{array}{ccc} X \otimes Y \boxtimes Y^\vee & \xrightarrow{1\boxtimes(1_{Y^\vee}\otimes \mathrm{coev}_X)} & (X \otimes Y) \boxtimes (X \otimes Y)^\vee \otimes X \\ {\scriptstyle 1\otimes i_Y}\downarrow & & \downarrow{\scriptstyle i_{X\otimes Y}\otimes X} \\ X_{1'} \otimes \mathbb{F}_{1''2} & \xrightarrow{\quad w \quad} & \mathbb{F}_{12'} \otimes X_{2''} \end{array}$$

induces the lower row.

Proof. Let $Y \in \mathcal{C}$. The considered isomorphism is

$$w : \int^{Z\in\mathcal{C}} Y\otimes Z\boxtimes Z^\vee \to (\mathbb{1}_1\boxtimes \mathrm{Hom}_{2,3}\boxtimes \mathbb{1}_4)\Big(\int^{X\in\mathcal{C}} X\boxtimes X\Big)\boxtimes \int^{Z\in\mathcal{C}} (Y\otimes Z)\boxtimes Z^\vee$$
$$\to \int^{X\in\mathcal{C}} X \boxtimes X^\vee \otimes Y.$$

When Y is injective, $\mathrm{Hom}(X, Y \otimes Z)$ is exact in X and Z, and the above isomorphism coincides with

$$w : \int^{Z\in\mathcal{C}} Y \otimes Z \boxtimes Z^\vee \to \int^{X,Z\in\mathcal{C}} X \boxtimes \mathrm{Hom}(X, Y \otimes Z) \otimes Z^\vee$$
$$\simeq \int^{X,Z\in\mathcal{C}} X \boxtimes \mathrm{Hom}(Z^\vee, X^\vee \otimes Y) \otimes Z^\vee$$
$$\simeq \int^{X,W\in\mathcal{C}} X \boxtimes \mathrm{Hom}(W, X^\vee \otimes Y) \otimes W \simeq \int^{X\in\mathcal{C}} X \boxtimes X^\vee \otimes Y.$$

The above morphism can be presented in another form via the exterior of the diagram

$$\begin{array}{ccccc} \int^{Z\in\mathcal{C}} Y\otimes Z\boxtimes Z^\vee & \xleftarrow{\sim} & \int^{X,Z\in\mathcal{C}} X\otimes \mathrm{Hom}(X,Y\otimes Z)\boxtimes Z^\vee & \xrightarrow{\sim} & \int^{Z,X\in\mathcal{C}} X\boxtimes \mathrm{Hom}(Z^\vee,X^\vee\otimes Y)\otimes Z^\vee \\ {\scriptstyle 1\otimes \mathrm{coev}_Y}\downarrow & & {\scriptstyle \alpha}\downarrow & & \downarrow\wr \\ \int^{Z\in\mathcal{C}} Y\otimes Z\boxtimes Z^\vee\otimes Y^\vee\otimes Y & \xrightarrow{\sim} & \int^{Z\in\mathcal{C}} (Y\otimes Z)\boxtimes (Y\otimes Z)^\vee\otimes Y & \xrightarrow{\beta} & \int^{X\in\mathcal{C}} X\boxtimes X^\vee\otimes Y \end{array} \tag{C.2.1}$$

Here α and β are defined in such a way that the diagrams

$$\begin{array}{ccc}
X \otimes \mathrm{Hom}(X, Y \otimes Z) \boxtimes Z^\vee \otimes 1 & \xrightarrow{\mathrm{ev} \boxtimes Z^\vee \otimes \mathrm{coev}_Y} & (Y \otimes Z) \boxtimes Z^\vee \otimes Y^\vee \otimes Y \\
\downarrow i_{X,Z} & & \downarrow i'_Z \\
\int^{X,Z \in \mathcal{C}} X \otimes \mathrm{Hom}(X, Y \otimes Z) \boxtimes Z^\vee & \xrightarrow{\alpha} & \int^{Z \in \mathcal{C}} (Y \otimes Z) \boxtimes (Y \otimes Z)^\vee \otimes Y
\end{array}$$

$$\begin{array}{ccc}
(Y \otimes Z) \boxtimes Z^\vee \otimes Y^\vee \otimes Y & \xrightarrow{\sim} & (Y \otimes Z) \boxtimes (Y \otimes Z)^\vee \otimes Y \\
\downarrow i'_Z & & \downarrow i_{Y \otimes Z \otimes Y} \\
\int^{Z \in \mathcal{C}} (Y \otimes Z) \boxtimes (Y \otimes Z)^\vee \otimes Y & \xrightarrow{\beta} & \int^{X \in \mathcal{C}} X \boxtimes X^\vee \otimes Y
\end{array}$$

are commutative.

The left square of diagram (C.2.1) is commutative by the definition of α. Commutativity of the right square of (C.2.1) follows by the equation

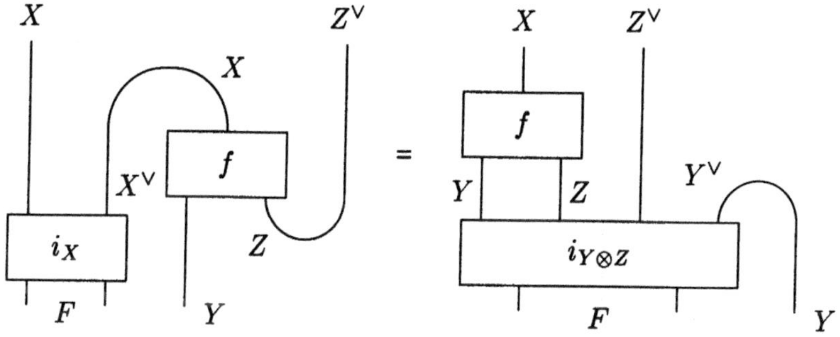

valid for any $f : X \to Y \otimes Z \in \mathcal{C}$. Indeed, by the definition of the coend $\mathbb{F} = \int^{X \in \mathcal{C}} X \boxtimes X^\vee \in \mathcal{C} \boxtimes \mathcal{C}$ we have

$$\big(X \boxtimes (Y \otimes Z)^\vee \xrightarrow{X \boxtimes f^t} X \boxtimes X^\vee \xrightarrow{i_X} \mathbb{F}\big)$$
$$= \big(X \boxtimes (Y \otimes Z)^\vee \xrightarrow{f \boxtimes (Y \otimes Z)^\vee} Y \otimes Z \boxtimes (Y \otimes Z)^\vee \xrightarrow{i_{Y \otimes Z}} \mathbb{F}\big).$$

Commutativity of Diagram (C.2.1) implies the lemma.

Lemma C.2.3. *The following two arrows* $(\mathbb{1} \boxtimes \mathrm{Inv} \boxtimes \mathbb{1})(\mathbb{F}_{12'} \otimes \mathbb{F}_{2''3}) \to \mathbb{F}$ *coincide:*

$$\mathbf{z} \boxtimes \mathbb{1} = \mathbb{1} \boxtimes \mathbf{n} : (\mathbb{1} \boxtimes \mathrm{Inv} \boxtimes \mathbb{1}) \int^{X,Y \in \mathcal{C}} (X \boxtimes X^\vee \otimes Y \boxtimes Y^\vee) \to \int^{Z \in \mathcal{C}} Z \boxtimes Z^\vee.$$

Proof. The isomorphisms $\mathbf{z}$ and $\mathbf{n}$ are made explicit in Lemma 5.1.11. Since we deal with left exact functors, it suffices to prove for projective X and injective Y that the following diagram is commutative:

$$
\begin{array}{ccccc}
X\boxtimes\mathrm{Inv}(X^\vee\otimes Y)\boxtimes Y^\vee & \xrightarrow{\sim} & X\otimes\mathrm{Hom}(X,Y)\boxtimes Y^\vee & \xrightarrow{\mathrm{ev}\boxtimes Y^\vee} & Y\boxtimes Y^\vee \\
\downarrow \wr & & & & \downarrow i_Y \\
X\boxtimes\mathrm{Hom}(Y^\vee,X^\vee)\otimes Y^\vee & \xrightarrow{X\boxtimes\mathrm{ev}} & X\boxtimes X^\vee & \xrightarrow{i_X} & \int^{Z\in\mathcal{C}} Z\boxtimes Z^\vee
\end{array}
\tag{C.2.2}
$$

The object $X \boxtimes \mathrm{Inv}(X^\vee \otimes Y) \boxtimes Y^\vee$ is a direct sum of $\dim_{\Bbbk} \mathrm{Inv}(X^\vee \otimes Y)$ copies of $X \boxtimes Y^\vee$. To prove the above diagram, it suffices to pick up an element $\hat{f} : 1 \to X^\vee \otimes Y$ and represent it in the form $f : X \to Y$ and $f^t : Y^\vee \to X^\vee$. Then Diagram (C.2.2), restricted to the subobject $X \boxtimes \hat{f} \boxtimes Y^\vee : X \boxtimes Y^\vee \to X \boxtimes \mathrm{Inv}(X^\vee \otimes Y) \boxtimes Y^\vee$, is nothing else but the definition of the coend in (5.2.1).

Proof (Theorem C.2.1). The constructed functor is fixed on 1-morphisms by the following requirement: it strictly commutes with the tensor product and the composition of 1-morphisms. We have to prove diagram (C.1.5) for w using the calculation made in Lemma C.2.2. That is,

$$
\Big(X \otimes Y \otimes \int^Z Z \boxtimes Z^\vee \xrightarrow{1\boxtimes 1_{Z^\vee}\otimes \mathrm{coev}_Y} X \otimes \int^Z Y \otimes Z \boxtimes Z^\vee \otimes Y^\vee \otimes Y \to
$$

$$
\to X \otimes \int^U U \boxtimes U^\vee \otimes Y \xrightarrow{1\boxtimes 1_{U^\vee}\otimes \mathrm{coev}_X \otimes 1_Y} \int^U X \otimes U \boxtimes U^\vee \otimes X^\vee \otimes X \otimes Y \to
$$

$$
\to \int^W W \boxtimes W^\vee \otimes X \otimes Y\Big) = \Big((X \otimes Y) \otimes \int^Z Z \boxtimes Z^\vee \xrightarrow{1\boxtimes 1_{Z^\vee}\otimes \mathrm{coev}_{X\otimes Y}}
$$

$$
\to \int^Z (X \otimes Y \otimes Z) \boxtimes Z^\vee \otimes Y^\vee \otimes X^\vee \otimes X \otimes Y \to \int^W W \boxtimes W^\vee \otimes X \otimes Y\Big).
$$

This is obvious since

$$
\big(1 \xrightarrow{\mathrm{coev}_{X\otimes Y}} (X \otimes Y)^\vee \otimes (X \otimes Y) \simeq Y^\vee \otimes X^\vee \otimes X \otimes Y\big)
$$

$$
= \big(1 \xrightarrow{\mathrm{coev}_Y} Y^\vee \otimes Y \simeq Y^\vee \otimes 1 \otimes Y \xrightarrow{1_{Y^\vee}\otimes \mathrm{coev}_X \otimes 1_Y} Y^\vee \otimes X^\vee \otimes X \otimes Y\big).
$$

Lemma C.2.3 contains the proof of Relation (C.1.4). Relations (C.1.2) and (C.1.3) are obviously satisfied. This proves Theorem C.2.1.

Remark C.2.4. If a thick graph is composed of subgraphs, the corresponding functor is also the composition of relevant functors.

Bibliography

[Abr96] Abrams L., *Two-dimensional topological quantum field theories and Frobenius algebras*, J. Knot Theory Ramifications **5** (1996), no. 5, 569–587.

[Ale23] Alexander J., *On the deformation of an n-cell*, Proc. Nat. Acad. Sci. **9** (1923), 406–407.

[And92] Andersen H. H., *Tensor products of quantized tilting modules*, Commun. Math. Phys. **149** (1992), no. 1, 149–159.

[APS75a] Atiyah M., Patodi V., Singer I. M., *Spectral asymmetry and Riemannian geometry. I*, Math. Proc. Camb. Phil. Soc. **77** (1975), 43–69.

[APS75b] Atiyah M., Patodi V., Singer I. M., *Spectral asymmetry and Riemannian geometry. II*, Math. Proc. Camb. Phil. Soc. **78** (1975), no. 3, 405–432.

[Ati88] Atiyah M., *Topological quantum field theories*, Inst. Hautes Études Sci. Publ. Math. **68** (1988), 175–186.

[Ati90] Atiyah M., *On framings of* 3*-manifolds*, Topology **29** (1990), 1–7.

[Bae97] Baez J. C., *Higher-dimensional algebra II: 2-Hilbert spaces*, Adv. Math. **127** (1997), 125–189.

[BD95] Baez J. C., Dolan J., *Higher-dimensional algebra and topological quantum field theory*, Jour. Math. Phys. **36** (1995), 6073–6105.

[Bel98] Beliakova A., *Spin topological quantum field theories*, Internat. J. Math. **9** (1998), no. 2, 129–152.

[Bén67] Bénabou J., *Introduction to bicategories*, Lecture Notes in Math., vol. 47, Springer, New York, 1967, pp. 1–77.

[BHMV95] Blanchet C., Habegger N., Masbaum G., Vogel P., *Topological quantum field theories derived from the Kauffman bracket*, Topology **34** (1995), no. 4, 883–927.

[Bir69a] Birman J., *On braid groups*, Commun. Pure and Appl. Math. **22** (1969), 41–72.

[Bir69b] Birman J., *Mapping class groups and their relations to braid groups*, Commun. Pure and Appl. Math. **22** (1969), 213–238.

[Bir74] Birman J., *Braids, links and mapping class groups*, Ann. of Math. Studies, vol. 82, Princeton Univ. Press, 1974.

[BJ73] Bröcker T., Jänich K., *Einführung in die Differentialtopologie*, Heidelberger Taschenbücher, vol. 143, Springer-Verlag, 1973.

[BK01] Bakalov B., Kirillov A., Jr., *Lectures on tensor categories and modular functors*, University Lecture Series, vol. 21, Amer. Math. Soc., Providence, RI, 2001.

[BKLT00] Bespalov Yu. N., Kerler T., Lyubashenko V. V., Turaev V. G., *Integrals for braided Hopf algebras*, J. Pure and Appl. Algebra **148** (2000), no. 2, 113–164.

[BL98a] Baez J. C., Langford L., *2-tangles*, Lett. Math. Phys. **43** (1998), no. 2, 187–197.

[BL98b] Baez J. C., Langford L., *Higher-dimensional algebra IV: 2-tangles*, math.QA/9811139, 1998.

[BN96] Baez J. C., Neuchl M., *Higher-dimensional algebra I: Braided monoidal 2-categories*, Adv. Math. **121** (1996), no. 2, 196–244.

[Bre94] Breen L., *On the classification of 2-gerbes and 2-stacks*, Astérisque, Soc. Math. de France **225** (1994), 1–160.

[BW89] Birman J. S., Wenzl H., *Braids, link polynomials and a new algebra*, Trans. Amer. Math. Soc. **313** (1989), no. 1, 249–273.

[BW95] Barrett J. W., Westbury B. W., *The equality of 3-manifold invariants*, Math. Proc. Camb. Phil. Soc. **118** (1995), 503–510.

[Cer70] Cerf J., *La stratification naturelle de espace des fonctions différentiables réelles et le théorème de la pseudoisotopie*, Publ. Math. I.H.É.S. **39** (1970).

[CF94] Crane L., Frenkel I. B., *Four dimensional topological quantum field theory, Hopf categories, and the canonical bases*, J. Math. Phys. **35** (1994), no. 10, 5136–5154.

[CKY97] Crane L., Kauffman L. H., Yetter D. N., *State-sum invariants of 4-manifolds*, J. Knot Theory Ramifications **6** (1997), 177–234.

[CRS97] Carter J. S., Rieger J. H., Saito M., *A combinatorial description of knotted surfaces and their isotopies*, Adv. Math. **127** (1997), no. 1, 1–51.

[CS98] Carter J. S., Saito M., *Knotted surfaces and their diagrams*, Math. Surveys and Monographs, vol. 55, Amer. Math. Soc., Providence, RI, 1998, xii+258 pp.

[CY93] Crane L., Yetter D. N., *A categorical construction of 4D topological quantum field theories*, Quantum topology, Ser. Knots Everything, no. 3, World Sci. Publishing, River Edge, NJ, 1993, pp. 120–130.

[Del91] Deligne P., *Catégories tannakiennes*, The Grothendieck Festschrift, Progress in Mathematics, no. 87, Birkhäuser, Boston, Basel, Berlin, 1991, pp. 111–195.

[DJ94] Durhuus B., Jonsson T., *Classification and construction of unitary topological quantum field theories in two dimensions*, J. Math. Phys. **35** (1994), 5306–5313.

[DM82] Deligne P., Milne J. S., *Tannakian categories*, Hodge cycles, motives, and Shimura varieties, Lecture Notes in Math., no. 900, Springer-Verlag, Berlin, Heidelberg, New York, 1982, pp. 101–228.

[Don99] Donaldson S. K., *Topological field theories and formulae of Casson and Meng-Taubes*, Proceedings of the Kirbyfest (Joel Hass and Martin Scharlemann, eds.), Geometry and Topology Monographs, no. 2, 1999, pp. 87–102.

[DPR90] Dijkgraaf R., Pasquier V., Roche P., *Quasi-quantum groups related to orbifold models*, Nucl. Phys. B. Proc. Suppl. **18** (1990), 60–72.

[Dri87] Drinfeld V. G., *Quantum Groups*, Proceedings of the International Congress of Mathematicians (Berkeley 1986), Vol. 1 (Providence, RI) (A. Gleason, ed.), Amer. Math. Soc., 1987, pp. 798–820.

[Dri90] Drinfeld V. G., *On almost cocommutative Hopf algebras*, Leningrad Math. J. **1** (1990), no. 2, 321–342.

[DS97] Day B. J., Street R. H., *Monoidal bicategories and Hopf algebroids*, Adv. Math. **129** (1997), no. 1, 99–157.

[DW90] Dijkgraaf R., Witten E., *Topological gauge theories and group cohomology*, Commun. Math. Phys. **129** (1990), 393–429.

[Ehr63a] Ehresmann C., *Catégories doubles et catégories structurées*, C. R. Acad. Sc. **256** (1963), 1198–1201.

[Ehr63b] Ehresmann C., *Catégories doubles des quintettes, applications covariantes*, C. R. Acad. Sc. **256** (1963), 1891–1894.

[EK95] Evans D. E., Kawahigashi Y., *From subfactors to 3-dimensional topological quantum field theories and back. A detailed account of Ocneanu's theory.*, Internat. J. Math. **6** (1995), 537–558.

[Eps66] Epstein D. B. A., *Curves on 2-manifolds and isotopies*, Acta Math. **115** (1966), 83–107.

[ES52] Eilenberg S., Steenrod N. E., *Foundations of algebraic topology*, Princeton Mathematical Series, vol. 15, Princeton University Press, 1952.

[Fi94] Fischer J. E., *2-categories and 2-knots*, Duke Math. J. **75** (1994), no. 2, 493–526.

[FK93] Fröhlich J., Kerler T., *Quantum Groups, Quantum Categories and Quantum Field Theory*, Lect. Notes Math., vol. 1542, Springer, Berlin, 1993.

[FKB96] Frohman C., Kania-Bartoszyńska J., *SO(3)-topological quantum field theory*, Comm. Anal. Geom. **4** (1996), no. 4, 589–679.

[FN62] Fadell E., Neuwirth L., *Configuration spaces*, Math. Scand. **10** (1962), 111–118.

[FN91] Frohman C., Nicas A., *The Alexander polynomial via topological quantum field theory*, Differential geometry, global analysis, and topology (Halifax, NS, 1990) (Providence, RI), CMS Conf. Proc., no. 12, Amer. Math. Soc., 1991, pp. 27–40.

[FN94] Frohman C., Nicas A., *An intersection homology invariant for knots in a rational homology 3-sphere*, Topology **33** (1994), no. 1, 123–158.

[FQ90] Freedman M., Quinn F., *Topology of 4-manifolds*, Princeton Mathematical Series, vol. 39, Princeton University Press, 1990.

[FQ93] Freed D. S., Quinn F., *Chern–Simons theory with finite gauge group*, Commun. Math. Phys. **156** (1993), no. 3, 435–472.

[FR79] Fenn R., Rourke C., *On Kirby's calculus of links*, Topology **18** (1979), 1–15.

[Fre64] Freyd P. J., *Abelian categories*, Harper and Row, New York, 1964.

[Fre94] Freed D. S., *Higher algebraic structures and quantization*, Commun. Math. Phys. **159** (1994), 343–398.

[Fuk99] Fukaya K., *Floer homology for 3 manifold with boundary, I*, `http://www.kusm.kyoto-u.ac.jp/~fukaya/bdrt1.pdf`, March 1999.

[FV62] Fadell E., Van Buskirk J., *The braid groups of E^2 and S^2*, Duke Math. J. **29** (1962), 243–257.

[FY92] Freyd P. J., Yetter D. N., *Coherence theorem via knot theory*, J. Pure and Appl. Algebra **78** (1992), 49–76.

[Gel98] Gelca R., *$SL(2,C)$-topological quantum field theory with corners*, J. Knot Theory Ramifications **7** (1998), no. 7, 893–906.

[GK96] Gelfand S. I., Kazhdan D., *Invariants of three-dimensional manifolds*, Geom. Funct. Anal. **6** (1996), 268–300.

[GL98] Garoufalidis S., Levine J., *Finite type 3-manifold invariants and the structure of the Torelli group. I.*, Invent. Math. **131** (1998), 541–594.

[GPS95] Gordon R., Power A. J., Street R. H. , *Coherence for tricategories*, vol. 117, Memoirs Amer. Math. Soc., no. 558, Amer. Math. Soc., Providence, Rhode Island, September 1995.

[GS99] Gompf R. E., Stipsicz A. I., *4-manifolds and Kirby calculus*, Graduate Studies in Mathematics, vol. 20, Amer. Math. Soc., Providence, Rhode Island, 1999, xvi+558 pp.

[Hem76] Hempel J., *3-manifolds*, Ann. of Math. Studies, vol. 86, Princeton University Press, 1976.

[Hen96] Hennings M., *Invariants of links and 3-manifolds obtained from Hopf algebras*, J. London Math. Soc. (2) **54** (1996), no. 3, 594–624.

[Hir76] Hirsch M. W., *Differential topology*, GTM, vol. 33, Springer, Berlin, 1976.

[HW73] Hatcher A., Wagoner J., *Pseudo-isotopies of compact manifolds*, Astérisque, vol. 6, Soc. Math. de France, 1973.

[Jon87] Jones V. F. R., *Hecke algebra representations of braid groups and link polynomials*, Ann. Math. **126** (1987), no. 2, 335–388.

[JS91] Joyal A., Street R. H. , *Tortile Yang–Baxter operators in tensor categories*, J. Pure Appl. Algebra **71** (1991), 43–51.

[Kar71] Karoubi M., *K-théorie*, Les Presses de l'Université de Montréal, 1971.

[Kel82] Kelly G. M., *Basic concepts of enriched category theory*, London Math. Soc. Lecture Notes, vol. 64, Cambridge Univ. Press, Cambridge, 1982.

[Ker92] Kerler T., *Non-Tannakian categories in quantum field theory*, New Symmetry Principles in Quantum Field Theory (Cargése, 1991). NATO Adv. Sci. Inst. Ser. B Phys., vol. 295, Plenum Press, New York, 1992, pp. 449–482.

[Ker94] Kerler T., *Mapping class group actions on quantum doubles*, Commun. Math. Phys. **168** (1994), 353–388.

[Ker97] Kerler T., *Genealogy of nonperturbative quantum-invariants of 3-manifolds: The surgical family*, Geometry and Physics, (Aarhus, 1995) (New York), Lecture Notes in Pure and Appl. Math., no. 184, Marcel Dekker, 1997, pp. 503–547.

[Ker98a] Kerler T., *Equivalence of a bridged link calculus and Kirby's calculus of links on non-simply connected 3-manifolds*, Topology Appl. **87** (1998), 155–162.

[Ker98b] Kerler T., *On the connectivity of cobordisms and half-projective TQFT's*, Commun. Math. Phys. **198** (1998), no. 3, 535–590.

[Ker99] Kerler T., *Bridged links and tangle presentations of cobordism categories*, Adv. Math. **141** (1999), 207–281.

[Ker00] Kerler T., *Homology TQFT via Hopf algebras*, math.GT/0008204, 2000.

[Kir78] Kirby R. C., *A calculus for framed links in S^3*, Invent. Math. **65** (1978), 35–56.

[Kir89] Kirby R. C., *The topology of 4-manifolds*, Lect. Notes in Math., vol. 1374, Springer–Verlag, Berlin, 1989.

[KM91] Kirby R. C., Melvin P., *The 3-manifold invariants of Witten and Reshetikhin–Turaev for $sl(2, C)$*, Invent. Math. **105** (1991), 473–545.

[Kne26] Kneser H., *Die Deformationssätze der einfach zusammenhängenden Flächen*, Math. Zeit. **25** (1926), 362–372.

[Kon] Kontsevich M., *Rational conformal field theory and invariants of 3-dimensional manifolds*, Marseille preprint CPT-99/P2189.

[KR93] Kauffman L. H., Radford D. E., *A necessary and sufficient condition for a finite-dimensional Drinfel'd double to be a ribbon Hopf algebra*, J. Algebra **159** (1993), no. 1, 98–114.

[KR95] Kauffman L. H., Radford D. E., *Invariants of 3-manifolds derived from finite dimensional Hopf algebras*, J. Knot Theory Ramifications **4** (1995), no. 1, 131–162.

[KRT97] Kassel C., Rosso M., Turaev V., *Quantum groups and knot invariants*, Panoramas et Synthèses de la S.M.F., vol. 5, Soc. Math. de France, Paris, 1997.

[KS74] Kelly G. M., Street R. H. , *Review of the elements of 2-categories*, Category Seminar (Proc. Sem., Sydney, 1972/1973), Lecture Notes in Mathematics, vol. 420, Springer-Verlag, 1974, pp. 75–103.

[KS93] Karowski M., Schrader R., *State sum invariants of three-manifolds: a combinatorial approach to topological quantum field theories*, J. Geom. Phys. **11** (1993), 181–190.

[KSS97] Kauffman L. H., Saito M., Sullivan M. C., *Quantum invariants of templates*, available at `http://www.math.usf.edu/~saito/preprints.html`, 1997.

[Kup91] Kuperberg G., *Involutory Hopf algebras and three-manifold invariants*, Internat. J. Math. **2** (1991), 41–66.

[Kup96] Kuperberg G., *Non-involutory Hopf algebras and 3-manifold invariants*, Duke Math. J. **84** (1996), no. 1, 83–129.

[KV94] Kapranov M. M., Voevodsky V. A., *2-categories and Zamolodchikov tetrahedra equations*, Algebraic groups and their generalizations: quantum and infinite-dimensional methods, Proc. Symp. Pure Math., Vol. 56, Part 2 (Providence, RI) (William J. Haboush et al., eds.), Summer Research Institute on algebraic groups and their generalizations, July 6-26, 1991, Pennsylvania State University, University Park, PA, USA., Amer. Math. Soc., 1994, pp. 177–260.

[Law93] Lawrence R. J., *Triangulation, categories and extended topological field theories*, Quantum Topology (R. A. Baadhio and L. H. Kauffman, eds.), Ser. Knots Everything, no. 3, World Sci. Publishing, River Edge, NJ, Singapore, 1993, pp. 191–208.

[Lic62] Lickorish W. B. R., *A representation of orientable 3-manifolds*, Ann. Math. **76** (1962), 531–540.

[LM94] Lyubashenko V. V., Majid S., *Braided groups and quantum Fourier transform*, J. Algebra **166** (1994), 506–528.

[LMMO95] Le T. Q. T., Murakami H., Murakami J., Ohtsuki T., *A three-manifold invariant derived from the universal Vassiliev–Kontsevich invariant*, Proc. Japan Acad., Ser. A **71** (1995), 125–127.

[LS69] Larson R. G., Sweedler M. E., *An associative orthogonal bilinear form for Hopf algebras*, Amer. J. Math. **91** (1969), no. 1, 75–94.

[Lyu95a] Lyubashenko V. V., *Tangles and Hopf algebras in braided categories*, J. Pure and Applied Algebra **98** (1995), no. 3, 245–278.

[Lyu95b] Lyubashenko V. V., *Modular transformations for tensor categories*, J. Pure and Applied Algebra **98** (1995), no. 3, 279–327.

[Lyu95c] Lyubashenko V. V., *Invariants of 3-manifolds and projective representations of mapping class groups via quantum groups at roots of unity*, Commun. Math. Phys. **172** (1995), no. 3, 467–516.

[Lyu96] Lyubashenko V. V., *Ribbon abelian categories as modular categories*, J. Knot Theory Ramifications **5** (1996), no. 3, 311–403.

[Lyu99] Lyubashenko V. V., *Squared Hopf algebras*, Mem. Amer. Math. Soc. **142** (1999), no. 677, 184 pp.

[Mac88] Mac Lane S., *Categories for the working mathematician*, GTM, vol. 5, Springer Verlag, New York, 1971, 1988.

[Maj93] Majid S., *Braided groups*, J. Pure Appl. Algebra **86** (1993), no. 2, 187–221.

[Mas97] Masbaum G., *Introduction to spin TQFT*, Geometric topology (Athens, GA, 1993) (Providence, RI), AMS/IP Stud. Adv. Math., no. 2.1, Amer. Math. Soc., 1997, pp. 203–216.

[Mil65] Milnor J. W., *Lectures on the h-cobordism theorem*, Princeton Mathematical Notes, Princeton Univ. Press, Princeton, 1965.

[Mil69] Milnor J. W., *Morse theory*, Annals of Mathematical Studies, vol. 51, Princeton Univ. Press, Princeton, 1969.

[Mil86] Miller E. Y., *The homology of the mapping class group*, J. Diff. Geom. **24** (1986), 1–14.

[MK89] Melvin P., Kazez W., 3-*dimensional bordism*, Michigan Math. J. **36** (1989), no. 2, 251–260.

[Moi52] Moise E. E., *Affine structures in 3-manifolds II. positional properties of* 2-*spheres*, Ann. of Math. (2) **55** (1952), 172–176.

[MP94] Matveev S., Polyak M., *A geometrical presentation of the surface mapping class group and surgery*, Commun. Math. Phys. **160** (1994), no. 3, 537–550.

[MS89] Moore G., Seiberg N., *Classical and quantum conformal field theory*, Commun. Math. Phys. **123** (1989), 177–254.

[Mur87] Murakami J., *The Kauffman polynomial of links and representation theory*, Osaka J. Math. **24** (1987), no. 4, 745–758.

[Mur94] Murakami H., *Quantum $SU(2)$-invariants dominate Casson's $SU(2)$-invariant*, Math. Proc. Camb. Phil. Soc. **115** (1994), 83–103.

[NV00] Nikshych D., Vainerman L., *A characterization of depth 2 subfactors of II_1 factors*, J. Funct. Anal. **171** (2000), no. 2, 278–307.

[Oht96] Ohtsuki T., *Finite type invariants of integral homology* 3-*spheres*, J. Knot Theory Ramifications **5** (1996), 101–115.

[PR68] Ponzano G., Regge T., *Semiclassical limit of Racah coefficients*, Spectroscopic and Group Theoretical Methods in Physics, North-Holland, Amsterdam, 1968, pp. 1–58.

[Qui95] Quinn F., *Lectures on axiomatic topological quantum field theory*, Geometry and Quantum Field Theory (Park City, UT, 1991), IAS/Park City Math. Ser., no. 1, Amer. Math. Soc., Providence, RI, 1995, pp. 323–453.

[Rad76] Radford D. E., *The order of antipode of a finite-dimensional Hopf algebra is finite*, Amer. J. Math **98** (1976), 333–335.

[Res88] Reshetikhin N. Yu., *Quantized universal enveloping algebras, the Yang–Baxter equation and invariants of links, I and II*, Preprints, LOMI, E-4-87, E-17-87, Leningrad, 1988.

[Res90] Reshetikhin N. Yu., *Quasitriangular Hopf algebras and invariants of tangles*, Leningrad Math. J. **1** (1990), no. 2, 491–513, Russian: **1** (1989), no. 2, 169-188.

[Rob95] Roberts J., *Skein theory and Turaev–Viro invariants*, Topology **34** (1995), no. 4, 771–787.

[Rob97] Roberts J., *Kirby calculus in manifolds with boundary*, Turkish J. Math. **21** (1997), no. 1, 111–117.

[Roh51] Rohlin V. A., *A three dimensional manifold is the boundary of a four dimensional one*, Dokl.Akad. Nauk. SSSR (N.S.) **114** (1951), 355–357.

[Rol76] Rolfsen D., *Knots and links*, Mathematics Lecture Series, vol. 7, Publish or Perish, 1976, 439 p.

[RT90] Reshetikhin N. Yu., Turaev V. G., *Ribbon graphs and their invariants derived from quantum groups*, Commun. Math. Phys. **127** (1990), no. 1, 1–26.

[RT91] Reshetikhin N. Yu., Turaev V. G., *Invariants of 3-manifolds via link polynomials and quantum groups*, Invent. Math. **103** (1991), no. 3, 547–597.

[Saa72] Saavedra Rivano N., *Catégories Tannakiennes*, Lecture Notes in Math., vol. 265, Springer, Berlin, Heidelberg, New York, 1972, 420p.

[Saw99] Sawin S., *Three-dimensional 2-framed TQFTs and surgery*, math.QA/9912065, 1999.

[Sch92] Schauenburg P., *Tannaka duality for arbitrary Hopf algebras*, Algebra Berichte, vol. 66, Verlag Reinhard Fischer, München, 1992.

[Seg88] Segal G. B., *The definition of conformal field theory*, Differential geometrical methods in theoretical physics (Como, 1987) (Dordrecht), NATO Adv. Sci. Inst. Ser. C: Math. Phys. Sci., no. 250, Kluwer Acad. Publ., 1988, pp. 165–171.

[Seg98] Segal G. B., Lecture Notes (unpublished) 1998.

[Shu94] Shum M. C., *Tortile tensor categories*, J. Pure Appl. Algebra **93** (1994), no. 1, 57–110.

[Sma59] Smale S., *Diffeomorphisms of the 2-sphere*, Proc. Amer. Math. Soc. **10** (1959), 621–626.

[Swe69a] Sweedler M. E., *Hopf algebras*, W. A. Benjamin, New York, 1969.

[Swe69b] Sweedler M. E., *Integrals for Hopf algebras*, Ann. of Math. **89** (1969), no. 2, 323–335.

[Tho54] Thom R., *Quelques propriétés globales des variétés différentiables*, Comment. Math. Helv. **28** (1954), 17–86.

[Til95] Tillman U., *On the homotopy of the stable mapping class group*, Oxford Preprint, 1995.

[Tur88] Turaev V. G., *The Yang–Baxter equation and invariants for links*, Invent. Math. **92** (1988), 527–553.

[Tur94] Turaev V. G., *Quantum invariants of knots and 3-manifolds*, de Gruyter Stud. Math., vol. 18, Walter de Gruyter & Co., Berlin, New York, 1994, 588 pp.

[TV92] Turaev V. G., Viro O., *State sum invariants of 3-manifolds and quantum 6-j-symbols*, Topology **31** (1992), 865–902.

[TW93] Turaev V. G., Wenzl H., *Quantum invariants of 3-manifolds associated with classical simple Lie algebras*, Internat. J. Math. **4** (1993), no. 2, 323–358.

[Wal] Walker K., *On Witten's 3-manifold invariants*, UCSD preprint, 1991.

[Wal60] Wallace A. H., *Modifications and cobounding manifolds*, Cam. J. Math. **12** (1960), 503–528.

[Wal68] Waldhausen F., *On irreducible 3-manifolds which are sufficiently large*, Ann. of Math. **87** (1968), 57–88.

[Wal69] Wall C. T. C., *Non-additivity of signature*, Invent. Math. **7** (1969), 269–274.

[Wit89] Witten E., *Quantum field theory and the Jones polynomial*, Commun. Math. Phys. **121** (1989), no. 3, 351–399.

[Xu] Xu F., *3-manifold invariants from cosets*, Preprint available as math.GT/9907077.

[Yet] Yetter D. N., *Coalgebras, comodules, coends and reconstruction*, Preprint.

[Yet97] Yetter D. N., *Portrait of the handle as a Hopf algebra*, Geometry and Physics, (Aarhus, 1995) (New York), Lecture Notes in Pure and Appl. Math., no. 184, Marcel Dekker, 1997, pp. 481–502.

Index

Vol. 1676: P. Cembranos, J. Mendoza, Banach Spaces of Vector-Valued Functions. VIII, 118 pages. 1997.

Vol. 1677: N. Proskurin, Cubic Metaplectic Forms and Theta Functions. VIII, 196 pages. 1998.

Vol. 1678: O. Krupková, The Geometry of Ordinary Variational Equations. X, 251 pages. 1997.

Vol. 1679: K.-G. Grosse-Erdmann, The Blocking Technique. Weighted Mean Operators and Hardy's Inequality. IX, 114 pages. 1998.

Vol. 1680: K.-Z. Li, F. Oort, Moduli of Supersingular Abelian Varieties. V, 116 pages. 1998.

Vol. 1681: G. J. Wirsching, The Dynamical System Generated by the 3n+1 Function. VII, 158 pages. 1998.

Vol. 1682: H.-D. Alber, Materials with Memory. X, 166 pages. 1998.

Vol. 1683: A. Pomp, The Boundary-Domain Integral Method for Elliptic Systems. XVI, 163 pages. 1998.

Vol. 1684: C. A. Berenstein, P. F. Ebenfelt, S. G. Gindikin, S. Helgason, A. E. Tumanov, Integral Geometry, Radon Transforms and Complex Analysis. Firenze, 1996. Editors: E. Casadio Tarabusi, M. A. Picardello, G. Zampieri. VII, 160 pages. 1998

Vol. 1685: S. König, A. Zimmermann, Derived Equivalences for Group Rings. X, 146 pages. 1998.

Vol. 1686: J. Azéma, M. Émery, M. Ledoux, M. Yor (Eds.), Séminaire de Probabilités XXXII. VI, 440 pages. 1998.

Vol. 1687: F. Bornemann, Homogenization in Time of Singularly Perturbed Mechanical Systems. XII, 156 pages. 1998.

Vol. 1688: S. Assing, W. Schmidt, Continuous Strong Markov Processes in Dimension One. XII, 137 page. 1998.

Vol. 1689: W. Fulton, P. Pragacz, Schubert Varieties and Degeneracy Loci. XI, 148 pages. 1998.

Vol. 1690: M. T. Barlow, D. Nualart, Lectures on Probability Theory and Statistics. Editor: P. Bernard. VIII, 237 pages. 1998.

Vol. 1691: R. Bezrukavnikov, M. Finkelberg, V. Schechtman, Factorizable Sheaves and Quantum Groups. X, 282 pages. 1998.

Vol. 1692: T. M. W. Eyre, Quantum Stochastic Calculus and Representations of Lie Superalgebras. IX, 138 pages. 1998.

Vol. 1694: A. Braides, Approximation of Free-Discontinuity Problems. XI, 149 pages. 1998.

Vol. 1695: D. J. Hartfiel, Markov Set-Chains. VIII, 131 pages. 1998.

Vol. 1696: E. Bouscaren (Ed.): Model Theory and Algebraic Geometry. XV, 211 pages. 1998.

Vol. 1697: B. Cockburn, C. Johnson, C.-W. Shu, E. Tadmor, Advanced Numerical Approximation of Nonlinear Hyperbolic Equations. Cetraro, Italy, 1997. Editor: A. Quarteroni. VII, 390 pages. 1998.

Vol. 1698: M. Bhattacharjee, D. Macpherson, R. G. Möller, P. Neumann, Notes on Infinite Permutation Groups. XI, 202 pages. 1998.

Vol. 1699: A. Inoue,Tomita-Takesaki Theory in Algebras of Unbounded Operators. VIII, 241 pages. 1998.

Vol. 1700: W. A. Woyczyński, Burgers-KPZ Turbulence,XI, 318 pages. 1998.

Vol. 1701: Ti-Jun Xiao, J. Liang, The Cauchy Problem of Higher Order Abstract Differential Equations, XII, 302 pages. 1998.

Vol. 1702: J. Ma, J. Yong, Forward-Backward Stochastic Differential Equations and Their Applications. XIII, 270 pages. 1999.

Vol. 1703: R. M. Dudley, R. Norvaiša, Differentiability of Six Operators on Nonsmooth Functions and p-Variation. VIII, 272 pages. 1999.

Vol. 1704: H. Tamanoi, Elliptic Genera and Vertex Operator Super-Algebras. VI, 390 pages. 1999.

Vol. 1705: I. Nikolaev, E. Zhuzhoma, Flows in 2-dimensional Manifolds. XIX, 294 pages. 1999.

Vol. 1706: S. Yu. Pilyugin, Shadowing in Dynamical Systems. XVII, 271 pages. 1999.

Vol. 1707: R. Pytlak, Numerical Methods for Optimal Control Problems with State Constraints. XV, 215 pages. 1999.

Vol. 1708: K. Zuo, Representations of Fundamental Groups of Algebraic Varieties. VII, 139 pages. 1999.

Vol. 1709: J. Azéma, M. Émery, M. Ledoux, M. Yor (Eds), Séminaire de Probabilités XXXIII. VIII, 418 pages. 1999.

Vol. 1710: M. Koecher, The Minnesota Notes on Jordan Algebras and Their Applications. IX, 173 pages. 1999.

Vol. 1711: W. Ricker, Operator Algebras Generated by Commuting Projections: A Vector Measure Approach. XVII, 159 pages. 1999.

Vol. 1712: N. Schwartz, J. J. Madden, Semi-algebraic Function Rings and Reflectors of Partially Ordered Rings. XI, 279 pages. 1999.

Vol. 1713: F. Bethuel, G. Huisken, S. Müller, K. Steffen, Calculus of Variations and Geometric Evolution Problems. Cetraro, 1996. Editors: S. Hildebrandt, M. Struwe. VII, 293 pages. 1999.

Vol. 1714: O. Diekmann, R. Durrett, K. P. Hadeler, P. K. Maini, H. L. Smith, Mathematics Inspired by Biology. Martina Franca, 1997. Editors: V. Capasso, O. Diekmann. VII, 268 pages. 1999.

Vol. 1715: N. V. Krylov, M. Röckner, J. Zabczyk, Stochastic PDE's and Kolmogorov Equations in Infinite Dimensions. Cetraro, 1998. Editor: G. Da Prato. VIII, 239 pages. 1999.

Vol. 1716: J. Coates, R. Greenberg, K. A. Ribet, K. Rubin, Arithmetic Theory of Elliptic Curves. Cetraro, 1997. Editor: C. Viola. VIII, 260 pages. 1999.

Vol. 1717: J. Bertoin, F. Martinelli, Y. Peres, Lectures on Probability Theory and Statistics. Saint-Flour, 1997. Editor: P. Bernard. IX, 291 pages. 1999.

Vol. 1718: A. Eberle, Uniqueness and Non-Uniqueness of Semigroups Generated by Singular Diffusion Operators. VIII, 262 pages. 1999.

Vol. 1719: K. R. Meyer, Periodic Solutions of the N-Body Problem. IX, 144 pages. 1999.

Vol. 1720: D. Elworthy, Y. Le Jan, X-M. Li, On the Geometry of Diffusion Operators and Stochastic Flows. IV, 118 pages. 1999.

Vol. 1721: A. Iarrobino, V. Kanev, Power Sums, Gorenstein Algebras, and Determinantal Loci. XXVII, 345 pages. 1999.

Vol. 1722: R. McCutcheon, Elemental Methods in Ergodic Ramsey Theory. VI, 160 pages. 1999.

Vol. 1723: J. P. Croisille, C. Lebeau, Diffraction by an Immersed Elastic Wedge. VI, 134 pages. 1999.

Vol. 1724: V. N. Kolokoltsov, Semiclassical Analysis for Diffusions and Stochastic Processes. VIII, 347 pages. 2000.

Vol. 1725: D. A. Wolf-Gladrow, Lattice-Gas Cellular Automata and Lattice Boltzmann Models. IX, 308 pages. 2000.

Vol. 1726: V. Marić, Regular Variation and Differential Equations. X, 127 pages. 2000.

Vol. 1727: P. Kravanja, M. Van Barel, Computing the Zeros of Analytic Functions. VII, 111 pages. 2000.

Vol. 1728: K. Gatermann, Computer Algebra Methods for Equivariant Dynamical Systems. XV, 153 pages. 2000.

Vol. 1729: J. Azéma, M. Émery, M. Ledoux, M. Yor, Séminaire de Probabilités XXXIV. VI, 431 pages. 2000.

Vol. 1730: S. Graf, H. Luschgy, Foundations of Quantization for Probability Distributions. X, 230 pages. 2000.

Vol. 1731: T. Hsu, Quilts: Central Extensions, Braid Actions, and Finite Groups,. XII, 185 pages. 2000.

Vol. 1732: K. Keller, Invariant Factors, Julia Equivalences and the (Abstract) Mandelbrot Set. X, 206 pages. 2000.

Vol. 1733: K. Ritter, Average-Case Analysis of Numerical Problems. IX, 254 pages. 2000.

Vol. 1734: M. Espedal, A. Fasano, A. Mikelić, Filtration in Porous Media and Industrial Applications. Cetraro 1998. Editor: A. Fasano. 2000.

Vol. 1735: D. Yafaev, Scattering Theory: Some Old and New Problems. XVI, 169 pages. 2000.

Vol. 1736: B. O. Turesson, Nonlinear Potential Theory and Weighted Sobolev Spaces. XIV, 173 pages. 2000.

Vol. 1737: S. Wakabayashi, Classical Microlocal Analysis in the Space of Hyperfunctions. VIII, 367 pages. 2000.

Vol. 1738: M. Émery, A. Nemirovski, D. Voiculescu, Lectures on Probability Theory and Statistics. XI, 356 pages. 2000.

Vol. 1739: R. Burkard, P. Deuflhard, A. Jameson, J.-L. Lions, G. Strang, Computational Mathematics Driven by Industrial Problems. Martina Franca, 1999. Editors: V. Capasso, H. Engl, J. Periaux. VII, 418 pages. 2000.

Vol. 1740: B. Kawohl, O. Pironneau, L. Tartar, J.-P. Zolesio, Optimal Shape Design. Tróia, Portugal 1999. Editors: A. Cellina, A. Ornelas. IX, 388 pages. 2000.

Vol. 1741: E. Lombardi, Oscillatory Integrals and Phenomena Beyond all Algebraic Orders. XV, 413 pages. 2000.

Vol. 1742: A. Unterberger, Quantization and Non-holomorphic Modular Forms. VIII, 253 pages. 2000.

Vol. 1743: L. Habermann, Riemannian Metrics of Constant Mass and Moduli Spaces of Conformal Structures. XII, 116 pages. 2000.

Vol. 1744: M. Kunze, Non-Smooth Dynamical Systems. X, 228 pages. 2000.

Vol. 1745: V. D. Milman, G. Schechtman, Geometric Aspects of Functional Analysis. VIII, 289 pages. 2000.

Vol. 1746: A. Degtyarev, I. Itenberg, V. Kharlamov, Real Enriques Surfaces. XVI, 259 pages. 2000.

Vol. 1747: L. W. Christensen, Gorenstein Dimensions. VIII, 204 pages. 2000.

Vol. 1748: M. Růžička, Electrorheological Fluids: Modeling and Mathematical Theory. XV, 176 pages. 2001.

Vol. 1749: M. Fuchs, G. Seregin, Variational Methods for Problems from Plasticity Theory and for Generalized Newtonian Fluids. VI, 269 pages. 2001.

Vol. 1750: B. Conrad, Grothendieck Duality and Base Change. X, 296 pages. 2001.

Vol. 1751: N. J. Cutland, Loeb Measures in Practice: Recent Advances. XI, 111 pages. 2001.

Vol. 1752: Y. V. Nesterenko, P. Philippon, Introduction to Algebraic Independence Theory. XIII, 256 pages. 2001.

Vol. 1753: A. I. Bobenko, U. Eitner, Painlevé Equations in the Differential Geometry of Surfaces. VI, 120 pages. 2001.

Vol. 1754: W. Bertram, The Geometry of Jordan and Lie Structures. XVI, 269 pages. 2001.

Vol. 1755: J. Azéma, M. Émery, M. Ledoux, M. Yor, Séminaire de Probabilités XXXV. VI, 427 pages. 2001.

Vol. 1756: P. E. Zhidkov, Korteweg de Vries and Nonlinear Schrödinger Equations: Qualitative Theory. VII, 147 pages. 2001.

Vol. 1757: R. R. Phelps, Lectures on Choquet's Theorem. VII, 124 pages. 2001.

Vol. 1758: N. Monod, Continuous Bounded Cohomology of Locally Compact Groups. X, 214 pages. 2001.

Vol. 1759: Y. Abe, K. Kopfermann, Toroidal Groups. VIII, 133 pages. 2001.

Vol. 1760: D. Filipović, Consistency Problems for Heath-Jarrow-Morton Interest Rate Models. VIII, 134 pages. 2001.

Vol. 1761: C. Adelmann, The Decomposition of Primes in Torsion Point Fields. VI, 142 pages. 2001.

Vol. 1762: S. Cerrai, Second Order PDE's in Finite and Infinite Dimension. IX, 330 pages. 2001.

Vol. 1763: J.-L. Loday, A. Frabetti, F. Chapoton, F. Goichot, Dialgebras and Related Operads. IV, 132 pages. 2001.

Vol. 1764: A. Cannas da Silva, Lectures on Symplectic Geometry. XII, 217 pages. 2001.

Vol. 1765: T. Kerler, V. V. Lyubashenko, Non-Semisimple Topological Quantum Field Theories for 3-Manifolds with Corners. VI, 379 pages. 2001.

Recent Reprints and New Editions

Vol. 1200: V. D. Milman, G. Schechtman, Asymptotic Theory of Finite Dimensional Normed Spaces - Corrected Second Printing 2001. X, 156 pages. 1986.

Vol. 1618: G. Pisier, Similarity Problems and Completely Bounded Maps - Second, Expanded Edition VII, 198 pages. 2001.

Vol. 1629: J. D. Moore, Lectures on Seiberg-Witten Invariants - Second Edition. VIII, 121 pages. 2001.

Vol. 1638: P. Vanhaecke, Integrable Systems in the realm of Algebraic Geometry - Second Edition. X, 256 pages. 2001.

Vol. 1702: J. Ma, J. Yong, Forward-Backward Stochastic Differential Equations and Their Applications - Corrected Second Printing 2000. XIII, 270 pages. 1999.

4. Lecture Notes are printed by photo-offset from the master-copy delivered in camera-ready form by the authors. Springer-Verlag provides technical instructions for the preparation of manuscripts. Macro packages in TEX, LATEX2e, LATEX2.09 are available from Springer's web-pages at

http://www.springer.de/math/authors/b-tex.html.

Careful preparation of the manuscripts will help keep production time short and ensure satisfactory appearance of the finished book.

The actual production of a Lecture Notes volume takes approximately 12 weeks.

5. Authors receive a total of 50 free copies of their volume, but no royalties. They are entitled to a discount of 33.3 % on the price of Springer books purchase for their personal use, if ordering directly from Springer-Verlag.

Commitment to publish is made by letter of intent rather than by signing a formal contract. Springer-Verlag secures the copyright for each volume. Authors are free to reuse material contained in their LNM volumes in later publications: A brief written (or e-mail) request for formal permission is sufficient.

Addresses:

Professor J.-M. Morel
CMLA, Ecole Normale Supérieure de Cachan
61 Avenue du Président Wilson
94235 Cachan Cedex France
E-mail: Jean-Michel.Morel@cmla.ens-cachan.fr

Professor B. Teissier
Université Paris 7
UFR de Mathématiques
Equipe Géométrie et Dynamique
Case 7012
2 place Jussieu
75251 Paris Cedex 05
E-mail: Teissier@ens.fr

Professor F. Takens, Mathematisch Instituut,
Rijksuniversiteit Groningen, Postbus 800,
9700 AV Groningen, The Netherlands
E-mail: F.Takens@math.rug.nl

Springer-Verlag, Mathematics Editorial, Tiergartenstr. 17
D-69121 Heidelberg, Germany
Tel.: *49 (6221) 487-701
Fax: *49 (6221) 487-355
E-mail: lnm@Springer.de

GPSR Compliance

The European Union's (EU) General Product Safety Regulation (GPSR) is a set of rules that requires consumer products to be safe and our obligations to ensure this.

If you have any concerns about our products, you can contact us on ProductSafety@springernature.com

In case Publisher is established outside the EU, the EU authorized representative is:

Springer Nature Customer Service Center GmbH
Europaplatz 3
69115 Heidelberg, Germany

Batch number: 09624486

Printed by Printforce, the Netherlands